W0106950

Techniques and Applications of Fast Reactions in Solution

NATO ADVANCED STUDY INSTITUTES SERIES

*Proceedings of the Advanced Study Institute Programme, which aims
at the dissemination of advanced knowledge and
the formation of contacts among scientists from different countries*

The series is published by an international board of publishers in conjunction
with NATO Scientific Affairs Division

A	Life Sciences	Plenum Publishing Corporation
B	Physics	London and New York
C	Mathematical and Physical Sciences	D. Reidel Publishing Company Dordrecht, Boston and London
D	Behavioral and Social Sciences	Sijthoff International Publishing Company Leiden
E	Applied Sciences	Noordhoff International Publishing Leiden

Series C – Mathematical and Physical Sciences

Volume 50 – Techniques and Applications of Fast Reactions in Solution

Techniques and Applications of Fast Reactions in Solution

Proceedings of the NATO Advanced Study Institute on New Applications of Chemical Relaxation Spectrometry and Other Fast Reaction Methods in Solution, held at the University College of Wales, Aberystwyth, September 10–20, 1978.

edited by

W. J. GETTINS AND E. WYN-JONES

University of Salford, Department of Chemistry and Applied Chemistry, Salford, England

D. Reidel Publishing Company

Dordrecht : Holland / Boston : U.S.A. / London : England

Published in cooperation with NATO Scientific Affairs Division

Library of Congress Cataloging in Publication Data

Nato Advanced Study Institute, Aberystwyth, 1978.
    Techniques and applications of fast reactions in solution.

    (NATO advanced study institutes series : Series C, Mathematical and physical
sciences ; v. 50)
    Includes index.
    1. Chemical reaction, Rate of–Congresses.   2. Solution (Chemistry)–
Congresses.   I. Gettins, W. J.   II. Wyn-Jones, Evan. III. Title.   IV. Series.
QD502.N37 1978   541'.39   79–16650
ISBN-13: 978-94-009-9492-8   e-ISBN-13: 978-94-009-9490-4
DOI: 10.1007/978-94-009-9490-4

Published by D. Reidel Publishing Company
P.O. Box 17, Dordrecht, Holland

Sold and distributed in the U.S.A., Canada, and Mexico
by D. Reidel Publishing Company, Inc.
Lincoln Building, 160 Old Derby Street, Hingham, Mass. 02043, U.S.A.

All Rights Reserved
Copyright © 1979 by D. Reidel Publishing Company, Dordrecht, Holland
Softcover reprint of the hardcover 1st edition 1979
No part of the material protected by this copyright notice may be reproduced or utilized
in any form or by any means, electronic or mechanical, including photocopying,
recording or by any informational storage and retrieval system,
without written permission from the copyright owner

CONTENTS

PREFACE

As a result of the pioneering efforts of Eigen, de Maeyer, Norrish and Porter, the kinetics of fast reactions in solution can now be studied using chemical relaxation methods, as well as many other fast reactions techniques. These methods have been applied successfully in many branches of the natural sciences. The simultaneous growth in the number of investigators and the diversity of their research interests has inevitably led to communication problems. The purpose of the NATO Advanced Study Institute entitled "New Applications of Chemical Relaxation Spectrometry and Other Fast Reaction Methods in Solution", was to create a forum so that research scientists working in different areas concerned with fast reactions could interact. This meeting was held at the Llandinam Building, University College of Wales, Aberystwyth from September 10th-20th, 1978. In addition to lectures on techniques and theory, two days of the NATO Advanced Study Institute, were spent discussing the current state of the art in this field. This two day meeting was also run under the auspices of the Chemical Society, Fast Reactions in Solution Group.

The papers in this volume are the result of the contributions given in the Aberystwyth meeting. We have attempted to make this volume useful for the non-expert and a comprehensive introduction to theory, as well as the instrumentation used in the studies are discussed in detail. The application of fast reaction methods to a wide range of topics are described and in addition, new developments and perspectives have been covered by several authors. This volume reflects the current state of the art of fast reactions in solution and it is hoped that it will serve as an incentive for further studies, particularly with reference to new areas of research.

In the organisation of the NATO meeting we would like to acknowledge the assistance of the Co-Directors Dr. W. Knoche, Professor J. E. Rassing and Dr. B. Robinson. We would also like to thank Mrs. P. Cassell and Mrs. G. Wyn-Jones for assistance in organising the meeting. We are also grateful to Miss M. Cradden for invaluable assistance in the preparation of this manuscript. Last but not least the financial assistance is gratefully acknowledged from: ICI Ltd., Esso Petroleum Company, Dia-log Ltd., Durram Instrument Corporation, V. A. Howe and Co. Ltd. and Nortech Laboratories Ltd.

SALFORD
MARCH 1979

W. J. Gettins
E. Wyn-Jones

THE STUDY OF FAST REACTIONS IN SOLUTION

E.F. Caldin

University of Kent at Canterbury, U.K.

1. Introduction.
2. Methods and techniques.
3. Characteristics of some fast-reaction techniques.
4. The choice of method and technique.
5. Fields of application.

1. INTRODUCTION

The study of reaction kinetics is our main way of investigating reaction mechanisms and the energetics of the actual path. Its data are measurements on the kinetic effects of systematically varying the concentration, temperature, pressure, solvent, and (where possible) the substituents and isotopic composition of the reactants. Accurate determinations of rate constants are therefore fundamental. Many familiar reactions, however, are too fast to be followed by conventional means, such as mixing the reactant solutions by hand and following the reaction by an ordinary spectrophotometer. Such reactions used to be written off as 'instantaneous' and their kinetics could not be investigated. This was unfortunate, since fast reactions (i.e. those with half-lives less than a few seconds) have widely varying rate constants, distributed over a range covering as many powers of ten as the range of conventional rates. The development of special methods has now made possible the study of even the fastest reactions.

Fast reactions in solution are the subject of a number of books (1-11) and innumerable papers. The early experiments of Hartridge and Roughton which laid the foundation for modern flow

1

W. J. Gettins and E. Wyn-Jones (Eds.). Techniques and Applications of Fast Reactions in Solution. 1–11.
Copyright © 1979 by D. Reidel Publishing Company.

methods were published in 1923 (12), and up to 1940 it was mainly
flow methods that were developed (13), apart from some work on
fluorescence quenching and photochemical methods, and on the use
of low temperatures. When the Faraday Society held a discussion
on fast reactions (14) in 1954, however, not only had the
stopped-flow technique reached essentially its present form but
relaxation methods, flash photolysis, and the use of n.m.r. and
e.s.r. had been developed. A conference in Germany in 1959 (15)
showed that all these methods were in full and systematic use.
In 1967 the fifth Nobel Symposium (3) was devoted to fast
reactions, and the Nobel Prize for chemistry was awarded to
Professors Eigen, Norrish and Porter jointly for their pioneer
work on the newer methods. From being the concern largely of
biochemists and a few physical chemists, fast-reaction
techniques have become major tools for the mechanistic
investigation of organic and inorganic reactions, and have
begun to generate chemical information of new types. From
being difficult techniques requiring special experimental skill,
some of them have developed into almost routine methods, not
much more difficult to learn than conventional spectrophotometry.

In this introductory lecture we consider the use of fast-
reaction methods in the measurement of the effects on reaction
rates of variations in the concentration, temperature, pressure,
and solvent, as required for the purposes of reaction kinetics.
The aim is to show the characteristic virtues and defects of the
various methods and techniques, so that an informed choice may
be made for a given investigation.

2. METHODS AND TECHNIQUES

Four Basic Methods

In conventional methods, where the reactant solutions are mixed
by shaking together and the progress of the reaction is
subsequently followed by chemical or physical means, the time
for mixing is short compared with the much longer time required
for observing the course of the reaction. For fast reactions,
ordinary mixing methods are too slow. There are various ways
of overcoming the difficulty; they may be classified according
to four basic principles which lead to four possible types of
procedure, as follows.

1. Bring the rate down to the conventional range, by the
use of low concentrations, or low temperatures.
2. Mix the solutions rapidly, by flow methods, as in the
continuous-flow and stopped-flow methods.
3. Disturb a system at equilibrium, and follow the
subsequent chemical change. There are two variants:-

3(a) Flash photolysis and pulse radiolysis produce large changes, e.g. formation of excited states or free radicals.

3(b) Small perturbations are brought about by the temperature-jump and other 'relaxation' methods; the rate of the shift of a chemical equilibrium to its new position is observed.

4. Make the reaction compete with some other fast process. In fluorescence quenching, for example, the reaction between the fluorescent molecule and some added reagent competes with the fluorescence of the molecule, which takes about 10^{-8} seconds. In nuclear magnetic resonance and electron-spin resonance methods, the reaction competes with spin relaxation. In the various electrochemical methods, reaction competes with diffusion towards an electrode.

Some representative techniques based on these four basic principles are summarised in Table 1. For descriptions of these and other techniques the reader should consult references 1-11. We now consider their applicability to the objectives of chemical kinetics, and their advantages and limitations from various points of view. The kineticist's concern is to select the technique which best suits his problem, as defined by the nature of the reactants, the solvent, temperature, rate constant, etc.

3. CHARACTERISTICS OF SOME FAST REACTION TECHNIQUES

(a) What is the maximum rate constant that can be measured? Most of the methods, if pushed, allow the determination of second-order rate constants approaching 10^{10} M^{-1} s^{-1}, which is about the diffusion-controlled limit in ordinary solvents. Such high rate constants are relatively easy to determine by temperature-jump, electric-field jump, ultrasonic absorption, dielectric absorption, flash, fluorescence-quenching, and e.s.r. methods. The stopped-flow method is not commonly used to determine rate constants above about 10^5 M^{-1} s^{-1}.

(b) What concentration range can be used? With most methods there is no problem. Fluroscence quenching and e.s.r. methods allow the use of exceptionally low concentrations. Relatively high concentrations are needed for n.m.r. and ultrasonics measurements as commonly practised.

(c) Can the temperature be varied, and can low temperatures be used? Temperature-control over a range sufficient for the determination of activation parameters presents no problem with most of the techniques; it is of course less precise where it is done by circulation of fluid rather than by immersion in a bath. Low-temperature work (to - 100°C or below) is not difficult for n.m.r., e.s.r., fluorescence quenching, and stopped-flow measurements.

Method	Time resolution, sec	Comments
1. Reduce rate to conventional.		
Low concentrations	—	Accurate estimation needed.
Low temperatures	—	Non-aqueous solvent needed
2. Rapid mixing by flow.		
Continuous flow	10^{-3}	Large volume needed.
Stopped flow	10^{-3}	Any solvent; versatile.
3. Disturb equilibrium.		
Flash photolysis	$<10^{-8}$	(Any solvent.
Pulse radiolysis	$<10^{-8}$	(Free radicals.
Photostationary methods	10^{-2}	(Excited states.
Temperature-jump, etc	$<10^{-5}$	Versatile. Equilibrium needed.
Ultrasonic absorption	10^{-9}	Solvent restricted.
5. Competition with fast process.		
Fluorescence quenching	10^{-8}	Reagents restricted.
Electrochemical methods	10^{4}	Ionising solvent, etc.
N.m.r. and e.s.r. methods	$<10^{-4}$	Infn. on particular atoms.

Table 1

Some Techniques for Fast Reactions in Solution

(d) <u>Can high pressure be used?</u> High-pressure forms of temperature-jump (16-19), pressure-jump (20), n.m.r. (21), and ultrasonic-absorption apparatus have been constructed. These can be pressurised to several kilobars. A stopped-flow apparatus working at pressures up to 1 kilobar has been constructed (22). Components for high-pressure work are available commercially, and these techniques seem likely to come into wider use (23).

(e) <u>Can the solvent be varied?</u> The choice of solvent is unrestricted for several of the techniques, including stopped-flow. Some techniques require an electrolyte solution, notably the commonest form of temperature-jump apparatus, which uses joule heating. With ultrasonic relaxation the solvent must be one which does not itself have a high absorption. With n.m.r. methods the signals due to the solvent must be considered; they may indeed be the important ones.

<u>Characteristics of the most-used techniques.</u> Some special benefits and restrictions of some of the most popular techniques are as follows (see Table 2). The use of low concentrations or low temperatures avoids expense but requires some time and ingenuity. The stopped-flow method is very adaptable to a variety of reactions (whether or not they go to completion; whether or not the product is stable; whether or not they involve ions or radicals) and to a range of solvents, detection methods, and temperatures; commercial apparatus is available relatively cheaply; but the fastest reactions are not accessible. The temperature-jump method is suitable for a wide variety of reactions in aqueous solution, and can handle very fast reactions; small volumes of solutions are enough, since the reactant solution is not lost but reverts to its original state; and commercial apparatus is available, which gives good results with aqueous solutions, though not with non-electrolyte or non-aqueous solutions, which require specially built laser or microwave apparatus (24). Flash-photolysis machines are commercially available and are especially useful for the study of excited states and of radical reactions; the reaction must of course be capable of photochemical initiation. Fluorescence quenching is perhaps the easiest method of studying very fast (diffusion-controlled) reactions, in any solvent, down to low temperatures; the main limitation is that one at least of the molecules taking part in the reaction must be fluorescent. The n.m.r. line-broadening technique can be used with good commercial apparatus, and gives uniquely detailed information, especially about the participation of the solvent (25); low temperatures are easily obtained; but the concentrations normally required are relatively high.

Principle	Method	Method requires:			Method Permits:				Apparatus can be bought
		Fast mixing	Fast obsn	Equilm.	Small volume	Low temp	High pressure	Very fast≠ reactions	
1	Low concentrations					*			
	Low temperatures					*			
2	Stopped flow	*	*			*	*		*
3	T-jump		*	*	*	*	*	*	*
	P-jump		*	*	*	*	*		*
	Ultrasonics			*	*	*	*	*	*
	Flash		*		*	*		*	*
	Pulse radiolysis		*			*		*	
4	Fluorescence				*	*		*	*
	Electrochemical			*				*	
	N.m.r.			*	*	*	*	*	*

≠ Those with second-order rate constants approaching 10^{10} M^{-1} s^{-1}.

Table 2

Characteristics of Some Fast-Reaction Techniques

3. THE CHOICE OF METHOD AND TECHNIQUE

The experimental kineticist's choice between various courses of action, when confronted with a particular kinetic problem, will depend on several factors. He will first choose a method that seems appropriate to the nature of the reaction, - according to whether it involves ions, or free radicals, for example; whether it goes to completion or to a balanced equilibrium; whether the product is stable or not - bearing in mind the expected rate of the reaction. He will then choose one of the particular techniques, depending on such factors as the type of solvent (ionising, polar, non-polar), the possible ways of monitoring the reaction (spectrophotometric, conductimetric, etc), the expected rate of the reaction, and whether variation of concentration and temperature will suffice or whether variation of solvent or pressure is also desirable. The scheme in Table 3 (which is not exhaustive) may serve to summarise some of the points that have to be considered.

Considerations of cost and of workshop facilities are also important. Commercial instruments are well developed for stopped-flow, pressure-jump, temperature-jump (by joule heating), fluorescence, flash photolysis, pulse radiolysis, n.m.r. and e.s.r. Apparatus can also be built for most of these, given the necessary man-hours and workshops.

The techniques that appear to be most popular at present among kineticists in general are stopped-flow, temperature-jump, flash photolysis, and n.m.r. The stopped-flow apparatus is the most generally useful for times down to a few milliseconds, and spectrophotometric detection allows the use of concentrations down to about $10^{-5}M$ or below; several hundred such machines are now in existance. Temperature-jump and flash-photolysis apparatus extend the time-range down to microseconds or below; the former is used when a well-characterised equilibrium is to be studied, the latter when free radicals or excited states are to be generated. N.m.r. methods give more detailed information than others, for instance on the role of hydroxylic solvents in proton-transfer reactions (25).

5. FIELDS OF APPLICATION

The widespread use of fast-reaction methods has opened up the study of new types of reaction and new chemical species, and has raised new questions about physical aspects of mechanisms. Examples of types of reaction which could not be studied without fast-reaction techniques include the following: among reactions of labile metal ions, ligand substitution, solvent exchange, and electron-transfer; among organic reactions, many proton-transfer

Reactants	Is an equilm. set up?	Methods available	Techniques available
Molecules or ions	no	Flow	(Stopped-flow (Continuous flow
	yes	Relaxation Single pulse	(T-jump: Joule heating; (laser; microwave (P-jump: E-jump.
		Periodic disturbance	(Ultrasonic absorption: (Pulse; resonance etc.
Free radicals or excited states		Above methods plus:-	
		Flash photolysis	(Flashlamp (Laser
		Pulse radiolysis	(Accelerator (Radiation
		Photostationary methods	(Rotating sector etc.

Table 3

Factors in the Choice of Method and Technique

reactions (26, 27), hydrogen bonding, radical reactions, and fast
steps in polymerisation reactions; among reactions of biological
importance, those of haemoglobin with oxygen, helix-coil
transitions in macromolecules, and fast steps in enzyme
reactions, photosynthesis and reactions in micelles. Among new
chemical species studied by these methods are short-lived free
radicals, excited states and the solvated electron. Examples
will be found in this volume and in references 1-11. Among new
questions about physical aspects of mechanisms are the role of
hydroxylic solvents as bridge molecules in proton transfer[25],
the question whether solvent motions accompany or precede or
follow changes of bonding, and problems about diffusion effects
which are important when the activation energy is small.

CONCLUSION

Fast-reaction methods have developed rapidly and continuously
over the last 25 years and are still being improved. Their
applications have likewise proliferated. They are indispensable
in the study of the kinetics and mechanisms of the numerous
reactions which, though not essentially different from more
familiar reactions, are too fast for conventional methods.
They permit investigation of the reactions of short-lived species -
free radicals and excited states - which would otherwise be
inaccessible. They throw light on physical phenomena such as
diffusion control of very fast reactions, the role of the
solvent, and solvation changes. They are valuable both in the
more intensive study of familiar types of reaction and in the
exploration of new ones such as helix-coil transitions and
enzyme reactions, which are of great biological importance.
They appeal alike to chemists who, being romantic by temperament,
are attracted to novel or complex situations, and those whose
more classical bent leads them to enquire more deeply into
subjects where there is already a modicum of understanding.
The kinetics and mechanisms of fast reactions do not constitute
a subject; they form part of the general subject of kinetics
and mechanisms, and cannot be isolated. The techniques and
methods, however, do have sufficient unity to constitute a
recognisable field, as has been shown by successive meetings of
the Fast Reactions in Solution discussion group of the Chemical
Society; there is a degree of common approach that encourages
cooperation and emulation. This unity of approach, coupled
with the diversity of chemical problems, is the reason and
justification for a book such as the present one.

REFERENCES

1.   Caldin, E.F.:   1964, "Fast Reactions in Solution",
     Blackwell, Oxford.
2.   Britton Chance, Eisenhardt, R.H., Gibson, R.H., and
     Lonberg-Holm, K.K. (eds.): 1964, "Rapid Mixing and
     Sampling Techniques in Biochemistry", Academic Press,
     New York.
3.   Claesson, S.; (ed.): 1967, "Fast Reactions and Primary
     Processes in Chemical Kinetics", Interscience.
4.   Czerlinski, G.: 1966, "Chemical Relaxation", Arnold.
5.   Kustin, K., (Ed.): 1969, "Fast Reactions" (Methods in
     Enzymology, Vol. XVI), Academic Press, New York.
6.   Hague, D.N.: 1971, "Fast Reactions", Wiley.
7.   Hammes, G.G., (ed.): 1974, "Investigation of Rates and
     Mechanisms of Reactions", 3rd edition (Techniques of
     Chemistry, vol. VI, part II), Interscience, New York, 1974.
8.   Bradley, J.N.: 1975, "Fast Reactions", Clarendon Press,
     Oxford.
9.   Wyn-Jones, E., (ed.): 1975, "Chemical and Biological
     Applications of Relaxation Spectrometry", Reidel, Dordrecht.
10. Bernasconi, C.F.: 1976, "Relaxation Kinetics", Academic
     Press, New York.
11. Strehlow, H., and Knoche, W.: 1977, "Fundamentals of
     Chemical Relaxation", Verlag Chemie, Weinheim, 1977.
12. Hartridge, H., and Roughton, F.J.W.: 1923, Proc. Roy.
     Soc., A, 104, p. 376; B, 94, p. 336. For a historical
     account, see Roughton, F.J.W., in Ref 2, pp. 5-13.
13. Britton Chance: 1940, J. Franklin Inst., 229, pp. 455,
     613, 737.
14. Disc. Faraday Soc., 1954, 17, pp. 114-234.
15. Z. Elektrochem, 1960, 64, pp. 1-204.
16. Caldin, E.F., Grant, M.W., Hasinoff, B.B., and Tregloan
     P.A.: 1973, J. Physics E., Sci. Instrum., 6, p. 349.
17. Yu, A.D., Waissbluth, M.D., and Grieger, R.A.: 1973,
     Rev. Sci. Instr., 44, p. 1390.
18. Jost, A.: 1974, Ber. Bunsenges. Physik. Chem., 78, p. 300.
19. Heremans, K. Snauvaert, J., Vandersypen, H.A. and van
     Nuland, Y.: 1974, Proc. 4th Int. Conf. High Pressure,
     Kyoto, p. 623.
20. Brower, K.R.: 1968, J. Amer. Chem. Soc.,   90, p. 5401.
21. Merbach, A.E., et al.: 1977, Helv. Chem. Acta, 60,
     p. 1124; Inorg. Chem. Acta, 25, p. L91.
22. Heremans, K., Snauvaert, J., and Rijckenberg, J.: 1977,
     Proc. 6th Int. Conf. High Pressure, Boulder.
23. For a review see Heremans, K.: in Proceedings of N.A.T.O.
     Advanced Study Institute, Corfu, 1977, on "High-Pressure
     Chemistry", forthcoming.
24. Caldin, E.F.: 1975, Temperature-jump Techniques, in "Chemistry
     in Britain", 11, p. 4.

25. Caldin, E.F., and Gold, V. (ed.): 1975, "Proton-transfer Reactions", London, article by Grunwald, E., and Eustace, D., chap. 4, pp. 103-120.
26. Robinson, B.H.: in ref 25, chap. 5.
27. Crooks, J.E.: in ref. 25, chap. 6.

FAST CONTINUOUS FLOW

Joseph F. Holzwarth

Fritz-Haber-Institut der Max-Planck-Gesellschaft,
Faradayweg 4-6, D-1000 Berlin 33, Germany

The "Continuous Flow Method with Integrating Observation",
CFMIO, is described, which uses a very fast jet mixer and incorpo-
rates this mixing chamber into the observation tube. To increase
the detection path length up to some cm and to bring the mixing
area into the detection region, the steady state of reaction is
followed in the direction of flow. The mixing process is examined
and integrated rate equations are given to determine the rate
constants of first and second order reversible, as well as irreversible, reactions. The time resolution of the CFMIO is demonstrated by measuring half lives of very fast electron transfer reactions as low as 5×10^{-6} s.

1. INTRODUCTION

The continuous flow method provides an excellent example of
the way in which a fast chemical reaction can be measured, using
a slow detection system and observing the steady state inside a
flow tube in which turbulent flow occurs at a constant rate.
Rutherford in 1897 (1) was the first who applied this principle
to observe the recombination of photolyzed substances in the gas
phase. He achieved a time-resolution of 10^{-1} s. In 1923 Hartridge
and Roughton employed the same method for reactions in solutions
after developing a jet mixer which allowed the observation of reactions occurring in 10^{-2} s. Since that time great efforts have
been made to improve the mixing process and to lower the consumption of reactants and solvent. As a result stopped-flow (2) arrangements with mixing times of 10^{-3} s and solution volumes of 1 ml
are commercially available (3).

W. J. Gettins and E. Wyn-Jones (Eds.). Techniques and Applications of Fast Reactions in Solution. 13–24.
Copyright © 1979 by D. Reidel Publishing Company.

In contrast to the relaxation technique, the flow methods can be used for reactions with large equilibrium constants (irreversible reactions) as well as with small ones (reversible reactions). The shortest half life that can be measured in flow systems depends on the mixing time. In stopped flow arrangements, the stopping time sets a barrier at 5.10^{-4} s. In a continuous flow system, only the mixing process is limiting.

We have therefore developed a novel, very fast jet mixer and incorporated this "mixing chamber" into the observation tube of a continuous flow cell with integrating observation (CFMIO) which allows mixing and reaction to be followed simultaneously.

2. FLOW METHOD WITH INTEGRATING OBSERVATION

The principle of our CFMIO measuring method is shown in Fig. 1. To increase the detection path length and to bring the complete mixing area into the detection beam, the steady state of reaction is followed in the direction of flow. This is possible because two quartz windows are situated at the beginning and end of the observation tube. Light guides are inserted perpendicular to the flow tube, to achieve additional information about the concentration profile inside the tube or to be used as stopped-flow detection channels (4). The integral of the extent of reaction inside the observation tube (Fig. 1), which is measured in the CFMIO, remains constant at constant flow rate.

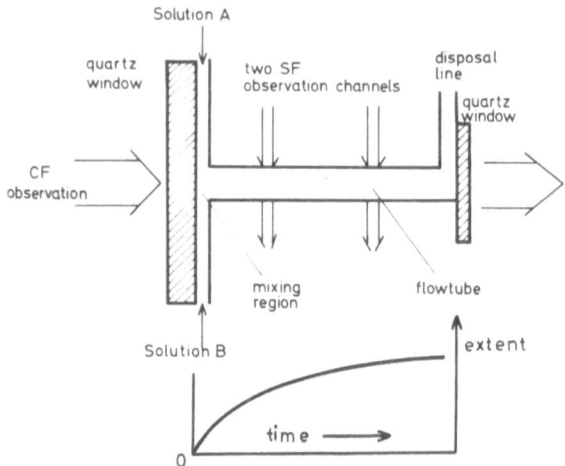

Fig. 1: Principle of the CFMIO: CF, continuous flow; SF, stopped flow; together with a plot of the extent of reaction against time inside the flow tube.

By changing the initial concentrations or the flow rate, its value can be changed and complicated profiles are observable. It is therefore not necessary to use measuring cells with different length of the observation tube though this would be possible. To calculate the rate constant, the absorbance inside the observation tube, of the reactants E_o and the products E_∞ as well as the absorbance during the course of reacton E has to be known. The quotient $M = (E - E_\infty).(E_o - E_\infty)^{-1}$ is then obtained. M is connected with the rate constant k, the length of the observation tube l, the initial concentrations of the reactants inside the flow tube c_o and the flow velocity v, by the parameter ξ. In Table 1 this relationship is given for different rate laws (5,6).

Table 1: Some examples of the theoretical equations for the calculation of the rate constant k from the measured quotient M using the initial conc. c_o, the flow velocity v, the tube length l and the equilibrium constant K.

Type of Reaction	$M=(E-E_\infty).(E_o-E_\infty)^{-1}=f(\xi)$	$k=f(\xi,l,v,c_o,k)$
irrev. first order $A \xrightarrow{k_{12}} B$	$M=\xi^{-1}(1-\exp(-\xi))$	$k_{12}=\xi.v.l^{-1}$
irrev. pseudo-first order $A+B \xrightarrow{k_{12}\cdot c_{oA}} C$ $c_{oA} \gg c_{oB}$	$M=\xi^{-1}(1-\exp(-\xi))$	$k_{12}=\xi.v.l^{-1}.c_{oA}^{-1}$
irrev. second order $A+B \xrightarrow{k_{12}} C+D$ $c_{oA}=c_{oB}$	$M=\xi^{-1}.\ln(\xi+1)$	$k_{12}=\xi.v.l^{-1}.c_{oA}^{-1}$
rev.second order second order $A+B \underset{k_{21}}{\overset{k_{12}}{\rightleftharpoons}} C+D$ $c_{oA} = c_{oB}$	$M=1-\xi^{-1}\left[\ln(\exp(\xi)+\alpha)\right.$ $+\alpha^{-1}\ln(1+\alpha\exp(-\xi))$ $\left.-(1+\alpha).\alpha^{-1}\ln(1+\alpha)\right]$ $\alpha =(1-K^{1/2}).(1+K^{1/2})^{-1}$ $K=k_{12}/k_{21}$	$k_{12}=0.5.\xi.v.K^{1/2}.l^{-1}.c_{oA}^{-1}$

Fig. 2 shows the schematic diagram of the complete experimental arrangement. Four syringes are connected to the measuring cell and the solution containers via three-way stop-cocks, so that E, E_0 and E_∞ can be measured. The measuring cell and all parts containing solution are made out of teflon or glass and the whole flow system can be evacuated and kept free of oxygen. A slow spectral photometric detection system is used with an integration time of 3s and an accuracy of 0.1 % for each measured absorbance. The design of the measuring cell is shown in Fig. 3. 10 to 20 slits are radially arranged around the top of a 20 or 30 mm long 2 mm diameter observation tube which itself is enclosed by two quartz windows. The slit dimensions at the entrance of the observation tube are 0.1 x 0.2 mm^2. A uniform turbulent character of flow at any cross section is obtained with flow velocities between 1.5 and 5 m s^{-1} using aqueous solutions. To avoid cavitation, the complete flow system is kept under a pressure of 10 bar.

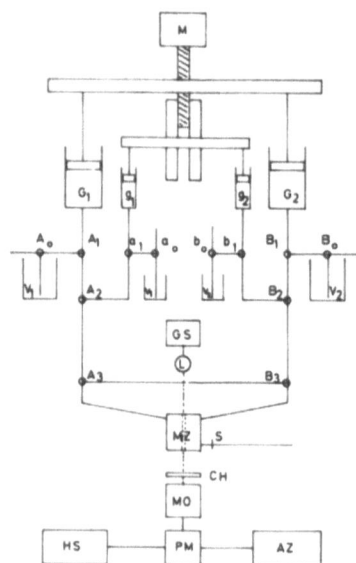

Fig. 2: Experimental arrangement of the CFMIO: M, motor. Flow system: V, v solution container; G,g, syringes; A,a,B,b, three-way stop-cocks; MZ, measuring cell; S, clamp. Detection system: GS, power supply; L, lamp with lenses and filters; CH, lens; MO, monochromator; PM, photomultiplier; HS, high voltage supply; AZ, unit for registration.

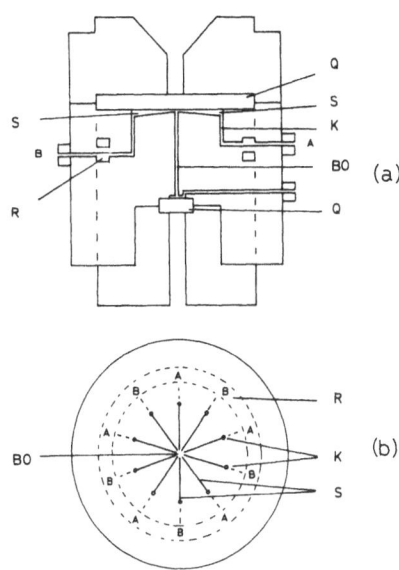

Fig. 3: Measuring cell of the CFMIO:
(a) longitudinal section: Q, quartz windows; s, inlet
jets; K, bores to the ring channel, R; A and B, inlet
of the solutions into R; BO, observation tube.
(b) cross-section showing the inlet jets, S, and the
bores, K, to the ring channels R.

3. RESULTS

     In the CFMIO, the complete mixing process occurs inside the
observation tube, it is therefore necessary to achieve detailed
information about mixing in order to understand its influence on
the reaction profile. Unfortunately, mixing inside a jet mixer
with turbulent flow is such a complicated kinetic problem that
it was not possible until now to derive an exact mathematical
description. A phenomenological description is attempted in Fig.
4. On the left side of the figure, the decay of the normalized
value of the intensity of segregation S is shown. S(x) represents
the mean square deviation form homogeneity or simply the part of
unmixed solution volumes at different distances from the beginning of the observation tube (7); S = 1 means complete separation
of the two solutions and S = O complete mixing. At a certain
cross-section of the observation tube, a distribution of differently mixed volume elements is found. These volume elements are the
turbulence eddies, their size distribution ranges from 1 µm to

1 mm and their lifetime is around 1 ms (8). In a good mixer, like
the one here described, mainly small eddies (20 μm) with a short
lifetime are created. This is demonstrated for different cross-
sections x along the observation tube at the right side of Fig.
4. At the entrance of solution A and B into the observation tube
(jets) new eddies are built by mixing. These new volume elements
contain mainly large volume ratios, V_A/V_B if we start with volume
elements of solution A, and V_B/V_A if we start with volume elements
of solution B. That means a small part of the volume of solution
A is badly mixed with solution B and vice versa. If we go further
down the observation tube, the better mixed parts of the solutions
are increasing and the less mixed parts are decreasing. Only at
the end of the mixing process do all volume elements contain 1/1
volume ratios of the initial solutions. It is extremely difficult
to measure the exact size-distribution function, the actual life
time and the distribution of the volume ratios V_A/V_B and V_B/V_A
inside the turbulence eddies at a certain cross section of the
flow tube because the whole mixing process is occurring in the
first 2 mm of the observation tube (9). However, exactly this
would be necessary to formulate a mathematical description of the
kinetics of mixing. We have therefore used an empirical method
to account for the influence of mixing in the calculation of rate
constants.

Mixing Process

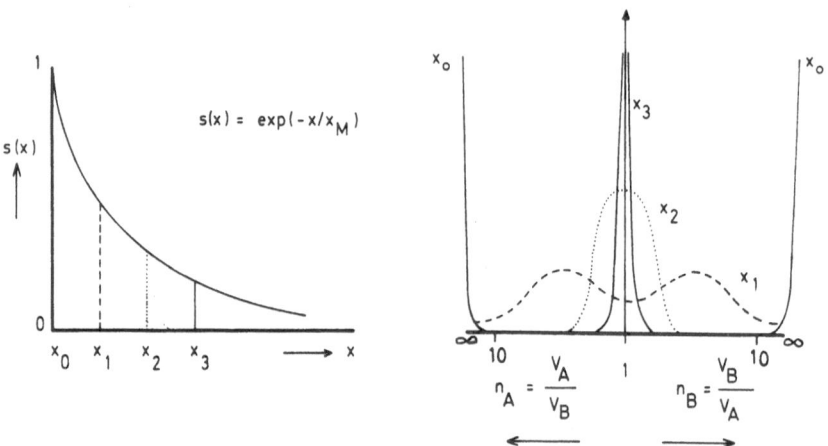

Fig. 4: Mean square deviation from homogeneity of mixing s(x) at different cross-sections of the observation
tube together with the distribution function of the
volume ratios n_A and n_B inside turbulence eddies at
the cross-sections: x_0, beginning of flow tube, and
further down the tube x_1, x_2 and x_3.

   In Fig. 5 the relationship between the observed and the real
half lives of three different second order irreversible electron
transfer reactions with equal initial concentrations measured in
our flow cell is shown. Below 2×10^{-4} s the observed half life
τ_R^* is greater than its real value but, even at half lives around
10 µs an exact correlation between the half lives observed and
their real values exists and the behaviour of all three reactions
is the same. This proves that the influence of mixing on reaction
rates in the CFMIO is a function of the half lives. The rate law
and the initial concentrations together with the rate constant,
define the half life; no simple dependence of M on the value of
the rate constant exists.

   In Fig. 6, the general dependence of the measured quotient
M on the half life of a second order irreversible reaction with
equal initial concentration obtained in the CFMIO is given.
In Fig. 7, a similar plot for pseudo-first order reactions is
shown. The lower values of M, observed at short half lives (compare Fig. 6 and Fig. 7), are not surprising if we remember the
model of mixing described in the previous part. In pseudo-first
order reactions, all turbulence eddies, with a volume ratio V_A/V_B
up to the excess ratio of the higher concentrated reactant (B),
can react completely; this is not possible in a second order
reaction with equal concentrations of the reactants. This is the
reason that the M values of the latter are higher at the same half
lives. The time resolution for pseudo-first order reactions in
Fig. 7 is again better than 10 µs.

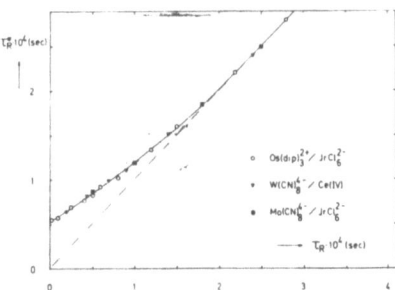

Fig. 5: Plot of the observed half-lives τ_R^* of three
electron transfer reactions ($k_o = 3.7 \times 10^9$ M^{-1}s^{-1},
$k_\nabla = 1.9 \times 10^8$ m^{-1}s^{-1} and $k_\bullet = 1.9 \times 10^6$ M^{-1}s^{-1};
T = 297 K in 0.5 M H$_2$SO$_4$) against the real half lives
τ_R, obtained with equal initial concentrations.

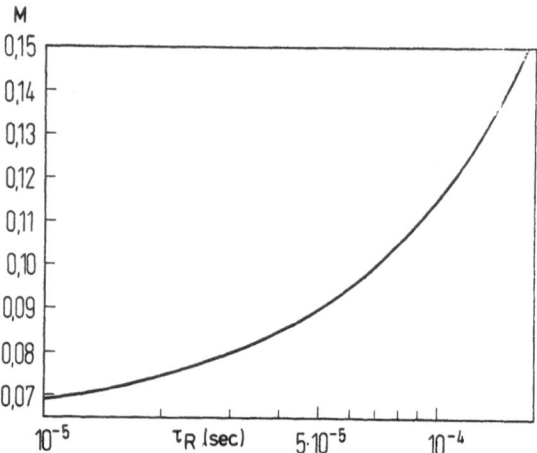

Fig. 6: Experimental dependence of the measured dimen-
sionless quotient M on the real half lives τ_R of second
order reactions, with equal initial concentrations, in
the mixing influenced range.

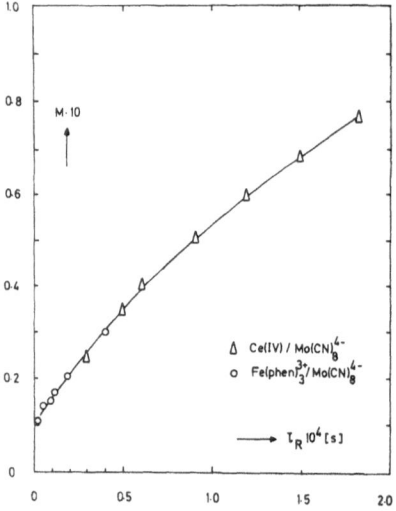

Fig. 7: Plot of the quotient M against the real half
lives τ_R for two electron transfer reactions (k_Δ = 1.1
x 10^7 $M^{-1}s^{-1}$ in 0.5 M H_2SO_4, k_o = 3.9 x 10^9 $M^{-1}s^{-1}$ in
0.5 M $HClO_4$, T = 296 K) measured in a pseudo-first-
order mode with an excess ratio of at least 10:1 of
one of the initial concentrations.

Single measurements of the forward rate constants of electron transfer of an irreversible second order reaction and a reversible reaction of the same type in dependence on the reciprocal
initial concentrations are shown in Fig. 8. Both reactions are
completely diffusion-controlled and were measured with equal
initial concentrations of the reactants. At higher initial concentrations, the influcence of the mixing process on the results,
is observed as an apparent decrease in the measured value of the
rate constant. Mixing was neglected in the calculations leading
to these results. The earlier decrease of the rate constant, in
the case of the second order reversible reaction, is caused by
the shorter half-lives in comparison to the irreversible reaction
with the same forward rate and equal initial concentration. In
reversible second order reactions, the back-reaction keeps the
reactant concentrations higher during the course of reaction.
This causes shorter half lives. The results of Fig. 8 again prove
that the influence of mixing is a function of the actual half
lives during a reaction and not of the rate constants.

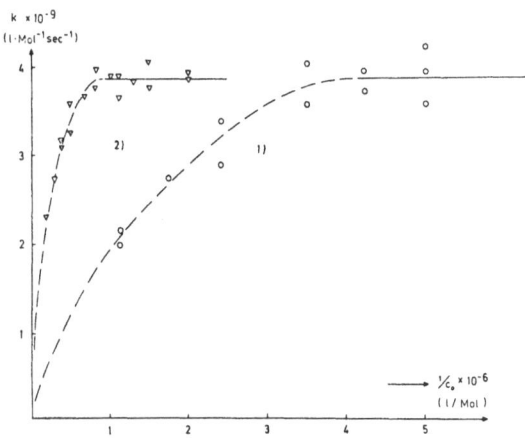

Fig. 8: Dependence of the forward rate constant of a
reversible reaction (1) and an irreversible reaction
(2) on the reciprocal initial concentration c_o (both
reactant conc. were equal)

(1) $Mo(CN)_8^{4-} + Os(dipy)_3^{3+} \underset{k_{21}}{\overset{k_{12}}{\rightleftharpoons}} Mo(CN)_8^{3-} + Os(dipy)_3^{2+}$

$K = k_{12}/k_{21} = 6$, in 0.5 M H_2SO_4 at T = 293 K

(2) $W(CN)_8^{4-} + Fe(5-CH_3-phen)_3^{3+} \xrightarrow{k_{12}} W(CN)_8^{3-} +$

$+ Fe(5-CH_3-phen)_3^{2+}$

$\Delta G^o = 44.5$ kJ, in 0.5 M $HClO_4$ at T = 293 K.

In Table 2 a number of outer sphere electron transfer reactions between substitution-inert transition metal complexes are summarized. The rate constants of these reactions are equal within the accuracy of measurement. Different charge products are not reflected in the results because electrostatic interactions are shielded by the high ionic strength of added acids. Also the change in the free energy of these redox couples does not influence the rates, even if the same reducing agent is reacting with almost similar oxidizing complexes ($Mo(CN)_8^{4-}$ + $Fe(X-phen)_3^{3+}$) or vice versa ($Ru(dipy)_3^{3+}$ + $Mo(CN)_8^{4-}$, $W(CN)_8^{4-}$ ....). The only possible explanation for this behaviour is that all reactions are completely diffusion controlled. The rate constants are reflecting the encounter rates and not the much faster electron transfer rates (10). The average rate constant of 3.2×10^9 $M^{-1}s^{-1}$ agrees with the Smoluchowski theory (11) for diffusion controlled reactions of such complexes.

Red	Ox	k_{exp} [$l \cdot Mol^{-1}sec^{-1}$]	Redox potential diff. ΔE [eV]	Medium [$Mol \cdot l^{-1}$]
	$Fe(DM-phen)_3^{3+}$	3.1×10^9	0.17	
	$Fe(M-phen)_3^{3+}$	3.2×10^9	0.22	
$Mo(CN)_8^{4-}$	$Fe(phen)_3^{3+}$	3.2×10^9	0.26	1 $HClO_4$
	$Fe(Cl-phen)_3^{3+}$	3.1×10^9	0.31	
	$Fe(NO_2-phen)_3^{3+}$	3.3×10^9	0.45	
$W(CN)_8^{4-}$	$Fe(DM-phen)_3^{3+}$	3.2×10^9	0.41	1 $HClO_4$
	$Fe(NO_2-phen)_3^{3+}$	3.2×10^9	0.69	
$Mo(CN)_8^{4-}$		3.2×10^9	0.44	
$Os(dipy)_3^{2+}$		3.3×10^9	0.44	
$Mn(CN)_5NO^{3-}$	$Ru(dipy)_3^{3+}$	3.3×10^9	0.64	1 H_2SO_4
$W(CN)_8^{4-}$		3.4×10^9	0.68	

Table 2: Second order rate constants of completely diffusion controlled electron transfer reactions measured with the CFMIO at 296 K.

4. CONCLUSION

The "Flow Method with Integrating Observation" is the only method which allows the measurement of reversible as well as irreversible reactons with half-lives down to 10 µs using reactants which are stable against decomposition (in contrast to pulsradiolysis and photochemistry where at least one reactant is unstable). This high time-resolution is comparable with that of a commercially available Joule heating apparatus, but such are restricted to reversible reactions. Only very small reactant concentrations down to 10^{-7} M, are necessary for measurements. A cheap conventional spectrophotometric detection system is used. All parts are easy to handle. The only remaining disadvantage is the high solvent consumption of 300 ml. This problem can be avoided if a continuous wave tunable dye laser is used as the detection light source. Such a laser allows a complex light path in our cell, through the inlet bores, the exit bore, and the observation tube using the same beam. In such an arrangement, all three absorbances necessary to calculate the rate constant could be measured inside 2 s, the solvent consumption would be below 20 ml and a time resolution of 10 µs would still be possible.

ACKNOWLEDGEMENT

I would like to thank my coworkers Dr. L. Strohmaier and Dr. D. Seifert who developed the CFMIO together with me, and who did most of the measurements mentioned in this article.

REFERENCES

(1) Rutherford, E.: 1897, Phil. Mag. 44, pp. 422.
(2) Chance, B.: 1974, in "Investigation of Rates and Mechanisms of Reactions" ed. G. G. Hammes, J. Wiley, N. Y. pp. 5.
(3) Robinson, B. H.: 1975, in "Chemical and Biochemical Applications of Relaxation Spectrometry" ed. E. Wyn-Jones, D. Reidel, Pub. Comp., Boston, USA, pp. 41.
(4) Bruhn, H., Westerhausen, J., J. H. Fuhrhop and Holzwarth, J. F.: 1979, this volume, "Combined Stopped-Flow/Continuous-Flow Arrangement".
(5) Gerischer, H., and Heim, W.: 1965, Zeitschr. Physik. Chem. Neue Folge, 46, pp. 345.
(6) Gerischer, H., Holzwarth, J. F., Seifert, D. and Strohmaier, L.: 1970, Ber. Bunsenges. Phys. Chem., 74, pp. 589.
(7) Danckwerts, P. V.: 1957, Chem. Engng. Sci., 7, pp. 116.
(8) Davies, J. T.: 1972, in "Turbulence Phenomena;, Acad. Press, London, pp. 51.
(9) Gerischer, H., Holzwarth, J. F. Seifert, D. and Strohmaier, L.: 1972, Ber. Bunsenges. Phys. Chem., 76, pp. 11.
(10) Holzwarth, J., and Jürgensen, H.: 1974, Ber. Bunsenges. Phys. Chem., 78, pp. 526.
(11) Smoluchowski, v., M.: 1917, Z. physik. Chem. 92, pp. 129.

COMPUTER-CONTROLLED REPETITIVE OPERATION OF A STOPPED-FLOW
SPECTROPHOTOMETER WITH A MAGNETIC VALVE IN PLACE OF THE
STOPPING SYRINGE

Raymond P. Cox,[*] Andrzej Ormicki[+] and Hans Degn[*]

[*]Institute of Biochemistry and [+]Department of
Computer Science, Institute of Mathematics, Odense
University, Campusvej 55, DK-5230 Odense M, Denmark.

INTRODUCTION

The stopped-flow technique (1,2) is the most flexible and
widely used method for the study of rapid biochemical reactions.
A major limitation, however, is the collection and analysis of
the results. The introduction of laboratory on-line computers
allowed a considerable increase in the accuracy and convenience
of data collection and processing (3). A logical further advance
is to allow the computer to control the operations of the apparatus. The major obstacle to this is the stopping syringe. We
describe here how this may be replaced by a suitable electromagnetic valve, allowing the whole measuring cycle to be controlled
by the computer.

DESCRIPTION OF THE APPARATUS

Stopped-flow spectrophotometer

A commercially available apparatus of the type developed by
Gibson and Milne (4) (Applied Photophysics, London, W.1.) was
modified by replacing the vertically arranged glass stopping
syringe by a 24 V electromagnetic valve (Type 65.231, H. Kuhnke,
Malente/Holstein, W. Germany). This is originally designed as a
two-way pneumatic valve. We have blocked the common opening per-
manently, and altered the port which is normally closed when the
valve is unpowered so that flow through it is always possible.
After these modifications the valve is closed when the power is
on and otherwise held open by the spring. It is mounted on an
adaptor so that the piston moves in the direction of fluid

25

W. J. Gettins and E. Wyn-Jones (Eds.). Techniques and Applications of Fast Reactions in Solution. 25–28.
Copyright © 1979 by D. Reidel Publishing Company.

flow during closing. The closing time quoted by the manufacturers
is 10-15 ms, which would provide a significant limitation to
stopped-flow measurements. However, the important consideration
in practice is the time during which the closing valve provides
a flow-limiting resistance. This occurs only when the valve is
almost shut. Since the liquid flow is undirectional, only spent
solution can come into contact with the valve, which has brass
parts. A second electromagnetic valve of the same type was placed
in parallel with the manually operated valve of the pneumatic
drive to allow electrical triggering.

Computer and interface

A minicomputer (D116, Digital Computer Controls Inc.
Fairfield, N.J.) with 8K of store was linked to the stopped-flow
apparatus by an interface built in our workshops. This has an
analog-to-digital converter (Type DAS 1128, Analog Devices,
Norwood, Mass.), powered (24 V) logic outputs to control the
magnetic valves, analog output to allow the results to be plotted
on an X-Y recorder, logic inputs to allow control of the operations without using the teletype, and a real time clock. Fig. 1
shows the layout of the apparatus.

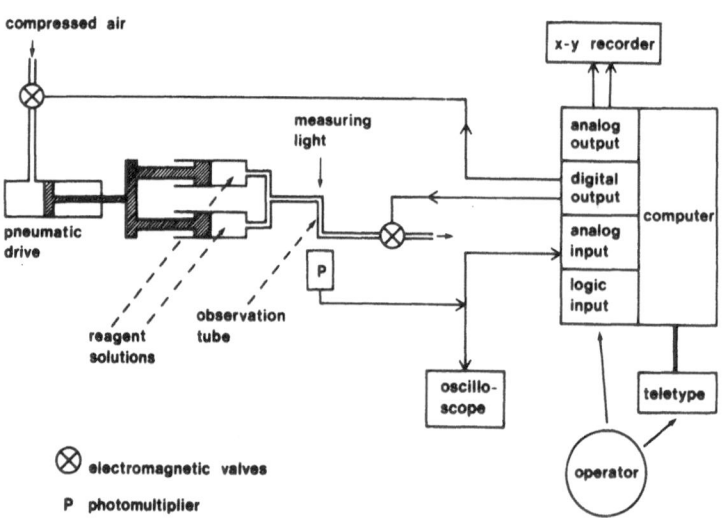

Fig. 1.   Block diagram of the apparatus

BASIC Statement	Result
550 CALL 4,0,1	Open flowline valve
560 CALL 4,1,0	Open pneumatic valve
570 FOR Y=1 TO 100	Delay loop
580 NEXT Y	
590 CALL 4,14,0	Trigger oscilloscope
600 CALL 4,14,1	
610 CALL 5,3	Start clock
620 CALL 4,0,0	Close flowline valve
630 CALL 10,A(1),A(N),Q	Take in N data points
640 CALL 6,T	Read clock

Fig. 2.   Program segment controlling start and stop of flow

OPERATION AND PERFORMANCE

A program in BASIC controls the whole sequence of operations via the interface using special CALL statements linked to the various inputs and outputs.

Before a series of experiments the computer is told the number of points to be recorded and the interval between them. The signal corresponding to zero transmittance (with the lamp shutter closed) is measured. The measuring cycle (Fig. 2) is started either by a teletype command or by a logic input switch. The computer then opens the valve controlling the pneumatic drive and after a suitable flow time, the magnetic valve replacing the stopping syringe closes and the signal from the photomultiplier is recorded by the computer. Once the predetermined number of points has been stored, the computer continues to sample the signal until there is no further change, to determine the end value. The results are then converted to changes in absorbance. These can be plotted on an X-Y recorder either directly or after further calculations. The whole sequence may be repeated automatically, the absorbance values from a number of runs being averaged to improve the signal-to-noise-ratio.

The "dead time" of the apparatus was estimated from measurements of the rate of reaction of Fe(III) myoglobin with different concentrations of azide. With a pneumatic drive pressure of 250 kPa the first valid data point (that falling on the linear portion of a semilogarithmic plot) was recorded 5 ms after the apparent start of the reaction. The dead time estimated from the maximum deflection before the flow stops was 4 ms, suggesting that the effective closing time of the valve is not much more than 1 ms.

Semilogarithmic plots of suitable test reactions (Fig. 3) were linear with no evidence of artefacts caused by the valve closing. It was, however, necessary to keep the drive syringes activated during the recording of data to prevent a small artefact when the pressure was released.

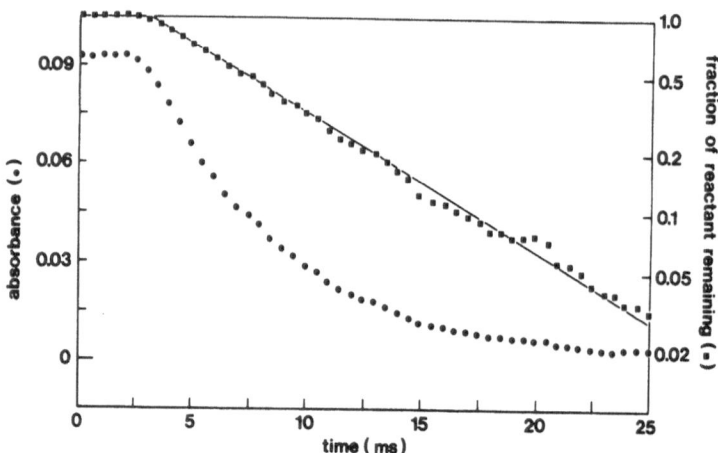

Fig. 3.   Example of the performance of the apparatus. The test reaction was the reduction by ascorbate of 2,6-dichlorophenol-indophenol, measured at 600 nm. The interval between points is 0.5 ms. The figure is redrawn from the computer output to an X-Y recorder.

DISCUSSION

This relatively simple modification provides a significant increase in the convenience of use of a stopped-flow spectrophotometer. It also greatly increases the practical value of repetitive stopped flow experiments, since it is only necessary to activate a single switch to average 8 runs. With a conventional apparatus manual intervention is necessary to refill the drive syringes after this, but modifications to avoid this limitation should pose no major problem. Already averaging 16-25 runs, corresponding to a theoretical improvement in the signal-to-noise-ratio of 4-5, is a much less tedious operation than with the conventional apparatus.

REFERENCES

1. Gibson, Q.H. (1969) Methods Enzymol. 16, 187-228
2. Chance, B. (1974) in Techniques of Chemistry, vol. 6 (Hammes, G.G. ed.) Part 2, pp. 5-62, Wiley-Interscience, New York.
3. DeSa, R.J. and Gibson, Q.H. (1969) Comput. Biomed. Res. 6, 216-227
4. Gibson, Q.H. and Milnes, L. (1964) Biochem. J. 91, 161-171

STOPPED-FLOW MEASUREMENTS IN THE INFRARED: SIMULATED RAPID SCAN
SPECTRA: SOME OBSERVATIONS ON THE REACTIONS OF METAL CARBONYL
COMPOUNDS WITH HALOGEN.

John P. Maher

School of Chemistry, University of Bristol, Bristol, U.K.

The design and operating details for a proto-type
infrared stopped-flow (IRSF) machine were previously described
(S.E. Brady, J.P. Maher, J. Bromfield, K. Stewart, and M. Ford,
J.Physics(E), Sci.Instr., 1976, $\underline{9}$, 19). This machine was based
upon a Perkin Elmer 237 infrared spectrophotometer; a novel
feature of the flow system was that it incorporated a new form
of 'diaphragm syringe' for injecting the solutions into the
mixing chamber. A new machine has been built which, whilst
maintaining the electronics and basic working principles of the
original machine, has an improved performance and ease of
operation.

The new machine uses the same optical system to focus
the infrared beam on the observation cell and detector, and the
same 2 kHz radiation chopper system. The PE 237 monochromator
has been replaced by a Grubb-Parsons M2 monochromator with the
optical filters and diffraction grating retained from the PE 237.
This new system gives and absolute wavelength accuracy of
$\pm$ 1.0 cm^{-1} at 2000 cm^{-1}, together with a similar 'reset'
accuracy. Because large changes in the kinetics signals (e.g.
from a positive-going exponential to a negative-going exponential)
are observed in IRSF kinetics over regions of a few wavenumbers,
such accuracy is necessary. The effect is especially pronounced
when product and reactant absorptions are narrow and close
together. With this wavenumber accuracy it is now possible to
use the machine to simulate rapid-scan infrared spectra in the
region 1800 cm^{-1} to 2400 cm^{-1}.

In the new machine the mixing and observation chamber
have been rebuilt on the lines of an infrared cell low temperature

29

W. J. Gettins and E. Wyn-Jones (Eds.). Techniques and Applications of Fast Reactions in Solution. 29–34.
Copyright © 1979 by D. Reidel Publishing Company.

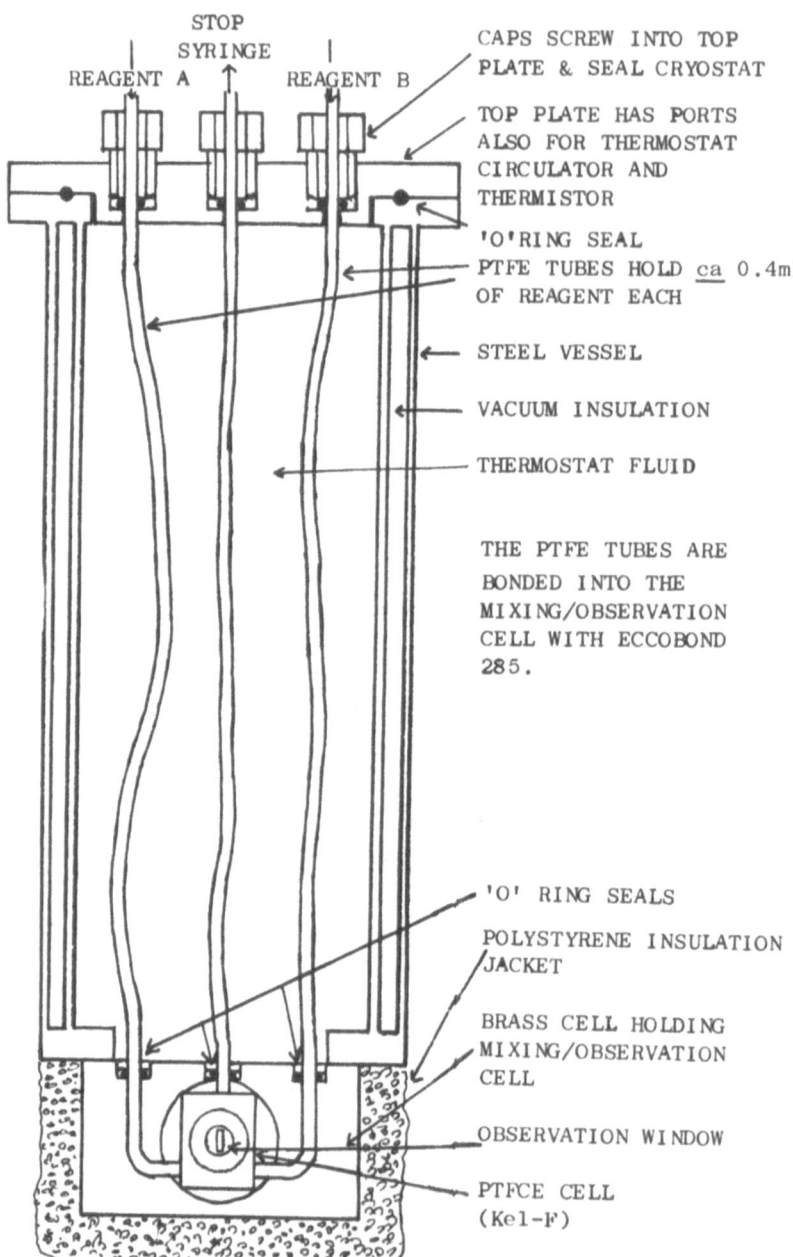

FIGURE 1   The Cryostat for the Mixing and Observation Cell.

cryostat (Figure 1). This enables the solutions to be thermostated to within ± 0.1 C, in the range −10 to +50 C, the lower
limit being presently set by the thermostat bath. The mixing
chamber has four tangential jets and is separated from the middle
of the observation window by 7 mm. The path−length in the
observation cell is 2.5 ± 0.02 mm, the observation window is
1 x 6 mm. and 1 mm inlet and outlet tubes lead to and from the
observation chamber. The whole mixing/observation cell is
machined from a single piece of Kel-F. The cell windows are of
CaF_2 and are held in place by rubber and p.t.f.e. gaskets and
screw caps (Figure 2). The dead−time has not been measured but
is estimated to be less than 5 msecs; mixing is complete before
the observation window.

The diaphragm syringes, solution taps and stopping
syringe are machined from a single slab of Kel-F (150 x 150 x
12 mm). Solutions are led to the mixing chamber, and then back
into this block to the stopping syringe through lengths of about
30 cm of thick walled (1.6 mm i.d.) p.t.f.e. tubing. The tubing
is passed through the thermostating fluid in the cryostat, to the
mixing and observation cell. (Similar to the principle used in
the Canterbury machines). The cell cryostat and the optics and
monochromator are covered by thermostated sealed boxes purged of
CO_2 and H_2O to facilitate the single beam infrared measurements.
The syringes and taps are thus mounted outside, and above the
apparatus making access to taps, stopping syringe, etc., easy.
Solutions are input from standard volumetric flasks with special
attachments for working anaerobically when necessary (Figure 3).

CAPS SCREW INTO THE BRASS CELL AT
THE BASE OF THE CRYOSTAT & HOLD
THE CaF_2 WINDOWS & THE VARIOUS
GASKETS IN BY PRESSURE

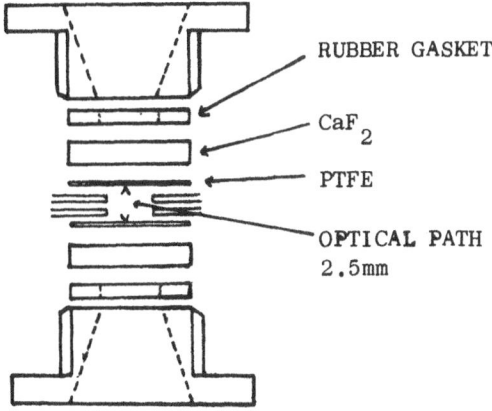

RUBBER GASKET

CaF_2

PTFE

OPTICAL PATH
2.5mm

FIGURE 2 Section through the Observation Cell.

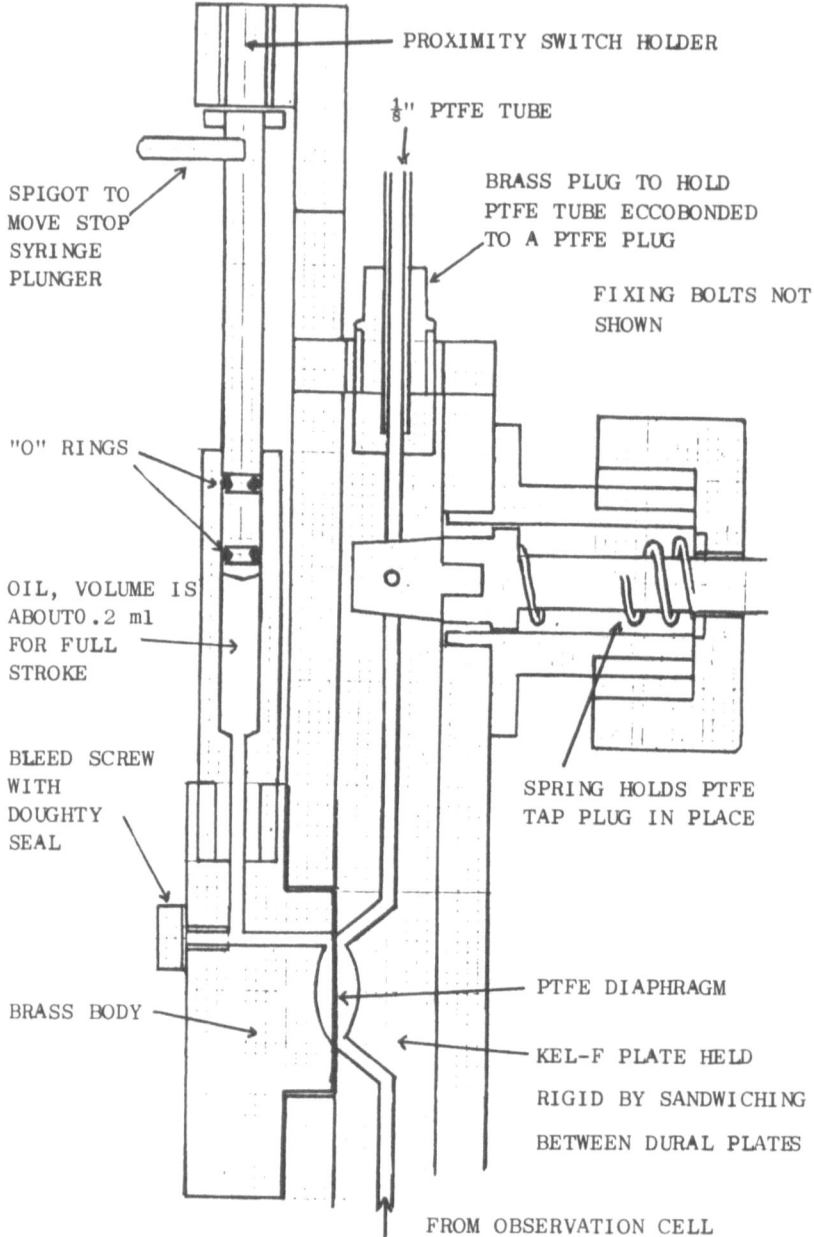

The two solution syringes are machined out of the Kel-F plate
on the opposite side to the stop-syringe, they hold approximately 5ml each with the diaphragms flat.

FIGURE 3   Diaphragm and Tap Block . Showing the Stopping Syringe
           and one of the Taps.

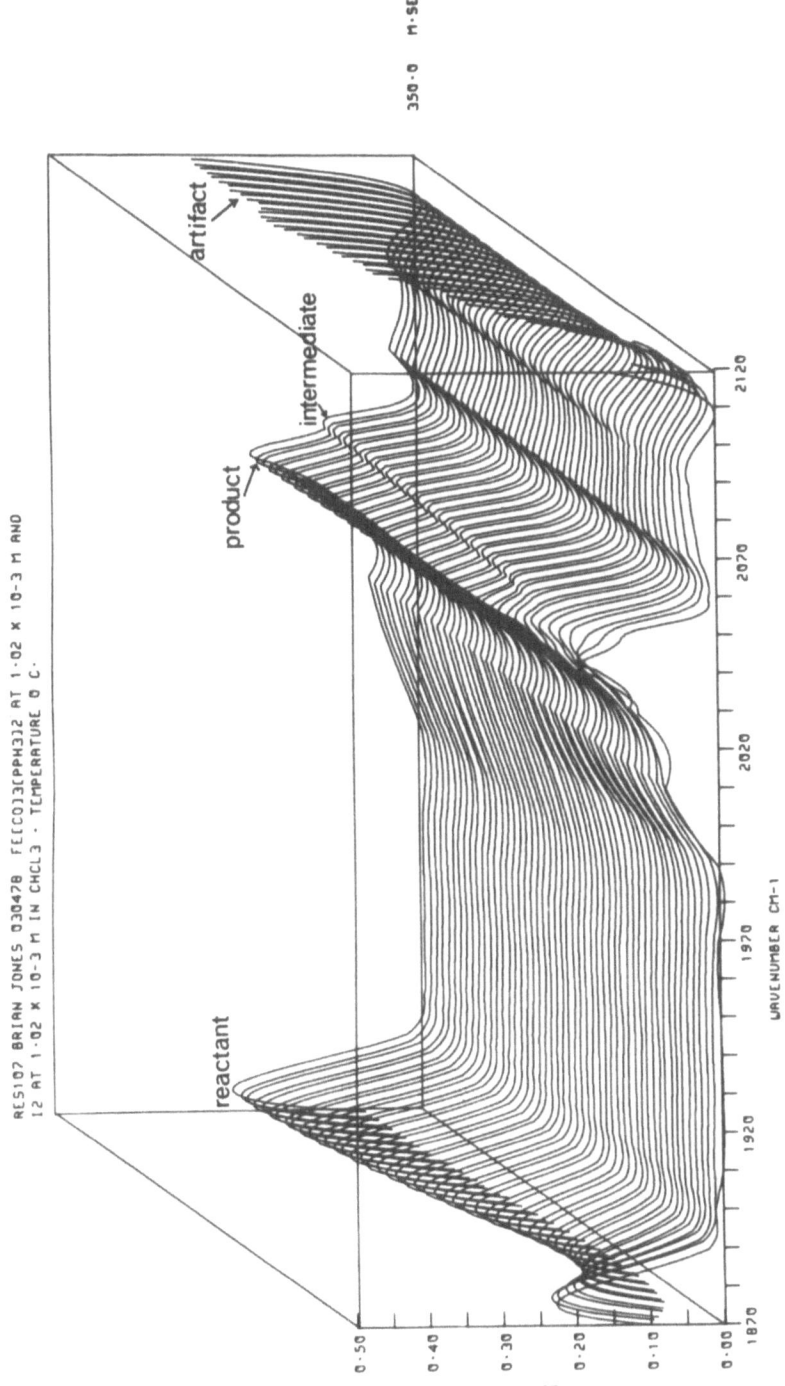

FIGURE 4   Transient Spectra observed in the reaction of $Fe(CO)_3(PPh_3)_2$ with iodine in chloroform at 0 C.

Rapid scan infrared spectra may be simulated by observing the kinetic changes in intensity at a series of different wavelengths, typically separated by 1 to 10 cm^{-1}, and concentrating on those regions of the spectrum corresponding to the absorption of reactants, any intermediates, and products. The data is collected on a Datalab DL 905 transient recorder (1024 points in 8-bits precision) and the separate runs are transferred to a digital cassette tape recorder (designed and built by J.S. Littler and K. Stewart at Bristol). Up to 50 runs can be recorded on a tape which is then read by the University computer (an ICL 4-75) where AlgolW programs interpret and edit the data. A picture of the reactions at 1 cm^{-1} intervals is then interpolated from the data, this can be used to generate diagrams of the spectra:- intensity versus wavelength versus time. Whilst this method is time consuming (a set of runs taking 2-3 hours) the data are accurate, giving rate constants and intensity data and some smoothing of the data. There is of course no distortion from the time required to sweep through the spectrum. Individual kinetic runs may be analysed by the micro-processor-based system described by J.S. Littler (page

Amongst the reactions which have been studied on this system the reaction of $Fe(CO)_3(PPh_3)_2$ with I_2 in $CHCl_3$ provides a good example of the type of information which can be obtained by this form of infrared spectroscopy. At a 1:1 ratio of reactants a very curious feature of the reaction was the regeneration of 50% of the starting Fe(0) complex after an initial reaction in which it nearly completely disappears. In Figure (4) the first 350 msecs of the reaction is shown for a spectrum between 1870 and 2120 cm^{-1}. Note (a) the decrease and then increase at 1878 cm^{-1} for the reactant, (b) the intermediate behaving in the same fashion, only with an initial increase and then decrease at ca 2050 cm^{-1}, (c) the product peak increase at ca 2043 cm^{-1}. Other features of the spectrum are observable at 2010 cm^{-1}, 2080 and 2100 cm^{-1}. At 2120 cm^{-1} the 'turning-up' effect for the plots is due to an artifact generated from the interpolation procedure. (The ALGOLW program uses the NAG E01AAF AITKEN interpolation), care has to be taken to provide the program with sufficient data points to provide an accurate interpolation.

Our full interpretation of these observations will be published elsewhere (submitted to J.Chemical Society,Dalton, P.K. Baker, N.G. Connelly, B.M.R. JOnes, J.P. Maher, and K. Somers); the proposed mechanism involves the formation and subsequent disproportionation reaction of an unusual 19 electron Fe(I) intermediate.

I thank Dr. John S. Littler for valuable discussions and Mr. B.M.R. Jones who carried out the chemical study leading to Figure (4).

SOLVENT-JUMP METHOD

Z. A. Schelly

Department of Chemistry
University of Texas at Arlington, TX 76019, USA

The principles and the experimental aspects of the solvent-jump (or concentration-jump) relaxation method are discussed.

INTRODUCTION

Chemical equilibria can be disturbed by changing intensive or extensive thermodynamic variables of the system. For experimental reasons, usually intensive variables are changed (1). From an operational point of view, one can distinguish perturbations by external and internal intensive variables. The external parameters usually altered are the temperature T, the pressure P, or the electric field strength E and the internal ones are the chemical potentials μ_i of the species present, i.e. the concentration of solutes and/or the composition of the solvent. A perturbation of the total solute concentration can be achieved by sudden dilution (concentration-jump) of the equilibrium system using rapid mixing devices (2). Similarly, the composition of the solvent can be abruptly changed by fast mixing of two equilibrium solutions of different solvents, or one equilibrium solution with another solvent. Since the total concentration of the solute is not necessarily changed in the experiment, the name solvent-jump is more generally descriptive of the method than concentration-jump. This type of relaxation experiment was first suggested by Ljunggren and Lamm (3) and has been applied to slow (4) as well as to fast reactions (5,6) using transient and also non-transient (7) observation in stopped- and continuous flow systems, respectively.

W. J. Gettins and E. Wyn-Jones (Eds.), Techniques and Applications of Fast Reactions in Solution. 35–39.
Copyright © 1979 by D. Reidel Publishing Company.

The difference between the regular flow methods (2) and the solvent-jump method is manifested in experimental procedures and in the evaluation of the data as well.

EXPERIMENTAL

In contrast to regular flow experiments where the reactants are mixed and the rate of reaction is measured, in solvent-jump experiments an equilibrium solution is perturbed, typically by a sudden dilution or by a sudden mixing with another solvent (Figure 1) and the relaxation time τ is measured. Besides the greater simplicity associated with the interpretation of relaxation times instead of rates, the solvent-jump method has obvious advantages in studying association equilibria where the reactants cannot be mixed.

The monitoring of the progress of the reaction after perturbation is usually done optically. However, conductometric, thermometric, CD, NMR, etc. detections can also be used.

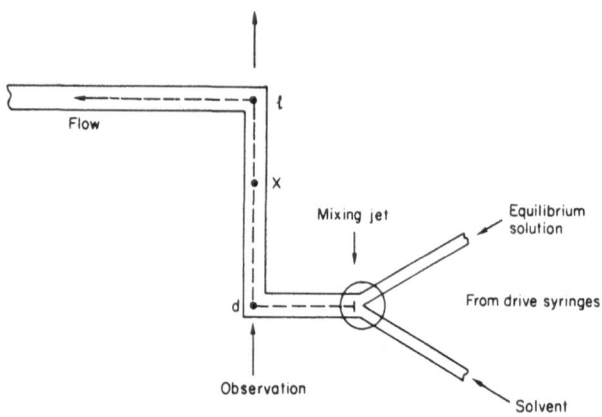

Figure 1.   Schematic diagram of a solvent-jump experiment using a flow system (stopped or continuous) with uniform cross section. Observation (optical detection) is done along the distance coordinate x between d and 1, all counted from the mixing jet.

Since the duration of the perturbation is the time required for complete and homogeneous mixing (which is typically one millisecond), relaxation times τ as short as a few milliseconds can be measured. If the flow velocity and the mixing jet design are optimized to reduce cavitation (8), the measurable limit can be reduced to half a millisecond when large perturbations are used.

The flow system can be operated either in the stopped or continuous mode. In the first case, as in other jump experiments, τ is obtained from the transient change of the concentration of the species that is being observed .

If continuous flow is used, the relaxation time is determined from the integrated relaxation signal amplitude A measured (on the steady state flow system during the flow) coaxially with the flow direction (7), using eq. (1)

$$(A-\bar{A})t_s/(\bar{A}-A_o/n) = \tau \exp(-1/v\tau)-\exp(-d/v\tau). \tag{1}$$

With reference to Fig. 1, A is the constant steady state absorbance (integrated signal amplitude) between d and 1. $\bar{A}$ and A_o are the absorbances of the final and initial (prior to mixing) equilibrium solutions, respectively, of path length s = 1-d. The residence time of the solution between d and 1 is symbolized by t_s, n stands for the dilution ratio and v for the constant flow velocity.

RELAXATION AMPLITUDES, $\Delta\xi$

The overall mixing-relaxation process can be subdivided as a transition from state-1 to state-3 over a distinct intermediate state-2:

$$\text{state-1} \longrightarrow \text{state-2} \longrightarrow \text{state-3} \tag{2}$$

State-1 is the initial equilibrium state at T and P. State-2 arises shortly after the sudden n-fold dilution with the original or with another solvent. At this point all fast physical processes (e.g. the redistribution of the solvent and ionic atmospheres) have been completed and the system is ready for the slower chemical relaxation. This is a non-equilibrium state with $\Delta G_2 \neq 0$ from which the system irreversibly (1) approaches the final equilibrium state-3.

For a single equilibrium present, the thermodynamic analysis (9) of the process state-2 $\longrightarrow$ state-3 results in an expression for the relaxation amplitude $\Delta\xi$ (i.e. the change of the extent of reaction during relaxation) as a function of the dilution ratio n, the equilibrium constants K_1 and K_3 in the initial and final media, respectively, of given dielectric constants ε_1 and ε_3, as well as the stoichiometric coefficients ν_i, activity coefficients γ_i, mole numbers n_i, and charges z_i of the species:

$$\Delta\xi = \frac{\ln\dfrac{K_1}{K_3} - \ln n \, \Sigma\nu_i - BI_1^{1/2} (\varepsilon_3^{-3/2} n^{-1/2} - \varepsilon_1^{-3/2}) \Sigma\nu_i z_i^2}{\Sigma\nu_i^2/n_{ii} - \Sigma \ln \gamma_{12} \dfrac{\Sigma\nu_i^2 z_i^2}{2 \Sigma n_{ii} z_i^2}} \tag{3}$$

The numeral subscripts refer to the states of eq. (2), and the summations are taken over all species i present. B is a function of the temperature only (9) which is kept constant during the experiment. For nonionic species ($z_i = 0$) in equilibrium diluted with the original solvent ($K_1 \equiv K_3$), eq. (3) simplifies to

$$\Delta\xi = -\ln n \, \Sigma\nu_i/(\Sigma\nu_i^2/n_{i1}) \tag{4}$$

In such a case, $\Sigma\nu_i \neq 0$ is the necessary requirement for the applicability of solvent-jump perturbation, and the dilution ratio n that results in a certain relaxation amplitude can be estimated from eq. (4). Typically, $\Delta\xi$ should be kept around $.05n_{i1}$ for the species i that is present in limiting amount.

For systems with coupled equilibria, an overall relaxation amplitude $\Delta\xi_o$ can be defined (10). $\Delta\xi_o$ can be both measured and calculated for an assumed mechanism, and the comparison of the two values can be used in mechanistic decision making.

Acknowledgement. Financial support by the R. A. Welch Foundation and the Organized Research Fund of The University of Texas at Arlington is greatfully appreciated.

REFERENCES

1.   See article on "Thermodynamics of Relaxation Methods" in this volume.
2.   See article on "Flow Methods" in this volume.
3.   S. Ljunggren, O. Lamm, Dcta Chem. Scand., 12, 1834 (1958).
4.   J. H. Swinehart and G. W. Castellan, Inorg. Chem., 3, 278 (1964).

5.   Z. A. Schelly, R. D. Farina and E. M. Eyring, J. Phys. Chem.,
     74, 617 (1970).
6.   M. M. Wong, Z. A. Schelly, J. Phys. Chem., 78, 1891 (1974).
7.   Z. A. Schelly and M. M. Wong, Intern. J. Chem. Kinet., 6,
     687 (1974).
8.   M. M. Wong and Z. A. Schelly, Rev. Sci. Instr., 44, 1226
     (1973).
9.   D. Y. Chao and Z. A. Schelly, J. Phys. Chem., 79, 2734 (1975).
10. Z. A. Schelly and D. Y. Chao, Adv. Mol. Relax. Processes,
     1979, in press.

CONVENTIONAL AND U. V. FLASH TEMPERATURE-JUMP

Z. A. Schelly

Department of Chemistry
University of Texas at Arlington, TX 76019, USA

The principles and some of the experimental aspects of the temperature-jump relaxation method are discussed.

METHODS OF HEATING

The T-jump method is probably the most versatile and popular of the jump-techniques. For perturbation (1), it utilizes the temperature dependence of the equilibrium constant K, $\partial \ln K / \partial T = \Delta H^o / RT^2$, of those chemical (and physical) equilibria which have a non-zero standard enthalpy change, $\Delta H^o \neq 0$. Initially, the system under investigation is at thermodynamic equilibrium. Its total (internal) energy U is stored in the form of electronic U_e, vibrational U_v, rotational U_r and translational U_t energy: $U = U_e + U_v + U_r + U_t$. The additivity of the different energy modes is also true for the description of the total energy (levels) ε_j of the individual molecules

$$\varepsilon_j = \varepsilon_e + \varepsilon_v + \varepsilon_r + \varepsilon_t \tag{1}$$

The relative population n_j/n of the energy levels follows the Boltzmann distribution

$$n_j/n = g_j \exp(-\varepsilon_j/kT)/\Sigma g_j \exp(-\varepsilon_j/kT) \tag{2}$$

where g_j are the degeneracies of the quantum states j. Eq. (2)

41

W. J. Gettins and E. Wyn-Jones (Eds.). Techniques and Applications of Fast Reactions in Solution. 41–45.
Copyright © 1979 by D. Reidel Publishing Company.

can be used as the statistical thermodynamic definition of the
temperature T. However, the Boltzmann distribution also applies
to each of the energy modes separately. E.g., equation (2)
written for the rotational levels only

$$_r n_j / n = {} _r g_j \exp(-_r \varepsilon_j / kT_r) / \Sigma_r g_j \exp(-_r \varepsilon_j / kT_r) \tag{3}$$

defines the rotational temperature T_r of the system. In a similar
fashion, the electronic T_e, vibrational T_v and translational T_t
temperatures can be defined as well. At thermodynamic equilibrium of the system:

$$T_e = T_v = T_r = T_t = T \tag{4}$$

In general, the different temperatures will deviate from one
another only if the population distribution of the different modes
(e.g. the rotational distribution, as given in eq. 3) is suddenly
and selectively changed. However, due to rapid internal relaxation and collisional energy exchange between molecules in the
liquid phase, temperature equilibrium (eq. 4) is usually reestablished in less than a nanosecond. Thus, the temperature T of a
system can be rapidly and effectively raised by funneling energy
into the system in either of the four modes.

Accordingly, four types of T-jump techniques have been
developed: i) U. V. flash T-jump (heating by electronic excitation) (2), ii) laser T-jump (vibrational excitation) (3), iii)
microwave T-jump (rotational excitation) (4) and iv) Joule
heating (5) and heat-exchange (6) T-jumps (maybe classified as
mainly translational excitation).

In this paper only i) and iv) will be discussed.

U. V. FLASH T-JUMP

Using a short intense light pulse in the ultra violet range,
the solvent and the solute are electronically excited, thus, first
raising the electronic temperature T_e of the system. Then, thermodynamic equilibrium is rapidly reestablished and the chemical
equilibration (relaxation) takes place at the new constant temperature T.

The improved version (7) of the original (2) flash T-jump
apparatus (Figure 1) can produce a temperature rise of 1° C in
about 5 μsec in aqueous solutions. The subsequent chemical
relaxation can be followed by electrical conductivity detection.

Advantages of the method are the lack of necessity of supporting electrolytes, and the possibility of conductometric detection. As disadvantages one should list the smallness of the temperature rise that can be achieved, the non-uniform heating of the solution, and the possibility of unwanted photochemical reactions caused by the U. V. flash.

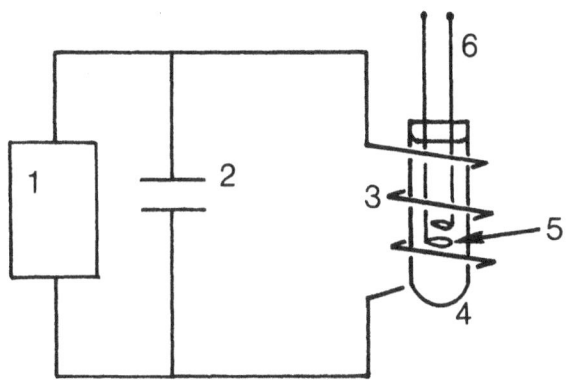

Figure 1. Schematic diagram of a U. V. flash T-jump apparatus. 1-power supply (15 kV), 2-discharge capacitor ($\sim$ 10 μF), 3-helical U. V. flash lamp, 4-quartz sample cell, 5-electrodes for conductivity detection, 6-leads to high frequency Wheatston bridge.

JOULE HEATING T-JUMP

The original version of the Joule heating T-jump apparatus (5) has gone through a spectacular series of improvements, with respect to heating time, sensitivity, and versatility of detection. All have been described in great detail in the literature (8-10). The last developments incorporate fluorescence detection with excellent S/N ratio (11) and electronic compensation for cooling through the observation windows (12). The temperature-jump is brought about by discharging a previously charged capacitor through the suitably conducting equilibrium solution. Typically jumps of a few degrees can be achieved in 1 to 5 μsec. The method is limited to solutions of high ionic strength, and in most instruments to the measurement of relaxation times between 2 μsec and $\sim$ 0.1 sec. The 0.1 sec upper limit is the time usually required by the electrolysis products to diffuse from the electrodes into the observation area.

HEAT EXCHANGE T-JUMP

In spite of its great simplicity and potential, the heat
exchange T-jump method (6) has not been used sufficiently. The
method is based on the fast switching of the circulating fluids
of two thermostats around the equilibrium solution located in a
micro-cell (Figure 2). Since it uses heat conduction for the
transfer of energy, it allows the solution to be heated or cooled
by an arbituary amount, which is given by the temperature settings
of the two water baths. Since the heat conduction controlled
heating time is around 0.5 sec, this "slow" method can be used
for measuring relaxation times in the seconds to hours range.
Usually absorbance, fluorescence and conductometric detections
are used.

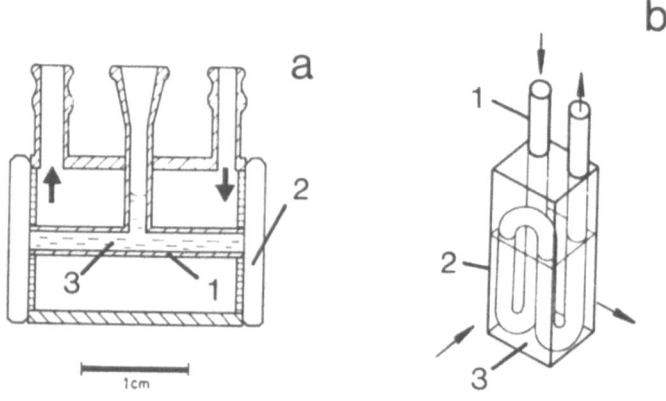

Figure 2. Schematic diagram of microcells used in "slow" T-jump
experiments (From references 6a and 6b). 1-heat exchange Pt
tube, 2-quartz windows, 3-sample. Arrows indicate the flow of
the thermostating liquid and in b) also the paths of the excitation and emission lights in case of fluorescence detection.

Acknowledgement. Support by the R. A. Welch Foundation and the
Organized Research Fund of The University of Texas at Arlington
is greatfully acknowledged.

REFERENCES

1.   See article on "Thermodynamics of Relaxation Methods" in
     this volume.

2. L. S. Nelson and J. L. Lundberg, Nature, 179, 367 (1957).

3. W. H. Inskeep, D. L. Jones, W. T. Silfvast and E. M. Eyring, Proc. Natl. Acad. Sci. USA, 59, 1027 (1968); see also article on "Laser T-Jump" in this volume.

4. G. Ertl and H. Gerischer, Ber. Bunsenges., 65, 629 (1961).

5. G. H. Czerlinski and M. Eigen, Z. Elektrochem., 63, 652 (1959).

6. a) F. M. Pohl, Europ. J. Biochem., 4, 373 (1968); b) F. M. Pohl in "Chemical Relaxation in Molecular Biology", I. Pecht and R. Rigler, eds., Springer-Verlag, Berlin, 1977, p. 282 ff.

7. H. Strehlow and S. Kalarickal, S. J., Ber. Bunsenges, 70, 139 (1966).

8. Z. A. Schelly and E. M. Eyring in "Chemical Instrumentation", Vol. 2, G. W. Ewing, ed., American Chemical Society, Washington, DC, 1977.

9. T. C. French and G. G. Hammes in "Methods of Enzymology" Vol. 16, K. Kustin, ed., Academic Press, New York, NY, 1969, p. 3.

10. G. G. Hammes in "Techniques of Chemistry" Vol. 6, Part 2, 3rd ed., G. G. Hammes, ed., Wiley-Interscience, New York, NY, 1974, p. 147.

11. R. Rigler, C. R. Rabl and T. M. Jovin, Rev. Sci. Instr., 45, 580 (1974).

12. See C. R. Rabl's article in this volume.

LASER TEMPERATURE JUMP

Joseph F. Holzwarth

Fritz-Haber-Institut der Max-Planck-Gesellschaft
Faradayweg 4-6, D-1000 Berlin 33

After a short description of the various methods of producing
temperature jumps in aqueous solutions attention is focussed on
the laser temperature jump techniques which have been built so
far. It is shown that the iodine laser, which is described in this
paper, having an emission wavelength of 1.315 µm, an energy output
of 1-20 Joule, and characteristic pulse lengths of 2.4 µs or 3 ns
is the most versatile arrangement for the direct heating of water.
Its application in a T-jump experiment has already been realized
and is described in this publication. The system was tested by investigating the dissociation-association reaction of water using
a conductance bridge and the protonation-deprotonation reaction
of the pH indicator dyes tropoeolin O and phenolphthalein using a
spectral photometric detection system. Due to the advantageous
wavelength of the laser shock waves and different relaxation amplitudes inside the measuring cell could be completely avoided. Measurements at the same very short heating time over the complete
conductivity range of water are possible.

1. INTRODUCTION

Among the relaxation techniques (1) such as pressure-jump,
electrical field-jump and ultrasonic absorption the temperature
jump method is most widely used because almost every chemical
equilibrium shows a variation with temperature.

In commercially available instruments a high-voltages capacitor is discharged through an electrolyte solution. It is therefore not possible to achieve heating times shorter than RC/2 of
the measuring cell and the discharge circuit (1). To reduce the
electrical resistance of the sample salt concentrations around

W. J. Gettins and E. Wyn-Jones (Eds.). Techniques and Applications of Fast Reactions in Solution. 47–59.
Copyright © 1979 by D. Reidel Publishing Company.

0.1 mol/l are necessary (2). If relaxation signals in the nanosecond region are measured using coaxial cable discharge even
higher concentrations of electrolyte are required (3).

Direct optical heating of the solution or the solvent, however, is not limited by the electrical resistance of a discharge
circuit and the sample especially. Here the heating time is go-
verned by the emission of the pulsed light source and the lifetime
of the excited states. In the case of water the relaxation time of
the vibrational rotational excited states in the near-IR is about
10^{-12} s(4). Pulsed microwave sources (2,5) and flashlamps (6) have
been successfully used in the microsecond range. But only mode-
locked lasers are able to emit an energy of about 1 Joule at a
suitable wavelegnth in a time below one nanosecond; here the optical damage thresholds of the sample and the measuring cell are
the limiting parameters. Power densities of 10 GW cm^{-2} should be
possible in aqueous solutions (7). The following section will concentrate on laser-temperature jump techniques only.

2. LASER-T-JUMP ARRANGEMENTS

We can distinguish between two processes for transferring
the laser pulse energy into thermal energy of the solution: either
via the solute or the solvent. Because only ruby (614 nm) and neodymium (1060 nm) lasers were available at the wavelength of which
water shows weak absorption, a dye or a coloured transition metal
complex was initially used to absorb the energy and transmit it
to the solvent. A serious problem in these experiments is the
interference of the energy transferring substance with the reaction under investigation via the excited electronic states. This
causes side-reactions and decomposition of the dye, which changes
the intensity of the detection light beam so that the reaction of
interest can not be analyzed. Another disadvantage is the long
lifetime of excited electronic states which is in most cases
10^{-6} - 10^{-8} ns. For literature see reference (2). A more elegant
method is available if direct absorption of laser pulse energy
into vibrational or rotational states of the solvent with lifetimes around 10^{-12} s is possible. We focuss our attention now
on the most important solvent water.

We can neglect the visible range of the spectrum because of
extremely weak absorptions as well as the UV region because of
the complicated UV-laser technique and the very high excitation
energy which certainly would cause serious problems in normal
kinetic experiments. The absorption spectrum in the near IR-region
is shown in Fig. 1 (for wavelengths above 2 μm see Falk and Ford
(8)). The wavelength at which laser T-jump experiments have been
reported using heating pulses shorter than 30 ns are indicated
(9). At 1.06 μm water has a very weak absorption (1.3 % mm^{-1}),

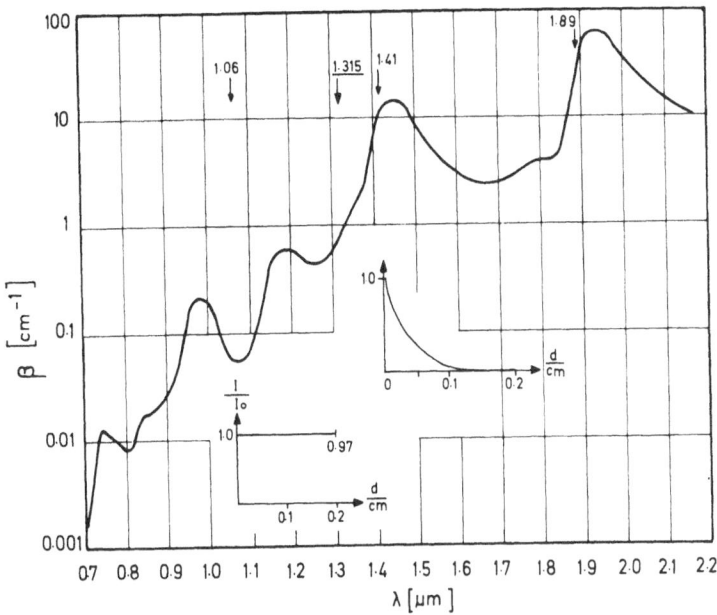

Fig. 1: Absorption spectrum of water in the near IR region. The wavelengths of laser T-jump experiments already performed are indicated by ↓ ; the change of incident laser energy I/I$_o$ with absorption path length for two extreme cases is included.

lasers with an energy output of 50 J or more would be necessary. 1.41 µm and 1.89 µm can be reached by stimulated Raman scattering of the neodymium laser in liquid N$_2$ or H$_2$ under high pressure (9). Both wavelengths are strongly absorbed in water (Fig. 1); this causes serious temperature inhomogeneities inside the measuring cell, which disturb the relaxation signal in the microsecond time regime and creates different relaxation amplitudes over the whole time scale. Furthermore only very thin layers for the detection beam can be used, which results in a small detection sensitivity. Another disadvantage is the Raman conversion rate of 10 - 20 %, initiated by nanosecond pulses. Only picosecond laser pulses (<100 ps) show a better conversion efficiency (10). The time region in between from 10 ns to 100 ps is very dangerous for solid state lasers because Brillouin scattering might easily destroy the laser rod. At 1.34 ± 0.2 µm the neodymium laser shows a second emission, the wavelength of which would be very useful for temperature jumps in H$_2$O. It appears, however, that nobody has succeeded

in achieving an energy of 1 J in a time of 10 ns or less. Future
development might solve this problem.

    Taking the above mentioned difficulties into account we have
constructed an iodine laser with a wavelength of 1.315 μm. When an
energy of some J is delivered a waterlayer of 2 mm can be heated
by some degrees allowing a variable detection light path length
up to 10 mm perpendicular to the heating beam.

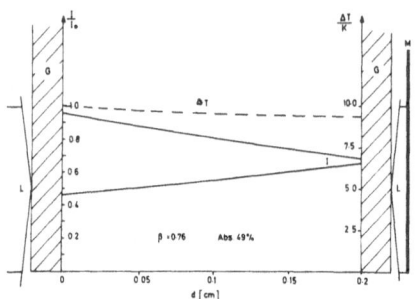

    Fig. 2: Dependence of temperature jump, ΔT, (right ver-
    tical scale) and initial laser intensity, I, (left ver-
    tical scale) on the position, d, inside the measuring
    cell (heated layer = 2 mm); L, laser beam; M, reflecting
    mirror; β, absorption coefficient of H_2O at 1.315 μm.

In Fig. 2 the energy and temperature profiles inside such a measuring cell are shown, with the laser beam passing twice through
the sample. The energy absorption is 49 % and the gradient of the
temperature jump ΔT is only 6 %, changing from 10 K at the entrance
window to 9.4 K at the opposite window. A general treatment of the
dependence, absorption coefficient of solvent times thickness of
heated layer versus laser energy absorption and temperature inhomogeneity can be found elsewhere (11).

3. THE IODINE LASER

    This photochemical gas laser was first mentioned by Pimentel
(12). Its chemical principle is shown in Fig. 3. Perfluoroisopro-
pyliodide (i-C_3F_7J) gas is flashed by UV-light resulting in electronically excited iodine atoms; their stimulated emission produces laser pulses. The main part of the initial compound is rebuilt after deexitation of iodine atoms and the C_3F_7-radicals.

    A schematic diagram of the laser oscillator is given in Fig. 5.
A low inductance capacitor (2 F), charged up to 40 kV, is discharged
over linear Xenon flash lamps by triggering the spark gap. The
flash lamps (rise time 2 μs) are situated inside an aluminium

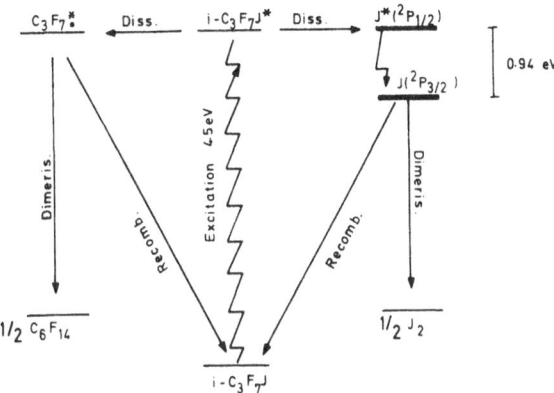

Fig. 3.: Principle of the light-induced chemical processes leading to laser action of i-C$_3$F$_7$J. Laser transition at 1.315 µm with a photonenergy of 0.94 eV.

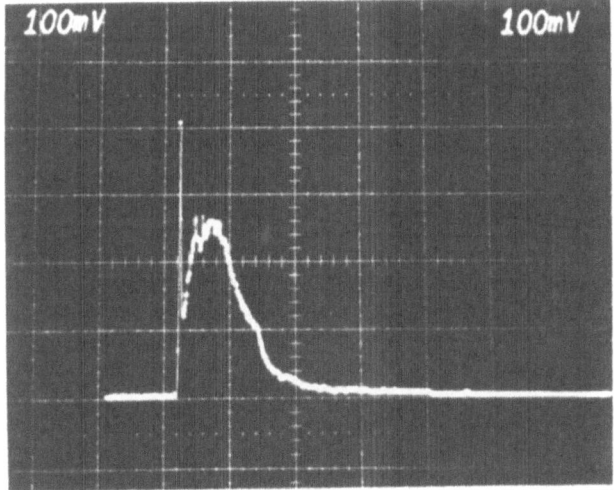

Fig. 4: Dependence of the intensity of an iodine laser
oscillator pulse in the multi-mode version on time;
horizontal scale 2 µs/div. τ_L = 2.4 µs.

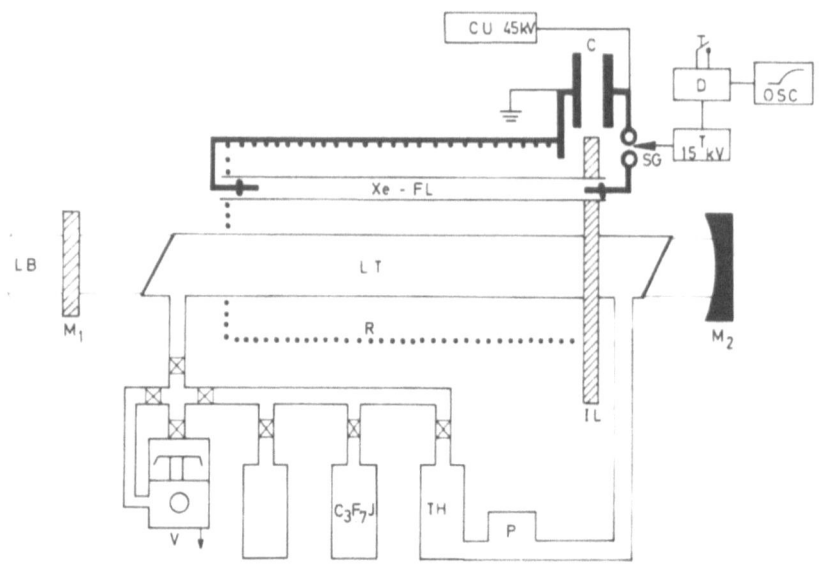

Fig. 5: Schematic diagram of the iodine laser oscillator:
CU, charging unit; C, capacitor; SG, spark gap; T, trigger circuit; D, delay unit with starter.
Xe-Fl, xenon flash lamps; R, reflector; IL, electrical isolation; LT, laser tube connected with the gas recycling and cleaning system consisting of the pump, P, the thermostat, TH, and the vacuum device, V; M_1 and M_2 are mirrors and LB is the laser beam.

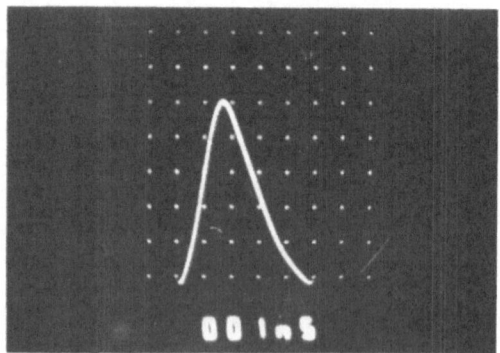

Fig. 6: Single pulse of a mode-locked iodine laser selected by an optical switch; time scale 1 ns/div, measured with the Thomson CSF-TSN 660 4 GHz oscilloscope.

reflector around the laser tube (1m x 1 cm), made from quartz
suprasil with Brewster windows on both ends. The laser cavity
is semi-confocal and consists of two dielectric mirrors, M_2
(R = 99,9 % at 1.315 μm, f = 2.5 m) and M_1 (plane, R = 40 %),
2.5 m apart. The laser tube is filled with pure C_3F_7J. The gas
pressure is controlled by a thermostat filled with liquid C_3F_7J.
The oscillator is connected to a gas recycling system which re-
moves iodine molecules and replaces the decomposed part of the
laser gas by circulating C_3F_7J through a thermostat held at 260 K.
This ensures that the output energy remains constant from shot
to shot. The vacuum system allows cleaning of the laser gas circuit from oxygen. This laser oscillator emits 2 J in a characteristic time of 2.4 μs (Fig. 4). The energy output shows a simple
exponential decay with time. If higher energies are required a
laser tube larger in size can be used or an amplifier device, described elsewhere in this volume (13). When the very intense spike
appearing in the first 100 ns of the oscillator emission is actively mode-locked, pulses with a duration of 0.7 to 3 ns can be
produced (13). Such a single pulse selected by a Pockels cell is
shown in Fig. 6. Its energy can be boosted by an amplifier system
up to more than 100 J if desired (14). This is much more energy
than necessary for temperature-jump experiments.

4. T-JUMP DETECTION SYSTEM

     The complete detection system is schematically shown in
Fig. 7; to avoid noise from the electrical part of the laser it
is placed inside a Faraday cage. Triggering of the recording circuit is made by a part of the laser beam via a vacuum photodiode
(Valvo XA 1003, jitter ∿1 ns) or externally via an opto-coupler
(jitter ∿50 ns).

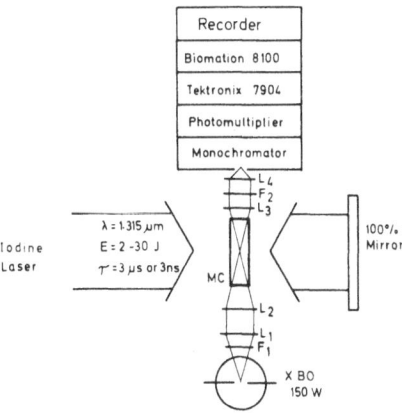

Fig. 7: Schematic diagram of the T-Jump detection system:
XBO, Xenon arc lamp;L, lenses; F, filters; MC, measuring
cell.

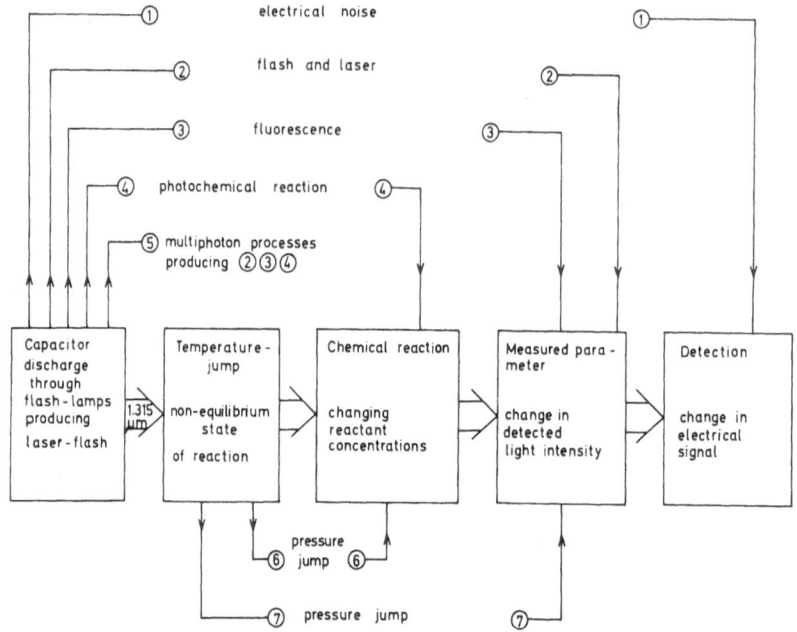

Fig. 8: Signal and noise sources in a laser T-Jump arrangement. ① - ⑦ indicate the origin of noise and its consequences; ⟹ shows the intended signal and reaction path.

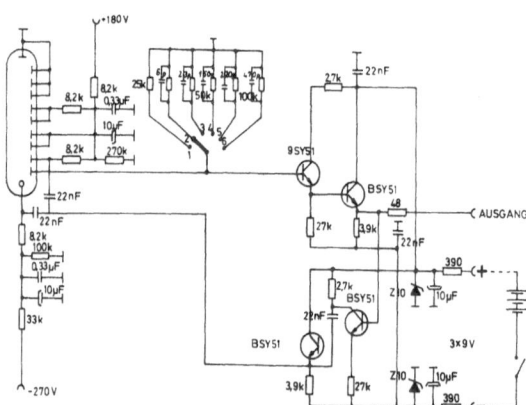

Fig. 9: Photomultiplier circuit using a 1P28 RCA tube; amplification adapted to a CW Xenon arc lamp (15).

In Fig. 8 the most important effects, which may disturb the relaxation signal under investigation are schematically indicated. Great effort has been made to minimize these noise sources, and their pick-up in the measuring circuit. The correct series of physical and chemical reactions is indicated by broad horizontal arrows. A photomultiplier circuit using a 1P28 tube designed for variable rise times between 1 µs and 40 ns is shown in Fig. 9 (15). The amplification of this photomultiplier is adapted for a xenon arc lamp, which is not pulsed (CW). Other multiplier circuits with different rise times and different light sensitivity are found in the literature (13).

5. RESULTS

The rate constants for the recombination of H^+ and OH^- in pure water over a wide temperature range measured with a conductance bridge are given in Table 1. The activation energy $E_a = 11.3 \pm 0.5$ kJ mol^{-1} is in excellent agreement with literature values (16). The experimental setup is shown elsewhere in this volume (17). The relaxation trace of a single measurement at 305.2 K is also shown in the table to demonstrate the accuracy of the experiments.

Table 1: Rate constants, k_{12}, and activation energy E_a, of the recombination of H^+ and OH^- in pure water at different temperatures; the measured relaxation trace at 305.2 K with $\tau = 24.5$ µs is included in the table.

$T \; [K]$	$k_{12} \; \left[\dfrac{dm^3}{mol \; s}\right]$	relax. trace
285.1	0.98×10^{11}	$\tau = 24.5$ µs
293.3	1.13×10^{11}	
300.9	1.27×10^{11}	
305.2	1.34×10^{11}	
311.8	1.50×10^{11}	
321.0	1.65×10^{11}	
332.0	1.90×10^{11}	
338.8	2.10×10^{11}	

relax. signal
20 µs/div

$$E_a = 11.3 \pm 0.5 \; kJ \; mol^{-1}$$

In Fig. 10 a perfect relaxation signal (τ = 4.9 μs) of a single measurement observed between 1 ms and 1 μs is shown.

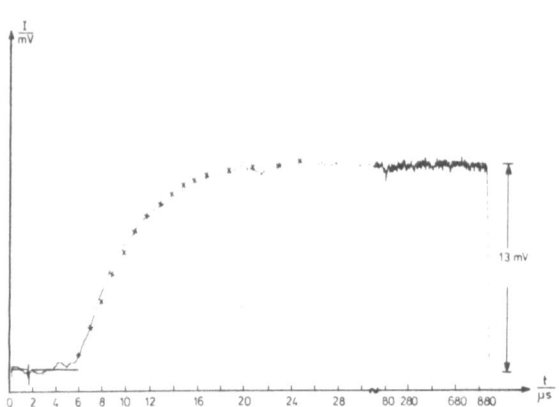

Fig. 10: Relaxation signal of a single measurement of phenolphthalein in H_2O at pH 9.6, temperature 298 K, registered with a Biomation 8100, (x) calculated points of a perfect exponential trace.

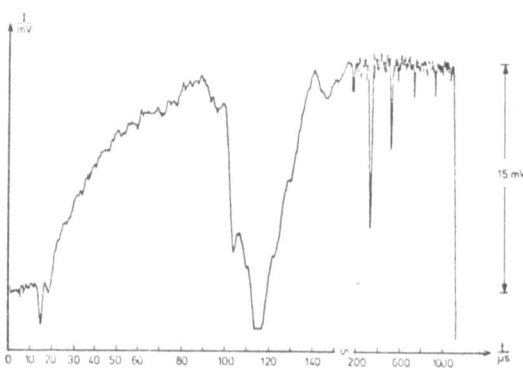

Fig. 11: Disturbance of an optical relaxation measurement by shock-waves, demonstrated from the relaxation signal of Tropaeolin O in H_2O (pH 11.5, temperature 298 K), occuring in our measuring cell, after the voluntary creation of a temperature gradient of 25 %.

If the laser beam is focussed into the measuring cell so that
inhomogeneous heating of the sample occurs the detection channel
is completely disturbed in definite time regions depending on the
size of the measuring cell. This is demonstrated with the relaxation signal of Tropaeolin O in Fig. 11. The use of smaller sample
volumes than in Fig. 11 having higher temperature gradients would
disturb the relaxation signal completely in the time region longer
than 1 μs (9). The limit·in time resolution of relaxation signals
measured with an iodine laser oscillator (τ heating = 2.4 μs) is
demonstrated in Fig. 12 by calculating the signal which results
from superimposing the heating and relaxation of the sample (18).
Other T-jump measurements performed with the iodine laser amplifier system are reported elsewhere in this volume (13).

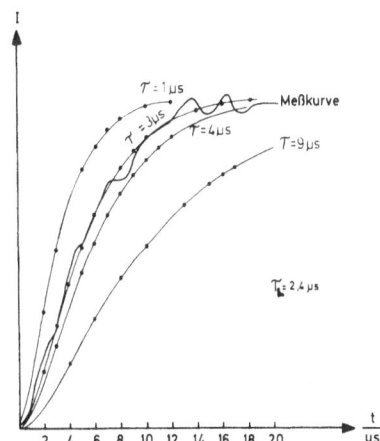

Fig. 12: Time resolution limit of the iodine laser
oscillator shown by calculating the relaxation trace
resulted from superimposing a heating time of 2.4 μs
(τ_L) with the relaxation times (τ) of chemical equilibria. Lower limit τ = 1μ s; Meßkurve means a measured
signal, I the measured parameter.

6. CONCLUSION

The iodine laser presents the most flexible heating source
to perform T-jump experiments in aqueous solutions:

Direct and homogeneous heating of the solvent is achieved.

The heating time is independent of the ionic strength of
solutions.

Sample volumes as small as 10 µl or as large as 1 ml can
be used.

The laser oscillator offers a simple system which can be
used for relaxation experiments with a time resolution down
to 1µs. If a mode-locking device is inserted into the oscillator
and one or more amplifiers are added (11,13) heating times down
to 700 ps can be achieved. Further developments such that a heating
time of 100 ps is reached are also possible.

The iodine laser can therefore be used as a modular device
starting with a laser oscillator and ending up with an extremely
fast T-jump source, which covers the complete time range between
seconds and picoseconds.

ACKNOWLEDGEMENTS

I like to thank my coworkers H. Wolff, A. Schmidt and
W. Frisch for their patience during the development of the laser
system and R. Volk of the Projektgruppe für Laserforschung, Garching bei München, for fruitful cooperation as well as Thomson
CSF Paris for the loan of their superfast 4 GHz oscilloscope
TSN 660. We wish to thank the DFG Bonn-Bad Godesberg for a research grant.

REFERENCES

(1) Hammes, G. G.: 1974 in "Techniques of Chemistry", Vol. VI,
     Part II, A. Weissberger, Ed., Interscience, N. Y.
(2) Caldin, E. F.,: 1975, Chem. Britain, 11, pp. 4
(3) Pörschke, D.: 1976, Rev. Sci. Instrum. 47, pp. 1363
(4) Goodall, D. M., and Greenhow, R. C.: 1971, Chem. Phys. Lett.,
     9, pp. 583
(5) Ertl, G., and Gerischer, H.: 1961, Z. Elektrochem., Ber.
     Bunsenges. physik. Chem., 65, pp. 629
(6) Strehlow, H., and Kalarickal, S. J.: 1966, Ber. Bunsenges.
     physik. Chem., 70, pp. 139
(7) Smith, W. L., Liu, P., and Bloembergen, N.: 1977, Physical
     Review A, 15, pp. 2396
(8) Falk, M., and Ford, T. A.: 1966, Can. J. Chem., 44, pp. 1699
(9) Turner, D. H., Flynn, G. W., Sutin, N., and Beitz, J. V.:
     1972, J. Am. Chem. Soc., 94, pp. 1554
     Flynn, G. W., and Sutin, N.: 1974, in "Chemical and Biochem-
     ical Applications of Lasers", ed. C. B. Moore, Acad. Press,
     N. Y., pp. 309
     Ameen, S., and De Maeyer, L.: 1975, J. Am. Chem. Soc.,
     97, pp. 1590
     Holzwarth, J. F., and Wolff, H.: 1975, Ber. Bunsenges. physik.
     Chem., 73, pp. 952
(10) Maier, M.: 1976, Appl. Phys. 11, pp. 209
(11) Holzwarth, J. F., Schmidt, A., Wolff, H., and Volk, R.: 1977,
     J. Phys. Chem., 81, pp. 2300
(12) Kapser, J. V. V., and Pimentel, G. C.: 1964, Appl. Phys. Lett.,
     5, pp. 231
(13) Frisch, W., Schmidt, A., Volk, R., and Holzwarth, J. F.: 1979,
     "Laser T-Jump Arrangement with Time Resolution in the Second
     to Picosecond Range", this volume
(14) Witte, K. J. Brederlow, G., Volk, R., et al.: 1978, in
     "High-Power Lasers and Applications", ed. Kompa, K. L., and
     Walther, H., Springer Series in Optical Sciences, Springer-
     Verlag, Berlin, pp. 142
(15) Rabl, C. R.: 1973, in Dechema-Monographie, Bd. 71, Techni-
     sche Biochemie, "Relaxationsmeßtechnik", Frankfurt
(16) Barker, G. M., Fowles, P., Sammon, D. C., and Stringer, B.:
     1970, Trans. Faraday Soc. 66, pp. 1498
     Eigen, M., and De Maeyer, L.: 1958, Proc. Roy. Soc.
     A 247, pp. 505
(17) Goodall, D. M., Holzwarth, J. F., et al.: 1979, this volume,
     "Single-Photon Infrared Photochemistry"
(18) Czerlinski, G., and Eigen, M.: 1959, Zeitschr. Elektrochem.,
     Ber. Bunsenges. physik. Chem., 63, pp. 652

LASER T-JUMP ARRANGEMENT WITH TIME RESOLUTION IN THE SECOND TO PICOSECOND RANGE

W. Frisch, A. Schmidt, J. F. Holzwarth and R. Volk*

Fritz-Haber-Institut der Max-Planck-Gesellschaft,
Faradayweg 4-6, D-1000 Berlin 33, West Germany

*Projektgruppe für Laserforschung der MPG, D-8046
Garching, West Germany

A laser temperature jump arrangement is described which uses a versatile iodine laser with a wavelength of 1.315 µm as an energy source for the direct heating of water. The energy output of the laser-amplifier system varies between 1 and 3 J. Characteristic heating times of 700 picoseconds to 3 nanoseconds are reported depending on the mode of operation. Two different photomultiplier circuits were used to take optimum advantages of the heating time. Due to the advantageous absorption of water at the wavelength of this type of laser, homogeneous temperature profiles inside an aqueous solution of a thickness of 2 mm have been produced, thus avoiding shock-waves in the microsecond and dielectric breakdown in the nanosecond regions. The proton transfer reaction of phenolphthalein was used to test the complete system.

1. INTRODUCTION

For the study of very fast reversible reactions in solution the T-Jump method is the most universal technique because almost every chemical reaction is accompanied by a change in reaction enthalpy.

W. J. Gettins and E. Wyn-Jones (Eds.). Techniques and Applications of Fast Reactions in Solution. 61–70.
Copyright © 1979 by D. Reidel Publishing Company.

If optical heating of the solvent is applied there are no restrictions on the composition of solutions. The limiting parameters are the emission time of the light source, the relaxation
times of the excited solvent molecules, and the optical damage
thresholds of both solution and measuring cell (1). In the case
of an aqueous solution in an Infrasil glass cell the optical
damage threshold is above 10^{11} J cm^{-2} s^{-1} (1). The relaxation time
of the vibrational-rotational excited states of H_2O in the near
IR is shorter than 10^{-12} s (1). Bearing these last two points in
mind we have developed an iodine-laser T-Jump arrangement with a
heating time shorter than 1 nanosecond. The reason for choosing
a laser with an emission wavelength of 1.315 µm can be seen from
the absorption spectrum of water (2,3) and from Fig. 3. Details
of the iodine-laser oscillator and a comparison with other laser
T-Jump arrangements are reported in another article in this
volume (2).

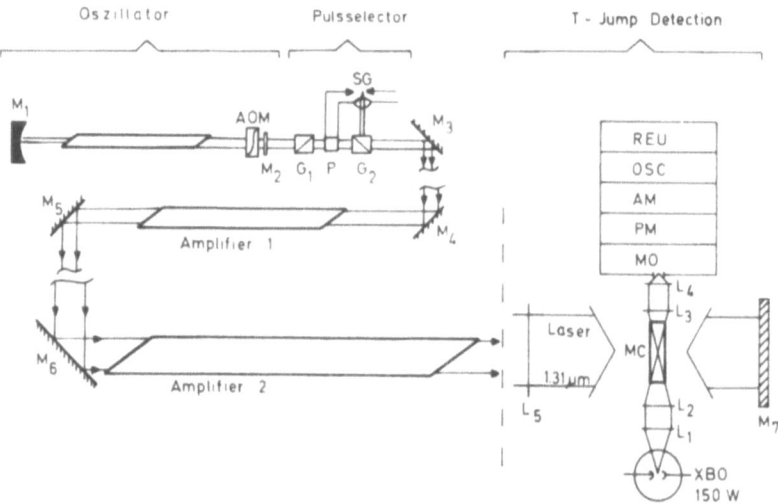

Fig. 1: Experimental arrangement of the iodine laser
amplifier T-jump: M, mirrors; AOM, acousto-optical
mode-locking device; G, Glanprism; P, Pockels cell;
SG, spark gap; L, lenses; MC, measuring cell; MO,
monochromator; PM, photomultiplier; AM, amplifier;
OSC, oscilloscope; and REU, registration unit.

2. EXPERIMENTAL

A schematic diagram of the whole experimental arrangement is shown in Fig. 1. The laser itself consists of three main units, the oscillator, the pulse selector, and the amplifiers. The oscillator produces a pulse at 1.315 µm with a characterstic emission time of 2.4 µs in the multimode version. Details of this unit are given in reference (2). If an acousto-optical mode locker (AOM) with a mode selecting diaphragm of 2-3 mm is inserted into the laser cavity near mirror M_2 a train of short mode locked pulses is obtained as shown in Fig. 2. Our acousto-optical mode

$\lfloor 20\,ns \rfloor$

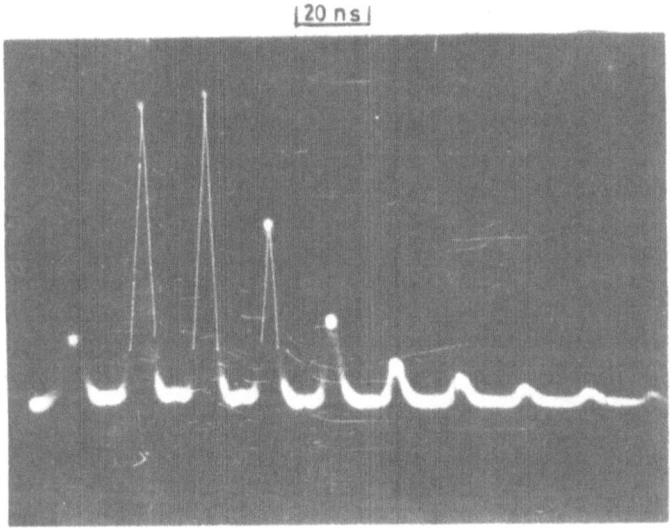

Fig. 2: Pulse train of an acousto-optical mode-locked iodine laser-oscillator; pulse period 16.6 ns.

locking device consists of a quartz-infrasil block with the dimensions 15 x 15 x 15 mm^3 which is placed on top of a quartz disc with a resonance frequency at 30 Megahertz. This system is driven by a pulsed transmitter of 30 MHz and 30 W. The standing wave generated inside the quartz block produces a modulation of the oscillator beam better than 95 %. In this way we achieve pulse trains of the TEM_{oo} mode with a distance of 16.6 ns. This is one round trip time between the mirrors M_1 and M_2 in Fig. 1.

The pulse train passes the pulse selector, which operates in the following way. The first Glanprism G_1 ensures that only linearly polarized laser light from the oscillator passes the deuterated coaxial KDP Pockels cell P and reaches the Glanprism G_2 which is perpendicular to G_1. In G_2 the pulse is directed to spark gap SG_1 The first intense pulse switches the spark gap so that a voltage pulse (10 kV, rise time <2 ns, duration 20 ns) generated by cable-discharge is applied to the Pockels cell. In this way only the second intense pulse from the pulse train changes its direction of polarization by 90° and can pass the Glanprism G_2 in the direction of the mirror M_3. This selected single pulse has an energy of 1 mJ. It passes the amplifiers 1 and 2 guided by the mirrors M_3, M_4, M_5 and M_6. The distance between the oscillator and amplifier 1 as well as amplifier 1 and 2 is 10 m. The cross section of the laser beam is dictated by its natural divergence of 0.8 mrad. The amplifiers are constructed similar to the oscillator but without mirrors for multiple reflections and are larger in size. The selected single pulse passes only once through the amplifier chain, which was brought into a stage of inversion just below self-emission before the single pulse reaches the chain. The amplified single pulse with an energy of 1-3 Joule can be focussed by a special surface mirror so that its size is compatible with the dimensions of the measuring cell.

We normally use a measuring cell of 10 x 10 x 2 mm^3 made from quartz-infrasil glass. The laser beam should have a cross-section of 8 mm and is reflected at the back of the measuring cell after passing a layer of 2 mm. This minimizes temperature inhomogeneities inside the heated water layer, which generate locally different relaxation amplitudes as well as unwanted shock waves. Shock waves may disturb the detection channel and locally diffe-

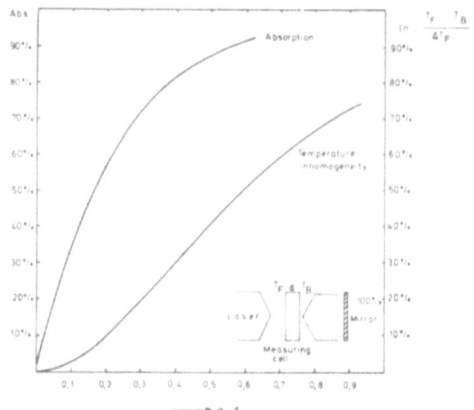

Fig. 3: Dependence of the absorption of initial laser energy and temperature inhomogeneity in the measuring cell on the product of absorption coefficient (β) times thickness of the heated layer (d) with the laser beam passing twice through the cell.

rent relaxation amplitudes make the calculation of relaxation
times difficult (2). In Fig. 3 we show the calculated dependence
of the product of absorption coefficient times thickness of the
heated layer as well as the temperature gradient for a single
reflection of the laser beam on the amount of laser energy absorption. If the detection light beam is directed perpendicular
to the heating laser beam as in our experiments temperature inhomogeneities inside the measuring cell should be below 10 %.
This allows a detection path length up to 10 mm which gives a
very high detection sensitivity. The T-Jump detection system consists of a XBO_2 150 W light source which has the highest light intensity per mm^2 of all XBO's. The XBO can be pulsed if appropriate
to reduce the signal to noise ratio (4). A lense and filter system
L_1 and L_2 as well as L_3 and L_4 provides maximum light intensity
of the selected wavelength at the entrance slit of the monochromator. The change of this light intensity down to 0.2 % is detected and converted into an electric current by the photomultiplier
PM, amplified or impedance matched by AM, and registered by a Tektronix 7904 oscilloscope (time resolution 500 MHz) or a transient
digitizer Biomation 8100 (time resoltuion 25 MHz). We have just
bought a Tektronix 7912 AD transient digitizer (time resolution
500 MHz) followed by a HP 9845 S calculator to use sampling techniques independent of signal repetition rate for achieving a better signal to noise ratio. This system will come into operation
in middle of 1979.

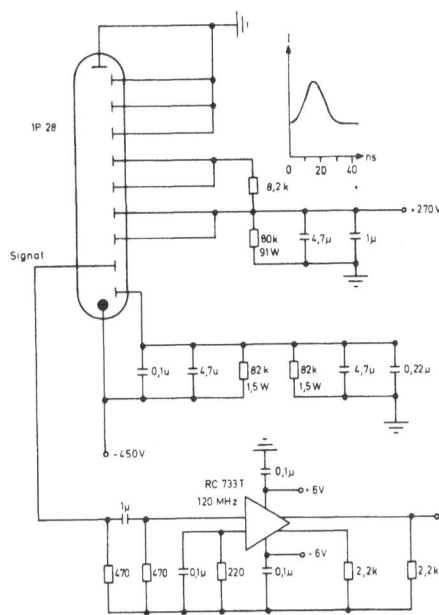

Fig. 4: Circuit used with a 1P28 photomultiplier tube
to achieve rise times of 9 ns.

We are using four different photomultiplier circuits with time resolution between 10 ms and 1ns (2,5). The circuit with a time resolution of 9 ns for an unpulsed XBO is shown in Fig. 4. This photomultiplier is designed to be used in connection with a Tektronix 7904 oscilloscope or the Tektronix transient digitizer 7912 AD together with a 7A13 differential comparator plug-in. This combination gives a sensitivity of 1mV into 1 M Ohm with an overall bandwidth of 100 MHz. If higher time resolution is necessary a pulsed xenon lamp and a less sensitive type of photomultiplier has to be used. Such a design can be found in the literature (4,5). The complete detection system is situated inside a Faraday cage to reduce the noise level. Triggering of the oscilloscope is initiated either by the heating laser beam itself via a XA1003 vacuum diode inside the Faraday cage or externally via an electrical pulse which is initiated outside the cage when the oscillator is fired. The latter is sent into the Faraday cage using an opto-coupler and has an accuracy of 30 ns, the former is correlated with the heating pulse inside 1 ns.

3. RESULTS

The quality of the actively mode-locked pulse trains and the smoothness of one single pulse depends on the acousto-optical modulator. In Fig. 5 a mode-locked pulse train and a single pulse of 3 ns halfwidth are shown. A further result of this measurement is that only one pulse is selected by the pulse cutting system,

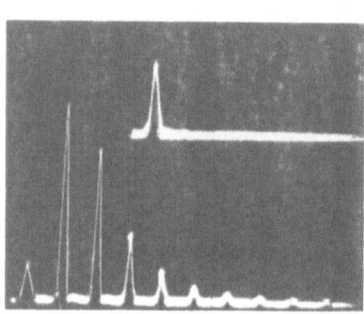
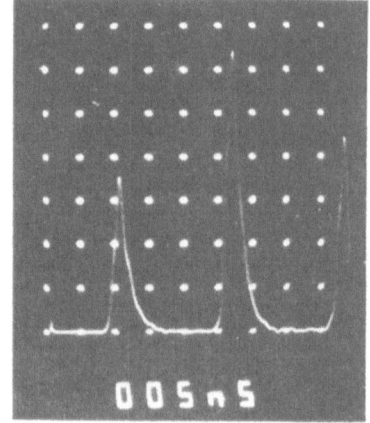

Fig. 5: a) Pulse train of the mode-locked iodine laser together with the optically switched single pulse; b) part of a pulse train showing its short rise time and perfect mode-locking as measured on a Thomson-CSF TSN 660 4GHZ oscilloscope, time base 5 ns/div.

the contrast ratio of which is around 10^3. These experiments are done at pressures of 50 Torr of the laser gas C_3F_7J. If the pressure inside the active part of the oscillator is increased, the pulse duration is decreased. In Fig. 6 the experimental results are given. Because of the finite time resolution of the detection system (Valvo vacuum diode XA 1003) it was not possible to measure pulse durations below 700 ps; therefore a further pulse shortening which might have occurred at pressures above 180 Torr of C_3F_7J could not be detected.

Two relaxation signals in aqueous solutions of the pH indicator phenolphthalein and the dye proflavin were measured. The experimental traces are shown in Fig. 7a, Fig. 7b and Fig. 8. The shock waves which might occur between 1 and 100 μs are not detectable. This proves that heating is very homogeneous. The relaxation times are in agreement with literature values (6). These results show that the above described arrangement can be used on the complete time scale between picoseconds and seconds.

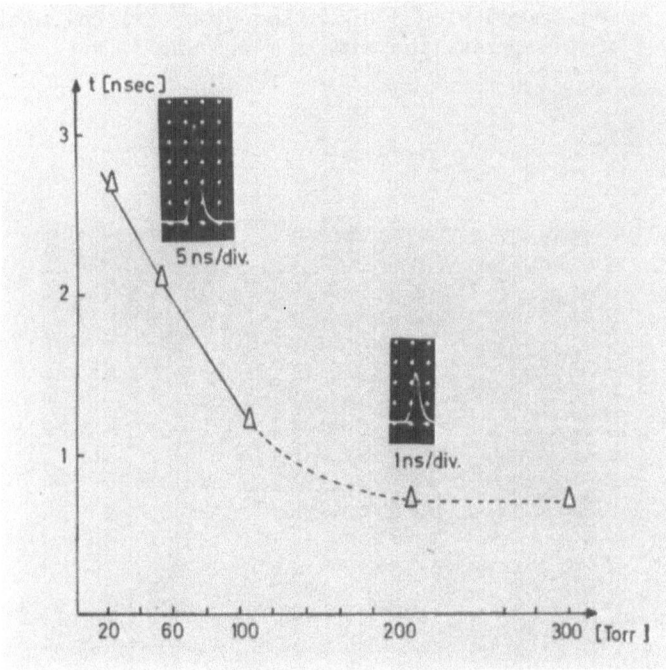

Fig. 6: Dependence of single pulse half width t on the pressure inside the active part of the iodine laser oscillator. Values below 0.7 ns are limited by the rise time of the vacuum photo diode (XA 1003) used.

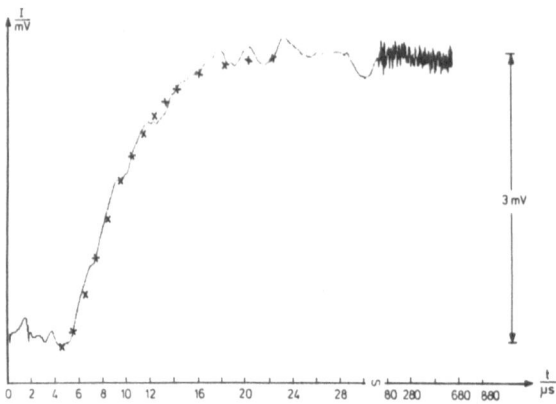

Fig. 7a: Relaxation signal of phenolphthalein in H_2O
at pH 9.1, concentration 5 x 10^{-5} M, temperature 297 K
registered with a Biomation 8100. (x) calculated points
of the relaxation time τ = 4.5 μs.

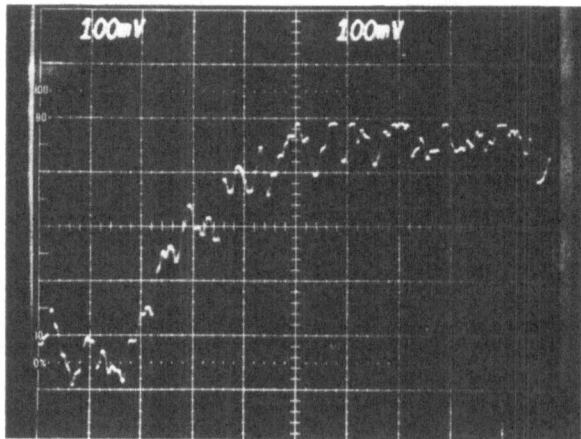

Fig. 7b: Relaxation signal of phenolphthalein in
glycine-buffered solution at pH 9.4, conc. = 10^4 M,
τ = 2.7 μs, registered with a Tektronix 7904 oscilloscope, time base 2 μs/div.

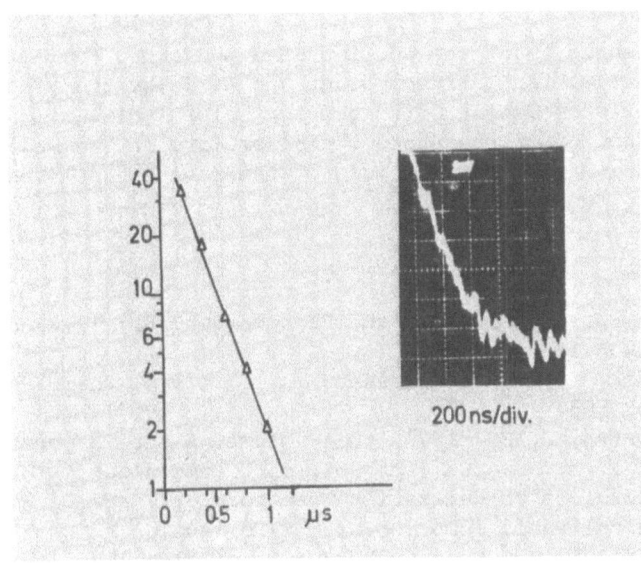

Fig. 8: Oscilloscope (Tek. 7904) trace and logarithmic plot of the relaxation of 5 x 10^{-4} M proflavin; temperature 300 K, relaxation time τ = 290 ns.

4. CONCLUSION

The following advantages of the iodine-laser T-Jump arrangement can be stated:

a) Direct and homogeneous heating of solvents such as H_2O, D_2O and alcohols is possible.

b) Very small volumes of the sample down to μl can be used.

c) Energy of some joules can be provided in a time short enough to measure relaxation signals from picoseconds to seconds.

d) Further development to provide more energy and shorter laser pulses is possible by scaling up the iodine laser amplifier system and using a high pressure oscillator.

The limiting time resolution of a laser T-Jump arrangement seems to be between 10 and 100 picoseconds. It is dictated by the optical damage threshold (1) of the sample, the state of the art of the very short laser pulse technique handling energies of more than 200 mJ (7), as well as the sensitivity and bandwidth of suitable detection systems.

ACKNOWLEDGEMENTS

We wish to thank the DFG Bonn-Bad Godesberg for a research grant, N.A.T.O. for a travel grant and R. Weise from the MPI für Biophysikalische Chemie Göttingen for helpful discussion during the construction of the photomultiplier circuit, as well as Thomson CSF, Paris, for the loan of their superfast 4 GHz oscilloscope TSN 660.

REFERENCES

(1) Smith, W. L., Liu, P., and Bloembergen, N.: 1977, Physical Review A 15, pp. 2396.
     Goodall, D. M., and Greenhow, R. C.: 1971, Chem. Phys. Lett. 9, pp. 583.
(2) Holzwarth, J. F.: 1979, this volume, Laser Temperature Jump.
     Schelly, Z. A.: 1975, pp. 61, in "Chemical and Biological Applications of Relaxation Spectrometry", ed. E. Wyn-Jones, D. Reidel Publ. Comp. Dordrecht-Holland.
(3) Goodall, D. M., Holzwarth, J. F., et al.: 1979, this volume, Single-Photon Infrared Photochemistry.
(4) Beck, G.: 1974, Rev. Sci. Instrum. 45, pp. 318.
     Frisch, W.: 1978, Master Thesis, Freie Universität Berlin
(5) Beck, G.: 1976, Rev. Sci. Instrum. 47, pp. 537.
(6) Turner, D. H., Flynn, G. W. Lundberg, S. K. Faller, L. D., and Sutin, N.: 1972, Nature 239, pp. 215.
     Rose, M. C., and Stuehr, J.: 1968, J. Am. Chem. Soc. 90, pp. 7205.
(7) Bradley, D. J.: 1977, "Methods of Generation of Ultra Short Light Pulses", pp. 17, in "Topics of Applied Physics", ed. Shapiro S. L., Springer-Verlag Berlin.

HIGHLY SENSITIVE AND VERY FAST CHEMICAL RELAXATION TEMPERATURE-JUMP APPARATUSES

J. Aubard, J.J. Meyer, J.M. Nozeran and P. Levoir.

Institut de Topologie et de Dynamique des Systèmes de l'Université Paris VII, associé au CNRS,
1, rue Guy de la Brosse, 75005 PARIS – France.

Within the last ten years, the time resolution and sensitivity of chemical relaxation apparatuses have markedly improved, and today these techniques are the best tools available for studying fast chemical and biological equilibria in solution.

Temperature-jump has proved to be the most versatile of these techniques, and the well-known Joule heating apparatus the one most widely used due to its convenience of use and commercial availability. However, this latter apparatus cannot be used to measure some highly important processes such as conformational equilibria in ribonucleic acids, enzymatic reactions, etc..., since the relaxation times range from 10^{-6} to 10^{-8}s. Furthermore, numerous very one-sided fast equilibria of biological interest are associated with only a small absorbance change (when optical detection is used) over a temperature variation of a few degrees. Tautomeric equilibria in purine and pyrimidine base solutions fall into this category.

To determine kinetic and thermodynamic parameters involved in these equilibria by direct relaxation measurements, very sensitive and rapid T-jump apparatuses appeared to be necessary. This objective can be attained by a fast Raman laser T-jump (time resolution about 10^{-8}s) and by a repetitive microwave T-jump (sensitivity better than 10^{-4} optical density units).

In this contribution we briefly describe the experimental set up in each case; kinetic results are then presented in order to show how these systems extend the field of study in fast kinetic experiments.

W. J. Gettins and E. Wyn-Jones (Eds.). Techniques and Applications of Fast Reactions in Solution. 71–75.
Copyright © 1979 by D. Reidel Publishing Company.

RAMAN LASER T-JUMP

The technical basis of this apparatus has been reported in detail[1,2], and is summarized in Figure 1. With such a system it is possible to carry out, in water, temperature-jumps up to 8°C in 20 ns; a very fast and sensitive pulsed spectrophotometric detection [30 MHz bandwidth, 10^{-3} optical Density (O.D) units sensitivity] makes it possible to detect rapid and small transient phenomena.

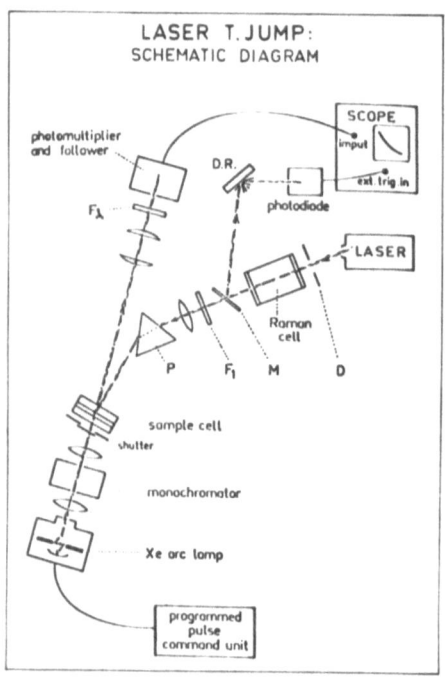

Fig. 1. Laser T-Jump: schematic diagram

We have first developed this apparatus to study very fast prototropic transformations in purine and pyrimidine aqueous solutions. For example Uracil with a pH above 9.5 exists in solution as a mixture of tautomeric anions[3]. The interconversion of these two species is always very fast (100 ns < τ < 3 μs) and must be studied with the laser T-jump technique. The study of the variation of the relaxation time with the pH and the concentration has allowed us to show that the N(1)H/N(3)H tautomeric interconversion of uracil monoanions proceeds via a dissociative mechanism[4,5], as it does for cytosine and isocytosine[6,7].

Moreover the performances achieved by our equipment have permitted
the direct study of certain major conformational equilibria. For
example in the 3'-5' dinucleosides, which are the simplest molecules that already have the helix structure characteristic of the
single stranded nucleic acids, the stacked ⇌ unstacked transition
is a very fast process[8] (always < 1 μs), associated with only a
very small variation in the molecular extinction coefficient.
Fig.2 shows one of the relaxation signals observed at wavelengths
λ > 275 nm (a second faster relaxation phenomenon with a relaxation
time about 40 ns at 25°C is observed at wavelengths λ < 275 nm),
for adenylyl 3'-5' adenosine (ApA).

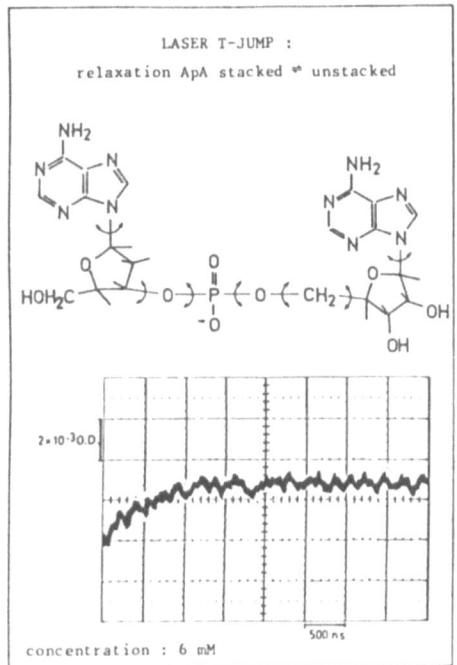

Fig. 2. ApA relaxation in aqueous solution (pH 6.86, 0.05 M ionic
strengh) performed at λ = 285 nm; t_i = 25°C.

It should be noted, first that the signals are very weak (about
2×10^{-3} O.D units), second that the relative "slowness" of the
conformational interconversion in these systems indicates the
existence of high energy barriers between the conformational
states.

REPETITIVE MICROWAVE T-JUMP

The analysis of the dynamics of very fast processes is
however only one of the aspects in relaxation measurements.
Another aspect is the problem of the detection of chemical or
biological equilibria even when they are very one-sided. So we
have developed a very sensitive T-jump apparatus using repetitive
heating generated by a 250 kW microwave source[9].

In this system the magnetron operating at 9.3 GHz delivers
up to 50 pulses per second; with water as a solvent, dielectric
losses cause a temperature rise of 1.5°C for each pulse of a
duration of 1.5 µs. The fast variations in composition of the
solution occuring after the perturbation are measured with
spectrophotometric detection using light pipe technology because
of the very small size of the T-jump cell. Periodical relaxation
signals delivered by the photomultiplier are amplified, converted
by a high speed A to D converter and summed in the memory of a
PDP 11 computer. The capabilities of this apparatus are illustrated by the following example.

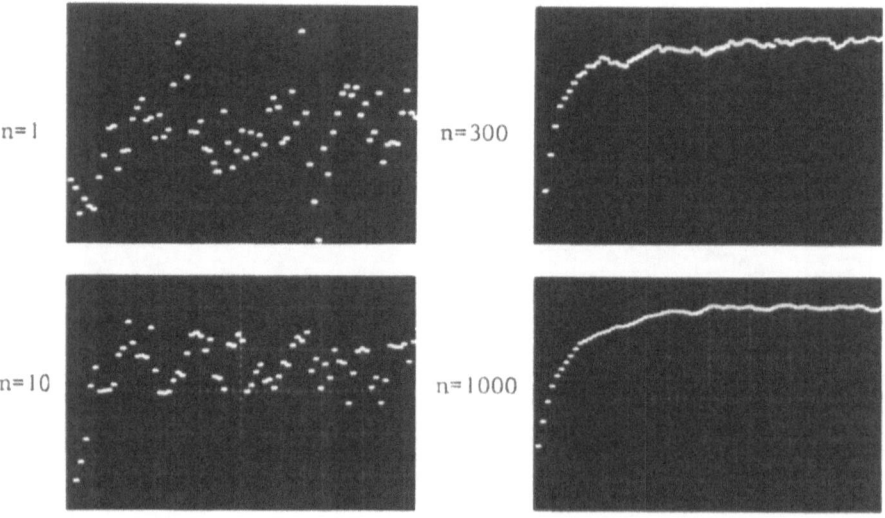

Fig. 3. Microwave T-jump relaxation of cytosine as a function of
the number, n, of accumulations; repetition rate: 20 Hz.
Experimental conditions: λ = 300 nm; t_i = 23°C; pH 6.4 ;
Cytosine concentration: 9.6 x 10^{-4} M; Horizontal scale: 20 µs/
major division; Vertical scale: 2 x 10^{-4} O.D units/major division.

The $N(1)H \rightleftharpoons N(3)H$ tautomeric equilibrium in cytosine is very one-sided in aqueous solution[6]. Detecting the very low proportion of the $N(3)H$ aminooxo form (2.5×10^{-3}) requires a highly sensitive method. Fig. 3 shows the improvement in the signal-to-noise ratio of the relaxation signal when the number of accumulations is increased. It should be noted that after 1000 accumulations, obtained in only 50 s recording time, the signal is well resolved and can be accurately analyzed.

REFERENCES

1.   J. Aubard, J.J. Meyer and J.E. Dubois, Chem.Instrum., 8, 1, (1977).
2.   J.J. Meyer and J. Aubard, Rev.Sci.Instrum., 48, 695,(1977).
3.   K. Nakanishi, N. Suzuki and P. Jamazaki, Bull.Chem.Soc.Jpn., 34, 53, (1961).
4.   J. Aubard, "Protons and Ions Involved in Fast Dynamic Phenomena", Elsevier Sci. Publ. Co., Amsterdam, 1978, p. 191.
5.   O. Bensaude, J. Aubard, M. Dreyfus, G. Dodin and J.E. Dubois. J.Am.Chem.Soc., 100, 2823, (1978).
6.   M. Dreyfus, O. Bensaude, G. Dodin and J.E. Dubois, J.Am.Chem. Soc., 98, 6338, (1976).
7.   O. Bensaude, M. Dreyfus, G. Dodin and J.E. Dubois, J.Am.Chem. Soc., 99, 4438, (1977).
8.   D. Pörschke, Eur.J.Biochem., 39, 117, (1973).
9.   J. Aubard, J.M. Nozeran, P. Levoir, J.J. Meyer and J.E. Dubois, Rev.Sci.Instrum., 50, 19, (1979).

HIGH-RESOLUTION TEMPERATURE-JUMP MEASUREMENTS USING COOLING-CORRECTION

Carl-Roland Rabl

Max-Planck-Institut für Biochemie,
D-8033 Martinsried / Munich, W.-Germany

ABSTRACT. Cooling of the sample after a temperature-jump results in a methodic error of the temperature-jump method in the ms- and sec-time range, especially when studying multiple relaxation phenomena and using micro or semi-micro sample cells. The cooling process has been studied both theoretically and experimentally using HV-conductivity heating and detection of optical absorption and fluorescence. An electronic signal correction circuit has been developed.

I. THEORY

Cooling starts immediately after the temperature-jump due to temperature equilibration of the sample with the cell body, the optical windows, and the electrodes. The apparatus detects cooling in the lightpath only. Thus the temperature gradient normal to the windows is most important and usually a one dimensional calculation will be sufficient for times up to several seconds. A further simplification can be obtained by assuming "thick" windows whereby thermal reflections from the exterior sides of the windows are neglected.

We assume the composite structure shown in Figure 1: fused silica glass (index 2) / water (index 1) / fused silica glass (index 2). Defining the optical axis as the x-axis and T_1 and T_2 as the temperatures of water and glass relative to the ambient temperature, respectively, diffusion of heat can be described by the differential equation

$$u_i^2 (\partial^2 T_i / \partial x^2) = \partial T_i / \partial t \qquad \begin{matrix} (i = 1: & \text{water,} \\ i = 2: & \text{glass)} \end{matrix} \qquad (1)$$

W. J. Gettins and E. Wyn-Jones (Eds.). Techniques and Applications of Fast Reactions in Solution. 77–82.
Copyright © 1979 by D. Reidel Publishing Company.

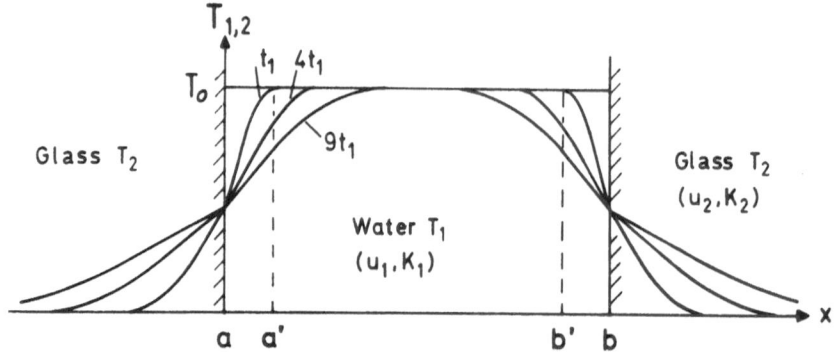

Fig. 1. Composite structure of fused silica glass windows and water. Temperature profile after T–Jump of size T_0. t_1 is of the order of 150 ms with a 7mm–cell, and 300 ms with a 10mm–cell.

where the initial condition at time $t = 0$ is $T_1 = T_0$ and $T_2 = 0$. The boundary condition at $x = a$ and $x = b$ is

$$K_1 \; (\partial T_1 / \partial x) \quad = \quad K_2 \; (\partial T_2 / \partial x) \qquad (t > 0) \; . \qquad (2)$$

Constants are (ρ_i = density, c_{pi} = specific heat; $i = 1, \; 2$):

K_i = conductivity of heat,

$u_i = \sqrt{K_i / (\rho_i c_{pi})}$ = propagation constant of temperature,

$Z_i = K_i / u_i = \sqrt{K_i \rho_i c_{pi}}$ = thermal impedance (see below).

Data obtained from reference [1] at $20°C$: $K_1 = 5.93$ and $K_2 = 13.6$ $[\text{W cm}^{-1} \text{grd}^{-1}]$, $u_1 = 0.0379$ and $u_2 = 0.092$ $[\text{cm sec}^{-1/2}]$, and $Z_1 = 156.7$ and $Z_2 = 147.5$ $[\text{W sec}^{1/2} \text{cm}^{-2} \text{grd}^{-1}]$.

Solving eq.(1) by Laplace transformation [2], $T_1(x,t)$ and $T_2(x,t)$ are obtained as infinite series of error function terms. For "short" times, i.e. $t \ll (D/u_1)^2$ where $D = b - a$ is the distance of the windows, only the first term of these series is important, resulting in a temperature profile as shown in Figure 1. Especially, the boundary condition (2) results in a partition of temperature: $T_1(x = a,b) = T_0 Z_1 / (Z_1 + Z_2)$.

The "cooling function F(t)" can now be defined by the average temperature $\overline{T}_{obs}$ of the observed sample volume as

$$F(t) \quad = \quad \overline{T}_{obs}(t) \, / \, T_0 \; . \qquad (3)$$

In measurements of absorption:

$$\overline{T}_{obs}(t) \quad = \quad \int_a^b T_1(x,t) \, dx \; \Big/ \; \int_a^b dx \qquad (4)$$

Eqs.(3) and (4) yield for short times a simple <u>square root law</u>:

$$F(t) \quad = \quad 1 - \alpha \sqrt{t} \quad = \quad 1 - k\sqrt{t}/D \qquad (D = b - a)$$

$$k \quad = \quad (4/\sqrt{\pi})\, u_1\, Z_2 \, / \, (Z_1 + Z_2) \qquad (\alpha = k/D) \tag{5}$$

The above data give $k = 0.0414$ cm sec$^{-1/2}$.

In fluorescence mode the integration limits in eq.(4) become modified because only part of the illuminated sample volume is observed. E.g., with reference to Figure 1, one may replace a and b by a' and b', respectively. For a general treatment the integrands must be multiplied by a weighting function depending on the optical parameters of observation. This results in a variety of cooling functions which can be classified into three groups: delayed cooling, time-linear cooling, and square root cooling similar to eq.(5) but with a smaller constant α.

II. MEASUREMENTS

Measurements of the cooling function have been performed with various sample cells using fast Tris-buffer pH-indicator systems. Previous results[3] have been improved.

In absorption mode results agree very well with eq.(5) up to several seconds (cf. Figure 4). Deviations at longer times are due to convection, due to temperature gradients perpendicular to x, and due to neglect of terms of higher order in eq.(5). Prior to the temperature-jump the initial temperature of sample and sample cell must be well equilibrated. Best experimental data with initial temperatures $T_a = 3 .. 51^{\circ}C$ yield

$$k \quad = \quad 0.042 + 6 \cdot 10^{-5} (T_a[^{\circ}C] - 20) \qquad [\text{cm sec}^{-1/2}] .$$

Fluorescence measurements[4], when using the appropriate weighting function, are also consistent with theory. However, problems become more complicated if the sample exhibits some photolytic sensitivity which is frequent with fluorescent samples and involves at longer times an irreducible non-linear superposition of cooling, photolysis, and convection.

III. COOLING CORRECTION CIRCUIT

Relaxation signals H(t) become convoluted by the cooling function F(t). Using the conventional definition of the convolution symbol * the actually measured signal becomes

$$H(t) * \dot{F}(t) \quad = \quad H(t) \cdot F(0) \quad + \quad \int_{+0}^{t} H(t - t') \cdot \dot{F}(t')\, dt' \tag{6}$$

where the point marks the derivative (linear relaxation assumed).

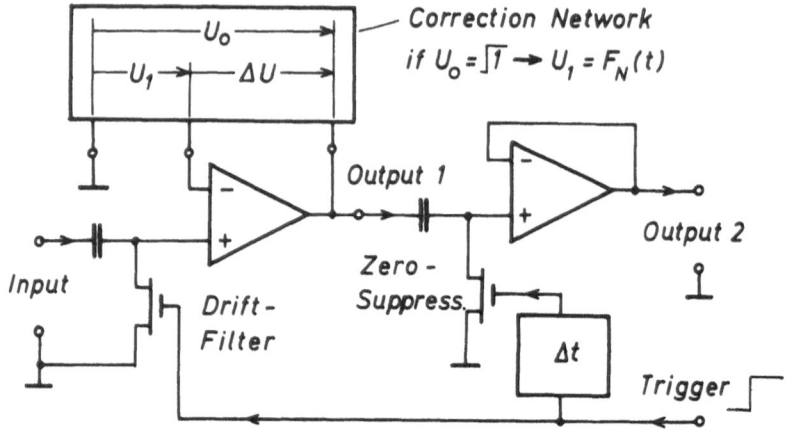

Fig. 2.   Cooling correction circuit integrated with drift-filter
(1.h.s.) and followed by zero-suppression circuit (r.h.s.) -
The photodetector signal is balanced by a reference signal, optio-
nally logarithmified, and applied to the input.   Drift-filter:
Mosfet switch opens at t = 0.   Static balancing offsets prior to
that are short-circuited but signal transients thereafter are by-
passed to the first operational amplifier. With the correction
network connected into the feedback loop of the amplifier, a temperature-jump signal will be corrected for at output 1 according
to eq.(8). - Zero-suppression circuit: Mosfet switch opens at t =
Δt. This yields a dynamic offset in order to separate slow signal
transients from preceeding fast ones.   Signal from output 2 is
filtered by a risetime filter and recorded.   Offset between outputs 1 and 2 is sampled for read out.

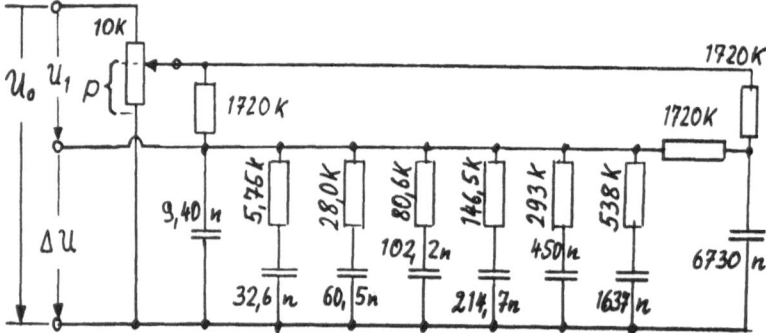

Fig.3.   Correction network of 8th order used with Fig.2. Coefficient of correction is adjusted by calibrated potentiometer. Unit
step response at ΔU-port is $0.25\,p\,\sqrt{t\,[sec]}$, $p = 4\alpha = 0..1$. Accuracy
is within $\pm 0.4\%$ from 1 ms to 1.5 s, $+2\%$ at 5 s, and -3% at 10 s.

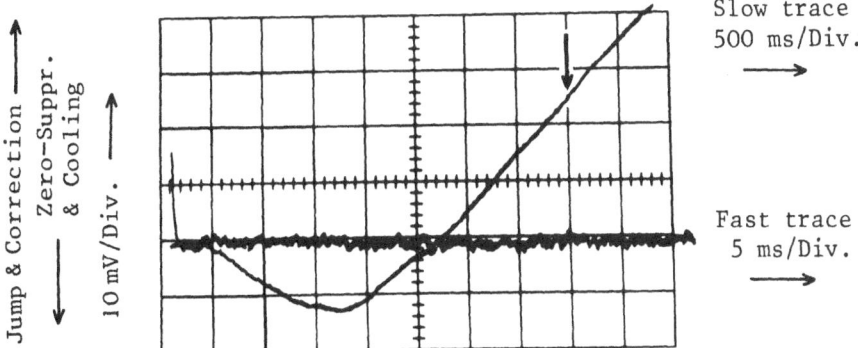

Fig.4.   Test of cooling correction with a 2 mm sample cell in absorption.   Sample cavity 2 x 8 x 9 mm, conical windows with diameter 6 x 12 mm, length 12 mm.   -   Tris-buffer Cresol Red system at pH 8.2 Tungsten lamp at 588 nm. T-Jump $20^\circ \rightarrow 24^\circ$ C.   Signal jump 2450 mV log nat scale of 25 V unity.   Zero-suppression $\Delta t = 0.32$ ms.   Dual trace record. Filter constant: 0.2 ms with fast trace (tail of zero-suppression transient at l.h.s. incompletely blanked out); 5 ms with consecutive slow trace. Cooling corrected for $\alpha = 21.1\%$ sec$^{-1/2}$. Signal deviation at 4 secs (marked by arrow) is $+1\%$ of jump instead of -42% without the correction.

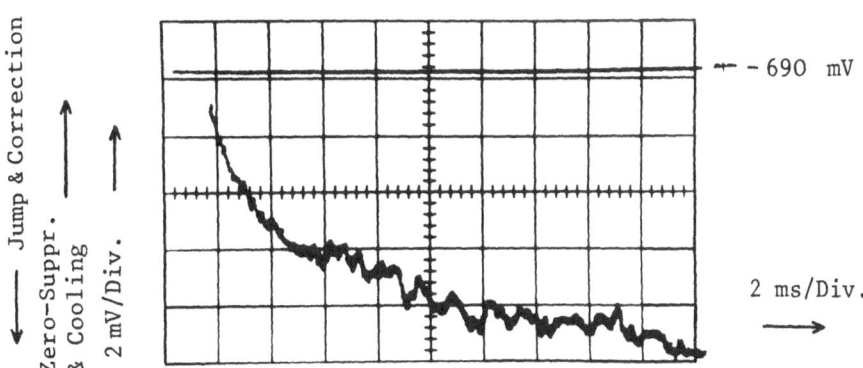

Fig.5.   Competitive binding of Acetylcholine and Ca·· to Acetylcholine Receptor using Ca-indicator and cooling correction. Input concentrations: $[\text{AcCh}] = 3 \cdot 10^{-6}$ M, $[\text{AcChR}] = 1 \cdot 10^{-6}$ M, $[\text{Ca}] = 1 \cdot 10^{-3}$ M, $[\text{Murexide}] = 5 \cdot 10^{-5}$ M, Tris buffer at pH 8.5 . - Sample cell D = 7 mm, T = 23.5° C. Vertical scale 2 mV/Div, horizontal 2 ms/Div. Full signal 20 V. Fast step of - 690 mV suppressed with $\Delta t = 1.4$ ms. Filter constant 0.5 ms. Tungsten lamp 480 nm. - Record gives a 3 ms relaxation term and the first part of a 30 ms relaxation term which become totally distorted without the correction. (Together with E.Neumann and H.W.Chang[5] .)

The correction circuit of Figure 2 uses an RC-network in the feedback loop of an operational amplifier. The network of Figure 3 has been calculated by Laplace transformation. The network function $F_N(t)$, which is the time response at the U_1-port to the unit step function applied at the U_0-port, simulates the square root decay function $F(t)$ eq.(5). The coefficient α can be adjusted by a calibrated potentiometer. The circuit has the unit step response $G(t)$ and performs a deconvolution for $F(t)$ according to

$$\dot{F}(t) * \dot{G}(t) \simeq \delta(t) \quad \text{where} \quad \dot{G}(t) = \mathcal{L}^{-1}(1/\mathcal{L}\,\dot{F}_N(t)) . \quad (7)$$

$\delta(t)$ denotes the delta-function and $\mathcal{L}$ and $\mathcal{L}^{-1}$ denote the Laplace transform and the inverse Laplace transform, respectively. (The circuit has the spectral response $1/\mathcal{L}\,\dot{F}_N(t)$ which has to be re-transformed in order to obtain $\dot{G}(t)$.) Because of the commutivity of the convolution we now have with any linear relaxation function:

$$\{H(t) * \dot{F}(t)\} * \dot{G}(t) = H(t) * \{\dot{F}(t) * \dot{G}(t)\} \simeq H(t) * \delta(t) = H(t) \quad (8)$$

A different network approximating a time-linear decay has been developed for special fluorescence measurements.

IV.   TEST OF COOLING CORRECTION AND APPLICATIONS

Figure 4 shows a test record using an indicator system in a 2 mm -sample cell and correction by the above circuit. The square root law is so precisely followed that the deviations of the slow trace from horizontal are mainly due to the small imperfections of the present network.

Cooling corrections must also be applied with larger cells if large fast signal steps are superimposed on to slow relaxation steps. Figure 5 shows an example where the signal becomes totally distorted even in the short msec-range without the correction.

ACKNOWLEDGEMENT. This work was carried out at the MPI für Biophysikal. Chemie at Göttingen. Part of the computer work was done at the UNIVAC 1108 of the GWDG Göttingen assisted by G.Striker.

REFERENCES:

1.   F.Kohlrausch, Praktische Physik, Vol.3, 22.ed. Stuttgart 1968
2.   H.S.Carslaw and J.C.Jaeger, Conduction of Heat in Solids, 2.ed., Clarendon Press, Oxford 1973
3.   C.R.Rabl, Lecture at 1.Meeting on "Fast Reactions in Solution", Cranfield / U.K.   September 1977
4.   R.Rigler, C.R.Rabl, and T.Jovin, Rev.Sci.Instr. 45(1974), 580
5.   E.Neumann and H.W.Chang, Proc.Nat.Acad.Sci.USA 73(1976), 3994

CIRCULAR DICHROISM MEASUREMENTS AT MILLISECOND TIME-RESOLUTION; STOPPED-FLOW CD AND TEMPERATURE-JUMP CD SYSTEMS

P. M. Bayley

Biophysics Division, N.I.M.R., Mill Hill, London
NW7 1AA

Circular Dichroism is an optical parameter uniquely sensitive to molecular conformation which finds widespread use in structural studies in many branches of chemistry. It is determined by the optical configuration at asymmetric centres, and in macromolecular systems by the overall conformation which derives from the stereospecific linkage together of asymmetric units such as peptides, nucleotides and carbohydrates. In biochemical systems, biological activity is generally dependent upon a specific conformation and modulation of this structure (for example by a conformational change in an enzyme following the binding of a substrate or coenzyme as ligand) is thought to play a central role in control of the conformational process. This provides a model for the regulation of biological function such as activation or inhibition of catalytic activity. Conformational effects are also thought to be involved in the sequential binding and release of reactants and products in ordered processes such as a dehydrogenase reaction.

Since catalytic processes generally occur at the rates of 1 to 1000 s^{-1}, methods are required for measuring CD and hence following conformational changes at the millisecond time range. Commercial instrumentation has generally been limited to measurements with time constants 10s, commensurate with the small signals generally observed in CD properties. The disymmetry ratio $g = \Delta A/A = \Delta\varepsilon/\varepsilon$ (where $\Delta A = A_L - A_R$ and $\Delta\varepsilon = \varepsilon_L - \varepsilon_R$) is frequently of the order 10^{-3} to 10^{-5}.

Taking as specification at 1 msec time constant, a sensitivity of $\delta\Delta A = 10^{-5}$, i.e. $\Delta A = 10^{-5}$ at $A = 1.0$ for $S/N = 1.0$ we have developed such a CD measurement system with the following

W. J. Gettins and E. Wyn-Jones (Eds.). Techniques and Applications of Fast Reactions in Solution. 83–85.
Copyright © 1979 by D. Reidel Publishing Company.

features:

    (1)   High speed electro-optic modulation using the Morvue
           50 kHz modulator;

    (2)   High pressure 100 watt Hg arc source with interference
           filters or 10 mwatt He Ne laser (λ = 632.8 nm);

    (3)   High optical aperture for rapid reaction system;

    (4)   Photodiode detectors taking 1-50 μwatt incident radia-
           tion;

    (5)   Highly tuned lock-in amplification, precision dividers
           for CD and source ratioing of TX channels giving over-
           all time constants as low as 300 μs.

    (6)   Simultaneous recording of CD and transmission (TX) on
           Storage oscilloscope, Transient Recorders (DL905 and
           DL901), and data transmission interface to HP3000
           computer for on line processing.

The desired sensitivity is attainable with currently available incident light power; thus the Hg lines 577, 546, 436, 405,
365 have been extensively used. Under conditions of shot-noise
limitation, the sensitivity approaches that evaluated theoretically. Overall limitations are source stability and intensity
and magnitude of spurious mixing or T-jump transients ($\delta\Delta A \sim 10^{-5}$).

The measurements have been applied to several problems using
a stopped flow system with dead time $\sim$1 msec or a conventional
T-jump cell (and have recently included 90° fluorescence (FLU)
measurements simultaneously with CD and TX in stopped flow).
Examples are as follows:

    (a)   DNA-ligands: the binding of triphenylmethane dyes
(crystal violet and methyl green) to double stranded DNA (by
SFCD) show multi-step reactions.

    (b)   Carbonic anhydrase-azosulphonamides: the binding and
displacement of the chromophoric inhibitors (by SFCD) follow
simple bimolecular kinetics.

    (c)   Tetracycline - Mg^{2+}: formation and displacement
reactions of the complex follow simple kinetics (by SF) identical
in CD, TX and FLU.

    (d) Ni^{2+} - Histidine: the dissociation of the (mono) his-
tidine complex in acid solution follows three steps (by SF TX),
the CD being lost in the initial fast (bimolecular) reaction with
H^+.

    (e)   Liver alcohol dehydrogenase - Auramine O: TJCD shows
the presence of more than one chiral complex of this diphenylmethane dye with the enzyme.

(f) Glutamate dehydrogenase - NADPH: formation of the abortive complex E.NADPH. L-Glutamate is followed (by SF) by CD and FLU properties of the bound coenzyme. At least a two-step mechanism operates and the kinetic parameters are not identical in the two measurements.

Careful choice of concentrations is required for bimolecular mechanisms, to produce observable CD signals in the time range employed. Further improvements require more intense sources particularly for λ<350 nm and laser systems are being evaluated.

REFERENCES

1. Bayley, P.M. & Anson, M. Biopolymers 13, 401-405 (1974).
2. Anson, M. & Bayley, P.M. J. Phys. E. Sci. Instrum. 7, 481-486 (1974).
3. Bayley, P.M., Martin, S.R. & Anson, M. Studia Biophysica 57, 53-58 (1976).
4. Anson, M., Martin, S.R. & Bayley, P.M. Rev. Sci. Instrum. 48, 953-962 (1977).

RECENT DEVELOPMENTS AND APPLICATIONS OF PRESSURE JUMP METHODS

Bernd Gruenewald and Wilhelm Knoche*

Biozentrum, University of Basel, Switzerland

*Max-Planck-Institut für biophysikalische Chemie,
Göttingen, West Germany

Introduction

Pressure shifts a chemical equilibrium in a manner which is described by van't Hoff's equation

$$(\frac{\partial \ln K}{\partial P})_T = - \frac{\Delta V^O}{RT} \tag{1}$$

This relation is applicable to elementary reaction steps with equilibrium constant K and reaction volume $\Delta V^O = \sum \nu_i V_i^O$ where "O" stands for standard conditions, ν_i is the stoichiometric factor and V_i the partial molar volume of reactant i. Most reactions in solution imply volume changes due to changes in conformation or solvation and can consequently be influenced by pressure.

The formalism of relaxation methods demands stationary variation of an external thermodynamic parameter at frequencies near the relaxation frequency of the equilibrium or a fast jump of this parameter (fast compared with the relaxation rate). In the case of pressure as external parameter all published techniques apply pressure changes at a rate at which the system is no longer isothermal, but isentropic. Instead of eqn. (1) we have to write

$$(\frac{\partial \ln K}{\partial P})_S = - \frac{\Delta V^O}{RT} + \frac{\alpha}{\rho c_p} \cdot \frac{\Delta H^O}{RT} \tag{2}$$

The second term respects the temperature effect accompanying the pressure variation and is small for many reactions in aqueous solutions since $(dT/dP)_S$ is only 0.07 K/(100 bar) at 20° C. Thus in these special cases $(\partial \ln K/\partial P)_S \approx (\partial \ln K/\partial P)_T$. On the other

87

W. J. Gettins and E. Wyn-Jones (Eds.), Techniques and Applications of Fast Reactions in Solution. 87–94.
Copyright © 1979 by D. Reidel Publishing Company.

hand for many organic solvents the second term may be comparable
to the first one or even larger (e.g. $(dT/dP)_S$ for benzene is
2.45 K/(100 bar) at 20° C).

Consequence of the discussed pressure dependent variation of
the equilibrium constant is a concentration shift

$$\delta c = (\Sigma \frac{v_i^2}{c_i})^{-1} \cdot (- \frac{\Delta V^{\circ}}{RT} + \frac{\alpha}{\rho c_p} \cdot \frac{\Delta H^{\circ}}{RT}) \cdot \delta P \tag{3}$$

As long as $\delta c \ll c$ the mathematical formalism of relaxation kinetics
can be applied. For stationary methods this restriction is usually
well fulfilled, but for pressure jump methods, which are to be
treated here, it should be controlled in each individual case.

The pressure jump methods available today offer a large
variety of detection possibilities. In gereral one only has to know
the dependence of the observed signal on the concentration in
order to interpret the amplitudes according to

$$\delta A = (\frac{\partial A}{\partial c})_{P,S} \cdot \delta c \tag{4}$$

Combination of (3) and (4) permits the evaluation of amplitudes
as a function of the equilibrium conditions and the pressure jump.

Methods

A number of different techniques for the creation of a pressure step has been developed and both pressure decrease and increase jumps are applied. The principle, which was mainly used, is
that a bursting membrane releases hydrostatic pressure to ambient
conditions as it will be discussed as an example in the next
chapter [1,2]. The alternative method of pressure release through
fast opening of a mechanical valve was described by Davis and
Gutfreund [3]. These techniques allow pressure drop times of
50-100 µs. An approach to the pressure increase jump by means of
a fast electromagnetic valve controlling the connection between
the pressure vessel and a gas bomb was reported by Kegeles and Ke
[4]. Based on the same principle Yasunaga described a valve method
with repetitive application of gas pressure jumps [5]. Pressure
rise times lie between 2 and 40 ms. A very elegant method for
small (0.01 - 5 bar) pressure perturbations was reported by Clegg
and Maxfield [6] who use voltage pulses of various shapes repe-
titively applied to a stack of piezoelectric crystals for the
compression of the solution. Relaxation times of 20 µs up to
several minutes can be measured. - Obviously the method of detection
is independent of the pressure perturbation technique. Electrical
conductivity mainly used for inorganic systems was combined with
the apparatuses reported in [1], [5], and [6]. This detection is
very sensitive, but unspecific. For many problems, especially

macromolecular, mostly biological ones optical detection methods are more useful, because they are more specific. Transmittance (optical absorption or turbidity) is applicable to the apparatuses described in [2], [3], [4], and [6]. It was shown recently that these techniques can also be combined with the very specific detection of circular dichroism (or optical rotation) [7]. Finally also fluorescence and light scattering detection was combined with pressure jump methods [3], [4], and [6]. Also worth mentioning is the thermometrical detection method [8].

Summarizing it should be mentioned that as compared to the Joule heating temperature jump the pressure jump technique offers a wider choice of solvent composition, an extended time range and shorter time intervals between repeats. And finally it has been shown that for organic solvents the temperature jump due to the adiabatic pressure jump causes an equilibrium shift sufficiently large for kinetic measurements [9].

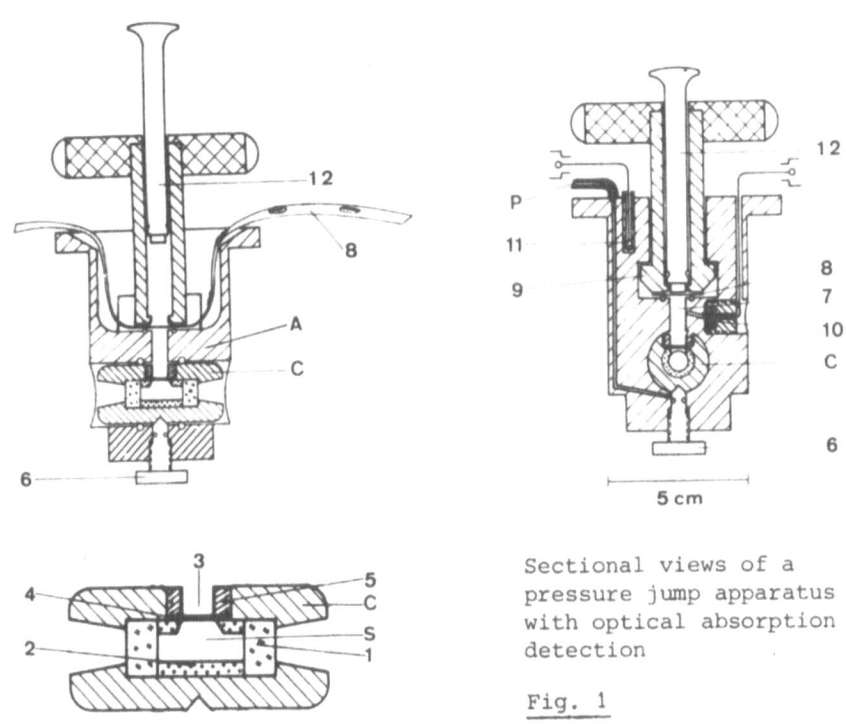

Sectional views of a pressure jump apparatus with optical absorption detection

Fig. 1

Jump Apparatus

Details of the pressure jump equipment with detection of electrical conductivity have been discussed in previous reviews

[10,11]; therefore we restrict ourselves to the description of
an apparatus with optical absorption detection [12] (Dia-Log,
Harffstr. 34, D-4 Düsseldorf), which is shown in fig. 1. The
apparatus consists of three parts: (i) the optical cell (C) containing the solution to be studied, (ii) the autoclave (A), in
which a fast pressure drop is achieved, (iii) a high pressure
manual water pump (not shown in fig. 1), which is connected to
inlet (P). The outer dimensions are the same as those of a conventional temperature jump cell, enabling the use of the optical
arrangement and thermostating system of temperature-jump equipment
without any modification.

The sample cell (C) is made of stainless steel with sapphire
windows (1) and has an insertion (2) made of ceramics (Macor,Corning)
which is inert against all solvents and solutes. The cell is taken
out of the autoclave for filling through inlet (3) which is closed
with a thin membrane of soft plastics (4) by means of screw (5). The
cell is then pushed into the autoclave and kept in position by
screw (6).

The autoclave (A) is made of bronze. The pressure chamber (7)
is filled with water, covered by a brass foil (thickness $\sim$ 0.1 mm)
(8) and rigidly sealed by putting the bayonet socket (9) into
position and turning it about 90°. The pump is connected to inlet
(P) and additional water is pressed into the autoclave to increase the pressure. The water passes around the cell to reach
the pressure chamber. The pressure sealing between autoclave and
cell is provided by O-rings. The hydrostatic pressure is transmitted to the sample volume through the thin plastic membrane (4).

The pressure in the autoclave is increased until the brass
foil ruptures spontaneously, whereupon the pressure drops to
1 atm within less than 10^{-4} s. The bursting pressure and to some
extent the pressure-drop time depends on thickness and type of
material used for the foil. The fast pressure drop causes a
piezoelectric capacitor (10) to generate a voltage pulse which
triggers oscilloscope and data-capturing device. The temperature
of the autoclave is determined with an accuracy of $\pm$ 0.1 K by
measuring the resistance of a Pt 100 resistor (11). In the com-
partment above the rupture disk the pressure can be reduced to
about 0.01 atm by a small pump (12) to avoid disturbances caused
by acoustical noise. The bursting membrane is exchanged by
loosening the bayonet socket (9) and moving forward the metal
band (8). This allows the measurements to be repeated every 30 s.
The equipment has been used to measure relaxation times between
10^{-4} s and 100 s, in the temperature range 0 to 60° C, and
pressure jumps between 30 and 200 atm.

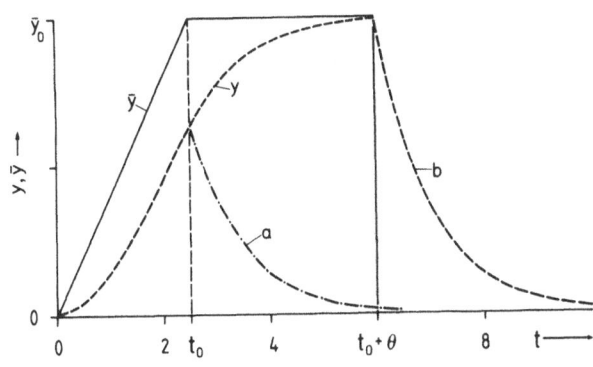

Fig. 2

Shift of equilibrium
concentration $\bar{y}$ and
actual concentration
y at a pressure jump
experiment, t in ar-
bitrary units. y is
calculated for a
single relaxation ef-
fect with $\tau = 1$ and
pressure drop at
a) $t = t_o$ and
b) $t = 6$.

Perturbation effects on the equilibrium

In pressure (-release) jump techniques the perturbation of
the chemical system is achieved by raising the pressure from the
ambient to some higher value and a subsequent sudden drop back
to the ambient value. To which extent the actual concentrations
can assume their equilibrium values at high static pressure before the jump depends on the ratio of the relaxation time of the
system and the pressure duration time θ, and also slightly on the
time t_o needed for the pressure increase (see fig. 2). Here y and
$\bar{y}$ refer to generalized concentration shifts where $\bar{y}$ is the equilibrium and y the actual shift. $\bar{y}_o$ is the equilibrium shift at high
static pressure. Obviously the choice of the pressure duration
time θ is critical for the relaxation amplitude which is measured
after the jump, because a jump before $t \sim 6\tau$ would lead to a reduction as compared to $\bar{y}_o$.

On the other hand this consideration may be a useful help
for the separation and the analysis of two relaxation effects
with relaxation times of which one is less than a factor of ~ 10
larger than the other, $\tau_2 = c\tau_1$. By choosing θ so that no complete
equilibrium is reached, $(y/\bar{y}_o)_2$ will always be more reduced than
$(y/\bar{y}_o)_1$ if τ_2 is larger than $\tau_1 (c > 1)$. This means that a short
(in relation to τ_2) pressure duration time θ always favors the
amplitude of the fast effect, or else: it always increases the
ratio $(y/\bar{y}_o)_1/(y/\bar{y}_o)_2$. Under the assumption of infinitely fast
pressure rise and drop this ratio is given by

$$\frac{(y/\bar{y}_o)_1}{(y/\bar{y}_o)_2} = \frac{1 - \exp(-\theta/\tau_1)}{1 - \exp(-\theta/\tau_2)} \tag{5}$$

For two special ratios of τ_1 and τ_2 this is plotted as a function
of θ/τ_1 or θ/τ_2 in fig. 3. Obviously the absolute maximum for the
relative amplitudes in eqn. (5) is reached for $\theta = 0$ and is given
by

$$\lim_{\theta \to 0} \frac{(y/\bar{y}_o)_1}{(y/\bar{y}_o)_2} = \frac{\tau_2}{\tau_1} \tag{6}$$

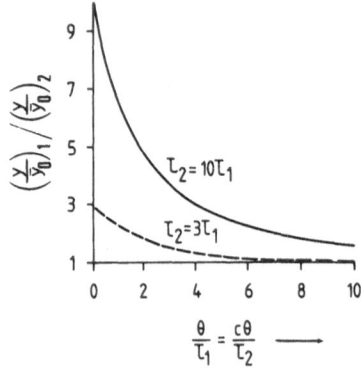

Fig. 3

The ratio $(y/\bar{y}_o)_1$ to $(y/\bar{y}_o)_2$ in dependence of the pressure duration time θ for $\tau_2 = 10\,\tau_1$ and $\tau_2 = 3\,\tau_1$

Very small θ favors the fast effect, but reduces also its amplitude. Therefore in each individual case an experimental optimum for θ has to be found. Amplitude analysis has to take this into account. Once relaxation time and amplitude of the fast effect are accurately known, it is easier to subtract it from the slow effect, for whose analysis the experiment has to be repeated with $\theta \gtrsim 6\tau_2$.

This consideration about the analysis of overlapping relaxation effects holds for all chemical relaxation methods with pulse shaped perturbations, it is not restricted to the pressure jump.

Applications

The pressure jump relaxation technique has first been applied with conductometric readout to study the formation of metal complexes involving Be^{2+}, Al^{3+}, Ga^{3+}, In^{3+}, Ni^{2+}, and Fe^{3+}, for which the reaction can be described by the Eigen-Tamm mechanism. These measurements have been summarized before. Recently examples have been found where in the case of bidentate ligands the formation of both monodentate and bidentate complexes could be observed separately [13]. Since the different complexes form at different rates, rate as well as equilibrium constants are obtained. In the case of carboxylate ligands it is difficult or even impossible to obtain these equilibrium constants by static measurements. Besides aqueous and nonaqueous solutions also mixed solvents have been used for the investigation of metal complexation [14,15]. In the case of the Be^{2+}/SO_4^{2-} system which shows a single relaxation effect in aqueous solution, up to five different relaxation effects are observed in mixed solvents, which refer to different composition of the solvation shell of the cation.

By several research groups the kinetics of micelle formation

has been studied using the pressure jump technique with conduc-
trometric readout for ionic surfactants and with spectrophotometric readout for uncharged surfactants. A summary of many results
is given in ref. [16]. Relaxation times have been determined as
a function of the hydrocarbon chain length and the polar head-
groups. Also the influence of the addition of hydrophobic solutes,
inert electrolyte and divalent counterions has been studied. The
kinetics of metal complex formation in the presence of micelles
has been investigated using the system Ni^{2+}-murexide-SDS [17].

   Further applications include the study of the hydration of
CO_2 [18] and carbonyl compounds [19], the dynamics of phase
separation in critical binary mixtures [20], [21], the kinetics
of heterogeneous catalytic processes [22], [23]. There is a
growing interest in applying the pressure jump technique to
problems of biological relevance: association behavior of ribo-
somal subunits [24], relaxation kinetics of helix-pomatia α-
hemocyanin [25], assembly of tubulin [3], [26] and myosin [3],
glutamate dehydrogenase self-association [26], folding of the
protein BSA [3], [7], the reaction between lactate and NAD [3],
phase transition of lipid bilayers [27], [28], and, as a truely
biophysical application, kinetics of frog skin open circuit voltage
and short circuit current [29].

References

[1]   W. Knoche and G. Wiese, Chem. Instrum., 5 (1973-74) 91.

[2]   W. Knoche and G. Wiese, Rev. Sci. Instrum., 47 (1976) 220.

[3]   J.S. Davis and H. Gutfreund, FEBS Lett., 72 (1976) 199.

[4]   G. Kegeles and Ch. H. Ke, Analyt. Biochem., 68 (1975) 138.

[5]   T. Yasunaga and N. Tatsumoto, Biophys. J., 24 (1978) 267.

[6]   R.M. Clegg and B.W. Maxfield, Rev. Sci. Instrum., 47 (1976)
      1383.

[7]   B. Gruenewald and W. Knoche, Rev. Sci. Instrum., 49 (1978)
      797.

[8]   J. Helisch and W. Knoche, Ber. Bunsenges. Phys. Chem., 75
      (1971) 951.

[9]   H.J. Buschmann, W. Knoche, R.A. Day, and B.H. Robinson,
      J. Chem. Soc., Faraday Trans. I   73 (1977) 675

[10] W. Knoche, in Techniques of Chemistry, Vol. VI, 2,
      G.G. Hammes, Ed., Wiley 1974.

[11] W. Knoche, in Chemical and Biological Applications of
      Relaxation Spectrometry, E. Wyn-Jones, Ed., Reidel 1975.

[12] W. Knoche and G. Wiese, Rev. Sci. Instrum., 47 (1976) 220.

[13] B. Gruenewald and W. Knoche, J. Chem. Soc., Dalton Trans.
      (1978) 1221.

[14] R. Lachmann, I. Wagner, D.H. Devia, and H. Strehlow, Ber.
      Bunsenges. Phys. Chem., 82 (1978) 492.

[15] D.H. Devia and H. Strehlow, this volume.

[16] J. Lang, C. Tondre, R. Zana, R. Bauer, H. Hoffmann, and
      W. Ulbricht, J. Phys. Chem., 79 (1975) 276.
[17] B.H. Robinson, J. McLagan Wedderburn, N.C. White, M. Fischer,
      and W. Knoche, this volume.
[18] R.C. Patel, R.J. Boe, and G. Atkinson, J. Sol. Chem., 2
      (1973) 357.
[19] H.J. Buschmann and W. Knoche, Ber. Bunsenges. Phys. Chem.,
      in press.
[20] D. Woermann, Z. Phys. Chem. (Frankfurt), 103 (1976) 219.
[21] N.C. Wong and C.H. Knobler, J. Chem. Phys., 66 (1977) 4707.
[22] J. Suzuki and Y. Kaneko, J. Catalysis, 36 (1973), 58.
[23] T. Yasunaga, this volume
[24] E. Schulz, R. Jaenicke, and W. Knoche, Biophys. Chem., 11
      (1976) 253.
[25] M.S. Tai, G. Kegeles, and C.H. Ke Huang, Arch. Biochem.
      Biophys., 180 (1977) 537.
[26] Y. Engelborghs, to be published.
[27] B.W. Maxfield, R.M. Clegg, L. Avers, and E.L. Elson, Biophys.
      J., 15 (1975) 104 a.
[28] B. Gruenewald, A. Blume, and F. Watanabe, to be published.
[29] J.P. Segal, Biochim. Biophys. Acta, 471 (1977) 453.

ELECTRIC FIELD METHODS

Paul Hemmes

Department of Chemistry, Rutgers University, Newark,
New Jersey 07102

Abstract: A brief survey of the uses of high electric fields for
kinetic measurements is presented. New experimental techniques
are also discussed.

Electric field methods for the study of fast kinetics encom-
pass a number of diverse experimental techniques. If the applied
field is small the techniques are usually considered under the
heading of dielectric relaxation techniques. These will be discussed elsewhere in this volume. At high field strengths there
is an increased dissociation of weak electrolytes, the so called
second Wien effect. Quantitatively this effect is given by the
equation of Onsager (1)

$$\frac{K(X)}{K(0)} = \frac{l_1(4\beta q)^{\frac{1}{2}}}{2\sqrt{\beta q}}$$

where $K(X)$ is the dissociation constant at high field strength
X, l_1 is a modified Bessel function of order one

$$\beta q = \frac{1}{2} \frac{X(e_j u_j - e_i u_i)}{kT\,(u_j + u_i)} \frac{-e_j e_i}{2DkT}$$

$e_j(i)$ is algebraic charge on the ion $j(i)$; $u_j(i)$ is mobility of
the ion $j(i)$; and D is the dielectric constant. For a very weak,
symmetrical electrolyte the conductance at high field is, to
first order terms,

$$\frac{\Lambda(X)}{\Lambda(0)} \quad 1 + \frac{9.695}{2} \frac{(1-\alpha)}{(2-\alpha)} \frac{z^3}{DT^2} X$$

W. J. Gettins and E. Wyn-Jones (Eds.). Techniques and Applications of Fast Reactions in Solution. 95–98.
Copyright © 1979 by D. Reidel Publishing Company.

with α the zero field degree of dissociation and Z the magnitude
of the valence. This relationship shows that for polyvalent
electrolytes or for any electrolyte in low dielectric media or
at low temperatures, the second Wien effect can be rather large.

It must be pointed out that high fields shift any equilibria
which involve a change in mobility and/or a change in dipole
moment. It can be shown using Maxwell's equations that even for
shifting dipolar equilibria without ion formation or combination
there is a conductance change associated with the process that
can be used for detection.

Transient Techniques: The most commonly used electric
field method is the E-jump technique. A high voltage pulse of
∿ 1 to more than 10 μsec duration with a rise time depending upon
the details of the apparatus but of the order of 50 nsec or less
is produced by a coaxial cable discharge (2) or twin spark gaps
(3) and a capacitor Conductimetric detection was used by Eigen
and DeMaeyer (4) for the kinetics of proton-hydroxide ion recombination, by Yasunaga et al (5) for helix coil transition studies,
the study of metal ion hydrolysis reactions (6)(7)(8) and dissociation of transition metal complexes in non-aqueous solvents by
McGarvey and coworkers (9)(10). The latter authors presented a
simplified conductimetric detection system which offers much
improvement over earlier bridge methods. Spectrophotometric
detection has been used by Ilgenfritz (11)(12) and Eyring (13)
(14) to study indicator dissociations including the use of
dynamic coupling between weak acids to determine the rate constants for dissociation of colorless acids (15). In a promising
application of field methods, the kinetics of proton transfers
in non-aqueous solution have been studied recently (16). Furthermore optical rotation changes have been used to study processes
induced by electric fields (17)(18). It is possible to eliminate
the effects of orientation of macromolecules on the observed
optical rotation changes by suitable experimental techniques.

Other applications of transient field methods include cooperative binding of dyes to macromolecules (19), studies of the
physical and chemical processes in proteins (20)(21), estimation
of equilibrium constants by field induced conductivity changes
(22) and some remarkable optical effects in dye solution under
high fields (23).

Frequency Domain Methods: Recently a number of new techniques have been developed for the study of molecular aggregation and ionic dissociation processes in very low dielectric
media. The first of these is chemically induced dielectric relaxation which has been theoretically treated by Schwarz (24)
and confirmed experimentally (25) for poly (γ-benzyl L-glutamate).
Under very high fields, equilibria involving changes in dipole

moment can be perturbed and the chemically induced change in
loss tangent can be measured as a function of frequency. Detailed
theory and experimental details have been covered in a number of
excellent reviews (26)(27). The method has been used to study
hydrogen bonded dimer formation (28). Still another technique
for the study of electrolytes is the field modulation method of
Persoons (29). While the theory is complex and the instrument
sophisticated, the experiment is simple and capable of producing
relaxation times of rather high precision, far better than the
transient methods. Thus more critical evaluation of the kinetics
of ionic processes in solution will be possible.

References

1. L. Onsager, J. Chem. Phys., 2, 599 (1934).
2. D. T. Rampton, L. P. Holmes, D. L. Cole, R. P. Jensen, and
    E. M. Eyring, Rev. Sci. Instrum., 38, 1637 (1967).
3. S. L. Olsen, R. L. Silver, L. P. Holmes, J. J. Auborn,
    P. Warrick, Jr. and E. M. Eyring, ibid, 42, 1247 (1971).
4. M. Eigen and L. DeMaeyer, Z. Electrochem., 59, 986 (1955).
5. T. Yasunaga, T. Sano, K. Takahaski, H. Takenaka, and S. Ito,
    Chem. Lett. (Japan), 405 (1973).
6. D. L. Cole, L. D. Rich, J. D. Owen, and E. M. Eyring,
    Inorg. Chem., 8, 682 (1969).
7. P. Hemmes, L. D. Rich, D. L. Cole, and E. M. Eyring, J.
    Phys. Chem., 74, 2859 (1970).
8. P. Hemmes, L. D. Rich, D. L. Cole, and E. M. Eyring, ibid,
    75, 929 (1971).
9. H. Hirohara, K. J. Ivin, and J. J. McGarvey, J. Am. Chem.
    Soc., 96, 3311 (1974).
10. H. Hirohara, K. J. Ivin, J. J. McGarvey, and J. Wilson,
    ibid, 96, 4435 (1974).
11. G. Ilgenfritz, Ph.D. Thesis, Georg August Univ., Gottingen
    (1966).
12. G. Ilgenfritz in "Probes of Structure and Function of Macro-
    molecules and Membranes". (B. Chance, ed.) Vol 1, p. 505,
    Academic Press, New York, 1971.
13. J. J. Auborn, P. Warrick, Jr., and E. M. Eyring, J. Phys.
    Chem., 75, 3026 (1971).
14. S. L. Olsen, L. P. Holmes, and E. M. Eyring, Rev. Sci.
    Inssum., 45, 859 (1974).
15. J. J. Auborn, P. Warrick, Jr., and E. M. Eyring, J. Phys.
    Chem., 75, 2488 (1971).
16. F. Strohbush, D. B. Marshall, and E. M. Eyring, ibid, 82,
    2447 (1978).
17. D. Porschke, Biopolymers, 15, 1917 (1976).
18. A. Revzin and E. Neumann, Biophys. Chem., 2, 144 (1974).
19. T. Yasunaga, H. Tahenaka, T. Sano and Y. Tsuji in "Chemical
    and Biological Application of Relaxation Spectrometry".
    (E. Wyn-Jones, ed.), p.467, Reidel, Dordrecht-Holland, 1975.

20. G. Ilgenfritz and T. M. Schuster in "Probes of Structure
     and Function of Macromolecules and Membranes" (B. Chance,
     ed.) Vol. 11, p. 399, Academic Press, New York, 1971.
21. G. Ilgenfritz and T. M. Schuster, _ibid_, p. 299.
22. P. Hemmes and J. J. McGarvey, submitted for publication –
     see also this volume.
23. Z. A. Schelley, K. Lundy-Douglas, and H. Eyring, Proc. Natl.
     Acad. Sci., U.S.A., 75, 2549, 1978.
24. G. Schwarz, J. Phys. Chem., 71, 4021, (1967).
25. G. Schwarz and J. Seelig, Biopolymers, 6, 1263 (1968).
26. L. DeMaeyer and A. Persoons in "Techniques of Chemistry",
     (G. G. Hammes, ed) Vol. VI, part 2, p. 211, Wiley (Inter-
     science), New York, 1973.
27. A. P. Persoons, J. Phys. Chem., 78, 1210 (1972).
28. K. Bergmann, M. Eigen, and L. DeMaeyer, Ber. Bunsenges.
     Phys. Chem., 67, 819 (1963).
29. A. Persoons, J. Phys. Chem., 78, 1210 (1974).

DESTRIAU-EFFECT IN AQUEOUS SOLUTION

Z. A. Schelly, G. Sumdani and Henry Eyring*

Department of Chemistry, University of Texas at
Arlington, TX 76019, USA

*Department of Chemistry, University of Utah, Salt Lake
City, UT 84112, USA

Transient luminescence induced by the high electric field present
in E-jump experiments led to the first observation of the
Destriau-effect in liquid solution.

INTRODUCTION

The phenomenological description of the Destriau-effect (1)
is illustrated in Figure 1. If a high voltage square pulse is
applied to a substance, transient light emission takes place both
at the leading and trailing edges of the perturbing square pulse.
Its most significant feature is that luminescence is induced also
by turning off the field. It is a high-field electroluminescence
that has previously been observed only in solid state phosphors.

SUMMARY

The Destriau-effect in liquid solution was first observed
(2) in E-jump experiments on aqueous solution of the sodium salt
of 6,8-dihydroxypyrene-1,3-disulfonic acid (DPD). The E-jump
apparatus used has already been described (3). The oscilloscope
traces of the signals have the feature of the curve shown in
Figure 1b. Luminescence is induced only at high field strengths
(> 2.5 x 10^4 V/cm), and the emitted light intensity decays with
a time constant of 0.5 μsec.

W. J. Gettins and E. Wyn-Jones (Eds.), Techniques and Applications of Fast Reactions in Solution. 99–101.
Copyright © 1979 by D. Reidel Publishing Company.

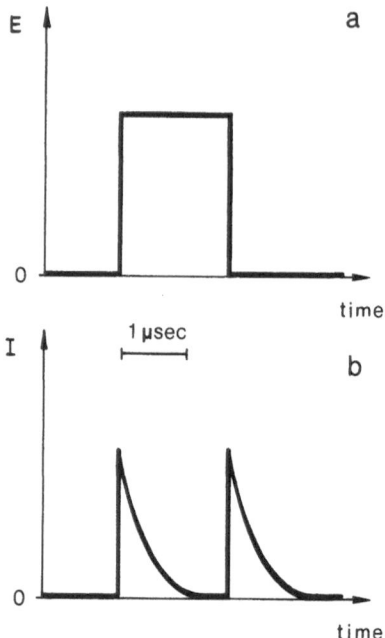

Figure 1   Schematic representation of the Destriau-effect.
(a) Perturbation high field pulse. E, electric field strength.
(b) Light emission from sample. I, intensity of luminescence.

The mechanism of the excitation and emission can be explained
by the electron injection model (2), in which the time constant
of the luminescence decay is associated with the rate of formation of a charged double layer in front of the cathode, which is
created and then stabilized by the applied potential.

The use of the Destriau-effect has great potential in the
investigation of the little known dynamics of the formation of
double layers.

ACKNOWLEDGEMENT

Financial support by the R. A. Welch Foundation and the
Organized Research Fund of the University of Texas at Arlington
is greatfully appreciated.

REFERENCES

1.   G. Destriau, J. Chem. Phys., 33, 587 (1936); Philos. Mag.,
      38, 700 and 774 (1947).

2.   Z. A. Schelly, K. Lundy-Douglas and Henry Eyring, Proc. Natl.
     Acad. Sci. USA, 75, 2549 (1978).
3.   M. W. Massey and Z. A. Schelly, J. Phys. Chem., 78, 2450
     (1974).

DIELECTRIC RELAXATION

R.A. Pethrick

Department of Pure and Applied Chemistry
Thomas Graham Building
University of Strathclyde
Glasgow G1 1XL.

Historically, dipole reorientation was one of the first molecular relaxation processes to be systemmatically investigated. The permittivity of a molecular system can be described in terms of three contributions:
(1) electronic polarization-(α_e)-associated with motion of the electrons independent of the nuclei and gives rise to the high frequency value of the dielectric constant (refractive index of a material-n).
(2) atomic polarization-(α_a)-associated with the vibrational motion of covalent molecules and gives rise to absorption and dispersion in the infrared spectral region.
(3) dipolar polarization-(α_d) associated with the reorientational motion of polar groupings.

The total polarization is the sum of all three contributions:

$$\alpha_T = \alpha_e + \alpha_a + \alpha_d \tag{1}$$

The relative permittivity ε_0 may be defined by

$$\varepsilon_0 = C/C_o = (Q + P)/Q \tag{2}$$

The permittivity of the material will be

W. J. Gettins and E. Wyn-Jones (Eds.). Techniques and Applications of Fast Reactions in Solution. 103–113.
Copyright © 1979 by D. Reidel Publishing Company.

higher the greater the polarizability of the
molecules contained in the capacitor. Alternatively
we can define(1,2) ε_0 as follows:

$$\varepsilon_0 = 1 + 4 \gamma \ P/\varepsilon E \tag{3}$$

where P is the permittivity of the material, E is the
applied field and ε is a constant whose value depends
on the system of units used. The polarization due to
a dipole reorientation is expressed by

$$P = N_1 \mu \langle \cos \Theta \rangle = N_1 \ \mu^2 F/3kT \tag{4}$$

where N_1 is the number of dipoles per unit volume, μ
is the dipole moment, F is the effective local field,
k and T have their usual meaning. Adding to this the
distortion polarization $N_1 \ a \ F$, we obtain for the
total polarization

$$P = N_1 (a + \mu^2/3kT) \ F_1 \tag{5}$$

The principle problem which remains is relating the
applied field E to the effective or loca field F.

The field at any point rather within a sphere
in a dielectric will be made up of three components
(3).

(i) the field due to the charge on the plates themselves

(ii) the field due to the
polarization charges on the
spherical surface surrounding
defined by the angles Θ and Θ
+ dΘ. This has area 2γ r^2
sin Θ dΘ and a surface density
of charge equal to the normal
component of the polarization, namely P cos Θ. By
symmetry it produces no field at A perpendicular to
the direction of P, while in the direction parallel
to P it produces a field

$$\frac{\text{area x surface density x cos } \Theta}{\varepsilon \text{ x (distance)}^2} =$$

$$\frac{2\gamma r^2 \sin \Theta \ d \ \Theta P \cos \Theta \cos \Theta}{\varepsilon \ r^2}$$

Integrating over all values of Θ, we have

$$F_2 = \frac{2\gamma}{\varepsilon} \int_0^\gamma P \cos^2 \Theta \sin \Theta \, d\Theta$$

$$= \frac{4\gamma P}{3\varepsilon} \tag{6}$$

(iii) the field due to molecules within the sphere. When molecules are distributed on a cubic lattice or form an ideal gas, this contribution is zero. In other systems this contribution is introduced as the Kirkwood 'g' factor(4)

The total field at a point within the sphere will be

$$F = F_1 + F_2 = E + 4\gamma P/3\varepsilon$$

$$4 \quad P = (\varepsilon_0 + 1)\varepsilon E$$

$$F = (\varepsilon_0 + 2)E/3\varepsilon$$

Combining these results yields:

$$P = N_1 (\alpha + \mu 2/kT)F = N_1 (\alpha + \mu 2/3kT) (\varepsilon_0 + 2)E/3 \tag{7}$$

we obtain

$$\frac{\varepsilon_0 - 1}{\varepsilon_0 + 2} = 4 N(\alpha + \mu^2/3kT)/3\varepsilon \tag{8}$$

This is the Debye equation for the static permittivity and can be rewritten in terms of the density (ρ) and molar weight (M) as

$$\frac{(\varepsilon_0 - 1) M}{(\varepsilon_0 + 2)} = 4 N(\alpha + \mu^2/3kT)/3\varepsilon \tag{9}$$

In practice this equation is not completely adequate and alternative forms have been proposed (3,5,6) which allow for the effects of the internal field on the field at the surface of the sphere. The Kirkwood form of the equation is one of the most general treatments (4).

$$\frac{(\varepsilon_0 - n^2) (2\varepsilon_0 + n^2)}{\varepsilon_0 (n^2 + 2)^2} = \frac{4 N g \mu^2}{9V \varepsilon kT} \tag{10}$$

here n is the refractive index and allows for the distortion polarization contribution to the permittivity; g is the correlation factor which allows for the effects orientation of neighbouring dipolar groups and V is molar volume.

Dynamic Polarizability

The permittivity is a complex quantity and exhibits a frequency dependence characterized by its real and imaginary parts. Rearranging equation (9) and putting the distortion polarization equal to the square of the refractive index yields

$$\frac{\varepsilon^{\ast} - n^2}{\varepsilon_0 - n^2} = \frac{1}{1 + jw\tau} \tag{11}$$

The real and imaginary parts of the permittivity will have the form

$$\frac{\varepsilon' - n^2}{\varepsilon_0 - n^2} = \frac{1}{1 + w^2\tau^2} , \frac{\varepsilon''}{\varepsilon_0 - n^2} = \frac{w\tau}{1 + w^2\tau^2} \tag{12}$$

Complex Plane Plots

By combining the above equations and eliminating $w\tau$ it is possible to obtain the equation of a circle,

$$\varepsilon' - \frac{\varepsilon_0 + n^2}{2} + \varepsilon''^2 = \left(\frac{\varepsilon_0 - n^2}{2}\right)^2 \tag{13}$$

The circle has as its centre $(\varepsilon_0 + n^2)/2, 0$ and radius $(\varepsilon_0 - n^2)/2$ an ideal relaxation curve.

Cole-Cole Plots (7)

For a relaxation process in which the mechanism is constant but the energy states between which exchange occurs can have different values of Cole-Cole analysis is appropriate. This type of analysis applies to relaxation in polymer molecules and hydrogen bond exchange

$$\varepsilon^{\ast} - n^2 = \frac{\varepsilon_0 - n^2}{1 + (jw\tau)(1 - a)} \tag{14}$$

Cole-Davidson Plots (8)

This form of equation applies to a situation where we believe that a particular relaxation corresponds to an overlap of a number of simple Debye

type relaxation processes. This is the sort of
situation which is encountered when librational and
reorientational relaxation are observed in close
proximity

$$\frac{\varepsilon - n_2}{\varepsilon_0 - n^2} = \frac{1}{(1 + j\omega\tau)^\beta}$$

Both of these forms of the equation can be put into
the complex plane notation and analysed to give both
an indication of the mean relaxation frequency and
the distribution parameters α and β. In both cases
in the limit i.e $\alpha \rightarrow 0$ and $\beta \rightarrow 1$ the equation equals
that of the ideal Debye relaxation, equation (11).
A combined complex plane plot has been proposed
however its application is somewhat limited and will
not be discussed here.

General Considerations

    For dielectric relaxation to be observed there
has to be a change in the polarizability of the media
under the influence of an applied field. This
selection rule implies that dispersion will only be
detected in polar materials. However, in certain
cases contributions to the polarizability have been
detected in non polar materials and are ascribed to
collisional polarization effects. If two non polar
molecules collide there is the possibility that
distortion of the electron density and atomic positions
may result in the formation of a transient dipole or
multipole. The induced polarization can be destroyed
by further collision with other non activated
molecules. The lifetime of these collisionally
activated molecules may be many times the collision
frequency and it then becomes possible to observe the
reorientational motion of the induced dipoles, which
act as though they were permanent dipoles.

    The total dielectric dispersion spectrum can
be represented as follows,

(i) low frequency-dipolar
reorientation. In solids
additional polarization
phenomena associated with
charge migration can be
observed, these will not
be discussed here.

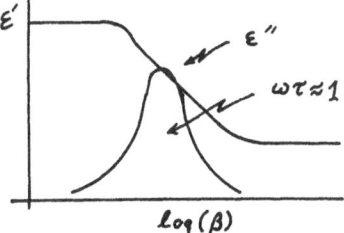

(ii)   intermediate frequency-collisional
polarization. The magnitude and position of the
relaxation depends on the type of molecule concerned
and the ease of reorientation of the induced dipole.

(iii) atomic distortion - this corresponds to the
real and imaginary parts of the permittivity changes
associated with molecular vibration and are observed
in the infrared region.

        Dielectric relaxation is somewhat limited
in its application to the study of chemical processes
because it reflects the reorientation of the dipole
which occurs with a time constant determined by
whole molecule motion rather than isomeric or
chemical change. For chemical data to be obtained
whole molecule motion has to be suppressed.
Dispersion of a polar molecule in a polymer matrix
can in certain instances lead to observations of
internal rotational isomerism rather than rotational
reorientation motion, however this is not the usual
situation.

        Dielectric relaxation is usually employed to
characterise the rate of whole molecule reorientation
in simple molecular fluids, segmental relaxation in
polymer molecules - here the size of the molecule
makes the whole molecule reorientation a slower
process than internal reorientation and structural
relaxation (rotation and translation) in associated
liquids such as water or alcohols.

Experimental Methods

        The dielectric spectrum ranges from 10^{-6} Hz
to above 10^{12} Hz, some eighteen decades of frequency.
It is impossible to describe in detail the techniques
used for dielectric relaxation. However we shall
consider the principles of the methods and see how
as we change frequency range so the methods used
have to be changed.

        The dielectric relaxation techniques can be
sub-divided into two groups: step response or
continuous wave methods.

(i)   Step Response

        The concepts of step response techniques have
been outlined in the proceeding lectures. In the

context of the dielectric experiment, observation of
the change in the permittivity or more correctly
the current or voltage with time following application
or removal of a step perturbation can be Laplace
transformed to yield the real and imaginary parts of
the permittivity as a function of frequency. This
method in suitable circumstances can be applied to
studies over the frequency range 10^{-6} to 10^9 Hz.

(ii) Continuous Wave Methods

Bridge or equivalent standing wave methods can
be used to cover the frequency range 10 Hz - 10^{12} Hz.
The method changes with frequency. At low frequency
10 Hz - 10^6 Hz bridge techniques are widely used, the
method resembles that of a Wheatstone resistance
bridge except that opposing arms contain capacitance
and resistance components and are coupled via a
radio frequency transformer. At frequencies above
10^6 Hz, the wavelength of the electro-magnetic
radiation becomes comparable to the physical size of
the components in the circuit, and they cease to
behave as either pure resistances, capacitances or
inductances. In the frequency range 10^6 to 10^8 Hz
resonance circuits are used in which the width and
position of resonance of an air and sample filled
cavity are used to probe the imaginary and real parts
of the permittivity. Above 10^8 Hz waveguide methods
are employed in which the wavelength and attenuation
of the signal as a function of distance are used to
obtain the dielectric parameters.

Due to the complexity of equipment it is simplest
to give reference to selected reviews of techniques,
Table (1).

Table (1) Methods of Measurement of the Complex Permittivity

Frequency Range (Hz)	Method	References
10^{-6} - 10^0	D.C. transient	9, 10, 11
10 - 10^6	Pulse methods	12
10^6 - 10^9	Time domain reflectrometry	13, 14
10 - 10^6	A.C. Bridge methods	9, 15
10^6 - 10^8	Resonators	16
10^8 - 10^{10}	Coaxial wave guide methods	17
10^{10} - 10^{12}	Microwave transmission methods	17, 18
10^{11} - 10^{13}	Lamellar and Michelson Interferometry	17, 19

Examples of Dielectric Relaxation Behaviour

(i)   Reorientational Motion in Simple Molecules

As an example of a typical dielectric relaxation
we shall consider the behaviour of propylene
carbonate[20,21]. The
relaxation spectrum is
dominated by peak centred
at 5×10^9 Hz, associated
with the rotational
motion of the C=O dipole
about the centre of gravity of the molecule. A
shoulder on the higher frequency side (9000 MHz) of
the main peak is ascribed to librational or
collisional polarization effects. Investigation of
the far infrared spectrum of this molecule[21] indicates
the existence of a loss feature which can be assigned
to collisional polarization indicating that the
shoulder is most probably librational in origin.
Librational relaxation arises from the fact that
liquid can have transient structure which leads to an
angular profile for the rotational energy. Reorientation of a dipole will require to overcome a
repulsive interaction from the surrounding molecules.
Partial randomization of the dipoles can occur
through a small angle but total randomization requires
dipoles to adopt all possible angular orientations.
Local randomization can occur more rapidly than large
scale randomization hence the observation of a high
frequency shoulder.

(ii)   Collisional Polarization

Benzene is a non polar molecule and as such
cannot exhibit dielectrically active reorientational
relaxation. Investigation of the microwave and far
infrared dielectric spectrum indicates that pure
benzene exhibits a distinct loss feature. It is well
known from ultrasonic studies
that molecules with a high
degree of symmetry can exhibit
translational-vibrational
relaxation[23-24]. If two
molecules collide inelasti-
cally part of their momentum can be used to excite
an internal vibrational mode to a higher state. In
the case of benzene it is assumed that this is one of
the low frequency ring vibrational modes[24,25].
Deactivation of this excited mode is not readily

achieved by direct radiation of energy and can only be efficiently deactivated by further inelastic collisions. The return of energy out of phase gives rise to a marked increase in the acoustic loss and can be ascribed to vibrational-translational relaxation. This excited state is in face a David's star structure[22,25] in which alternate carbon atoms at distorted out of the plane of the ring and assigned a partial positive charge. Close inspection of this activated form of benzene indicates that it carries a net dipole which can be aligned in an electric field. The observed dipole relaxation in benzene is therefore ascribed to this transient species. It is of interest that the ultrasonic and dielectric relaxation features in this instance reflect the same relaxation process[22]; this is not always the case.

(iii) **Structural Relaxation (breaking of hydrogen bonded structures) in Alcohols**

The magnitude of the dielectric relaxation process in alcohols is sensitive to both the orientation and extent of hydrogen bonding around a particular dipole. In water dispersion can be detected over a frequency range which extends from 10^6 to 10^{13} Hz. In normal alcohols the relaxation spectrum is slightly narrower indicative of narrower distribution of energy states contributing to the total relaxation spectrum. The relaxations are invariably broader than Debye and have half widths of several decades. The magnitude of the increment associated with the reorientational motion is indicative of the local structure around the OH group. The larger the value of 'g' the more structure of local organization occurring in the system. Structural relaxation in hydrogen bonding liquids is however rather complex since it corresponds to the motion and exchange of monomeric, dimeric, trimeric and higher oligomeric forms. The observed dielectric dispersion is the result of all possible exchange processes. In sterically hindered alcohols the higher oligomeric forms are not allowed and the observed dispersion approaches in width that a Debye relaxation and is associated with reorientation and exchange between monomeric, dimeric and possibly trimeric forms.

(iv) **Rotational Isomerism**

Studies of small moelcules dispersed in polymer matrices have allowed observation of the internal

rotational isomerism, the polymer in this case
effectively inhibits overall rotational motions as a
mechanism for relaxation of the dipole. Similarly
dipole relaxation in polymer molecules both in the
solid and solution phases occurs by segmental
(rotational) motion. In the case of a polymer overall
rotation may have a time constant of 10^{-4} secs or
longer, whereas rotational isomerism can occur with
rates of the order of 10^{-6} sec or shorter.

References

1. Debye, P.: 1929, Chemical Catalog, Co., New York.
2. Bottcher, C.F.J.: 1952, Theory of Electric
    Polarization, Elseveir, Amsterdam.
3. Frohlich, H.: 1952, Theory of Dielectrics,
    Elsevair, Amsterdam.
4. Kirkwood, J.G.: 1939, J. Chem. Phys., 7, pp. 911.
5. Onsager, L.: 1936, J. Amer. Chem. Soc. 58,
    pp. 1486.
6. Cole, R.H.: 1957, J. Chem. Phys., 27, pp.33.
7. Cole, K.S. and Cole, R.H.: 1949, J. Chem. Phys.
    9, pp. 341.
8. Davidson, D.W. and Cole, R.H.: 1951, J. Chem.
    Phys. 18, pp. 1417.
9. Vaughan, W.E.: 1969, Dielectric Properties and
    Molecular Behaviour, Van Nostrand, pp.108.
10. Dev, S.B., North, A.m. and Pethrick, R.A.:
    1972, Advances in Molecular Relaxation Processes,
    4, pp. 159.
11. Reddish, W.: 1962, Pure Appl. Chem., 5, pp. 723.
12. Hyde, A.J.: 1970, Inst. Elec. Eng., 117, pp. 1891.
13. Feuner-Feldegg, H.: 1969, J. Phys. Chem., 73,
    pp. 616.
14. Loeb, H.W., Young, G.M., Quickenden, P.A. and
    Suggett, A,: 1971, Br Bunsengesellschaft,
    Phys. Chem., 75, pp. 1155.
15. Roberts, S.: 1966, Report No. 66-C-333, General
    Electrical Research Development Centre Schenectady,
    New York.
16. Hartshorn, L. and Ward, W.H.: 1936, J. Inst.
    Elect. Engrs., 79, pp.597.
17. Hadi, Z.H.A., Hunter, W.N., North, A.M.,
    Pethrick, R.A. and Towland, M.: 1975, Advances
    in Mol. Relaxation Processes, 4, pp. 267.
18. Von Hippel, A.: 1954, Dielectric and Waves,
    Wiley, New York.
19. Chantry, G.W. and Gebbie, H.A.: 1965, Nature,
    London, 208, pp. 378.
20. Payne, R. and Theo Doru, I.E., 1972, J. Phys.
    Chem., 76, pp. 2892.

21. Masood, A.K.M., Pethrick, R.A., Barlow, A.J.,
    Kim, M.G., Plowiec, R.P., Barraclough, D. and
    Ladd, J.A.: 1976, Advances in Mol. Relaxation
    Processes, 9, pp. 29.

22. Masood, A.K.M., North, A.M., Pethrick, R.A.,
    Towland, M. and Swinton, F.1.: 1974, Advances
    in Molecular Relaxation Processes, 9, pp. 153.

23. Khabibullaev, P.K. and Khaliulin, M.G.: 1969,
    Akust Zh, 15, pp. 140.

24. Benedek, G.B., Lastovka, J.B., Fritsch, K.
    and Graytek, T.: 1964, J. Opt. Soc. Amer., 54,
    pp. 1284.

25. Varsanyi, G.: 1969, Vibrational Spectra of
    Benzene Derivatives, Academic Press, New York
    and London.

ULTRASONIC RELAXATION TECHNIQUES

R.A. Pethrick

Department of Pure and Applied Chemistry
University of Strathclyde
295 Cathedral Street
Glasgow G1 1XL

Chemical relaxations which occur with either a change in energy or volume and have a rate constants in the range 10^{-6} to 10^{-9} secs can in principle be studied using ultrasonic techniques (1). Ultrasonic relaxation is concerned with observation of the dispersion in either the velocity (V), the absorption coefficient (α/f^2) or the loss $(\alpha\lambda)$. An ideal relaxation is approximately 1.2 decades in width necessitating observation of the dispersion curves over as wide a frequency range as possible. The frequency range 1-1000 MHz can be adequately studied using a combination of three techniques.

(i) <u>Acoustic Resonator (1-10 MHz)</u> (2-4) The resonator, is constructed from two piezo-electric crystals separated by a spacer. One crystal acts as the transmitter, the second as the detector. The cavity is filled with either a reference fluid or with the unknown. Application of a radio frequency

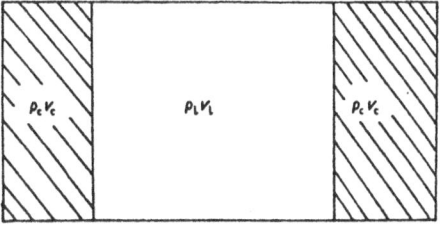

Schematic representation of acoustic cavity resonator. Note, the crystals and liquid have densities respectively designated ρ_c and ρ_l and a wave will have a velocity of propagation v_c and v_e

signal to the first transducer produces a displacement of the crystal-liquid interface relative to its

W. J. Gettins and E. Wyn-Jones (Eds.). Techniques and Applications of Fast Reactions in Solution. 115-122.
Copyright © 1979 by D. Reidel Publishing Company.

unperturbed position. The elastic displacement so
generated progates as a plane wave and will be
partially reflected at the second crystal-liquid
interface. The ratio of the amount of energy
transmitted to that reflected is determined by the
relative magnitudes of the acoustic impedances of the
two media involved. The detected electrical signal
generated in the second crystal is the sum of all
in-phase displacements generated by the time average
of the transmitted waves. The fundamental mode of
oscillation of a cavity constructed from two X-cut
quartz transducers of diameter 2 cm and separated by
1 cm will be approximately 175 kHz, when the coupling
media is water. Such a cavity will exhibit
resonances at approximately 100 kHz intervals
through the scanned frequency range 100 kHz - 30 MHz.
Resonances occur when the cell path length is an
integral number of half wavelengths of the
propagating wave.

If it is assumed that plane wave propagation
occurs, an ideal resonator (i.e. lossless reflection
at the quartz transducer and plane wave propagation)
can be described by the following simple analysis.

Velocity of propagation and resonance of the cavity(2)

The frequency of the jth resonance (f_j) will
be given by the solution of the transcendental
function (2).

$$\rho_Q v_Q \tan \left[\pi \frac{f_j}{f_Q} \right] = \begin{cases} -\rho_s v_s \tan \left[\pi \frac{f_j}{v} \ell \right] & \text{for even } j \\ \rho_s v_s \cos \left[\pi \frac{f_j \ell}{v} \right] & \text{for odd } j \end{cases}$$

$$(1)$$

where Q and s refer to the quartz and sample
respectively, ρ designates the density, 1 is the cell
path length and f_Q is the fundamental frequency of
the transducer. In practice the technique is used in
a comparative mode of operation and the velocity of
sound in a sample is referenced to that of a standard,
such as water or a suitable solvent. Equation (1) can
be rearranged to give

$$\frac{v_c}{v_r} = \frac{D_c}{D_r} \left[\frac{1 + 2 (D_c Z_c - D_r Z_r)}{f_q Z_q} \right] \tag{2}$$

where v_c and v_r are respectively the velocities in the sample and the reference media, Z_c and Z_r are respectively the acoustic impedances of unknown and reference media, D_c and D_r are the respective separations of adjacent resonances for the unknown and reference media and f_q and Z_q are respectively the frequency of the quartz and its acoustic impedance. It has been shown that the square bracket in equation (2) changes the ratio and hence the result by only a few parts per thousand for most cases. Equation (2) to a good approximation reduces to

$$\frac{\delta v}{v_r} = \frac{\delta f_j}{f_j} \tag{3}$$

where δv is the velocity difference between the unknown and reference media, δf_j is the difference between the separation of analogous resonances.

Energy Absorption in the Cavity

The energy loss due to interaction of the propagating plane wave with the coupling media can be calculated from the change of the quality factor (Q) of the resonance on substitution of the reference liquid by the unknown (2,5). The quality factor Q is defined as the frequency of the resonance f_j divided by the half power peak width Δf_j. The observed width is a function of both the mechanical damping in the cavity, the effects of diffraction due to the finite size of the transducer and the attenuation per wavelength of the elastic displacement $(\alpha\lambda)$. The measured quality factor Q_T is related to the mechanical (Q_M) and molecular (Q_L) by the expression

$$Q_T^{-1} = Q_L^{-1} + Q_M^{-1} \tag{4}$$

The mechanical quality factor will include losses arising from the effects of incomplete reflection of the sound beam in the quartz transducer, diffraction of the beam and dissipation of energy at the walls. For a simple harmonic oscillator (5), the quality can be related to the energy absorption per wavelength by the equation,

$$\Delta f_j / f_j \;=\; \alpha\lambda \tag{5}$$

If the quality factor of the unknown and reference are respectively Q_1 and Q_2 then

$$Q_1 - Q_2 = \pi/\alpha\lambda_{\text{unknown}} - \pi/\alpha\lambda_{\text{reference}} \tag{6}$$

This assumes that the value of Q_M^{-1} is unchanged by replacing the reference by the unknown.

Experiment

Acoustic propagation parameters are determined from observation of the position and half peak widths of the resonances of the cavity. The thermal stability of the cavity must be of the order of $0.01°C$ or better to give a 2% accuracy in the absorption coefficient. The electronic system (6), must be able to provide a signal stable to ± 0.01 dB and free from harmonic distortion to better than 80 dB. The detection system should have a gain of approximately 100 dB and an adjustable band width to enable optimization of the signal to noise ratio and reflection of spurious modulations from the detector. The signal used to excite the cavity is generated

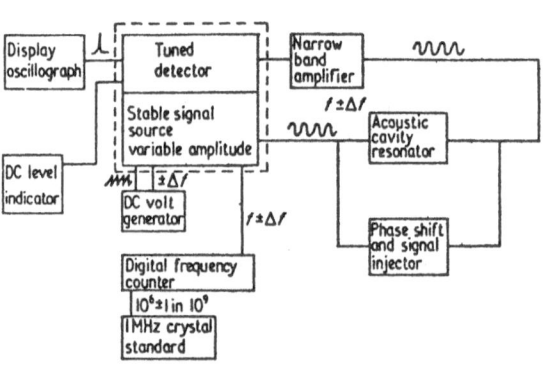

Block diagram of acoustic cavity resonance cavity. Note, the response of the cavity resonator may either be displayed on the oscillograph or as a voltage variation on the level meter. The basic swept level generator is shown inside the broken line box

by a stabilized RS oscillator, locked at 100 kHz intervals to a 1 MHz crystal standard of stability ± 1 part in 10^7. The stability of the detector is maintained by referencing the output to the receiver input. The receiver is tracked in phase with the swept frequency oscillator in the frequency range 100 kHz to 30 MHz with a resolution of 1 Hz. The overall accuracy of measurements for the system is estimated to be better than ± 2 parts in 10^6 for frequency and ± 0.01 dB for signal level.

A plot of the variation of the half peak width of the resonances with frequency is shown for water. The technique is sensitive to small changes in the absorption as shown by the acoustic loss curves for a solution of 0.005 molar 4 methyl piperidine in water.

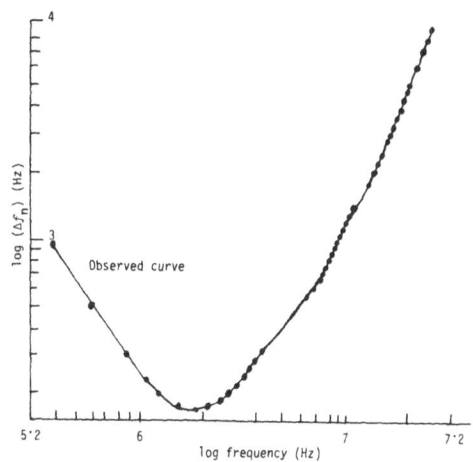

Plot of half-width vs. frequency for water

The above analysis is a simplification of the true picture $(7,8)$. An analysis of three dimensional wave propagation in the cell by solving the wave equation under appropriate boundary conditions allows a more precise analysis of both the absorption and velocity data.

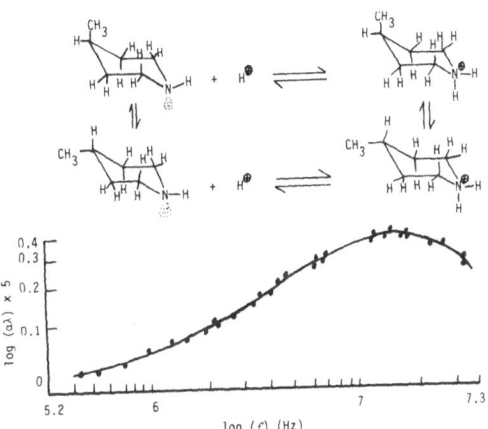

Plot of acoustic energy loss against frequency for a 0·005 molar aqueous solution of 4-methyl piperidine. Note, the stereochemical diagrams represent two of the possible conformations and their protonated structures

It has been pointed out (8) that not only plane wave but also radial mode propagation is possible in the cavity and this explains the satellite peaks which can under certain circumstances interfere with the main mode of propagation above 7 MHz $(7,8)$. In practice it is found that the simple theory outlined above is valid for a wide variety of systems

provided care is taken about matching impedances
of reference and unknown liquids.

(ii) **Pulse Technique** (5-200 MHz) This is the most
widely used method of studying sound propagation in
liquids. The method was originally developed from
the principles of radar by Pellam and Galt (9) and
Pinkerton (10). A d.c. pulse of some 10 μs duration
gates of oscillator to give a radio frequency pulse
typically of 100 V amplitude. This is applied to a
quartz X cut transducer which is usually attached to
a fused quartz delay line some 50 mm long. The pulse
of sound generated passes through the liquid and then
travels through the second delay line and is reconverted
into a radio frequency pulsed by a second transducer.
This pulse is amplified demodulated.

The amplitude of the received pulse as a
function of the separation between the two delay lines
gives the absorption coefficient of sound in the liquid.
The sound velocity may be found by measuring
electronically the time required for the ultrasonic
pulse to transverse the entire acoustic system on
changing the liquid path length: the velocity is
determined using standing wave methods associated
beating when the delay line separation is small
compared with the pulse length. The resultant beat
pattern can be used to estimate the wavelength and
hence the velocity of the sound wave.

(iii) **High Frequency Ultrasonic Measurements** (200-
100 MHz) In principle the technique resembles
the pulse method outlined above, however there are
several important differences. With a conventional
system problems arise at high frequency due to
attenuation in the delay lines and loss of efficiency
of transmission between the quartz and delay line.
These problems can be reduced by use of direct surface
excitation of the transducer (11-13).

The experimental system (14), is similar to
that used in (ii) except that the power levels
required are considerably greater and the precision of
construction of the cell has to be much better than
for lower frequency operation. A hydrogen thyrotron
line pulser is used to generate a 3 kV pulse of 2 μs
duration. A radio frequency pulse is generated with
several hundreds of watts of power and a repetition
frequency of about 1Hz. The spot frequencies (0.1 to
1.5 GHz) are applied to a lithium niobate crystal via

a tuned cavity. The crystals were polished and parallel to better than $\lambda/10$. A slow motion drive allows relative movement of the crystals with a precision of 0.1 μm: The alignment of the crystals being maintained to better than 1 μm over the total travel of 2.5 mm. The attenuation can range from 200-2000 dB mm^{-1} over the frequency range used. The temperature is controlled using a ducted air cooler and a proportional temperature controller.

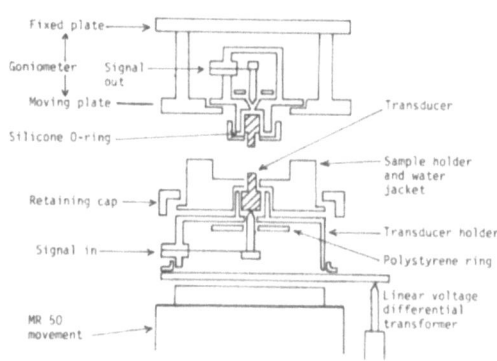

Mechanical system and acoustic cell

The method of measurement depends upon the attenuation-separation product-αr. At short distances the pulse will be reflected at the liquid-crystal interface and this leads to a standing wave pattern being generated between the crystals. If the reflection coefficient of the interface is $\emptyset$ then the pressure at a point x from the transmitting crystal is given by (15)

$$P_x = P_o \; (1-\emptyset) \; \exp \; (jwt)$$

$$\frac{\exp - (\alpha + j\beta)x + \emptyset \exp -(\alpha + j\beta)(2r - x)}{1 + \emptyset^2 \; \exp -(\alpha+j\beta) \; 2 \; r} \quad (7)$$

where α is the amplitude and β is the phase propagation constant (velocity) for the liquid, P_o is the incident pressure wave generated at the source crystal, r is the distance between the crystal and is the angular frequency of the signal. A detailed analysis of the operation of this system is presented elsewhere (11-18).

References

1. Matheson, A.J.: 1971, Wiley, pp. 19.
2. Eggers, F: 1968, Acoustica, pp.19.
3. Pethrick, R.A. and Wyn Jones, E.: 1972, Ultrasonics, pp. 228.

4. Pethrick, R.A.: 1972, J. Phys. E. 5, pp. 571.
5. Hunter, T.F. and Bolt, R.H.: 1955, Sonics, Wiley New York, pp. 321.
6. Wandel and Golterman, WM50.
7. Labhardt, A. and Schwarz, G.: 1976, Berichte der Bunsen Geselschaft, 80, pp. 83.
8. Sarker, S., Dev, S.B. and Pethrick, R.A.: 1973, J. Phys. E. 6, pp. 139.
9. Pellam, J.R. and Gait, J.K.: 1946, J. Chem. Phys. 14, pp. 60.
10. Pinkerton, J.M.M.: 1949, Proc. Phys. Soc. B, 62, 129, pp. 286.
11. Hunter, J.L. and Darby, H.D.: 1964, J. Acoust. Soc. Am., 36, pp. 1914.
12. Lamb, J. and Richter, J.: 1969, J. Acoust. Soc. Am., 41, pp. 1043.
13. Plass, K.G.: 1970, Ber. Bunsenges. Phys. Chem., 74, pp. 343.
14. Wright, T. and Campbell, D.D.: 1977, J. Phys. E. 10, pp. 1241.
15. Musa, R.S.: 1958, J. Acoust. Soc. Am. 30, pp.215.
16. Dunbar, J.H., Ph.D. Thesis, University of Strathclyde.

OPTICAL ULTRASONIC TECHNIQUES

Edward M. Eyring and Michael M. Farrow

Department of Chemistry, University of Utah, Salt Lake City, Utah 84112, U.S.A. and Office Products Division, IBM Corporation, Boulder, Colorado 80302, U.S.A.

ABSTRACT. A fully automated laser Debye-Sears ultrasonic absorption apparatus is briefly described and its relationship to Brillouin scattering is mentioned. The utilization of mini-computerized optical techniques at lower ultrasonic frequencies than the present lower limit of ~ 15 MHz is also considered.

The observation that chemical equilibria give rise to ultrasonic absorption is a very old one. For instance, as long ago as 1936, Bazulin (1) attributed an absorption of sound observed in acetic acid to hydrogen bonding. The discovery of an optical method of detecting ultrasonic absorption was reported independently even earlier (1932) by Debye and Sears (2) and by Lucas and Biquard (3). The essence of the method is suggested by the schematic of Figure 1 of one of its modern incarnations (4). A plane, traveling sound wave from a piezoelectric transducer passes through a condensed medium (in our case a liquid solution) at approximately right angles to the visible laser beam. The latter is diffracted by the alternating regions of compression and rarefaction in the liquid solution. The angle of diffraction is given by the Bragg relation and for first order diffraction is very small indeed, about $\sim 1°$ for $\lambda = 514$ nm and a frequency of sound $f = 100$ MHz. If a chemical equilibrium present in the sample solution absorbs energy at the incident sound frequency f, the regions of compression and rarefaction will be less well defined and the intensity of the diffracted laser beam will be diminished as a consequence. Thus in a sample solution in which several chemical equilibria are present their respective reciprocal relaxation times are given by $\tau_1^{-1} = 2\pi f_1$, $\tau_2^{-1} = 2\pi f_2$, etc. where f_1, f_2, etc. are those ultrasonic

W. J. Gettins and E. Wyn-Jones (Eds.). Techniques and Applications of Fast Reactions in Solution. 123–126.
Copyright © 1979 by D. Reidel Publishing Company.

frequencies at which a reduction in the intensity of the diffracted laser light beam are noted.

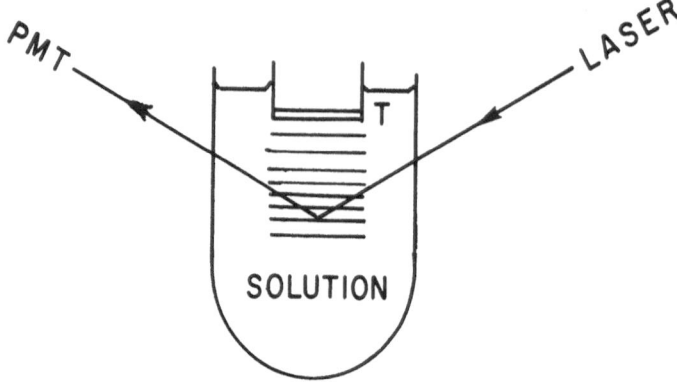

Fig. 1. Schematic of laser Debye-Sears apparatus for measuring
ultrasonic absorption (between ∿15 MHz and ∿300 MHz) in a sample
liquid. The piezoelectric (quartz) transducer T is cemented to
the bottom of a plastic rod that is driven up and down by a
stepping motor that rotates a micrometer. The stepping motor
is controlled by a minicomputer. The angle of diffraction of
the laser beam by the alternating regions of compression and
rarefaction in the liquid (suggested by the horizontal lines)
is exaggerated. (PMT denotes a photomultiplier tube.)

    Rather than a more formal title such as laser Debye-Sears
ultrasonic absorption spectroscopy we frequently refer to this
experiment as simply "looking at sound." This has the advantage
of drawing attention to the complementarity of this experiment
to photoacoustic spectroscopy (5) that is aptly described as
"listening to light." Piezoelectric transducers, lock-in amplifiers, and minicomputers play key roles in each of these two
experiments. The details of the minicomputerization of our
"looking at sound" experiment have been given elsewhere (6).
To date our applications have been to kinetic studies of aqueous
samarium sulfate (7), silver thiosulfate (8), and a variety of
macrocyclic systems (9). It can be persuasively argued that the
same systems could have been studied as effectively by other ultrasonic techniques, particularly the pulse method (10). Our
response would be that the "looking at sound" experiment lends
itself particularly well to minicomputerization and the latter
greatly speeds the data taking process (by a factor of more than
ten) and at the same time significantly reduces imprecision in
the experimental data.

The laser Debye–Sears ultrasonic absorption experiment provides an interesting insight to another, higher frequency ultrasonic technique that has come into vogue in recent years: laser Brillouin scattering (11). The similarity was brought to our attention through the examination of a colleague's Brillouin scattering studies of the flexing dynamics of polymeric chains in solution. In any liquid, fluctuations in density are occurring like those in the earth's atmosphere, which by scattering sunlight make the daytime sky blue. These density fluctuations in liquids correspond to an infinite number of plane waves of sound having all possible different frequencies and directions. By selecting the wavelength and direction of the incident laser beam and the angle to the laser beam at which the Brillouin scattering will be observed, the experimentalist has effectively chosen to look at Bragg diffraction of the laser light by just one of these infinitely many plane sound waves.

Our own chemical interests in complexation of inorganic ions by larger and larger ligands have prodded us to look for ultrasonic absorption techniques that would permit measurements below the low frequency end of the $\sim$15 MHz to $\sim$300 MHz range accessible to our laser Debye–Sears instrumentation rather than the high frequency side to which Brillouin scattering gives access. The very beautiful resonance ultrasonic instrumentation of Eggers and colleagues (12) is, of course, one answer. Raoul Zana graciously loaned us another type of resonance apparatus (13) that inspired true horror in our colleagues because of the virtuosity it demands from its operator in the manual adjustments that bring the two piezoelectric quartz transducers into parallelism.

Our recent experiences (14) with a minicomputerized Michelson interferometer for Fourier transform photoacoustic spectroscopy have suggested a speedier method of aligning a resonance ultrasound cell. Burleigh Instruments, Inc. (P. O. Box 270, East Rochester, NY 14445) manufactures a piezoelectric aligner-translator (PZAT 90) that under minicomputer control could hold a gold plated quartz transducer precisely parallel with a reflecting surface. The systematic search for parallelism in such a resonance interferometer (using either sound or light to monitor the approach to parallelism) is a task to which the mini-computer-PZAT combination is ideally suited. A number of interesting geometries for this type of sub-15 MHz ultrasonic absorption system can be imagined. It seems appropriate here to insist only on the great power of minicomputerization to speed up and refine the process of obtaining ultrasonic absorption data in liquids over an extended range of frequencies.

ACKNOWLEDGMENT. This work was sponsored by Grant AFOSR 77-3255 from the Directorate of Chemical Sciences, Air Force Office of Scientific Research.

REFERENCES

1. P. Bazulin, Compt. Rend. (Dokl.) Acad. Sci. URSS, $\underline{3}$, 285 (1936).

2. P. Debye and F. W. Sears, Proc. Natl. Acad. Sci. (U.S.), $\underline{18}$, 409 (1932).

3. R. Lucas and P. Biquard, Compt. Rend., $\underline{194}$, 2132 (1932).

4. M. M. Farrow, N. Purdie, A. L. Cummings, W. Hermann, Jr., and E. M. Eyring, in *Chemical and Biological Applications of Relaxation Spectrometry*, E. Wyn-Jones, ed., D. Reidel Publishing Co., Dordrecht-Holland, 1975, pp. 69-83.

5. M. M. Farrow, R. K. Burnham, M. Auzanneau, S. L. Olsen, N. Purdie and E. M. Eyring, Appl. Optics, $\underline{17}$, 1093 (1978).

6. M. M. Farrow, S. L. Olsen, N. Purdie, and E. M. Eyring, Rev. Sci., Instrum., $\underline{47}$, 657 (1976).

7. M. M. Farrow, N. Purdie, and E. M. Eyring, J. Phys. Chem., $\underline{79}$, 1995 (1975).

8. M. M. Farrow, N. Purdie, and E. M. Eyring, Inorg. Chem., $\underline{14}$, 1584 (1975).

9. See, for example, L. J. Rodriguez, G. W. Liesegang, M. M. Farrow, N. Purdie and E. M. Eyring, J. Phys. Chem., $\underline{82}$, 647 (1978) and references cited therein.

10. For references see J. Stuehr, in *Techniques of Chemistry, Vol. VI, Investigation of Rates and Mechanisms of Reactions, Part II*, G. G. Hammes, ed., Wiley-Interscience, New York, NY, 1974, Chap. VII, p. 256ff.

11. Reference 10, p. 260ff.

12. F. Eggers, Th. Funck, and K. H. Richmann, Rev. Sci. Instrum., $\underline{47}$, 361 (1976).

13. O. Funfschilling, Doctoral Thesis, Louis Pasteur University, Strasbourg, 1972.

14. M. M. Farrow, R. K. Burnham, and E. M. Eyring, Appl. Phys. Lett., $\underline{33}$, 735 (1978).

LIGHT SCATTERING STUDY OF ULTRASONIC RELAXATION

IN LIQUID FURAN-CYCLOHEXANE MIXTURES

K. Takagi, P.-K. Choi, and K. Negishi

Institute of Industrial Science,
University of Tokyo,
Minato, Tokyo, Japan

INTRODUCTION

Study of ultrasonic relaxation in pure liquids provides an useful means to observe vibration-translation (v-t) energy transfer in polyatomic molecules, but it is almost blind to vibration=vibration (v-v) transfer process. If experiments are made in mixtures, however, one can obtain the speed of v-v process between different molecules since the contribution of v-v coupling to the whole relaxation process would be dependent on the concentration. For this purpose, it is required to carry out measurement over a wide frequency range and to determine relaxation frequency at each concentration.

In this work, we measured ultrasonic velocity and absorption in furan and cyclohexane mixtures over a range from 60 to 700 MHz using high-resolution Bragg reflection technique.[1,2] Also used co-operatively were Brillouin scattering and pulse-echo-overlap techniques. In this binary system, small amount of cyclohexane increases substantially the relaxation frequency of furan.

HIGH-RESOLUTION BRAGG REFLECTION TECHNIQUE

Details of this new technique have been given in our previous papers.[1,2] Light scattered by sound waves excited by a ZnO film transducer is detected with an optical heterodyne system schematically shown in Fig.1. When the sound frequency is 700 MHz, for example, beat signal at 728 MHz is generated at the photodiode and recorded as shown in Fig.2. Angular width of the right-hand curve gives absorption coefficient, and the angle between the two peaks gives the sound velocity.

W. J. Gettins and E. Wyn-Jones (Eds.). Techniques and Applications of Fast Reactions in Solution. 127–130.
Copyright © 1979 by D. Reidel Publishing Company.

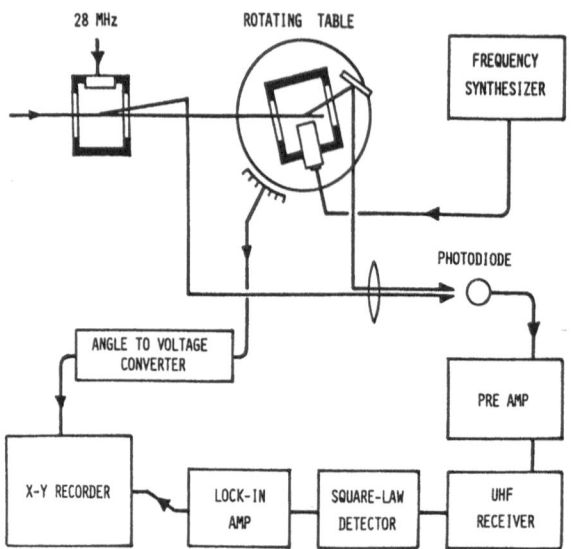

Fig.1 Blockdiagram of high-resolution Bragg reflection technique.

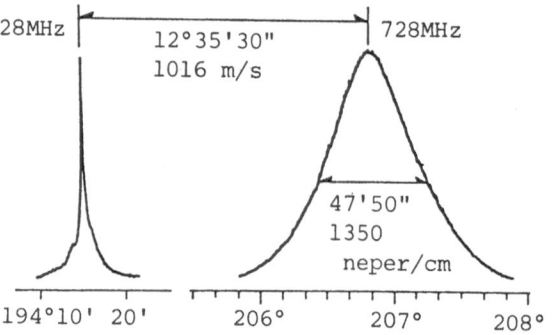

Fig.2 Typical recorder tracing obtained at f = 700 MHz.

VIBRATION-VIBRATION ENERGY TRANSFER

Velocity dispersion and absorption observed at 20°C are shown in Fig.3. The solid lines are the single relaxation curves fitted under the assumption that the relaxation of furan and that of cyclohexane are effectively isolated in the mixtures and independent on each other. This assumption might be invalid in the intermediate region of composition. Relaxation frequency of furan is obtained at each concentration and summarized in Fig. 4, which shows rapid increase at small percentages of cyclohexane. This suggests that collision between furan and cyclohexane (F-C collision) is more effective in energy transfer of furan than collision between furan and furan (F-F collision).

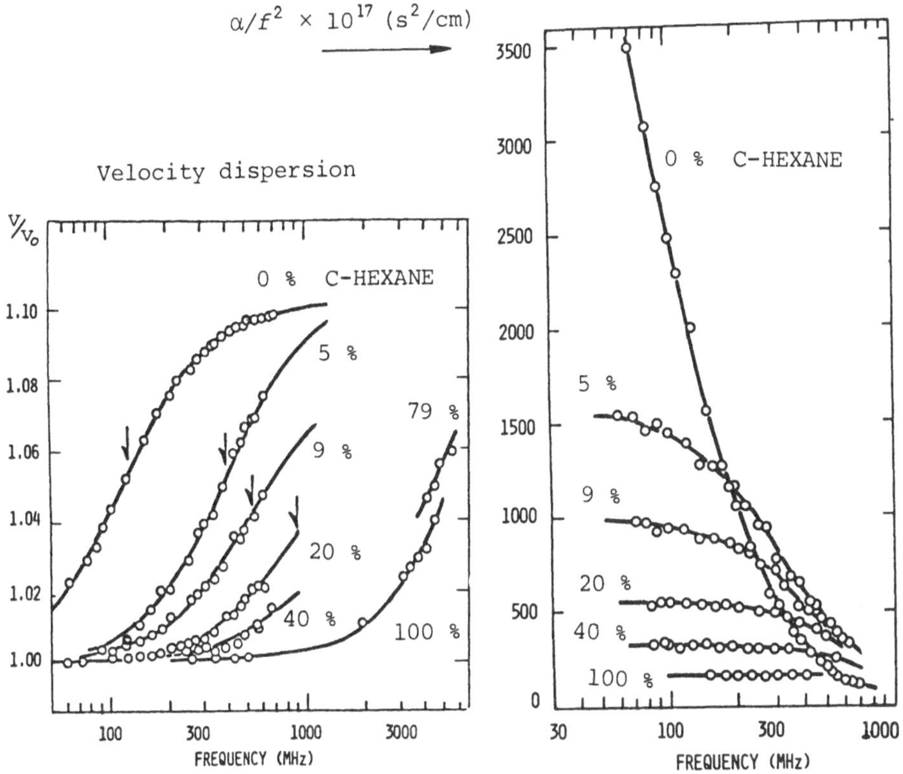

$\alpha/f^2 \times 10^{17}$ (s^2/cm)

Velocity dispersion

0 % C-HEXANE

5 %

79 %

9 %

20 %

40 %

100 %

Fig.3 Velocity dispersion and absorption observed at 20°C.

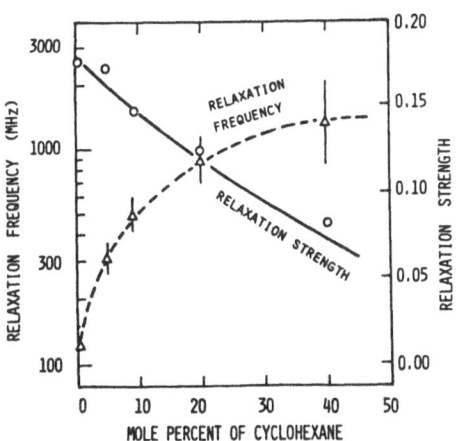

Fig.4 Fitted values of relaxation
frequency and strength. The solid
line shows the theoretical prediction of relaxation strength.

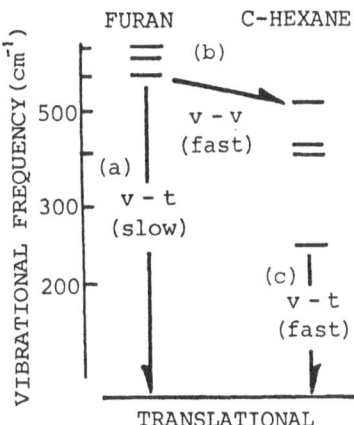

Fig.5 Fundamantal vibra-
tional modes of furan and
cyclohexane.

Following three reactions would be directly responsible for the relaxation of furan in mixtures:

$$F^* + F \rightleftharpoons F + F \ , \quad (v{-}t) \qquad 1)$$

$$F^* + C \rightleftharpoons F + C \ , \quad (v{-}t) \qquad 2)$$

$$F^* + C \rightleftharpoons F + C^* \ , \quad (v{-}v) \qquad 3)$$

where F and C denote furan and cyclohexane molecules, respectively, and * indicates vibrationally excited state. Equations 2) and 3) represent two types of energy transfer, v-t and v-v, in F-C collision. The following discussion will show that the high efficiency in F-C collision should be due to v-v process. Molecular factors which determine the efficiency in v-t process would be mass and repulsive term of intermolecular potential. Comparison of molecular constants of furan and cyclohexane shows that furan is lighter in weight and harder in collision, and must therefore has more ability in v-t transfer than cyclohexane. Then, v-t in F-C collision must be even slower than v-t in F-F collision. The high efficiency in F-C collision should be due to v-v process.

Figure 5 shows the fundamental vibrational modes of these molecules. The arrow (a) shows the v-t of furan. The rate of this process is derived from the result in pure furan and shown to be rather slow. In mixtures, however, close coupling between the lowest mode of furan($605 \ cm^{-1}$) and the nearest mode of cyclohexane ($522 \ cm^{-1}$) would arise. Vibrational energy of furan is transferred quickly in v-v process indicated by arrow (b) and then falls down in v-t of cyclohexane. This v-v process is expected to be very fast because the energy difference between these two levels is much smaller than the thermal energy: $h\Delta\nu \ll kT$. Further, the rate of v-t derived from the result in pure cyclohexane is also very fast. Thus, the indirect path (b)-(c), is much faster than the direct path(a). Analysis of relaxation frequencies[3] at small percentages of cyclohexane gives the rate constants of the v-v process in F-C collision. The results are listed in Table I.

Table I   Rate constants of vibrational energy transfer.

	v-t (F-F)	v-t (C-C)	v-v (F-C)	
k_f	4.4×10^8	2.1×10^{10}	3×10^{10}	(1/s)
k_b	—	—	2×10^{10}	

REFERENCES

1) K. Takagi and K. Negishi: Jpn.J.Appl.Phys. 16(1977) 1319.
2) K. Takagi and K. Negishi: Ultrasonics, 16(1978) 259.
3) A. A. Monkewicz: J.Acoust.Soc.Am. 42(1967) 258.

BRILLOUIN SCATTERING SPECTROMETRY

J. Gormally

Department of Chemistry and Applied Chemistry,
University of Salford, Salford M5 4WT, U.K.

Light is scattered from homogeneous materials due to microscopic local variations in the dielectric constant, and hence index of refraction within the bulk of the material. If these variations are the result of time dependent fluctuations, the scattering will generally be inelastic due to energy transfer between the incident photons and the medium. Inelastic scattering gives rise to scattered light which has a frequency spectrum different from that of the incident light and the object of many scattering experiments is to determine the form of this frequency spectrum.

In some experiments, time dependent variations in the refractive index are impressed upon the medium by passing a compressional sound wave through it. This is done in the Debye Sears technique and also in the more recent Bragg reflection technique which is described by K. Takagi elsewhere in this volume. In Brillouin scattering, the variations in refractive index are due to propagating fluctuations which have their origin in the random thermal motions of molecules within the medium (1). These random motions can be regarded as a superposition of well defined acoustic modes and the scattering experiment picks out one of these modes for investigation. The salient features of the scattering process are most easily appreciated by considering the inelastic interaction between a photon and a quantum of acoustic energy, a phonon. Conservation of energy requires that;

$$\hbar\omega_s = \hbar\omega_i \pm \hbar\Omega \tag{1}$$

where ω_s, ω_i, and Ω are the angular frequencies òf the scattered

W. J. Gettins and E. Wyn-Jones (Eds.). Techniques and Applications of Fast Reactions in Solution. 131–135.
Copyright © 1979 by D. Reidel Publishing Company.

photon, the incident photon, and the phonon respectively.
$\hbar = {}^{h}/_{2\pi}$ where h is Planck's constant. The positive sign corresponds to a scattering process in which a phonon is anni-
hilated and its energy carried away by the scattered photon.
The negative sign corresponds to the case in which a phonon is
created at the expense of the incident photon. Conservation of
momentum in the process requires that;

$$\hbar \underline{k}_s = \hbar \underline{k}_i + \hbar \underline{K} \tag{2}$$

where $\underline{k}_s$, $\underline{k}_i$, and $\underline{K}$ are the wavevectors of the scattered photon,
the incident photon and the phonon respectively. Inspection of
the corresponding vector diagram shows how the wavevector of the
phonon involved determines the angle of scattering, θ.

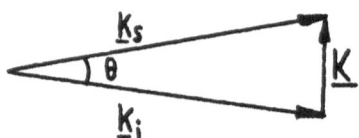

Figure 1

In practice, $k_s \simeq k_i = k$, and so we see from Fig. 1 that

$$K = 2k\mathrm{Sin}\ \theta/2 \tag{3}$$

Since $k = n\omega_i/c$ where n is the refractive index of the medium
and c is the speed of light in vacuum and $K = \Omega/c_s$ where c_s is
the phase velocity of the phonon we have;

$$\Omega = \frac{2n\omega_i c_s}{c}\cdot\ \mathrm{Sin}\ \theta/2 \tag{4}$$

Equation (4) specifies the relationship between the frequency,
Ω, of the propagating acoustic mode which interacts with light
scattered at the angle θ. Insertion of typical values into (4)
puts Ω in the range from 10^8 s^{-1} to about 5×10^{10} s^{-1} in the
usual Brillouin experiment. For this reason, Brillouin scatter-
ing has often been used as a high frequency extension to conventional ultrasonic techniques. The scattered light typically
consists of three frequency components as indicated in Fig. 2.
The central component has a centre frequency equal to the frequency of the incident light and is often referred to as the
Rayleigh component. This component is due to scattering from
non propagating fluctuations. The two Brillouin components show
frequency shifts dictated by (1). The lower frequency component

corresponds to phonon creation and is known as the 'Stokes' line.
The higher frequency component corresponds to phonon annihilation
and is known as the 'anti-Stokes' line.

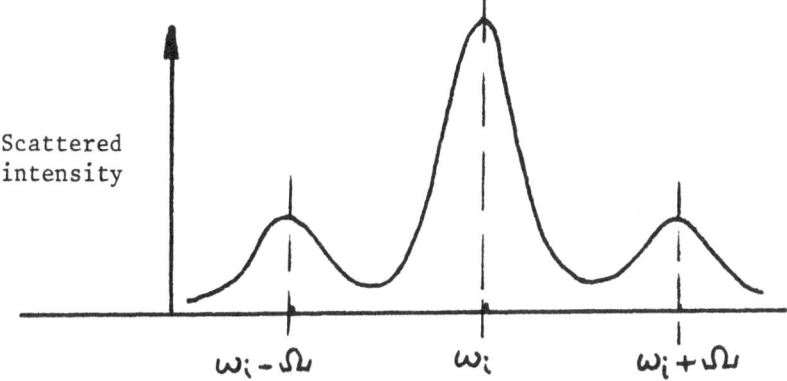

Frequency

Figure 2

Experimental technique

A typical Brillouin spectrometer is illustrated in Fig. 3.
A narrow light beam from a single mode laser enters the sample
cell C. Lenses L_1 and L_2 together with pinhole P serve to define the angle of scattering. The parallel beam emerging from
L_2 passes through a Fabry-Perot interferometer which acts as an
optical frequency filter, and the emerging light is allowed to
fall upon the cathode of a high gain photomultiplier tube. The
photomultiplier output can be processed by photon counting or by
using some form of lock-in amplifier technique. The Brillouin
spectrum is finally displayed on a chart recorder.

In order to obtain a spectrum of the type shown in Fig. 2
it is necessary to scan the Fabry-Perot interferometer in some
way. This is done by changing the optical path length between
the plates, either by moving the plates apart using a piezoelectric drive mechanism or by placing the instrument in a
sealed housing and varying the pressure of the air between the
plates. This latter method has the advantage of mechanical
stability but is slower than the piezoelectric method.

In experiments of this kind it is vital that careful consideration be given to the design of the sample cell and that
the liquid in it be as free from dust as possible. The reason
for this is that the Fabry-Perot interferometer is an instrument
of high resolution, but of limited contrast. This means that
any strong elastic scatterers in the cell will produce a central
peak in the spectrum which is so high and wide as to overlap the

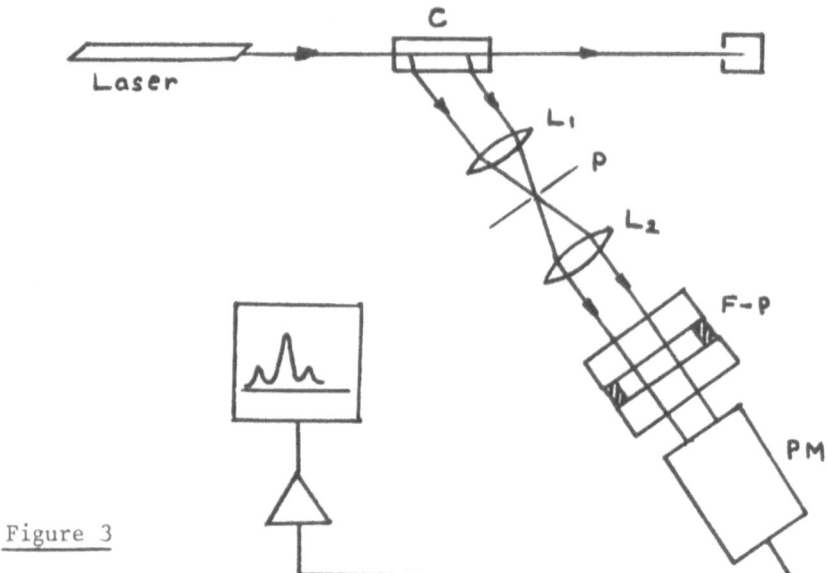

Figure 3

Brillouin components on each side. Problems of this sort become
particularly troublesome when the sample contains large molecules
or molecular aggregates which scatter strongly. The contrast of
the interferometer can be considerably improved by using it in a
multipass mode so that the scattered light passes through it
several times (2). In any event the interferometer must be made
to a high specification. The plates are usually polished flat
to within a hundredth of a light wavelength and they must be
maintained as parallel as possible during use (3). A method
which dispenses with the interferometer has been attempted by
Lastovka and Benedek (4). They mixed light scattered from toluene
with unshifted laser light on a high spped photodiode and observed
a beat signal. This technique does not seem to have been followed
up, but with the availability of very fast photomultiplier tubes
it should now be a more viable prospect.

Relaxation data from Brillouin spectra

The Brillouin components shown in Fig. 2 have a width which
is determined partly by the resolving power of the interferometer
and partly by mechanisms which lead to damping of acoustic modes
in the sample. The instrumental width is usually allowed for by
a convolution method so as to obtain the true line width. This
numerical procedure is one of the largest sources of inaccuracy
in this experiment. Each line in the pair has a true half width
at half maximum given by (5);

$$\Delta\omega \;=\; \alpha K^2 \tag{5}$$

where α is the acoustic attenuation coefficient. Relaxation processes are sought by measuring α as a function of Ω, the phonon frequency, as in ultrasonic experiments. It should be noted, however, that the Brillouin experiment is really an observation of the time decay of a phonon whereas conventional ultrasonic methods measure the spatial decay of acoustic waves. Relaxation times are obtained using the relation familiar in ultrasonics;

$$\frac{\alpha}{\Omega^2} = \frac{A}{1 + \Omega^2\tau^2} + B \qquad (6)$$

where τ is the relaxation time.

Most liquids exhibit vibrational and structural relaxation effects in the time scale accessible to the Brillouin technique. Structural relaxation is a phenomenon which will occur in all liquids and the time scale can vary from about 10^{-9}s to 10^{-12}s. Toluene has a structural relaxation spectrum which can be completely covered by the Brillouin technique (6). The relaxation of the vibrational specific heat of benzene has been extensively studied using ultrasonic methods and Brillouin scattering. With this liquid, there has been some disagreement as to whether the process can be accounted for using a single relaxation time or a distribution of relaxation times (7). It is with fast processes of this sort that Brillouin spectra have proved most useful. However, studies of chemical relaxation effects have been pursued from measurements made on the Rayleigh component using either Fabry-Perot techniques or optical mixing (8). It is doubtful, however, as to whether light scattering methods are the most appropriate in this case.

References

1. Fabelinskii, I.L. 1968. Molecular Scattering of Light (New York: Plenum).
2. Sandercock, J.R. 1971. Light Scattering in Solids, Edited by M. Balkanski, (Paris: Flammarion Press).
3. Sandercock, J.R. 1970, Opt. Comm., 2, 76.
4. Lastovka, J.B. and Benedek, G.3. 1965. Physics of Quantum Electronics. Edited by Kelley, Lax and Tannenwald. Page 231 (McGraw-Hill).
5. Berne, B.J. and Pecora, R. 1976. Dynamic Light Scattering (J. Wiley).
6. Chiao, R.Y. and Fleury, P.A. 1966. Physics of Quantum Electronics (New York, McGraw-Hill).
7. Nichols, W.H., Kunsitis-Swyt, C.R., and Singal, S.P. 1969. J. Chem. Phys., 51, 5659.
8. Yin Yeh and Keeler, R.N. 1969. J. Chem. Phys., 51, 1120.

DATA CAPTURE AND PROCESSING IN CHEMICAL RELAXATION MEASUREMENTS

W. Knoche and H. Strehlow

Max-Planck-Institut für biophysikalische Chemie,
D-3400 Göttingen, West Germany.

Non-Digital Techniques

The old method to evaluate relaxation jump experiments is
to photograph the trace on the oscilloscope screen and to plot
the measured amplitudes against time on semilogarithmic paper.
An improved evaluation technique introduced by Crooks et al. [1]
eliminates the amplitude measurements (and the error introduced
by the imperfection of the oscilloscope). The photographed relaxation trace is projected onto the same oscilloscope screen
which has been used for the actual experiments, and an exponentially decaying time function is generated with a suitable
RC network on the scope. The amplitude and time constant of this
function are varied until the projected curve on the scope is
matched. The method has been extended also for two relaxation
times. This technique is a definite improvement, though still
dark room work is necessary, and the results are not available
immediatedly after the experiment. Another more convenient version
of this technique consists in using a digital oscilloscope with
alternating points from the relaxation experiment proper and from
the simulated exponentially decaying curve. A faster matching
without photographic work can thus be realized. Both techniques,
however, do not allow signal averaging (see below). Also the
precision is not as good as that obtainable with the digital processing technique to be described now.

Digital Techniques

In jump-relaxation experiments we obtain a signal of the
form

$$y = A \exp(-t/\tau) + C \tag{1}$$

137

W. J. Gettins and E. Wyn-Jones (Eds.). Techniques and Applications of Fast Reactions in Solution. 137–142.
Copyright © 1979 by D. Reidel Publishing Company.

or, in the case of superposition of several relaxing effects

$$y = \sum_i A_i \exp(-t/\tau_i) + C \qquad\qquad (2)$$

C is the measured amplitude after complete equilibration.

In principle, if Eq. (1) applies, three values of y at
different times are sufficient to obtain the three unknowns A,
τ, and C. Of course, because of the limited precision of the
measurements, more values of y are needed to get the unknowns
with reasonable accuracy. Furthermore, it is not known whether
an experiment is described by Eq. (1) or by Eq. (2). To increase
the accuracy one may measure a large number of values y at different times t and fit these to Eq. (2). But since the general shape
of the signal is well known (i.e., the superposition of exponentially decaying curves), it is not necessary to accumulate an
exceedingly high number of amplitude measurements if the single
values of y have high precision. This can be achieved by proper
filtering to increase the signal-to-noise ratio. For an exponentially decaying signal, the nearly ideal filtering is the
integration of the signal during time intervals Δt with 0.2 τ $\lesssim$
$\Delta t \lesssim$ 0.3 τ. Thus, high frequency noise (compared to 1/τ) is eliminated to a large extent. Low frequency disturbances only change
the value of C. Of course, disturbances in the frequency range
of 1/τ cannot be reduced by filtering, since they have nearly the
same time constants as the relaxing effect. However, if these
disturbances are uncorrelated to the chemical relaxation, they
can be eliminated by signal averaging, i.e. by the superposition
of the signals of several experiments.

An exponentially decaying signal is reduced to 2 % of its
starting amplitude within a time t ≈ 4 τ. At longer times the
signal becomes too small to be detected. Therefore, if integration
is performed as discussed above, only 12 to 16 values of the
integrated signal have to be taken into account. According to
this consideration a simple on-line data processing system has
been developed which will be discussed now.

Hardware [2]

The output signal is amplified and then integrated over
time intervals Δt. At the end of each interval, the voltage on
the integrating capacitor is digitized to a number of three
digits (10 bits) and stored in a 16-word buffer memory. This high
precision is necessary, expecially if several relaxing effects
are superimposed. The capacitor is then discharged and the integrating process continued until the memory is filled. The integral
of the sum of exponentials is again a sum of exponentials with
the same time constants, though with changed amplitudes. The
data are transferred from the memory into a computer and are

further processed as discussed below. A dedicated computer is feasible, since only an inexpensive microcomputer is needed. Signal averaging is possible by adding the corresponding measured integrals of successive experiments in the computer.

Software [3]

The signal (as described by Eq. (2)) is characterized by $2n + 1$ parameters (the A_i, τ_i, and C) which must be obtained from the experiment. However, if only one of the superimposed exponentials has an appreciable amplitude, the other ones may be regarded as disturbances. In this case A and τ of the main relaxation effect are obtained by measuring accurately the values of 12 equidistant amplitudes y_i. The interval between successive measurements should be 0.25 to 0.3 τ. By a least sqare fit using eight amplitudes starting with the first and assuming a simple exponential (Eq. (1)) the time constant τ is evaluated. Then the calculation is repeated starting with the second amplitude. Thus, five time constants, τ_0 to τ_4, are obtained. Of course, if the curve is really a single exponential, the five τ_n values should be the same except for experimental error. If, however, two (or more) exponentials are superimposed, the τ_n values will change systematically with n.

The slopes and curvatures of the τ_n vs. n curves depend on the ratio of the time constant of the perturbing exponential to that of the main curve and the magnitude and relative sign of the two amplitudes as shown in Fig. 1.

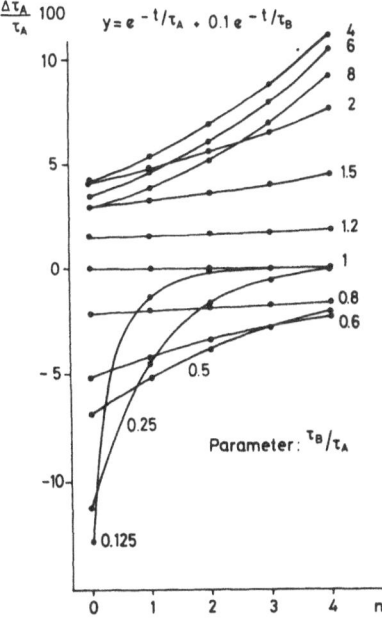

Fig. 1

Calculated τ_n-n curves for disturbed relaxation times. The amplitude of the disturbing exponential is assumed to be 10 % of the main amplitude. For other amplitude ratios (including sign), the ordinate must be changed accordingly.

The correct time constant τ_A of the main exponential is given by

$$\tau_A = \frac{\tau_1(2\tau_0\tau_2 - \tau_1\tau_2 - \tau_0\tau_1)}{\tau_0\tau_2 - \tau_1^2} \tag{3}$$

Other equidistant triples such as τ_1, τ_2, τ_3 or τ_0, τ_2, τ_4 may also be used in Eq. (3). (The derivation of Eq. (3) is given below.)

Also, if the amplitudes of two relaxations processes

$$y = A \exp(-t/\tau_A) + B \exp(-t/\tau_B) + C \tag{4}$$

are comparable, both time constants can be evaluated using Eq. (3). The procedure is demonstrated in Figure 2. For the sake of simplicity we assume $C = 0$ which, however, is not a necessary condition. After a rough estimation of the long relaxation time

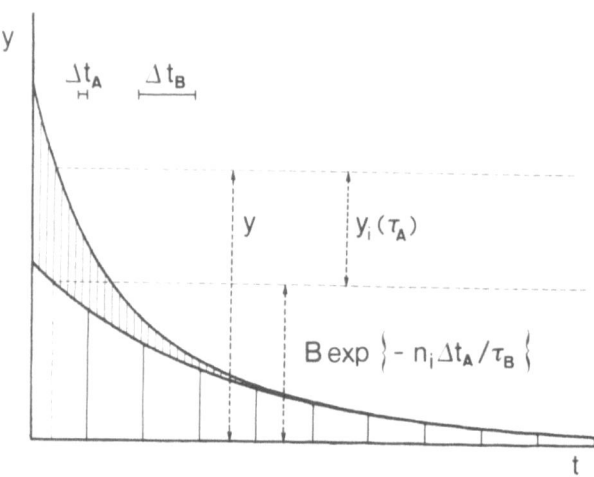

Fig. 2

Evaluation procedure of two relaxation processes with similar relaxation times and amplitudes [3].

τ_B we choose $\Delta t_B \approx \tau_B/3$ and apply the technique outlined above by starting with the second amplitude y_1. The fast relaxation has decayed to an extent that it is only a disturbance for the slow process and, therefore, Eq. (3) will supply the correct time constant τ_B. Then the partial initial amplitude B is given by

$$B = y_1 \exp(\Delta t/\tau_B) \tag{5}$$

(For the effect of integration which changes the correct relaxation amplitudes reference (3) should be consulted.)
At time $t_i = n_i \Delta t_A$ the corrected values for the fast relaxation process are given by

$$y_i(\tau_A) = y_i - B \exp(-n_i \Delta t_A/\tau_B) \tag{6}$$

and τ_A can be similarly obtained with 12 amplitudes with $\Delta t_A \approx$ $\tau_A/3$. If the value of B is not very exact, again the fast relaxation process is only weakly disturbed and is corrected by the use of Eq. (3). The indicated process has also been successfully used to separate 3 relaxation times. The relaxation times to be separated should differ by at least a factor of about 2.5. Otherwise, a fitting program should be used taking more points and processing the data in a minicomputer or a large computer using an appropriate 5 parameter fitting program. Also relaxation data with relatively strong low frequency noise are possibly better processed off-line with such programs or on-line if share time facilities or a large buffer memory is available.

The Derivation of Equation [3]

The basis of Eq. (3) is the fact that an average rate of decay for the superposition of two exponentials is given by the sum of the two rates weighted with the amplitude fraction (see [3])

$$y = A \exp(-t/\tau_A) + B \exp(-t/\tau_B) \approx A' \exp(-t/\tau) \tag{7}$$

$$\frac{1}{\tau} = \frac{A}{A+B}\frac{1}{\tau_A} + \frac{B}{A+B}\frac{1}{\tau_B} \tag{8}$$

(7) is, unfortunately for the experimenter, an excellent approximation, if $B \ll A$ and τ_A is not too different from τ_B. Using the procedure outlined above we obtain for the n th value of τ from (8)

$$\frac{1}{\tau_n} = \frac{A_n}{A_n + B_n}\frac{1}{\tau_A} + \frac{B_n}{A_n + B_n}\frac{1}{\tau_B}. \tag{9}$$

Introducing the abbreviation

$$z = \exp\left[-\frac{\Delta t}{\tau_A}\left(\frac{\tau_A}{\tau_B} - 1\right)\right] \tag{10}$$

we have

$$\frac{1}{\tau_n} = \frac{1}{1 + (B/A)z^n}\frac{1}{\tau_A} + \frac{(B/A)z^n}{1 + (B/A)z^n}\frac{1}{\tau_B}. \tag{11}$$

Since $(B/A)z^n \ll 1$, we approximate Eq. (11) by

$$\frac{1}{\tau_n} = \frac{1}{\tau_A} + \left(\frac{1}{\tau_B} - \frac{1}{\tau_A}\right)\frac{B}{A}z^n. \tag{12}$$

Eq. (3) is obtained from the observed values of τ_n with $n = 0, 1,$ and 2 by eliminating

$$\left(\frac{1}{\tau_B} - \frac{1}{\tau_A}\right) \frac{B}{A} \quad \text{and } z.$$

Results and Discussion

The program has been used extensively in this laboratory with chemical relaxation and NMR-spin-echo experiments. A long test series has been performed either with computer generated data or by using a special electronic generator for three super-imposed exponential decay curves with a precision in relaxation time and amplitude of about $\pm$ 0.5 % [3]. The conclusions are: Two relaxation processes with equal sign and similar magnitude of the amplitude may still be separated if $\tau_I/\tau_{II} \stackrel{>}{\sim} 2$. However, under these circumstances the accuracy of the evaluated parameters depends strongly on the precision of the experimental data.

If the relaxation times differ by a larger factor and/or if the amplitudes differ in sign the precision attainable is much better. Data with a good signal/noise ratio typically supply relaxation times with an error of $\leq \pm$ 2 %. Signal averaging gene-rally improves the accuracy of the data considerably even if only 5 - 10 experiments are averaged. The computing time needed for one relaxation time (and amplitude) using a microcomputer is about 20 seconds.

A BASIC computer program and programs for the microcumputers WANG 600, WANG 720, HP 9815 and for the pocket calculator HP 67/97 (computing time $\sim$ 5 minutes) are available. The hardware can be obtained from Dialog, Harffstr. 34, D-4000 Düsseldorf 13.

References

1. J.E. Crooks, M.S. Zetter, and P.A. Tregloan, J. Phys. E. Sci. Instr. 3, 73 (1970).

2. M. Krizan and H. Strehlow, Chem. Instr. 5, 99 (1973-1974).

3. H. Strehlow, Adv. Mol. Rel. Interact. Proc., 12, 29 (1978).

A MICROPROCESSOR BASED SYSTEM TO COLLECT, DISPLAY, AND EVALUATE STOPPED-FLOW AND OTHER KINETICS DATA

John S. Littler

School of Chemistry, University of Bristol,
Bristol   BS8 1TS

The evaluation of kinetic data has traditionally been carried out by first measuring chart recorder or photographic output by hand, and then using the data to calculate and plot various functions of concentration against time in order to establish reaction orders and evaluate rate constants. This is quite adequate for slow reactions using conventional analytical methods, where the time taken to evaluate the data is small relative to the time taken collecting it, but the operator is rapidly swamped by data when studying reactions by the stopped-flow or T-jump methods. One way of avoiding this is to add to the apparatus a data conversion unit which records the data in digital form suitable for reading into a computer. While this eases the operator's labour it often has the disadvantage of introducing a long delay before the results are available, and this may be a considerable disadvantage if quick results are needed to plan further experiments. A second approach has been to obtain rate constants by an analog curve fitting method, by which a curve, which in the simplest case is a single exponential decay, is derived from resistor and capacitor circuits.[1] These are adjusted by the operator, until the curve is superimposed, by eye, on the trace generated on an oscilloscope screen by the kinetic experiment. The rate constant is determined from the R and C values used. This method has all the advantages of speed, and is reproducible, but its accuracy depends on the accuracy with which the analog circuit time constants are known, and it does not in itself provide a permanent record of the experiment.

As cheap microprocessors are now available the obvious development is to bring the computer to the experiment, and so

W. J. Gettins and E. Wyn-Jones (Eds.). Techniques and Applications of Fast Reactions in Solution. 143–146.
Copyright © 1979 by D. Reidel Publishing Company.

to obtain results rapidly. At present these microprocessors are
still relatively limited in their arithmetic capabilities, so we
have combined the two approaches by using the microprocessor to
calculate the exponential decay curve. The numerical values it
uses are put in from a keyboard and from potentiometers via
analog to digital converters (ADC's) operating at 8 bit
precision. Such a simulation involves much less computational
labour than fitting a least squares line to a log function, so
a fresh curve can be calculated in ca. 1/20 sec. Both the data,
collected via a simple ADC, or at higher data rates via a
transient recorder, and the simulated curves are displayed in a
256 x 256 element area on the screen of a small commercial
television set.[2]   The operator can immediately verify that the
data collected are satisfactory, and so he does not need to use
a separate oscilloscope. He can then adjust the calculated
curve on the screen to a good fit to the data by adjusting the
potentiometers or by using the keyboard. He has a number of
options available at the keyboard. He can expand or contract
the horizontal axis of the display (the time coordinate) and
move the display window forwards or backwards along the reaction
curve. He can also expand the vertical axis so that a small
change can be examined in detail. Eventually when he has
selected the portion of the curve which interests him (avoiding,
for example, premixing transients in a stopped flow) he can
select the point from which the exponential is to begin to decay
by setting an initial x-value (time) and y-value on the two
potentiometers provided. He can then use them again to set up
the rate of decay of the exponential and the final value to
which it tends asymptotically. The rate potentiometer is
programmed to give a value between 0 and 256 which is used as a
fine adjustment; a coarse adjustment from the keyboard adds
multiples of 256 to this value. In practise it is easiest to
fit curves when the value lies between 256 and 768, so that the
rate constant can be set to an accuracy of about 0.2%. If the
analog input is noise-free and accurately exponential (e.g.
from a test RC decay circuit) operators easily get consistent
fittings for the rate constant to ±1 unit (i.e. a range of ca.
1/2%) even though the data has only been recorded to 8 bit
(1:256) precision. Noisy data or non-exponential curves
obviously introduce greater uncertainties. There is no
significant uncertainty in the time scale since it is derived
solely from the clock used in the original sampling and
digitisation of the data, and this is crystal controlled.
Uncertainties about the R and C values, or about the accuracy
of an oscilloscope time base, are thus avoided. When the
operator is satisfied with the fit he has achieved he enters,
via the keyboard, the original time-scale on which the data was
collected. If the internal ADC is used the processor will
already know at what time intervals it collected data. The
rate constant is then calculated and displayed numerically on

the screen, together with the infinity value chosen and the
coordinates of the start of the curve. A simple error function
is also calculated to give the operator a numerical estimate of
the differences between the calculated and exponential curves;
with noise free data the error should be no more than one part
in 256 at most positions on the screen.

Throughout this process the data is held in numerical form
in the memory of the microprocessor and the same system can
record the original data, and the results of the operator's
attempt to fit it, on a cassette tape, which can provide both
a permanent store and the means of transfer of the data to a
large computer for more sophisticated processing. The operator
can also choose to record only selected portions of the data, or
to add further identification information, etc. at this point.
The hardware for this digital processing and data transfer unit
was constructed for about £650, and it was demonstrated at the
conference collecting and fitting data, which was generated by
a student T-jump apparatus and collected and temporarily stored
by a transient recorder at 8 bits precision.

The same hardware, suitably reprogrammed, is capable of
fitting many different sorts of curves to their appropriate
data (e.g. for determination of areas under peaks in g.l.c. and
n.m.r. a standard peak shape could be fitted to actual traces).
It is in fact a cheap general purpose "intelligent" graphics
display unit and as such it has considerable possibilities for
use in displaying other chemical information such as structures,
or in computer aided learning situations where diagrammatic
formula, graphs, or animated pictures are useful. Programmes
already exist to display the stereochemistry of an cyclohexane
derivative, and to illustrate the Berry pseudo-rotation of
5-coordinate inorganic species. It is also easy to introduce
an element of experimental control, e.g. of the repetitive
operation of a T-jump apparatus to improve the signal to noise
ratio by averaging successive runs; and if the ADC used is
under programme control it is capable of collecting data at
intervals down to 250 μS with the computer specifying the
reading intervals and changing them during the run if required.
Faster data collection facilities (up to one 8 bit reading per
μS for limited periods) could be incorporated at reasonable
cost using additional hardware. It is thus inherently both a
programmable transient recorder and a signal averager, as well
as a display unit, a data logger, a curve fitting device, and
a small computer in its own right.

As a result of the practical experience at the Conference
a number of minor software modifications are being written to
increase the ease of use of the controls, to smooth noisy data,
to allow curves involving two successive exponential terms to

be simulated and fitted, and to provide a rapid logarithmic
routine to convert transmission values to optical densities.
A 12 bit (1:4096) ADC is being incorporated. This will enable
the unit to measure 0% and 100% transmission levels from a
photo detector without changing the range or offset of the ADC
and yet still record small changes in transmission at adequate
precision, so providing the necessary information to calculate
optical density, without introducing extra uncertainties. This
ADC will also be capable of reading data at intervals down to
50 µS under programme control. An improved format for transferring the data to cassette is also being developed, together
with the software to read it back again, and the software to
process the cassette data further in the University central
computer is being altered to take account of this. An
integrated circuit which can provide reasonably fast floating
point arithmetic is now available, and we intend to incorporate
this so that the arithmetic capabilities can be increased and
the unit made largely independent of the University computing
service. At present the apparatus has cost, in components,
bought-in systems etc. but not costing labour or design time,
about £650-£700, and the 12 bit ADC and faster arithmetic unit
may cost a further £200, but it can be seen that the cost is
quite reasonable compared with the cost of the apparatus used
in many fast-kinetics systems to generate the data, and indeed
it is competitive with the cost of many data collection systems
which do not have the versatility provided here by the
programmable microprocessor.[3] It is even cheap enough to be
used in a student practical course, where it has proved to be
usable by and acceptable to relatively inexperienced operators.
An earlier system using a similar curve-fitting technique made
use of spare time on a laboratory computer which had cost ca.
£18,000 in 1970.[4]

Acknowledgements

Dr. J.P. Maher of this University, for assistance in setting up
the demonstration and supervising the T-jump apparatus, and for
being a cooperative user when testing programmes.

References

1. E.F. Caldin, O. Rogne, and C.J. Wilson, J.C.S. Faraday I,
    1978, 74, 1796; C.J. Wilson personal communication.

2. K. Stewart and J.S. Littler, Microprocessors, 1978, 2, 139.

3. J.S. Littler and R.M. Reeves, Chem. Brit., 1978, 118.

4. S.E. Brady, R.L. Cleaver, R.A. Jewsbury, J.S. Littler, and
    J.P. Maher, Laboratory Practise, 1972, 21, 15.

AN AUTOMATIC SELF-SCALING TRANSIENT RECORDER

R.W. King, D.B. Everett, J.M. Sowden, J.H. Satchell
and J.E. Lewin.

Division of Molecular Pharmacology, and Computing
Laboratory, National Institute for Medical Research,
Mill Hill, London NW7 1AA.

A problem which all kineticists share is that of recording
their data, and importantly nowadays recording it in a form
suitable for subsequent computer analysis. This communication
reports on an ingenious solution to our own problems of data
recording, by building a Transient Recorder around the Motorola
M6800 microprocessor. The instrument is capable of recording
analogue data in a digitised form on a suitable time-base with
no instruction from the operator; of displaying the transient
on a screen in a convenient way for visual inspection; and of
adding together up to 16 transients for signal averaging before
transferring the data to main disc storage of a larger computer.

The requirements of a laboratory Transient Recorder may be
summarised as follows:

(a)   That it should sample as rapidly as possible.
(b)   That it should store only useful information.
(c)   That it should be easy to operate.

The instrument which we have built and are currently
operating satisfies all of these requirements to a greater or
lesser extent. Under heading (a) the microprocessor control
circuits can record one sampled voltage every 100 μseconds, i.e.
a 10 kHz frequency. This operational speed can be improved upon
by contemporary electronics, and an operational speed of 200 kHz
is now aimed at.

The requirements of heading (b) are satisfied by the method
in which samples are taken. The sampling algorithm is shown in
Figure 1 below. The voltage signal fed to the Recorder is

W. J. Gettins and E. Wyn-Jones (Eds.). Techniques and Applications of Fast Reactions in Solution. 147–149.
Copyright © 1979 by D. Reidel Publishing Company.

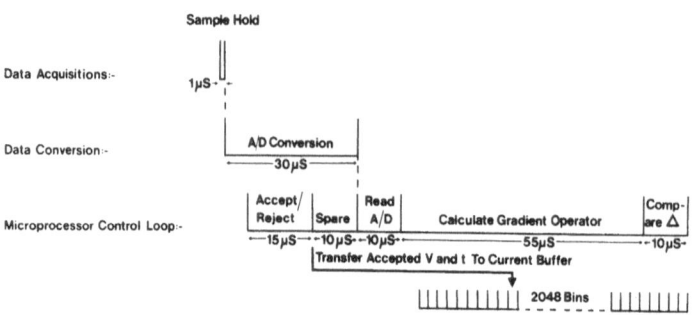

Figure 1. Transient Recorder sampling algorithm. The three
systems run in parallel under the control of a crystal clock
accurate to ± 0.01%.

sampled for 1 μs in each 100 μs. The sampled voltage is then
digitised and, after a synchronising interrupt, read by the
microprocessor. The microprocessor then calculates the
difference between the current value of the signal and the
previously-recorded value. This difference is then compared
with a pre-set significance parameter Δ. This parameter is
normally set at 5 mV in a total usable dynamic range of 10 V.
If this significance parameter is exceeded then the voltage and
elapsed time are stored sequentially in the 2048 bins of the
current buffer. Thus only a signal value which is significantly
different from the previously-recorded value is itself recorded.
If the difference is not significant the microprocessor does not
record in that cycle, and continues cycling until a significant
difference does occur, or until a pre-set truncation time is
exceeded.

Data stored as a series of digitised time/amplitude coordinates is ideal for transmission to a central computer and
for processing. However, it does have the drawback that the
time axis is not linear, and for display on the screen it has
to be relinearised.

Under heading (c), ease of operation, the instrument is
designed with only four operating controls which are push-button
switches. All that the operator needs to do, external to the
recorder, is to ensure that the maximum signal presented does
not exceed 10 V and that a suitable time constant is selected.

After the instrument has been switched on a self-checking
process begins which tests the memory in a complex pattern of
read/write routines. When this test is successfully completed

the truncation time (i.e. the pre-set limit of elapsed time
for sampling after cessation of flow) is entered using the
first push-button switch. A second switch initiates the
experiment and prepares the recorder to accept data arriving
after a trigger pulse returned from the stopped-flow instrument.
The recorded transient is displayed on the lower half of the
screen, suitably scaled to use as much as possible of the half-
screen area. After visual inspection the transient can be
transferred to the cumulative buffer using the third switch.
The retained transient is then displayed on the upper half of
the screen and the current buffer is cleared. A total of 16
transients can be added to the cumulative buffer. Transfer of
the contents of the cumulative buffer to main disc storage is
achieved through a resident program, and is initiated by the
fourth push-button switch. If any transfer difficulties are
encountered the information is not erased and is re-presented
on the screen after a few seconds delay.

APPLICATIONS OF KINETIC POLYCHROMATORS TO FAST REACTION TECHNIQUES

Dr. Paul Suppan

Nortech Laboratories Ltd., Brunel Road, Salisbury, Wilts. U.K.

INTRODUCTION

In kinetic spectrometry experiments of two types are normally considered:

1. Single wavelength kinetic photometry which describes the change of absorption at a narrow wavelength band in time.

2. Time-resolved spectroscopy which records a complete spectrum over a defined wavelength range at one set time after the chemical reaction has been started.

In general many experiments are needed with both techniques to obtain a comprehensive picture of a reaction. Kinetic photometry can be used in n experiments at different wavelengths, or time-resolved spectroscopy can be repeated in n experiments at different times and the data can be analyzed to provide a complete picture. Apart of being time-consuming and often expensive in chemicals, such repeated experiments present problems of reproducibility and accuracy which have so far severely limited their use.

The kinetic polychromator provides the answer for most applications. As the name implies, the optical system is similar to a monochromator or spectrograph but an array of detectors is used to scan n points of the spectrum simultaneously. In this way n kinetic traces are stored in n memories which can be accessed individually (to display the kinetic trace at any one wavelength) or in turn (to display the spectrum at a set time).

W. J. Gettins and E. Wyn-Jones (Eds.). Techniques and Applications of Fast Reactions in Solution. 151–155.
Copyright © 1979 by D. Reidel Publishing Company.

The advantages of the kinetic polychromator over other
systems such as fast scanning monochromators are:

A much faster response which makes them usable to microsecond scan times;

An absolutely simultaneous "scan" for time-resolved spectra
(unlike scanning monochromators which have a time spread
over each spectrum)

A better stability and signal to noise ratio due to the
absence of mechanically moving parts;

A much greater store of information;

An easier method of applying corrections to obtain true
spectra taking in account the wavelength dependence of
detectors and light sources;

An information easily transformed into digital form ready
for further processing.

General description.

The "white" light to be analysed enters the optical unit
of the polychromator through a vertical slit which can be selected for the resolution required. It is then dispersed by a
grating and focused by a spherical mirror which projects the
spectrum on to the photomultiplier array. Additional dispersing
optics can be used to increase the spectral spread for medium
or high resolution applications. The output of the photomulti-
pliers are individually stored in separate memories which share
a common time-base and trigger so that a sweep can be started
with the initiation of the experiment. At the end of the sweep
the information stored in the memories can be converted at
leisure into digital form for permanent storage and further
processing.

The memories can be read out in two ways:

In the single channel kinetic mode any one memory (corresponding to one wavelength of monitoring light) is read out on a
fixed time-base with display on an oscilloscope or chart recorder
(tape punch or magnetic tape recording optional).

In the multi-channel set time mode all the memories can be
read out in sequence at any one of the words which corresponds
to a set time following the start of the experiment. As the
word number is changed the read-out shows the change in the

whole spectrum in course of time; the display is an oscilloscope
or chart recorder, with the possibility of tape punch or magnetic
tape recording as well.

In all cases the gain of the detectors can be adjusted
individually to provide directly corrected spectra both for
absorption and emission work. For absorption detection the gain
of the detectors is adjusted in such a way as to obtain the same
output signal from each channel. For emission detection the
signals are set to correspond to a calibrating sample or light
source.

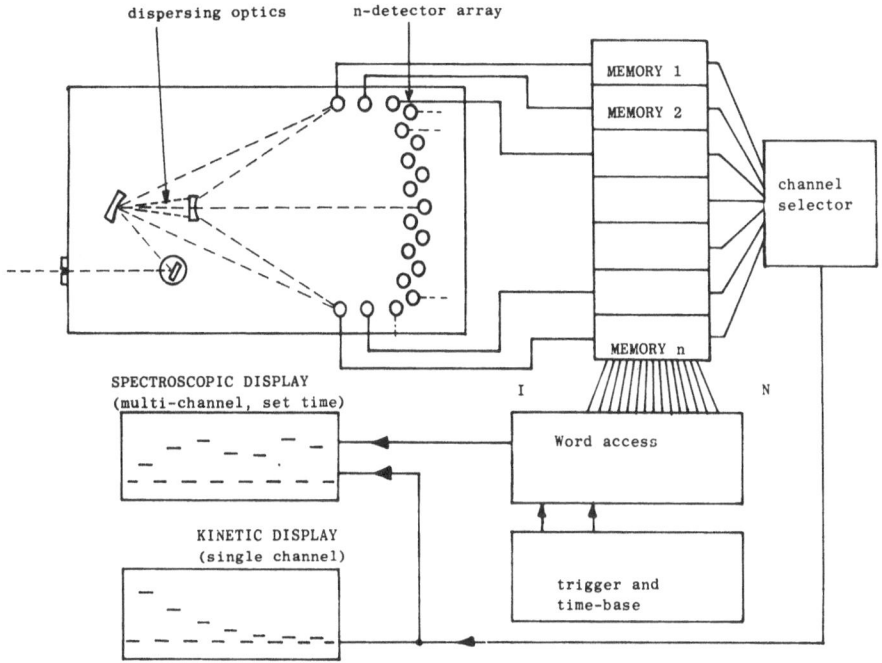

Memory systems.

Most modern transient recorders use a digital storage of
the signal. The analogue signal (defined for example as V Volts)
is converted into a binary number of 8 bits in general (256
points), sometimes of 10 or even 12 bits (4048). The process of
conversion takes a minimum time which defines the time resolution
of the memory system. The higher the signal resolution (e.g.
12 bits) the longer the conversion time. When simultaneous conversion must be performed on 10, 20, or 30 channels the cost
factor must also be considered. For single channel applications
an 8-bit resolution with a conversion rate of the order of 1 μs
is usually acceptable, as the amplification can be adjusted so

that the 256 available points cover nearly the whole signal or
the change in the signal.

For a multi-channel system such as a polychromator this
limited 8-bit resolution can lead to problems when the signal
amplitudes differ greatly from one channel to another. Consider
the case of an experiment where the absorbance change at one
wavelength defined by channel n is 0.5, and at another wavelength
defined by channel m is only 0.05. If the scaling factor is the
same on all channels, which is necessary for a direct comparison
of all channels (the "spectroscopic" display), the 256 points
may correspond to a change of absorbance of 1, which will give
an acceptable resolution of the order pf 1% for the n th channel;
however channel m uses only 12 points of the available 256 and
its resolution is therefore very low. This is overcome to some
extent by the use of individual scaling factors on each channel
in the KMD type memories. The scaling factors which multiply
the input signals are used also to divide the output signals to
restore them to a common level without loss of resolution. This
is not a completely satisfactory answer, however, since it
implies a prior knowledge of the scaling factors, that is of the
signal changes on each channel, which is the object of the
experiment. For this reason several experiments are needed with
the digital memory to obtain the best results.

A radically different approach uses analogue memories which
have a limited number of samples (in time) but theoretically
infinite signal resolution. The number of samples taken is
limited by the current output of the amplifier since each sample
("word") is in effect a small capacitor which holds the voltage
representing the signal at one instant in time. The main difficulty with this analogue storage method is the fact that the
memory is not a permanent type, as the charge on the capacitor
leaks slowly through any insulation imperfections of the circuit.
It is therefore necessary to restore the information from the
analogue memory fairly rapidly into digital form.

Diagram 2 illustrates the principle of the analog memory
for a 2-channel 8-word system. Each capacitor is connected to
the signal line via its individual switch which is open (0) or
closed (1) according to the logic output of the corresponding
bit of a parallel output shift register.

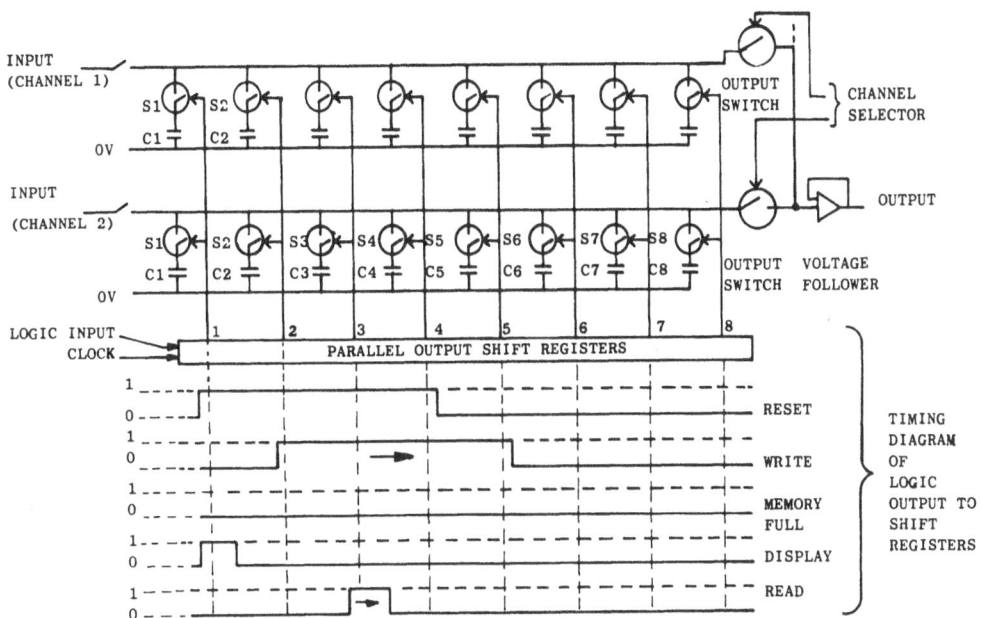

RESET.   Before the experiment the first n (in this case
                4) capacitors are connected to the output and
                track the signal.

WRITE.   When a trigger signal is given, the clock of the
                shift registers starts and the logic input goes
                to 0, A wave of n logic 1 signals travels through
                the shift registers so that the capacitors track
                and then hold the signal in sequence.

MEMORY FULL.   When the last shift register has a logic 1 to 0
                transition.

DISPLAY.   A single logic 1 input of the width of one clock
                pulse is given to the shift registers input.

READ.   The clock of the shift registers is started and
                a single logic 1. signal travels through the
                shift registers, connecting each capacitor in
                turn to the output line via a high-impedance
                voltage follower.

PICOSECOND TIME RESOLVED FLUORESCENCE SPECTROSCOPY

C.J. Tredwell, C.M. Keary and G. Porter

Davy Faraday Research Laboratory, The Royal Institution,
21 Albemarle Street, London W1X 4BS.

The time-resolved fluorescence emission from organic molecules can be used to elucidate a number of photophysical processes, such as heavy-atom enhanced intersystem crossing from the excited singlet state (S_1) to the triplet state (T_n), internal conversion from S_1 to S_0, singlet-singlet resonance energy transfer and molecular rotational diffusion. Since many of these processes can occur on a sub-nanosecond time scale, a source of picosecond duration light pulses and a detection system with a comparable time resolution are needed to obtain the necessary data.

A simplified diagram of the apparatus used to obtain picosecond time-resolved fluorescence decay curves, is shown in Figure 1 ; a more detailed diagram may be found in a recent publication (1). A mode-locked neodymium (Nd^{3+}):glass laser is used to generate a train of 150 pulses at a wavelength of 1060 nm; the pulses have a typical duration of 6 ps and a pulse separation time of 6.9 ns. The 1060 nm fundamental output is then converted to 530 nm by a temperature-tuned caesium dihydrogen arsenate (CDA) frequency-doubling crystal. Since a complete pulse train is unnecessary for excitation of the sample, a Pöckels cell electro-optic shutter selects a single pulse from the centre of the 530 nm pulse train. Fluorescence emitted by the sample is time-resolved with a picosecond streak camera; a diagram of the time-resolving element, the streak image tube, is shown in Figure 2. The incident light is focused onto a narrow slit whose image is subsequently focused onto the photocathode of the image tube. Photoelectrons emitted by the photocathode are accelerated by the intense electric field between the cathode and the extraction mesh in order to minimise the electron velocity distribution. The resultant electron beam is deflected to one side of the phosphor screen by the electric field between the two

W. J. Gettins and E. Wyn-Jones (Eds.). Techniques and Applications of Fast Reactions in Solution. 157–160.
Copyright © 1979 by D. Reidel Publishing Company.

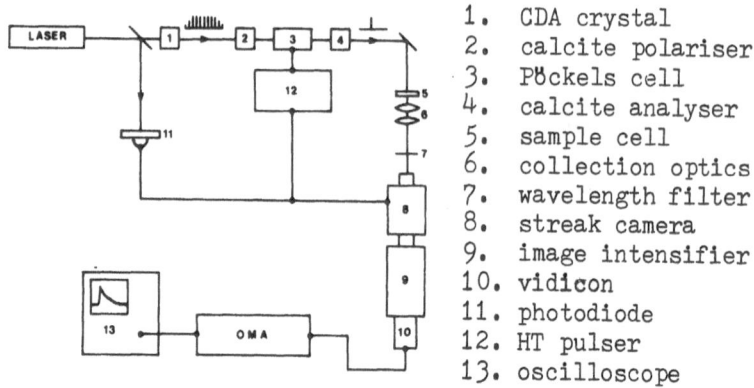

1. CDA crystal
2. calcite polariser
3. Pöckels cell
4. calcite analyser
5. sample cell
6. collection optics
7. wavelength filter
8. streak camera
9. image intensifier
10. vidicon
11. photodiode
12. HT pulser
13. oscilloscope

Figure 1. Picosecond laser and streak camera

deflection electrodes. When the streak camera is triggered, the
polarity of the deflection field is reversed and the slit image
is swept across the phosphor in a direction orthogonal to its
long axis. Time is therefore converted into the distance travelled
across the phosphor, and the incident light intensity is
proportional to the light emitted by the phosphor. Modern
streak cameras are capable of 0.5 ps to 5 ps time resolution on
their fastest sweep speed. After image intensification, the
streak trace is converted to an intensity versus time curve by
a vidicon optical multichannel analyser (OMA). Data stored in
the OMA memory can either be displayed on an X-Y oscilloscope or
be transferred to a computer for numerical analysis. Figure 3
shows an exponential fluorescence decay curve before and after

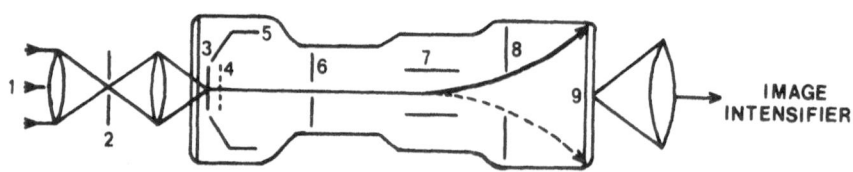

Figure 2. The streak image tube.
            1. photon input, 2. 50 μm slit, 3. photocathode,
            4. extraction mesh, 5. focus cone, 6. anode,
            7. deflection electrodes, 8. baffle plate,
            9. phosphor screen.

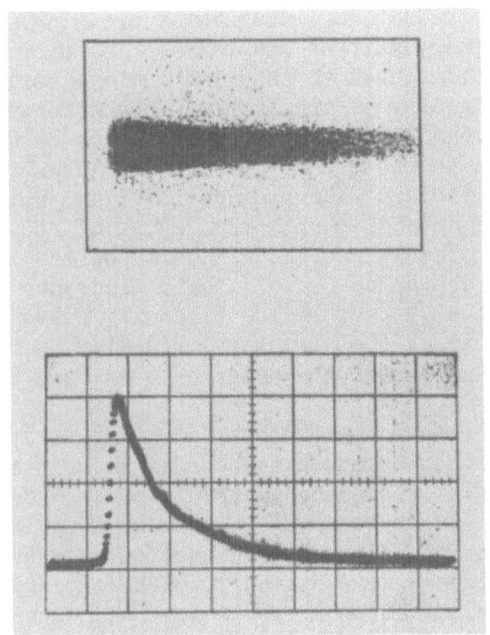

Figure 3. The fluorescence decay of tetrachloro-
tetraiodofluorescein (1 ns per major division):
Top - observed at the image intensifier.
Bottom - after conversion by the OMA.

conversion by the OMA; the fluorescence lifetime is 1.1 ns.

Fluorescence lifetime measurements provide invaluable
information about the processes involved in the dissipation of
excitation energy within organic molecules. Some of the
fastest excited state relaxation rates are observed in solvated
ionic dyes, such as the xanthene, triphenylmethane and polymethine
dyes. The stereochemistry of these dyes and the close proximity
of the ground and excited states make it possible for radiationless
transitions to occur at rates far in excess of those observable in
most aromatic molecules. Consequently, the fluorescence lifetimes
of these dyes are strongly dependent upon the rate constants of
the various radiationless relaxation processes.

Examples of heavy-atom enhanced intersystem crossing to the
triplet state can be found in the halogen substituted derivatives
of fluorescein (Fl) in aqueous solution. Using the streak camera
system, the fluorescence lifetimes of fluorescein (Fl, 4.6 ns),
eosin Y (Fl.Br$_4$, 1.24 ns), tetrachlorotetraiodofluorescein
(Fl.Cl$_4$.I$_4$, 1.10 ns) and erythrosin Y (Fl.I$_4$, 0.11 ns) have been

recorded. The marked reduction of the fluorescence lifetime with the increase in atomic weight of the substituent can be attributed to the enhancement of the rate of intersystem crossing.

Although the triphenylmethane and polymethine dyes have certain structural similarities to the xanthene dye series, they lack the structural rigidity of the latter. Consequently, intramolecular rotation allows excitation energy to be dissipated by internal conversion to the ground state. Our investigation of these dye systems has yielded fluorescence lifetimes which range from 1-2 ps up to 6 ns depending upon the degree of steric hindrance within the molecule and the viscosity of the solvent; many of the dyes with a high degree of steric hindrance exhibit fluorescence lifetimes of 1-2 ps in low viscosity solvents such as methanol. By increasing the structural rigidity of these dyes, fluorescence lifetimes of the order of 6 ns have been recorded.

Picosecond fluorescence spectroscopy can also be applied to the investigation of other molecular processes in solution including singlet-singlet resonance energy transfer and molecular rotational diffusion. In the case of resonance energy transfer, picosecond time resolution allows the measurements to be performed in a low viscosity solvent (e.g. ethanol) which is close to the situation envisaged by the Förster expression for resonance energy transfer. A picosecond time-resolved study of the donor fluorescence decay kinetics from the rhodamine 6G (donor) and malachite green (acceptor) energy transfer system has indicated that the theoretical Förster expression is valid to within 10 ps of excitation; the R_O value obtained for this system is 52.5 Å (2).

The technique for studying the rotational motion of molecules in solution by picosecond fluorescence spectroscopy can be used to time-resolve rotation times of less than 100 ps (3). The rotation correlation time is obtained from the rate of depolarisation of the fluorescence emission after excitation by a polarised laser pulse. The rotation times for tetrachloro-tetraiodofluorescein (TCTIF) and rhodamine 6G in various alcohols indicate that hydrogen bonding of the solvent to the dianionic dye, TCTIF, increases the rotation time by a factor of two over rhodamine 6G. The measurements also show that the Stokes-Einstein equation is obeyed at low viscosities, although the increase in molecular volume caused by hydrogen bonding with the solvent must be taken into account (3).

References:
1. M.D.Archer, M.I.C.Ferreira, G.Porter and C.J.Tredwell, Nouveau Journal de Chimie, 1 9-12 (1977).
2. G.Porter and C.J.Tredwell, Chem. Phys. Letters, 56 278-282 (1978).
3. G.Porter, P.J.Sadkowski and C.J.Tredwell, Chem. Phys. Letters, 49 416-420 (1977).

FOURIER TRANSFORM PHOTOACOUSTIC SPECTROSCOPY

Michael M. Farrow, Roger K. Burnham, and Edward M. Eyring

Office Products Division, IBM Corporation, Boulder, Colorado 80302, U.S.A. and Department of Chemistry, University of Utah, Salt Lake City, Utah 84112, U.S.A.

ABSTRACT. The construction and operation of a Michelson interferometer that permits Fourier transform photoacoustic spectroscopy of opaque and partially transparent samples at visible wavelengths is described. Multiplexing and throughput advantages are considered. A visible spectrum of Nd(III) doped laser glass is reproduced and potential kinetic applications are described.

INTRODUCTION

Photoacoustic spectrometers for recording visible spectra of opaque or practically transparent solid samples operate according to basic principles first described by Alexander Graham Bell (1,2). A visible wavelength is selected by a monochromator from the white light emitted by an arc or filament source, chopped with a rotating sector at an audio frequency ($\sim$50 Hz), and passed through a glass window onto the sample surface in an enclosed cell (3). If the surface absorbs light of the incident wavelength, electronic excitation rapidly degrades into heat by radiationless transitions, and the layer of gas in the cell in contact with the sample surface is intermittently heated at the frequency with which the light beam was chopped. The resulting sound wave in the gas is detected by a microphone located at a position in the cell where the scattered light beam cannot strike it. The signal voltage from the microphone is amplified by a lock-in amplifier that is coupled to the chopping wheel in such a way that only that part of the signal that bears a selected time relationship to the repetitive illumination intervals is amplified. Thus the lock-in amplifier can bring up to detectable levels, weak signals that would otherwise be buried in ran-

161

W. J. Gettins and E. Wyn-Jones (Eds.). Techniques and Applications of Fast Reactions in Solution. 161–174.
Copyright © 1979 by D. Reidel Publishing Company.

dom, environmental noise.

Photoacoustic spectroscopy (PAS) has found recent use in
determining the extinction coefficients of highly transparent
solids and gases (4), absolute fluorescence quantum yields in
liquids (5), lifetimes and mechanisms of radiationless transitions in solids (6), the band edge in semiconductors (7), the
visible absorption spectrum of highly scattering fluids such as
whole blood (7), the corrosion chemistry of copper intrauterine
devices (8), and the near infrared to ultraviolet absorption
spectrum of an inorganic linear-chain conductor (9). Other applications have included the measurement of thermal diffusivities
in solids (10), the measurement of the visible absorption spectrum of the waxy cuticle (surface layer) of a plant leaf (11),
the detection of tetracycline applied topically to human skin
(12), a similar study of newborn rat stratum corneum (the outer-
most layer of the epidermis) (13), and the detection of gases
evolving from the surface of a photoactivated heterogeneous catalyst (14).

None of the above noted studies touch on the use made in recent years of photoacoustic spectroscopy to investigate the properties of gases (15).

Karasek's investigation of two then commercially available
photoacoustic spectrometers for determining spectra of opaque
specimens led him to the pessimistic conclusion that PAS would
not become widely useful in studying solids until sensitive
measurements are routinely possible in the infrared (16).

Intense, broadly tunable infrared emitting lasers are still
not available, but even if they were, a much less expensive
method of enhancing the signal to noise ratio of an infrared
photoacoustic spectrometer is already available: the Michelson
interferometer and Fourier transform (FT) techniques. An alternative approach is the Hadamard transform spectroscopy technique
in which the spectral information is encoded in a binary code
based on Hadamard matrices (17). This corresponds to an algebraic permutation of the various spectral components. In Fourier
transform spectroscopy, the spectral information is encoded according to sines and cosines. Since both the Fourier transform
and Hadamard transform spectroscopic techniques are multiplexing
techniques, they should gain equally from any advantage in the
signal to noise obtained from multiplexing spectral information
in a situation where the system is detector noise limited (see
below). We applied the Fourier transform technique since the
cost of commercial Hadamard transform masks is of the order (18)
of $25,000, and an effective Michelson interferometer can be con-
structed for less than $10,000 exclusive of control electronics
and computer (19,20).

The fundamental components of a Michelson interferometer
are depicted schematically in Fig. 1. One possible misconcep-
tion regarding the use of a Michelson interferometer at visible
wavelengths is that it is difficult to obtain a sufficiently
precise movement for the scanning mirror, M_1, to provide useful
resolution in the absorption spectrum (Fourier transformed interferogram). This is definitely not the case. The actual reason
that FT techniques are not commonplace in visible spectroscopy
is that the usual visible spectrophotometer with a photomultiplier tube for a detector is source-noise-limited rather than
detector-noise-limited so that FT methodology does not improve
(and actually degrades) the quality of the measured spectrum.

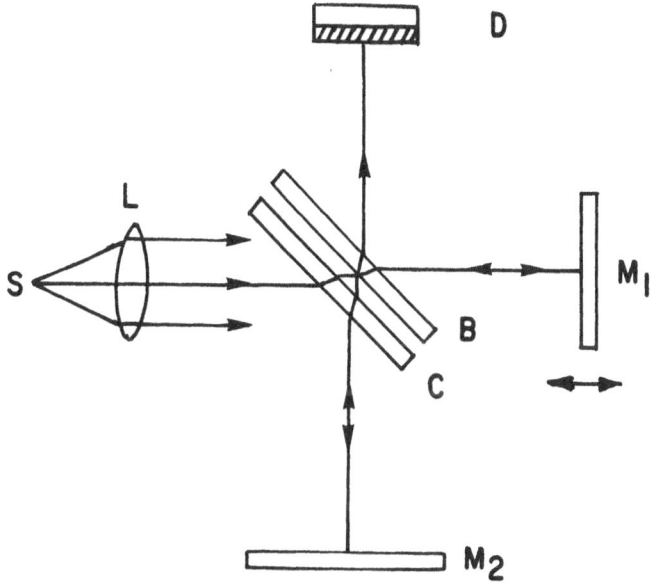

Fig. 1.   Schematic of a Michelson interferometer. S in the FTPAS
is a 100 watt tungsten iodide lamp. M_1 is a movable front surface mirror, L is a collimating lens, M_2 is a "stationary" front
surface mirror that flutters to provide the analog of beam chopping in ordinary PAS, B is a half silvered beam splitter, C is
a compensating plate, and D is the detector (a piezoelectric
transducer, microphone, photomultiplier tube, or the eye of an
observer).

Light from the source S is split into two beams of approximately equal amplitude. If the two beams have traversed precisely the same effective path length in returning from the mirrors
to the beamsplitter B, they will be in phase with one another
and will interfere constructively. In order to provide exactly

equivalent paths in both arms of the interferometer, a compensating plate of material identical to the beamsplitter substrate but without the beamsplitting coatings, is placed in the appropriate arm of the interferometer. This compensates for any wavelength dependent dispersion of the beam splitter substrate. Any difference in path length is referred to as retardation. If the effective path length on reaching B differs by a half integral wavelength distance, the beams from the two arms will be out of phase and will destructively interfere with each other. We will refer to the bright areas of constructive interference as "bright fringes".

The actual beam path for light passing through an interferometer may be more easily visualized with the aid of Fig. 2 which shows an equivalent arrangement for the mirrors in Fig. 1. The case illustrated is for one mirror, M_1, being nearer to the beam splitter than the other mirror, M_2. The virtual images of the source, S_i, appearing in each mirror are identified by the appropriate subscripts. If the path lengths in the two arms differ by distance d, then the virtual sources are separated by 2d, since the path is traversed twice by the beam on its way to and from the mirrors. The appropriate arrangement of the mirrors in the interferometer used in FTPAS is that in which the mirrors are perfectly perpendicular. This is equivalent to the parallel po-

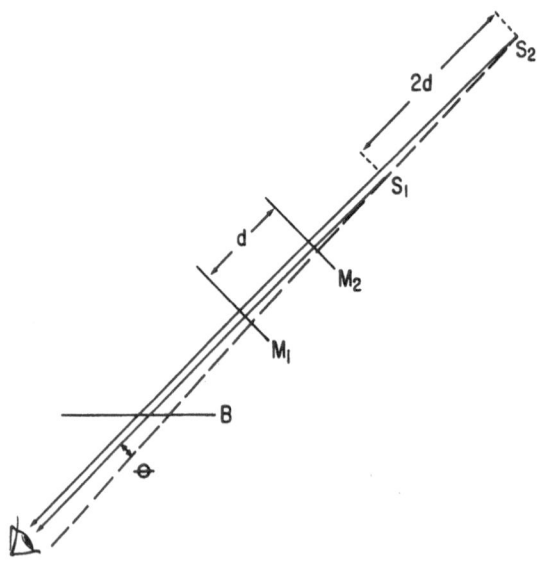

Fig. 2.   Schematic of light paths in a Michelson interferometer. Compensating plate has been omitted for clarity.   See text for complete description.

sitioning in Fig. 2 and contrasts with some other interferometric geometries in which one mirror is canted with respect to the other

If a perfectly aligned Michelson interferometer is illuminated with a monochromatic light source and an observer places his eye at the output of the interferometer, looking back toward the beam splitter and source (Fig. 2), he will observe circular interference fringes. Although all rays of light reflected normal to the mirrors will be in phase, rays reflected at any angle will not, in general, be in phase. Figure 2 shows that for any given angle θ the path difference is given by $2d \cos \theta$. Since the rays must be parallel for interference to occur, θ must be the same for both rays coming from either virtual source, S_1 or S_2. The rays will reinforce each other to produce maxima or bright fringes for those angles θ which satisfy the relationship:

$$2d \cos \theta = m \lambda \qquad (1)$$

Here d is the mirror separation, λ is the wavelength, m is an integer, and θ is a constant which produces a circular fringe. If m is half integer, then destructive interference occurs and a dark fringe results. At the center of the fringe, θ is zero and $\cos \theta = 1$, so that equation 1 becomes

$$2d = m \lambda \qquad (2)$$

In order for m to change by unity, d must change by λ. As the two virtual sources S_1 and S_2, approach each other, the angle θ must decrease and the fringes become more widely spaced. At a critical position, there is only one dark fringe visible. This happens when M_1 and M_2 are exactly coincident, since for this position the path difference is zero for all angles of incidence.

If white light is used to illuminate the interferometer, no fringes will be seen except very close to zero path difference between the two mirrors. At zero path difference there will be a dark central fringe, bordered by eight to ten colored fringes. That there are only a few fringes visible with white light is easily accounted for by recalling that all wavelengths between ∿400 and 750 nm are contained in visible light. There will be coincidence of various wavelength fringes only for d = 0, since the fringe spacing varies according to the wavelength with fringes for longer wavelengths being more widely spaced. Clearly the fringes of different colors will begin to separate on either side of zero producing impure colors. After eight or ten fringes so many colors are present that the resultant is essentially white. Interference is still occurring, but the eye is not sensitive enough to resolve the variations in intensity.

Two techniques are useful for locating the zero path difference position of the mirrors. One of these is to use a monochromatic light source and locate that position of the two mirrors for which the central fringe has expanded to fill the field
of view as mentioned above. Another is to use a monochromatic
light source that is actually a doublet such as the sodium D
line. The interference produced by two closely spaced spectral
lines will be a sine wave whose amplitude varies at some lower
sinusoidal frequency. The output of the interferometer is focused
on a photomultiplier and the path length difference of the two mirrors is varied in some uniform manner (i.e. a sawtooth motion).
If the output of the photomultiplier is displayed on an oscilloscope, the sinusoidal envelope which contains a high frequency
sinusoidal modulation is observed and as the path lengths of the
two arms approach equality, the modulation depth of the higher
frequency sinusoidal modulation in the envelope increases. If
the interferometer is adjusted to produce a maximum modulation
depth, zero path difference has been located. If a white light
is then used to illuminate the interferometer, the photomultiplier should detect the expected eight to ten fringes and present
the typical interferogram of a broadband source, Fig. 3.

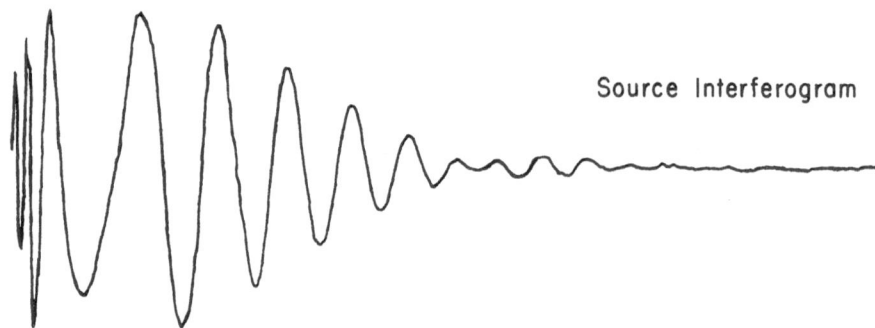

Source Interferogram

Fig. 3. Typical interferogram of a broadband light source. A
Schematic of the trace observed on an oscilloscope screen when
the PMT voltage (vertical) is displayed versus time (horizontal).
The time axis corresponds to changes in the mirror position M_1 of
Fig. 1 about equal path difference in the arms of the interferometer.

The amount of light passing through the interferometer will remain essentially a constant if averaged over the entire field of view. Only the central (Airy) disc will vary as the interferogram of the source. For this reason, one places an aperture at the focus of the interference pattern and only allows the central portion of the fringe pattern to fall on the detector.

Since Fourier transform photoacoustic spectroscopy (FTPAS) still requires a modulation of the optical signal at the detector, the source intensity must be modulated in some manner. This may be accomplished by a technique known as phase modulation (21). This method, also known as internal modulation, requires a slow scanning interferometer, usually employing the step and integrate method of traversing the moving mirror. At discrete positions of the moving mirror the beam is modulated not by a chopper, but by periodically varying the retardation of the moving mirror by a small amount. This small variation in the retardation or jitter, if made infinitesimally small, would result in the measured interferogram being the first derivative of the interferogram. In practice, even with finite jitter amplitudes, a phase modulated interferogram is a very good approximation to the first derivative of the amplitude modulated interferogram. An added benefit of a phase modulated interferogram is that the large DC offset which otherwise must be measured is eliminated. This DC offset corresponds to white light which does not interfere with itself in the interferogram and presents a large background. Since the first derivative of an amplitude modulated interferogram is a sine function, the spectrum can then be computed from the interferogram by sine transform (21).

In photoacoustic spectroscopy, signal amplitude increases only when absorption occurs. Thus, photoacoustic spectra will be absorption spectra as opposed to FTIR spectra which are transmission spectra. FTPAS interferograms should then show modulation at high values of retardation as well as near zero retardation. This comes about as a consequence of the properties of Fourier transforms which cause very broad general feature information to be collected at low retardation, with narrow and sharp features ("high frequency") to be collected at higher retardation.

EXPERIMENTAL

Figure 4 is a photograph of the working interferometer. The entire interferometer is mounted on a 96 x 80 x 20 cm granite slab which rests on a ∿15 cm thick bed of polyurethane foam. The combination of approximately 300 kilograms of granite and 15 cm of foam effectively decouples the interferometer from mechanical and seismic events in the building. The framework of the inter-

ferometer is one inch diameter stainless steel rod and is somewhat sensitive to thermal expansion caused by heating from the
light source. The 5 cm, $\frac{\lambda}{20}$ (full aperture) mirrors (Pyramid
Optical Corp.) are mounted in Star-Gimbal mounts (Burleigh).
The visible wavelength beam splitter and compensating plate are
both $\frac{\lambda}{20}$ (parallel to one second of arc) with 5 cm full aperture
(Pyramid Optical Corp.). The beam splitter is located in a 45°
optical mount (Burleigh) which is supported by a Star-Gimbal
mount for adjustment. The fixed mirror is also the phase modu-
lating mirror. In order to accomplish the phase modulation the
mirror is mounted in a PZAT-90 (Burleigh) piezoelectric mount.
There is sufficient linear motion available (∿10 micrometers/
1000 V) from this mount to drive several orders of visible
fringes.

Source   Sample   Beam Splitter

Ref. Laser   Fixed Mirror   Moving Mirror   PZAT-90
_Fig. 4.   Photograph of the FTPAS interferometer at Utah.

The moving mirror is mounted on a Velmax Corp. translation
stage (Model U 4006 GGMO). This specially selected stage provides more than 25 millimeters of highly precise motion without
appreciable side to side or vertical error in tracking. The
mirror and gimbal mount are located directly on the stage,
which is driven by a differential micrometer (Burleigh, DM-105)
rotated by a 2000 step per revolution stepping motor (USM Corp.,
Harmonic Drive Division, Model HDM-170-2000-8, with drive logic
321-3/3121-4).

Figure 5 is a schematic of the step and integrate FTPAS apparatus we have described in the preceding paragraphs.

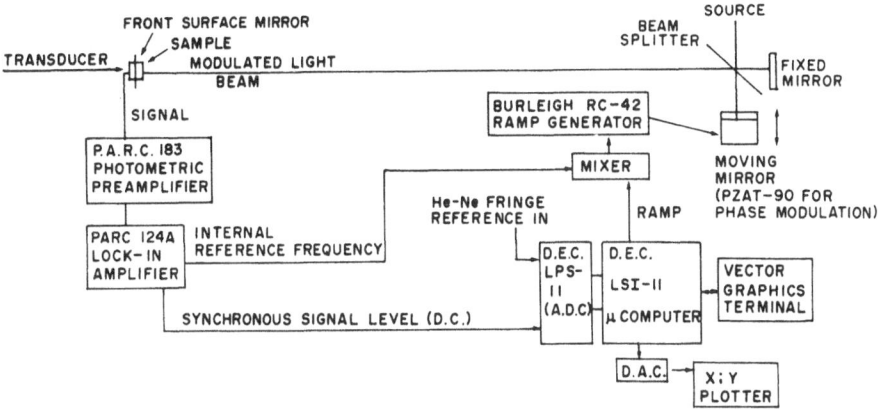

_Fig. 5.   Schematic of a step and integrate Fourier transform-visible-photoacoustic spectrometer.   The Michelson interferometer is in the top right-hand corner.   Reproduced with permission from Appl. Phys. Lett., _33_, 735 (1978).

RESULTS AND POTENTIAL APPLICATIONS

A representative interferogram (Fig. 6) and transformed absorbance spectrum (Fig. 7) of neodymium(III) doped laser glass suggest the potential utility of FTPAS for obtaining high resolution spectra in analytical chemistry.   In this case the FT feature of the experiment greatly speeds the accumulation of data (20).

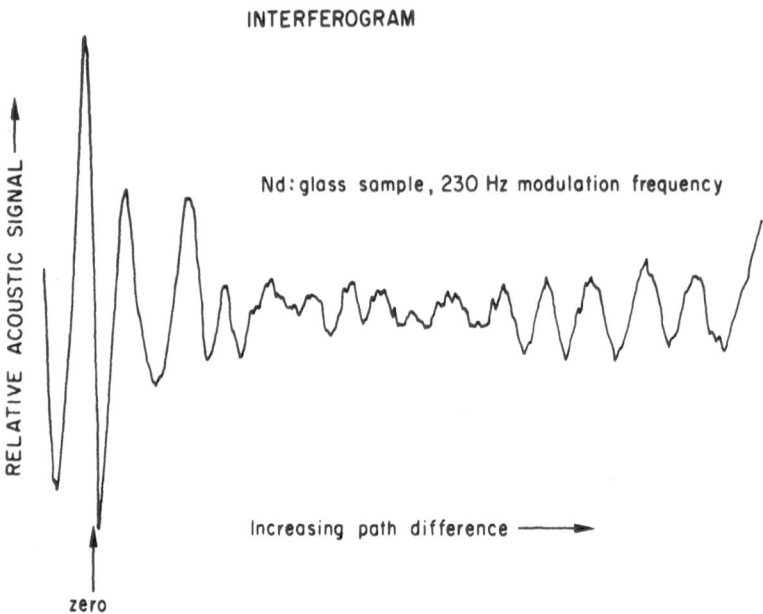

Fig. 6. Photoacoustic interferogram of Nd:glass sample. Reproduced with permission from Appl. Phys. Lett., 33, 735 (1978).

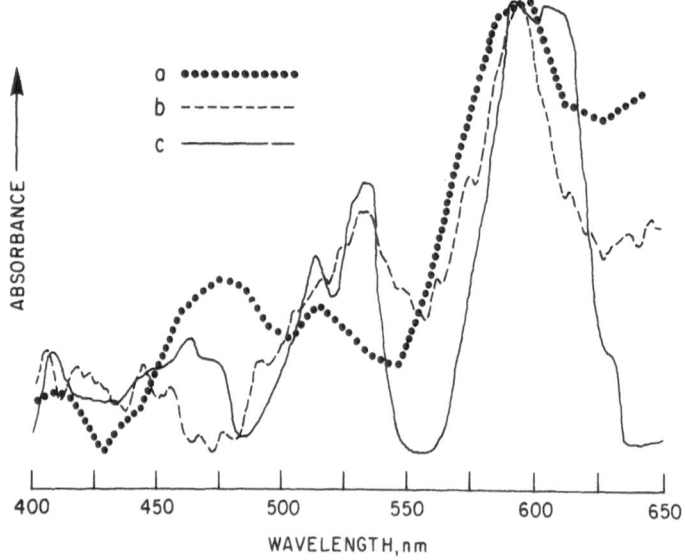

Fig. 7. Absorption spectra of Nd:glass samples. Curve a calculated from the interferogram of Fig. 6. The spectrum has not been source normalized (i.e. signal has not been divided by light source intensity at each wavelength). The interferogram was based on samples taken at 256 equally spaced mirror positions. The

spectral resolution is $\sim$850 cm^{-1} or $\sim$20 nm @ λ = 500 nm. Curve
b is a dispersive PAS spectrum[6e] of the same sample. Curve c
is a portion of a typical absorption spectrum of another Nd:glass
sample taken from the literature (K. Hauptmanova, J. Pantoflicek,
and K. Patek, Phys. Stat. Sol., 9, 525 (1965).) Reproduced with
permission from Appl. Phys. Lett., 33, 735 (1978).

    In addressing reaction kineticists, however, it is more inter-
esting to talk about potential application of FTPAS to the measurement of reaction rates occurring on opaque surfaces or in
liquids containing suspended particles that scatter light strongly. For instance, in medicine a real need exists for rapid and
precise information on the physiological state of available oxygen in whole blood. Although derivative measurements, such as
those provided by traditional spectrophotometric measurements
made with an oximeter do provide important information, the operator must rely on standard interpolations to provide understand-
able results. Since actual oxygen-hemoglobin affinities are not
easy to measure and to interpret, clinical correlations of physiological condition and exact oxy-hemoglobin affinity are difficult to make.

    An FTPAS spectrometer has, in principle, a capability for
sub-millisecond time resolution (>300 μsec) and is the instrument
of choice for obtaining optical absorbance spectra of opaque and
strongly scattering samples. Thus the optical spectra of whole
blood samples can be obtained without hemolysis. With the incorporation of P_{O_2}, pH and P_{CO_2} sensing electrodes directly in the

PAS cell one could measure oxy-hemoglobin saturation without alter-
ing the basic blood chemistry. Since routine oxygen-hemoglobin
affinity measurements may prove to be a useful diagnostic tool,
the potential for time resolution inherent in the basic FT-PAS
design to measure points relating the partial pressure of oxygen
and percent oxygen saturation of hemoglobin in the intact cell
should also be explored.

    Such an apparatus would have the capability of providing not
only a more exact measurement of the intraerythrocyte total hemoglobin, oxy-hemoglobin, deoxyhemoglobin and carboxyhemoglobin (as
well as methemoglobin and sulfhemoglobin (22)) concentrations but
also a measurement of P_{50}, the partial pressure of O_2 necessary
to cause 50% of the hemoglobin to be bound with oxygen, which is
actually based on a series of P_{O_2}-S_{O_2} (saturation partial pres-

sure) measurements. This information would allow determination
of the Hill coefficient (slope of the S_{O_2} vs. P_{O_2} plot), and per-

haps screen for mutant hemoglobins. Lichtman et al. (23) have

pointed out that the oxy-hemoglobin affinity is of clinical importance as it supplies a means of detecting the presence of a
mutant hemoglobin at high or low affinity for oxygen.

The interest in the role of the effect of the red blood cell
membrane on the rate of oxygenation of hemoglobin is of a more
fundamental nature. The variety of techniques brought to bear on
this problem have suffered from many of the same flaws as the
clinical instrumentation described above. The most severe restriction until now has been the inability to directly follow
the rate of oxygenation of hemoglobin within the intact erythrocyte. In order to evaluate the diffusion rate through the red
blood cell membrane, experimentalists were forced to rely on related measurements, such as those reported by Gros and Moll (24),
who measured diffusion through layers of intact erythrocytes. Obviously, the oxygen-hemoglobin affinity will affect this diffusion rate, at least until complete oxygen saturation is achieved.
However, if there is any hemoglobin facilitated O_2 transport (25)
in the erythrocyte, its physiological importance occurs not only
at 100% oxygen-hemoglobin saturation, but also at much lower
oxygen partial pressures.

Since the photoacoustic technique allows the effective
analytical penetration depth into the sample to be varied by
varying the optical modulation frequency (26), the actual hemoglobin oxygenation rates can, in principle, be followed at differing depths below the gas-blood interface and simultaneously
compared to the oxygen partial pressure as measured by the Clark
O_2 electrode. This new class of experiments could resolve the
disagreement noted by Kutchai (25) concerning the role of the red
blood cell membrane as the principal barrier to gas uptake:
"Perhaps this long-standing controversy can only be resolved by
direct determination of the permeability of the red cell membrane
to oxygen and carbon monoxide." (25)

Other interesting potential FTPAS kinetic studies that we
have planned include measurement of the rate of leaching of copper ions from suspensions of highly colored copper sulfide ores
in acidic ferric sulfate solutions and an infrared FTPAS study
of the epoxidation of ethylene on a solid silver-silver oxide
catalyst.

Since we have been working with PAS instrumentation for
three years and with a Michelson interferometer for almost one
year (20), it would be fair to ask why the present paper is not
more a report of chemistry accomplished rather than a prospectus
of things to come. Absorbance spectra for transmitting samples
such as neodymium(III) doped laser glass obtained on a Cary uv-
visible spectrophotometer are only roughly approximated by the
FTPAS spectra (Fig 7). It may turn out that the differences

provide important new information about the sample or it may just be a case of errors being introduced into the interferogram by phase lag problems. Until such questions about the FTPAS technique using a piezoelectric transducer detector are worked out, the interferograms and transformed spectra we obtain will be of uncertain value and the hot pursuit of more chemical goals will be premature.

ACKNOWLEDGMENT

Partial support of this work by the Chemical Directorate, Air Force Office of Scientific Research, Grant AFOSR 77-3255, is gratefully acknowledged.

REFERENCES

1. A. G. Bell, Proc. Am. Assoc. Adv. Sci., $\underline{29}$, 115 (1880).

2. A. G. Bell, Phil. Mag., $\underline{11}$, 5-10 (1881).

3. See, for example, J. F. McClelland and R. N. Kniseley, Appl. Optics, $\underline{15}$, 2967 (1976).

4. H. S. Bennett and R. A. Forman, Appl. Optics, $\underline{15}$, 1313, 2405 (1976); J. Appl. Phys., $\underline{48}$, 1217 (1977); A. Hordvik and H. Schlossberg, Appl. Optics, $\underline{16}$, 101 (1977); A. Hordvik and L. Skolnik, Appl. Optics, $\underline{16}$, 2919 (1977).

5. W. Lahman and H. J. Ludewig, Chem. Phys. Lett., $\underline{45}$, 177 (1977); M. J. Adams, J. G. Highfield, and G. F. Kirkbright, Anal. Chem., $\underline{49}$, 1850 (1977).

6. L. D. Merkle and R. C. Powell, Chem. Phys. Lett., $\underline{46}$, 303 (1977); J. C. Murphy and L. C. Aamodt, J. Appl. Phys., $\underline{48}$, 3502 (1977); M. G. Rockley and J. P. Devlin, Appl. Phys. Lett., $\underline{31}$, 24 (1977); R. G. Peterson and R. C. Powell, Chem. Phys. Lett., $\underline{53}$, 366 (1978); M. M. Farrow, R. K. Burnham, M. Auzanneau, S. L. Olsen, N. Purdie, and E. M. Eyring, Appl. Optics, $\underline{17}$, 1093 (1978).

7. A. Rosencwaig, Anal. Chem., $\underline{47}$, 592A (1975).

8. K. M. Lewis, R. D. Archer, A. P. Ginsberg, and A. Rosencwaig, Contraception, $\underline{15}$, 93 (1977).

9. A. Rosencwaig, A. P. Ginsberg, and J. W. Koepke, Inorg. Chem., $\underline{15}$, 2540 (1976).

10. M. J. Adams and G. F. Kirkbright, Analyst, 102, 281 (1977).

11. M. J. Adams, B. C. Beadle, A. A. King, and G. F. Kirkbright, Analyst, 101, 553 (1976).

12. S. D. Campbell, S. S. Yee, and M. A. Afromowitz, J. Bioengineering, 1, 185 (1977).

13. A. Rosencwaig and E. Pines, Biochim. Biophys. Acta, 493, 10 (1977).

14. R. C. Gray and A. J. Bard, Anal. Chem., 50, 1262 (1978).

15. M. B. Robin and N. A. Kuebler, J. Am. Chem. Soc., 97, 4822 (1975); L. A. Farrow and R. E. Richton, J. Appl. Phys., 48, 4962 (1977).

16. R. W. Karasek, Research/Development, Sept., 1977, p. 38.

17. A. G. Marshall and M. B. Comisarow, Anal. Chem., 47, 491A (1975); M. Harwit in *Transform Techniques in Chemistry*, P. R. Griffiths, ed., Plenum Press, New York, N.Y., 1978, Chapter 7.

18. A. J. Bard, private communication.

19. W. K. Yuen and G. Horlick, Anal. Chem., 49, 1446 (1977).

20. M. M. Farrow, R. K. Burnham, and E. M. Eyring, Appl. Phys. Lett., 33, 735 (1978).

21. J. Chamberlain, Infrared Phys., 11, 25 (1971).

22. V. F. Fairbanks in *Fundamentals of Clinical Chemistry*, N. Tietz, ed., Saunders, Philadelphia, PA, 1976, p. 401.

23. M. A. Lichtman, M. S. Murphy, and T. W. Adamson, Ann. Int. Med., 84, 517 (1976).

24. G. Gros and W. Moll, Pflügers Arch., 324, 249 (1971).

25. H. Kutchai, Resp. Physiol., 23, 121 (1975).

26. A. Rosencwaig and A. Gersho, Science, 190, 556 (1975).

A TEXTBOOK EXAMPLE OF COMPLEX KINETICS

Clauae F. Bernasconi and Hsien-chang Wang

Thimann Laboratories, University of California,
Santa Cruz, California 95064

Abstract. A kinetic scheme is analyzed which includes a number of
important features and special cases which are of interest to the
kineticist dealing with complex equilibrating systems. These
include (a) coupling of a slow reaction to a fast equilibrium,
(b) coupling of two reactions which equilibrate at similar rates,
(c) designing experiments which render a relaxation effect invis-
ible (zero amplitude) so that the remaining relaxation times can
be more easily evaluated, (d) evaluating two overlapping relaxation times by taking advantage of an isosbestic point, and (e)
converting an equilibrium reaction into an irreversible process
by "instantaneous" removal of one of the reagents, thereby making
an inaccessible rate constant measurable. Furthermore, our system permits a systematic discussion of situations where certain
reactions would escape kinetic detection even if they were present.

INTRODUCTION

We have been interested in the reactions of nucleophiles
with nitro-activated aromatics for some time (1). These reactions lead typically to complexes known as anionic σ-complexes
or Meisenheimer complexes (2, 3). One of our main purposes was
to use these reactions as models for elementary steps in
nucleophilic aromatic substitution and draw mechanistic conclusions from them (1, 4).

In some cases one observes the addition of two molecules of
nucleophile to the nitroaromatic, to form a 1:2 complex. A
particularly interesting aspect of the 1:2 complexes is the
possibility of detecting cis-trans isomerism. Such isomerism
has in fact been observed in the addition of sulfite ion to

175

W. J. Gettins and E. Wyn-Jones (Eds.). Techniques and Applications of Fast Reactions in Solution. 175–186.
Copyright © 1979 by D. Reidel Publishing Company.

Scheme I

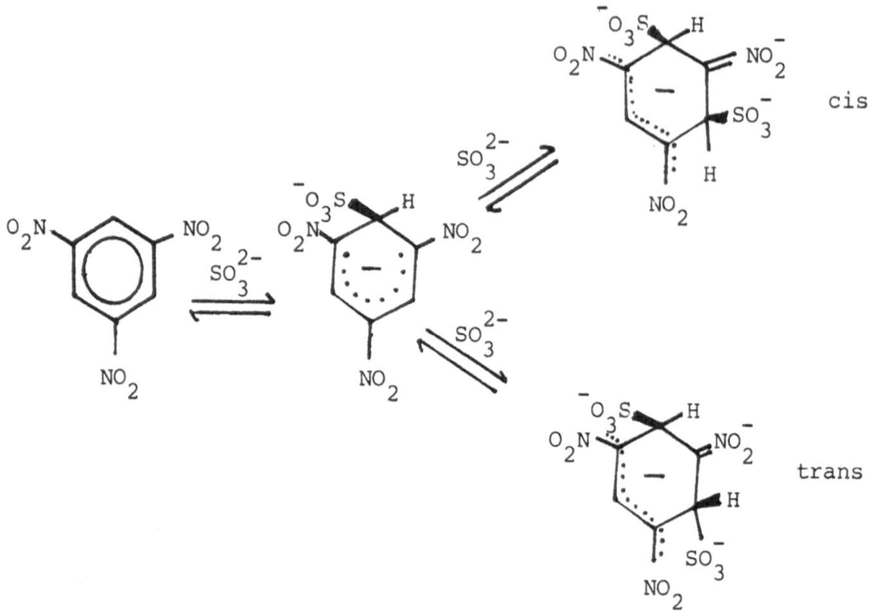

1,3,5-trinitrobenzene in aqueous solution (Scheme I) (5-7).
This system proved to be ideal for the detection of cis-trans
isomerism because (a) the trans isomer forms about 200 times
faster than the cis isomer, thus giving rise to two easily
detectable separate relaxation times for 1:2 complex formation
(5), (b) the thermodynamic stabilities of the cis and trans
isomer are virtually identical (5), which allows the detection,
by pmr spectroscopy, of both isomers after the system has reached
its final equilibrium (6, 7).

The search for cis-trans isomerism in other systems such
as the reaction of 1,3,5-trinitrobenzene with alkoxide ions
(8) or hydroxide ion (8), and the reaction of 1-X-2,4,6-tri-
nitrobenzene with sulfite ion where X = OMe, O⁻, NH_2, NHMe,
NMe_2 (9) has been unsuccessful. In continuing our search for
this kind of isomerism we decided to study the reaction of
sulfite ion with the spiro Meisenheimer complex which is formed
by base catalyzed cyclization of N-methyl-N-β-hydroxyethyl
picramide (10). As indicated in Scheme II, the two complexes
MS' and MS" formed by sulfite ion attack on M are geometric
isomers in a similar way as the cis and trans isomers in Scheme
I.

Since M is in equilibrium with its precursor, N-methyl-N-
β-hydroxyethyl picramide (P), competing sulfite ion attack on
P, to form PS and PS_2, had to be taken into account. Thus

Scheme II

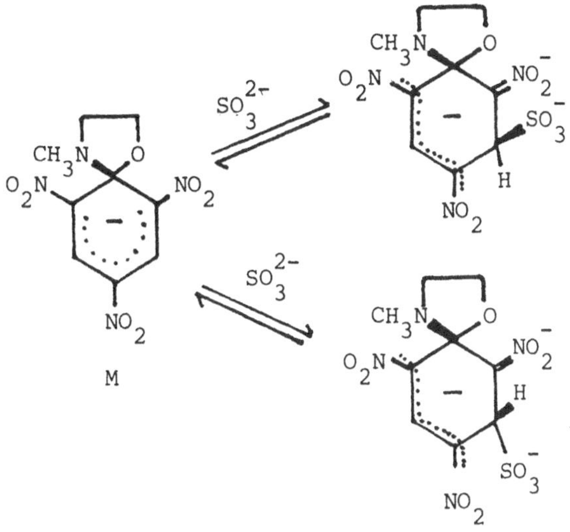

the various possible processes can be summarized in Scheme III.

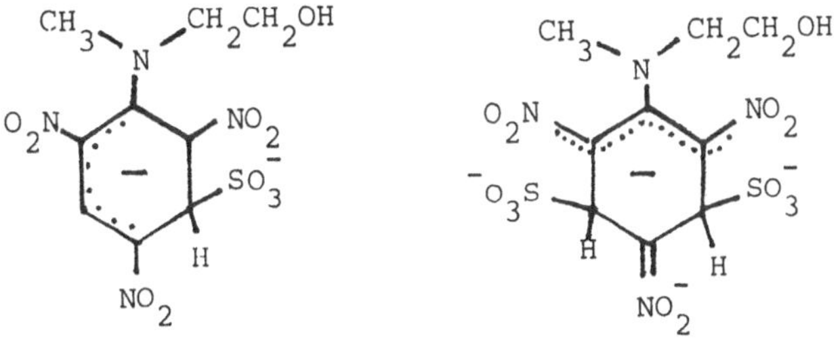

PS   PS$_2$ (cis and trans)

Note that, for the sake of simplicity, Scheme III does not
show the geometric isomers of PS$_2$ and MS because, as discussed
below, no evidence for isomerism could be found. Note also
that MS can be formed either by sulfite ion attack on M or by
cyclization of PS. Finally it should be realized that even
though the cyclization reactions are represented by one rate
constant only (k$_5$ and k$_4$, respectively), the reactions involve
a fast deprotonation equilibrium step followed by rate limiting
intramolecular nucleophilic attack (10).

Scheme III

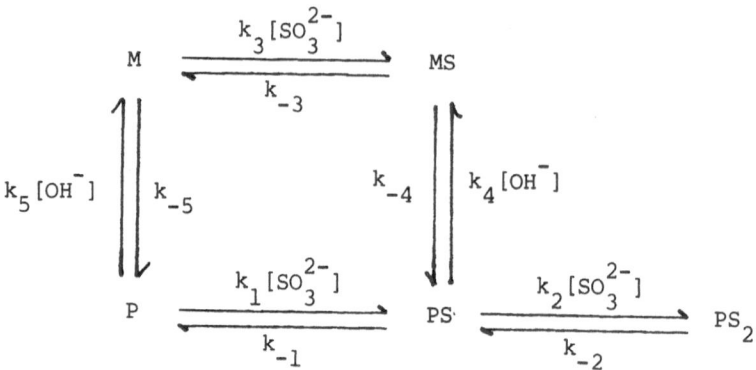

The kinetic analysis of Scheme III which leads to the evaluation of all 10 rate constants is the main topic of this paper. Although the scheme is kinetically fairly complex there are several features which, combined with an intelligent approach, make this a textbook example of complex kinetics (11).

RESULTS AND DISCUSSION

A. Experiments in the Absence of Sulfite Ions

In the absence of sulfite ion Scheme III reduces to

$$P \xrightarrow[\quad k_{-5} \quad]{\quad k_5 [OH^-] \quad} M \tag{1}$$

Since P and M have strongly different absorption spectra (10), the kinetics of reaction 1 could be followed spectrophotometrically in the stopped-flow spectrophotometer. Under pseudo first order conditions the relaxation time is given by

$$\tau_5^{-1} = k_5 [OH^-] + k_{-5} \tag{2}$$

where $k_5 = 1.30 \times 10^3 \ M^{-1} \ sec^{-1}$ and $k_{-5} = 3.50 \times 10^{-2} \ sec^{-1}$ were easily evaluated from a plot of τ_5^{-1} vs hydroxide ion concentration, as described in more detail elsewhere (10).

B. Sulfite Ion Addition in Weakly Basic Solution

If a weakly basic solution (pH 8.67) of P is mixed with a sulfite solution at the same pH one observes all the four relaxation times which characterize a five state system (12a)

such as Scheme III. Since three of the four relaxation times
overlap considerably it would be virtually impossible to determine them accurately. The three overlapping relaxation times
are the ones associated with the P $\rightleftharpoons$ PS $\rightleftharpoons$ PS$_2$ (τ_1 and τ_2) and
M $\rightleftharpoons$ MS (τ_3) processes. There is, however, a procedure by which
the amplitude of τ_3 can be reduced to zero. This procedure
takes advantage of the fact that even though there is a significant <u>equilibrium</u> concentration of M in a pH 8.67 solution of P,
the <u>rate</u> of formation of M from P and of MS from PS is slow
compared to that of the formation of PS and PS$_2$ at typical sulfite ion concentrations used in our study. Thus, if one mixes,
in the stopped-flow apparatus, a pH 6 solution of P (which does
<u>not</u> contain a significant equilibrium concentration of M) with a
buffered pH $\sim$ 8.7 sulfite solution (giving a pH 8.67 after mixing), one observes only the two relaxation times (τ_1 and τ_2)
associated with reaction 3. The values of τ_1^{-1} and τ_2^{-1}, deter-

$$P \xrightarrow[k_{-1}]{k_1 [SO_3^{-2}]} PS \xrightarrow[k_{-2}]{k_2 [SO_3^{-2}]} PS_2 \tag{3}$$

mined under pseudo first order conditions, have been reported
elsewhere (11) as a function of sulfite ion concentration. τ_1^{-1}
is only a factor of about 3 larger than τ_2^{-1} which indicates
that the rate of equilibration of the reaction P $\rightleftharpoons$ PS is similar to that of the reaction PS $\rightleftharpoons$ PS$_2$.

   In general it is difficult to evaluate two closely overlapping relaxation times unless conditions can be found where
the amplitude of one of them is very small or zero. In the
present case τ_1 could in fact be observed with almost no interference by τ_2 (negligible amplitude) by monitoring the reaction
at 420 nm which is close to the isosbestic point of PS and PS$_2$.
τ_2 was monitored at 470 nm where the extinction coefficiencts
of PS and PS$_2$ differ strongly; interference by τ_1 could be
minimized by evaluating only the last portion of the relaxation
curve.

   Due to the similar rates of the two reactions in eq 3 the
expressions for τ_1^{-1} and τ_2^{-1} become very complex and the evaluation of the individual rate constants has to be made from eq
4 and 5 (12a). A plot of $\tau_1^{-1} + \tau_2^{-1}$ <u>vs</u> [SO$_3^{2-}$] is linear as

$$\tau_1^{-1} + \tau_2^{-1} = (k_1 + k_2)[SO_3^{2-}] + k_{-1} + k_{-2} \tag{4}$$

$$\tau_1^{-1} \cdot \tau_2^{-1} = k_1 k_2 [SO_3^{2-}]^2 + k_1 k_{-2}[SO_3^{2-}] + k_{-1}k_{-2} \tag{5}$$

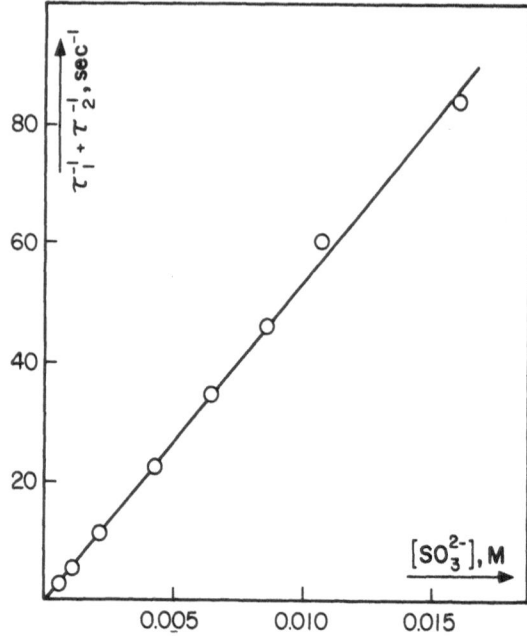

Figure 1. $\tau_1^{-1} + \tau_2^{-2}$ vs $[SO_3^{2-}]$, eq 4.

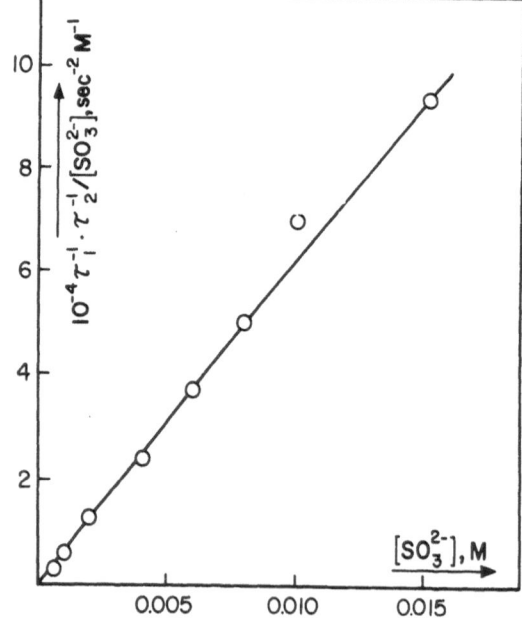

Figure 2. $\tau_1^{-1} \cdot \tau_2^{-1}[SO_3^{2-}]^{-1}$ vs $[SO_3^{2-}]$, eq 6.

shown in Fig. 1. A plot of $\tau_1^{-1} \cdot \tau_2^{-1}$ vs $[SO_3^{2-}]$ (not shown) is curved; it has a negligible intercept, hence $k_{-1}k_{-2} \approx 0$ and eq 5 can be rearranged to give

$$\tau_1^{-1} \cdot \tau_2^{-1} [SO_3^{2-}]^{-1} = k_1 k_2 [SO_3^{2-}] + k_1 k_{-2} \tag{6}$$

The linear plot according to eq 6 is shown in Fig. 2.

From the slopes of Fig. 1 and 2 one obtains $k_1 + k_2 = 5.60 \times 10^3$ M^{-1} sec^{-1} and $k_1 k_2 = 6.30 \times 10^6$ M^{-2} sec^{-2}, respectively. Solving the two simultaneous equations affords two sets of k_1 and k_2, viz. (1) $k_1 = 4.04 \times 10^3$ M^{-1} sec^{-1} and $k_2 = 1.56 \times 10^3$ M^{-1} sec^{-1}, or (2) $k_1 = 1.56 \times 10^3$ M^{-1} sec^{-1} and $k_2 = 4.04 \times 10^3$ M^{-1} sec^{-1}. We prefer the first set (11) based on a comparison of k_1 with the rate constant for sulfite ion attack on N,N-dimethylpicramide (13) which would be expected to have virtually the same reactivity as P.

From the intercept of Fig. 1 one obtains $k_{-1} + k_{-2} = 0.15$ sec^{-1}; the intercept of Fig. 2, $k_1 k_{-2}$, is indistinguishable from zero which indicates that k_{-2} must be very small and in fact much smaller than k_{-1}, so that we have $k_{-1} + k_{-2} \cong k_{-1} = 0.15$ sec^{-1}. k_{-2} can be obtained as follows. To an equilibrated solution, which contained a high enough sulfite ion concentration so that most of P was in the form of PS_2, was added an excess of H_2O_2 which very rapidly oxidizes all the sulfite ion. This effectively converts the equilibrium reaction 3 into the irreversible reaction 7, with k_{-2} as the rate determining step.

$$P \xleftarrow{\quad k_{-1} \quad} PS \xleftarrow{\quad k_{-2} \quad} PS_2 \tag{7}$$

The decay of PS_2 could be followed in a conventional spectrophotometer and gave a $k_{-2} = 4.8 \times 10^{-3}$ sec^{-1}.

C. Sulfite Ion Addition in Strongly Basic Solution

At high pH P is converted to >99% into M at equilibrium. When such a solution is mixed with a basic sulfite solution in the stopped-flow apparatus, two relaxation times (τ_3 and τ_4) are observed. The concentration dependence of τ_3^{-1} (Fig. 3) as well as the fact that its amplitude is largest around 420–430 nm where the difference in the extinction coefficients of M and MS is largest (11) is consistent with reaction 8

$$M + SO_3^{2-} \underset{k_{-3}}{\overset{k_3}{\rightleftharpoons}} MS \tag{8}$$

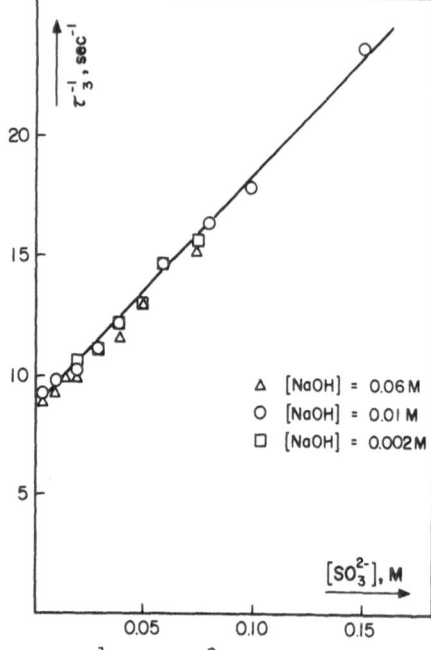

Figure 3. τ_3^{-1} vs $[SO_3^{2-}]$

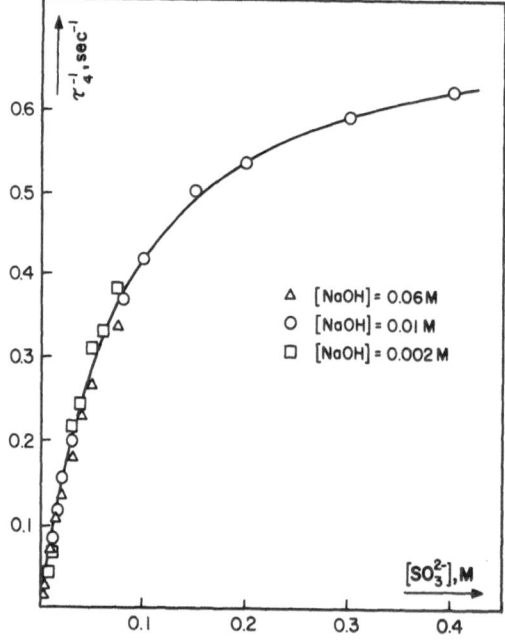

Figure 4. τ_4^{-1} vs $[SO_3^{2-}]$

with

$$\tau_3^{-1} = k_3 [SO_3^{2-}] + k_{-3} \tag{9}$$

As can be seen from Fig. 3 τ_3^{-1} is independent of base concentration which shows that there is no coupling with the reaction $P \rightleftharpoons M$ (since equilibrium greatly favors M over P), and no coupling with reaction $MS \rightleftharpoons PS$, indicating that the $MS \rightleftharpoons PS$ reaction is much slower. Evaluation of $k_3 = 94.0$ M^{-1} sec^{-1} and $k_{-3} = 8.45$ sec^{-1} from slope and intercept of Fig. 3 is straightforward.

The process associated with τ_4 must then be attributed to the reaction $MS \rightleftharpoons PS$ with the reactions $M \rightleftharpoons MS$ and $PS \rightleftharpoons PS_2$ acting as rapid pre- and postequilibria, respectively; the equilibrium position of reaction $P \rightleftharpoons PS$ strongly favors PS so that this reaction, even though it is a rapid postequilibrium, does not need to be considered. Thus τ_4^{-1} is given by

$$\tau_4^{-1} = \frac{K_3 k_{-4} [SO_3^{2-}]}{1 + K_3 [SO_3^{2-}]} + \frac{k_4 [OH^-]}{1 + K_2 [SO_3^{2-}]} \tag{10}$$

where K_3 and K_2 are the equilibrium constants of the $M \rightleftharpoons MS$ and $PS \rightleftharpoons PS_2$ reactions, respectively.

The concentration dependence of τ_4^{-1} is shown in Fig. 4. τ_4^{-1} is seen to be independent of base concentration and the intercept is indistinguishable from zero. This shows that the second term on the right side of eq 10 is negligible. Inverting the remainder of eq 10 then provides

$$\tau_4 = (k_{-4})^{-1} + (K_3 k_{-4} [SO_3^{2-}])^{-1} \tag{11}$$

A plot of τ_4 vs $[SO_3^{2-}]^{-1}$ (not shown) is a straight line from which $k_{-4} = 0.71$ sec^{-1} and $K_3 = 12.5$ M^{-1} are obtained. The agreement with $K_3 = k_3/k_{-3} = 11.1$ M^{-1} determined from τ_3 is excellent.

k_4, the only remaining unknown, can now be calculated as $k_4 = K_4 k_{-4} = 13.0$ M^{-1} sec^{-1} with $K_4 = (K_1)^{-1} K_5 K_3 = 16.7$ M^{-1}.

Some additional kinetic experiments, under conditions where PS and MS are steady state intermediates, were performed with results in agreement with the foregoing analysis (11).

The Question of Geometric Isomers for PS_2 and MS

We have no kinetic evidence, in the form of additional relaxation times, for such isomerism. Pmr experiments on equilibrated solutions did not indicate the presence of isomers either (11).

Let us briefly discuss possible reasons why no such isomerism was observed. We shall restrict ourselves to some qualitative arguments as they apply to PS_2; a more elaborate and quantitative discussion relating to both PS_2 and MS has been presented elsewhere (11).

There are essentially four possible different situations which might prevail.

Case I. One isomer is strongly favored over the other, both thermodynamically and kinetically, so that under no experimental conditions there is ever a detectable concentration of the disfavored isomer present. This would be the most trivial explanation why no isomer was detected, either kinetically or by pmr spectroscopy.

Case II. One isomer is thermodynamically strongly favored over the other but forms at a comparable rate. In such a situation there could not be any pmr evidence but there should be an additional relaxation time which essentially refers to the reaction $PS_2(cis) \rightleftharpoons PS_2(trans)$ via PS as steady state intermediate. This additional relaxation time would however only be detectable if the extinction coefficients of $PS_2(cis)$ and $PS_2(trans)$ were sufficiently different from each other as to produce a significant relaxation amplitude (12b). In view of the expected similarity of the spectra of the two isomers this may not be the case.

Case III. The isomers are of comparable thermodynamic stability but one forms faster than the other. This is similar to the reaction of 1,3,5-trinitrobenzene with sulfite ion (5) mentioned earlier where there was both kinetic and pmr evidence for the two isomers. It should be noted that even if the extinction coefficients of the two isomers were identical, an additional relaxation time must be observed in the H_2O_2 experiments (eq 7). Thus our results definitely exclude case III.

Case IV. The rate constant k_{-2} is the same for both isomers which may or may not have similar thermodynamic stabilities. In this case the additional relaxation time which, as in case II, corresponds to the reaction $PS_2(cis) \rightleftharpoons PS_2(trans)$ via PS as steady state intermediate, would have zero amplitude regardless of the extinction coefficients of the two isomers (12b). This is because the normal coordinate of this process is zero, due to the fact that, from the outset, the isomers are formed in concentrations equal to their final equilibrium values, thus making the subsequent equilibration $PS_2(cis) \rightleftharpoons PS_2(trans)$ redundant (12b). The H_2O_2 experiment would not reveal the presence of two isomers either because the two species decompose at the same rate, giving rise to a single exponential relaxation curve.

In conclusion: case III is definitely excluded, whereas cases I, II and IV are possible. Absence of pmr evidence make case IV for the situation where both isomers are of comparable thermodynamic stability unlikely.

REFERENCES AND NOTES

1. For a recent review, see Bernasconi, C. F.: 1978, Acc. Chem. Res. 11, p. 147.

2. Crampton, M. R.: 1969, Adv. Phys. Org. Chem. 7, p. 211.

3. Strauss, M. J.: 1970, Chem. Rev. 70, p. 667.

4. Bernasconi, C. F.: 1975, in "Chemical and Biological Applications of Relaxation Spectrometry," Wyn-Jones (ed.), Reidel, Dordrecht-Holland, p. 343.

5. Bernasconi, C. F. and Bergstrom, R. G.: 1973, J. Am. Chem. Soc. 95, p. 3603.

6. Crampton, M. R. and Willison, M. J.: 1973, J. Chem. Soc., Chem. Comm., p. 215.

7. Strauss, M. J. and Taylor, S. P. B.: 1973, J. Am. Chem. Soc., 95, p. 3813.

8. Bernasonconi, C. F. and Bergstrom, R. G.: 1974, J. Am. Chem. Soc., 96, p. 2397.

9. Crampton, M. R. and Willison, M. J.: 1976, J. Chem. Soc. Perkin 2, p. 160.

10. Bernasconi, C. F., Gehriger, C. L., and de Rossi, R. H.: 1976, J. Am. Chem. Soc. 98, p. 8451.

11. For a more detailed account of the results see Bernasconi, C. F. and Wang, H.-C.: 1979, Int. J. Chem. Kin., in press.

12. Bernasconi, C. F.: 1976, Relaxation Kinetics, Academic
     Press, New York, (a) Chapter 3; (b) Chapter 7.3.

13. Buncel, E., Norris, A. R., Russell, K. E., and Sheridan,
     P. J.: 1974, Can. J. Chem. 52, p. 25.

THERMODYNAMICS OF RELAXATION METHODS

Z. A. Schelly

Department of Chemistry
University of Texas at Arlington, TX   76019, USA

Using the framework of irreversible thermodynamics, the driving force of chemical relaxation and the thermodynamic meaning of the relaxation time τ are investigated.

INTRODUCTION

In discussing the thermodynamics of relaxation methods, the major aspects of consideration are usually i) the mode of perturbation of chemical equilibria (forcing parameters), ii) the enforced change of the concentrations (relaxation amplitudes) and iii) the chemical contributions to equations of state. In the present paper, after reviewing the pertinent basic definitions and forcing parameters, the driving force of relaxation and the thermodynamic meaning of the relaxation time τ are investigated, using the framework of irreversible thermodynamics.

DEFINITIONS AND PERTURBATION OF EQUILIBRIA

Consider the reaction

$$\nu_1 A_1 + \nu_2 A_2 + \cdots \rightleftharpoons \nu_i A_i + \nu_{i+1} A_{i+1} + \cdots \tag{1}$$

where the stoichiometric coefficients ν_i are positive for products and negative for reactants. Due to conservation of matter during chemical reactions, eq. (1) can also be written as $\sum_i \nu_i A_i = 0$.

W. J. Gettins and E. Wyn-Jones (Eds.). Techniques and Applications of Fast Reactions in Solution. 187–193.
Copyright © 1979 by D. Reidel Publishing Company.

The extent of reaction ξ at any time is defined according to De Donder by eq. (2)

$$n_i = {}_o n_i + \nu_i \xi \qquad (2)$$

where ${}_o n_i$ are the reference values of the mole numbers n_i present for each species A_i. The equilibrium constant and the rate of reaction (1) are defined as $K = \Pi a_i^{\nu_i}$ and rate $\equiv d\xi/dt$, respectively, where a_i are the equilibrium values of the activities of A_i.

If energy (in the form of heat or work) is exchanged reversibly between the system and the surrounding, the change of internal energy is given by the first law of thermodynamics

$$dU = \sum_j F_j d\Theta_j \qquad (3)$$

where F_j are intensive variables (generalized forces) and $d\Theta_j$ are differentials of extensive variables (generalized displacements). Examples for these conjugate variables are the following:

$$
\begin{array}{ll}
T & dS \\
P & dV \\
E & dD^* \\
\mu_i & dn_i \\
A & d\xi \\
H & dM
\end{array} \qquad (4)
$$

In addition to the familiar thermodynamic symbols (T, S, P, V, n_i, ξ), E stands for electric field strength, D^* for electric displacement (1), μ_i for chemical potential, A for affinity (defined by De Donder: see eq. 20), H for magnetic field strength and M for magnetization.

Due to the sum in eq. 3, indicating that there are several ways in which work can be done on the system, the differential of the Gibbs free enthalpy G becomes

$$dG = \sum_j \Theta_j dF_j - A d\xi \qquad (5)$$

Since the equilibrium constant K and ΔG^o of reaction (1) are related by $\Delta G^o = -RT \ln K$, the (usually sudden) change of any of the conjugate variables F_j or Θ_j may affect the equilibrium activities, i.e. may shift the equilibrium. The relaxation time τ of the shift is shorter if during equilibrium the extensive variables Θ_j instead of the intensive variables F_j are kept constant (2). Nevertheless, experimentally it is easier to vary the intensive variables F_j (see list (4)) which led to the development of the T-jump, E-jump, P-jump, solvent (μ_i)-jump, as well as the sound absorption or dispersion (oscillating P and T variation), and dielectric relaxation (oscillating variation of E) methods. Typically, one chooses such a forcing parameter F_j for perturbation to which the equilibrium is sensitive.

DRIVING FORCE OF RELAXATION

Once the value of a forcing parameter has been suddenly changed, the reaction starts to relax irreversibly to new equilibrium concentrations. Let us compare the relaxation with other irreversible processes, for the case where the systems are "not too far" removed from the final equilibrium (linear region).

The second law of thermodynamics ascertains that the entropy of an isolated system increases monotonically until it reaches its maximum at equilibrium,

$$dS/dt \geqslant 0 \qquad\qquad (6)$$

This formulation can easily be extended to open systems which exchange energy and matter with the surrounding. Here dS is composed of two, and only two, terms: dS_e, the transfer of entropy across the boundaries, and dS_i, the entropy produced within the system (3). The second law assumes that $dS_i \geqslant 0$.

If the assumption of "local equilibrium" (4) is accepted, one can obtain for the entropy production per unit time(5)

$$dS_i/dt = \sum_s J_s X_s \geqslant 0 \qquad\qquad (7)$$

where J_s are the rates of various processes s involved (chemical reactions, heat flow, diffusion) and the X_s are the corresponding generalized forces (affinities, gradients of temperature, gradients of chemical potential, etc.).

Near equilibrium, linear homogeneous relations exist between the fluxes J_s and the forces X_s (6). If several processes occur simultaneously, the associated fluxes J_s are given by

$$J_s = \sum_{s'} L_{ss'} X_{s'} \tag{8}$$

with $L_{ss'}$ being the phenomenological coefficients. One important result of linear thermodynamics of irreversible processes is the Onsager reciprocity principle (7) which states that the coefficients for coupled processes are symmetrical (if the fluxes J_s are rates of changes of state functions):

$$L_{ss'} = L_{s's} \tag{9}$$

In other words, if the flux J_s corresponding to the irreversible process s is influenced by the force $X_{s'}$ of another irreversible process s', then the flux $J_{s'}$ is also influenced by the force X_s through the same coefficient.

If only one irreversible process occurs in a system, $J = LX$ describes the flow. Well known examples in the linear region are the laws of heat conduction

$$J = -\kappa_T \text{ grad } T \qquad\qquad \text{Fourier} \tag{10}$$

where κ_T is the coefficient of thermal conduction; electric current J

$$J = \kappa \text{ grad } \phi \qquad\qquad \text{Ohm} \tag{11}$$

with the electric conductivity κ, and electric potential ϕ; fluid flow

$$J = -\gamma \text{ grad } P \qquad\qquad \text{Poiseuille} \tag{12}$$

where γ is the friction coefficient; and diffusion

$$J = -D \text{ grad } C \qquad\qquad \text{Fick} \tag{13}$$

with the diffusion coefficient D and concentration C.

In analogy with the physical processes described in (10) through (13), next we shall consider chemical reactions, where the rate of the reaction is the flux $J \equiv d\xi/dt$. We shall also identify the phenomenological coefficient L and the driving force X. Let us use an elementary reaction.

$$A + B \underset{k_{-1}}{\overset{k_1}{\rightleftharpoons}} C \tag{14}$$

for the derivation (3). The rate of reaction (i.e. the rate of equilibration or relaxation) is

$$dC_C/dt = k_1 C_A C_B - k_{-1} C_C \tag{15}$$

Since $C_i = n_i/V$, it follows from eq. (2) that $dC_i = \nu_i d\xi/V$. Hence eq. (15) can be written as

$$d\xi/Vdt = k_1 C_A C_B (1 - k_{-1} C_C / k_1 C_A C_B) \tag{16}$$

if the forward rate is factored out on the right hand side of eq. (15). But $k_{-1}/k_1 = K^{-1}$, and the non-equilibrium ratio of the concentrations $C_C/C_A C_B$ can be called Q. Since near equilibrium the forward rate will approach the forward equilibrium rate v_1

$$k_1 C_A C_B \longrightarrow k_1 \bar{C}_A \bar{C}_B \equiv v_1 \tag{17}$$

near equilibrium (16) becomes

$$d\xi/dt = Vv_1 (1 - Q/K) \tag{18}$$

Now, Q/K can be related to A, if one considers the total differential of G:

$$dG = \Sigma(\partial G/\partial n_i)dn_i = \Sigma\mu_i dn_i = \Sigma\nu_i\mu_i d\xi \tag{19}$$

where the last equality follows from eq. (2). From eq. (19), the differential quotient $dG/d\xi$ is

$$dG/d\xi = \Sigma\nu_i\mu_i = \Delta G \equiv -A \tag{20}$$

which is the definition of A. But $\Delta G = \Delta G^o + RT \ln Q$
$\Delta G^o = -RT \ln K$, hence

$$A = -RT \ln(Q/K) \quad \text{or} \quad Q/K = \exp(-A/RT). \tag{21}$$

Near equilibrium $A/RT \ll 1$, therefore $\exp(-A/RT)$ can be expanded into a series and terminated after the second term, resulting in $(1-A/RT)$. Substituting this for Q/K in eq. (18), we have

$$d\xi/dt = (Vv_1/RT)A \; . \tag{22}$$

Thus, L can be identified as Vv_1/RT and the driving force of relaxation as the affinity, $X = A$.

RELAXATION TIME

Next we shall focus on the time constant τ of the relaxation close to equilibrium. As a starting point, we write eq. (22) in the general form as

$$d\xi(T,P)/dt = \bar{\varepsilon} \, A(T,P,\xi) \tag{23}$$

indicating the functional dependencies of the variables. Also $\bar{\varepsilon}$ is a function of T,P and ξ, but close to equilibrium it can be considered as constant. If A is expanded in a Taylor series about its equilibrium value at constant T, and only the first order terms are kept, we have

$$A = A(\bar{\xi}, \; \bar{P}) + (\partial A/\partial \xi)_{T,P}(\xi - \bar{\xi}) + (\partial A/\partial \xi)_{T,\xi}(P - \bar{P}) \tag{24}$$

where the bars indicate equilibrium values. But at equilibrium the affinity is zero, $A(\bar{\xi}, \; \bar{P}) = 0$, and close to equilibrium $(\xi - \bar{\xi}) = d\xi$ and $(P - \bar{P}) = dP$. With these results, eq. (24) substituted into (23) yields

$$d\xi/dt = \bar{\varepsilon}\left[(\partial A/\partial \xi)_{T,P}d\xi + (\partial A/\partial P)_{T,\xi}dP\right] \tag{25}$$

or

$$= -\bar{\varepsilon}(\partial A/\partial \xi)_{T,P}\left[-d\xi - \frac{(\partial A/\partial P)_{T,\xi}}{(\partial A/\partial \xi)_{T,P}} \, dP\right] \tag{26}$$

The negative quotient of the partial differential quotients in the brackets is equal to $(\partial \xi/\partial P)_{T,A}$. However, constant A (as indicated by the subscript) means equilibrium conditions, with the equilibrium value of ξ. Therefore,

$$(\partial \xi/\partial P)_{T,A} \equiv (\partial \bar{\xi}/\partial P)_{T,A} = 0 \; , \tag{27}$$

since $\bar{\xi}$ is constant. Thus, eq. (26) becomes

$$d\xi/dt = -\bar{\epsilon}(\partial A/\partial\xi)_{T,P}(-d\xi) \tag{28}$$

or $\qquad d\xi/dt = -\tau_{T,P}^{-1} \; (\xi-\bar{\xi}) \tag{29}$

with $\tau_{T,P}^{-1} \equiv -\bar{\epsilon}(\partial A/\partial\xi)_{T,P} = (-Vv_1/RT)(\partial A/\partial\xi)_{T,P} \tag{30}$

if $\bar{\epsilon}$ is identified through eqs. (22) and (23).

The final conclusion is reached in the last two equations: i) the rate of approach to equilibrium (relaxation) is directly proportional to the distance $(\xi-\bar{\xi})$ from equilibrium (eq. 29); and ii) the rate coefficient τ^{-1} of the relaxation has a simple physical meaning, described by eq. (30).

Acknowledgement. The author is grateful for financial support by the R. A. Welch Foundation and the Organized Research Fund of The University of Texas at Arlington.

REFERENCES

1.   M. Eigen and L. DeMaeyer in "Techniques of Organic Chemistry", Vol. 8, Part 2, S. L. Friess, E. S. Lewis and A. Weisberger, editors, Interscience Publishers, New York, NY, 1963, p. 895 ff.
2.   R. Schlögl, Z. Physik.Chem., N. F., 9, 259 (1956).
3.   I. Prigogine, "Thermodynamics of Irreversible Processes", 3rd edition, Interscience Publishers, New York, NY, 1967.
4.   I. Prigogine, "Etude thermodynamique des phénomènes irreversibles", thesis, University of Brussels (1945).
5.   I. Prigogine, Science, 201, 777 (1978).
6.   S. R. de Groot and P. Mazur, "Non-Equilibrium Thermodynamics", North-Holland, Amsterdam, 1969.
7.   L. Onsager, Phys. Rev., 37, 405 (1931).

RELAXATION AMPLITUDES IN CONCENTRATION JUMP EXPERIMENTS. APPLICATION TO ANTHOCYANINS.

Raymond Brouillard and Bernard Delaporte

Institut de Topologie et de Dynamique des Systèmes
Université Paris VII - associé au C.N.R.S.,
1, rue Guy de la Brosse, 75005 PARIS - France.

The calculation of relaxation amplitudes by temperature jump is complex, except for very simple systems of the type $A \rightleftharpoons B$ or $A + B \rightleftharpoons C$.[1] This is unlike their calculation by concentration jump, since, in this case, the perturbation occurs at constant T and P, and the equilibrium constants remain unchanged. Moreover, we show in this paper, that by proper use of the concentration jump method, one is able to measure the equilibrium constants of complicated as well as simple chemical systems. Whenever possible, the perturbation should be performed by rapidly modifying the concentration of a species that is characteristic of the type of reaction studied: H^+ or OH^- for proton transfer, nucleophile for nucleophilic addition or substitution, etc... Thus, one can always write $\sum_k [X_k](t) = C_o$, where X_k represents the different chemical forms that can be adopted by the compound studied. C_o is the analytical concentration. To calculate the relaxation amplitudes, it suffices to set up the following relationships: $A_j = [X_i]_j - [X_i]_{j-1}$, where A_j is the amplitude at the jth mode of relaxation, and X_i represents a species involved in the jth mode of relaxation but not in any faster one. When the different steps are also kinetically decoupled, the expressions of A_j become remarkably simple, and it is easy to obtain the thermodynamic parameters of the different equilibria and the parameters concerning the type of detection used.

From an experimental standpoint, the observation of relaxation amplitudes by concentration jump requires only a classical spectrophotometer fitted with a stirred thermostatted cell. With

W. J. Gettins and E. Wyn-Jones (Eds.), Techniques and Applications of Fast Reactions in Solution. 195–199.
Copyright © 1979 by D. Reidel Publishing Company.

this set up, even the amplitudes of very rapid phenomena (whose kinetic study remains, of course, in the domain of specific relaxation techniques) can be obtained.

We now apply the method to the acid-base equilibrium and to the proton transfer, hydration and tautomeric equilibria of anthocyanins.

HOW TO MEASURE THE ACIDITY CONSTANT K'_a BY THE CONCENTRATION JUMP METHOD

Scheme I

$$AH \rightleftharpoons A^- + H^+ \qquad K'_a = ([A^-]/[AH]) \, a_{H^+}$$

It is important that the solution should be initially equilibrated at a given pH value ($pH_o = -\log a^o_{H^+}$). Thus $K'_a = ([A^-]_o/[AH]_o) a^o_{H^+}$.

$[AH]_o$ and $[A^-]_o$ are the equilibrium concentrations of the acid AH and its conjugate base A^- at pH_o, respectively. The pH jump is achieved by injecting into the solution a very small quantity of a more or less concentrated acidic or neutral solution. During the whole experiment the temperature is carefully kept at a constant value. The pH jump is from pH_o $(a^o_{H^+})$ to pH_f $(a^f_{H^+})$. Due to the effect of this perturbation the new equilibrium concentrations of AH and A^- are $[AH]_f$ and $[A^-]_f$ respectively. The theoretical amplitude of the relaxation process is simply given by :

$$A = [AH]_o - [AH]_f = -([A^-]_o - [A^-]_f)$$

The observable spectroscopic amplitude is:

$$D_o - D_f = \{(\varepsilon_{AH} [AH]_o + \varepsilon_A [A^-]_o) - (\varepsilon_{AH} [AH]_f + \varepsilon_A [A^-]_f)\} \ell$$

$$= (\varepsilon_{AH} - \varepsilon_A) . \ell . A$$

where D_o and D_f are the initial and final absorptions of the solution, respectively. ε_{AH} and ε_A are the molecular extinction coefficients of the acid and the base, respectively. It is easy to calculate A as a function of K'_a, C_o, $a^o_{H^+}$ and $a^f_{H^+}$.

Since $K'_a = ([A^-]_o/[AH]_o) a^o_{H^+} = ([A^-]_f / [AH]_f) a^f_{H^+}$ and

$C_o = [\bar{A}]_o + [AH]_o = [\bar{A}]_f + [AH]_f$ (no dilution), one obtains:

$$A = \frac{K_a' \, (a_{H^+}^o - a_{H^+}^f) \, C_o}{(K_a' + a_{H^+}^o)(K_a' + a_{H^+}^f)}$$

The observable spectroscopic amplitude becomes:

$$D_o - D_f = (\varepsilon_{AH} - \varepsilon_A) \, \ell \, \frac{K_a' \, (a_{H^+}^o - a_{H^+}^f) \, C_o}{(K_a' + a_{H^+}^o)(K_a' + a_{H^+}^f)}$$

This is the usual form for a relaxation amplitude expression. In particular $D_o - D_f$ may be positive, negative or nil, depending on the respective values of ε_{AH}, ε_A on one hand, and on the increasing or decreasing acidity of the solution on the other hand. Simplifications may lead to an easy and reliable determination of K_a'. For instance, assuming that the following conditions apply, $a_{H^+}^o \ll K_a'$ and $a_{H^+}^f \simeq K_a'$, one obtains:

$$\frac{- (\varepsilon_{AH} - \varepsilon_A).\ell.C_o - (D_o - D_f)}{D_o - D_f} = \frac{K_a'}{a_{H^+}^f}$$

or,

$$\log \left\{ \frac{- (\varepsilon_{AH} - \varepsilon_A).\ell.C_o - (D_o - D_f)}{D_o - D_f} \right\} + pK_a' = pH_f$$

$- (\varepsilon_{AH} - \varepsilon_A).\ell.C_o$ being equal to $D_o - D_f$ for a jump from $a_{H^+}^o \ll K_a'$ to $a_{H^+}^f \gg K_a'$. The term between brackets should always be positive. This is easy to demonstrate. If $\Delta\varepsilon = (\varepsilon_{AH} - \varepsilon_A) > 0$ and $a_{H^+}^f > a_{H^+}^o$, $D_o - D_f < 0$. Since $|- (\varepsilon_{AH} - \varepsilon_A).\ell.C_o| > |- (D_o - D_f)|$, one obtains:

$$\frac{- (\varepsilon_{AH} - \varepsilon_A).\ell.C_o - (D_o - D_f)}{D_o - D_f} > 0$$

Of course, the same inequality holds for $\Delta\varepsilon < 0$. In the case where $\Delta\varepsilon = 0$ (isosbestic point), the amplitude is nil whatever the pH jump is.

APPLICATION TO ANTHOCYANINS

Anthocyanins equilibrate in aqueous acidic solutions via the mechanism:[2]

Scheme II

$$AH^+ \underset{\rightleftharpoons}{\overset{-H^+}{}} A \qquad \text{proton transfer } K'_a \simeq 10^{-4}M; \ \tau \ \simeq 10^{-5}s$$

$$AH^+ \underset{\rightleftharpoons}{\overset{+H_2O/-H^+}{}} B \quad \text{hydration} \qquad K'_h \simeq 10^{-2}M; \ \tau \ \simeq 10 \ S$$

$$B \ \rightleftharpoons \ C \qquad \text{tautomerism} \qquad K_T < 1 \ ; \qquad \tau \ \simeq 10^3 \ S$$

The same principles as above apply to scheme II and one obtains for the observable spectroscopic relaxation amplitudes:[3]

$$D_0 - D_1 = \frac{-(\varepsilon_{AH^+} - \varepsilon_A)K'_a(a^f_{H^+} - a^o_{H^+}).\ell.C_o}{\delta_o(K'_a + a^f_{H^+})}$$

$$D_1 - D_2 = \frac{-(\varepsilon_{AH^+}a^f_{H^+} + \varepsilon_A K'_a)K'_h(a^f_{H^+} - a^o_{H^+}).\ell.C_o}{\delta_o(K'_a + a^f_{H^+})(K'_a + K'_h + a^f_{H^+})}$$

$$D_2 - D_f = \frac{-(\varepsilon_{AH^+}a^f_{H^+} + \varepsilon_A K'_a)K'_h K_T(a^f_{H^+} - a^o_{H^+}).\ell.C_o}{\delta_o \delta_f (K'_a + K'_h + a^f_{H^+})}$$

where : $\delta_o = K'_a + K'_h(1 + K_T) + a^o_{H^+}$

$\delta_f = K'_a + K'_h(1 + K_T) + a^f_{H^+}$

All of the thermodynamic constants $(K'_a, K'_h$ and $K_T)$, as well as

the values of ε_i (λ) and, consequently, the spectrum of each species (even those which cannot be isolated) can be calculated from the spectroscopic amplitudes measured by pH-jump. These amplitudes can be further simplified when the pH-jump conditions are carefully chosen. Beyond its simplicity, this method has the advantage of being very precise; we have shown, in the case of anthocyanins,[4] that even when large perturbations are induced by pH-jump, the linearization conditions of the rate equations are satisfied (in agreement with Bernasconi's remarks, see ref.1). We have therefore brought about large variations in the concentrations of species in equilibrium by this technique, and thereby observed large spectroscopic amplitudes which correspond to excellent signal-to-noise ratios.

REFERENCES

1. M. Eigen and L. De Maeyer, "Technique of Organic Chemistry", Vol.8, A. Weissberger, Ed., Interscience, New York, N.Y., 1963; R. Winkler, Doctoral Dissertation, Max Planck Inst., Göttingen, 1969; G. Czerlinski, "Chemical Relaxation", Dekker, New York, N.Y., 1966; D. Thusius, "Chemical and Biological Applications of Relaxation Spectrometry", ed., E. Wyn-Jones, Reidel, Dordrecht, Holland, p. 113, 1975; C.F. Bernasconi, "Relaxation Kinetics", Academic Press, New York, N.Y., 1976; H. Strehlow and W. Knoche, "Fundamentals of Chemical Relaxation", ed., H.F. Ebel, Verlag Chemie, New York, N.Y., 1977.

2. R. Brouillard and J.E. Dubois, J.Am.Chem.Soc., 99, 1359 (1977).

3. R. Brouillard, B. Delaporte and J.E. Dubois, ibid., 100, 6202 (1978).

4. R. Brouillard and B. Delaporte, ibid., 99, 8461 (1977).

EQUILIBRIUM STUDIES ON THE SELF-ASSOCIATION OF DRUGS IN AQUEOUS SOLUTION

David Attwood

Pharmacy Department, University of Manchester,
Manchester, M13 9PL

Equilibrium studies of aqueous solutions of amphiphilic drug
molecules have indicated a dependency of the mode of association
on the structure of the hydrophobic moiety. The association
behaviour of drugs possessing diphenylmethane hydrophobic groups
may be described using the mass action theory of micellization.
In contrast, some drugs with tricyclic hydrophobic moieties
exhibit continuous association with no apparent critical micelle
concentration. The association behaviour of these drugs may be
described by a stepwise association model. Apparent nonmicellar
association is observed with maleate salts of amphiphilic drugs
containing pyridine rings. Relaxation studies have revealed an
interionic proton transfer process in such systems.

INTRODUCTION

An important factor controlling the mode of association of
amphiphilic molecules in aqueous solution is the structure of the
hydrophobic moiety. In typical surfactants the hydrophobic
region is composed of a flexible hydrocarbon chain which can
intertwine during the micellization process to form approximately
spheroidal aggregates. Association commences at a critical concentration (the critical micelle concentration, cmc) and the
micelles, which are generally composed of between 30-100 monomers,
are of a narrow size distribution. In contrast, rigid planar
aromatic molecules, such as the cationic dyes and the purine and
pyrimidine bases of nucleosides can associate by a stacking
process.

W. J. Gettins and E. Wyn-Jones (Eds.). Techniques and Applications of Fast Reactions in Solution. 201–209.
Copyright © 1979 by D. Reidel Publishing Company.

Self-association is generally continuous i.e. there is no equivalent to the cmc and there is a wide range of aggregate sizes
in solution. The side chains are generally small with respect
to the hydrophobic ring and the self-association is controlled
by hydrophobic interactions; charge repulsion playing an insig-
nificant role in the aggregation process.

A large number of drug molecules are amphiphilic and associate in solution (1) but unlike typical surfactants, their hydrophobic groups are usually aromatic and may often be of limited
flexibility. Where the aromatic hydrophobic moiety is planar it
usually has side chains of appreciable length attached and in
this respect drug molecules differ from those of typical continuously associating systems. Drugs in general, represent an
interesting intermediate group of compounds between the two
extremes and it is of interest to study the effect of the structure of the hydrophobic group on the mode of association. For
this purpose it is convenient to consider the following classes
of compound.

DIPHENYLMETHANE DERIVATIVES

Typical examples of this class of drug are shown in Scheme 1.

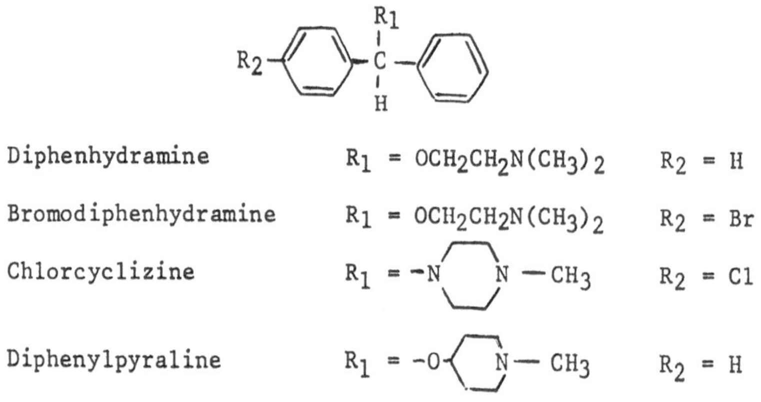

$$R_2 - \bigcirc - \underset{\underset{H}{\overset{\overset{R_1}{|}}{C}}}{} - \bigcirc$$

Diphenhydramine	$R_1 = OCH_2CH_2N(CH_3)_2$	$R_2 = H$
Bromodiphenhydramine	$R_1 = OCH_2CH_2N(CH_3)_2$	$R_2 = Br$
Chlorcyclizine	$R_1 = -N\bigcirc N - CH_3$	$R_2 = Cl$
Diphenylpyraline	$R_1 = -O\bigcirc N - CH_3$	$R_2 = H$

Scheme 1. Structures of some typical diphenylmethane derivatives

Light scattering measurements on such compounds (2) (3)
invariably produce apparently micellar plots (see Fig. 1) showing
distinct inflections, which for conventional surfactants are
identified with the cmc. Aggregation numbers are however much
lower (usually around 9-12 monomers per micelle) than those of
flexible chain surfactants and in view of this it is necessary
to critically examine the evidence for micellization as opposed
to continuous association.

The mass action model of micellization assumes that the cationic micelle, MP^+, is formed by an all-or-none process from N monomers, D^+, and N-P firmly bound anions, X^-.

$$ND^+ + (N - p)X^- \rightleftharpoons MP^+ \tag{1}$$

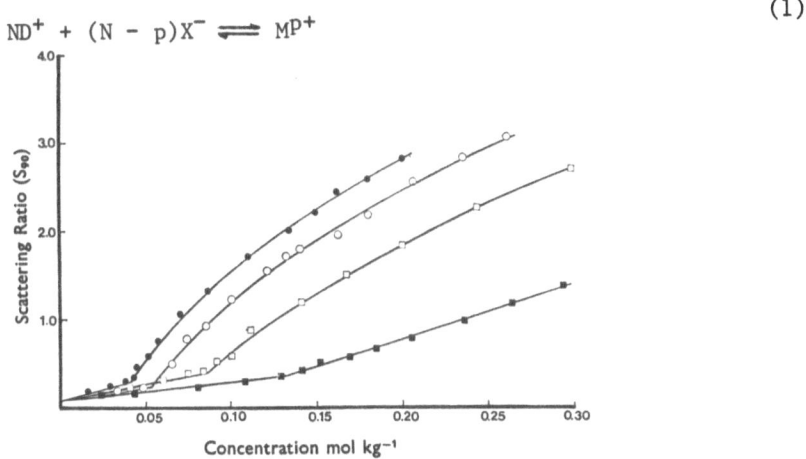

Fig. 1. Variation of the scattering ratio, S_{90}, with concentration for aqueous solutions of ●, chlorcyclizine hydrochloride; o, bromodiphenhydramine hydrochloride; □, diphenylpyraline hydrochloride; ■, diphenhydramine hydrochloride. (2)

In the absence of added electrolyte the micellar equilibrium constant, K_m, is given by

$$K_m = \frac{[MP^+]}{([X^-] - p[MP^+])^N [X^-]^{N-p}} \tag{2}$$

A computational procedure has been devised (4) for the simulation of the light scattering data for a wide range of combinations of K_m, N and p using eq. 2. The lines through the experimental scattering points in Fig. 1 have been computed in this manner showing clearly that the scattering behaviour of these compounds may be described by the mass action equation of micellization.

The possession of a cmc may be viewed as a criterion of micellar association. However, as discussed by Mukerjee (5), it is all too easy to imagine inflection points in experimental curves and it is preferable to examine the system using a variety of techniques and to compare cmc values. Table 1 shows such a comparison for antihistamine drugs containing a diphenylmethane hydrophobic group and shows reasonable agreement between values.

The addition of electrolyte to the diphenylmethane derivatives causes an increase in aggregation number as with typical surfactants. The effect of added electrolyte on the cmc of micellar systems is described by

$$\log \text{cmc} = -(1-\alpha)\log X^- + \Delta G_h^o/2.303RT + \frac{1}{N} \log F \ (M^{p+}) \qquad (3)$$

Table 1. Cmc values (mol kg^{-1}) of antihistamines at 303 K

	Light-scattering	Conductivity	Surface Tension
Diphenhydramine	0.132	–	0.122
Bromodiphenhydramine	0.053	0.049	0.041
Chlorcyclizine	0.040	0.040	0.039
Diphenylpyraline	0.086	0.094	–

where $\alpha = p/N$, ΔG_h^o is the hydrophobic contribution to the standard free energy of micellization and F is a term involving the activity coefficients of all species present in solution. Plots of log cmc as a function of log counterion concentration, X^-, for the antihistaminic diphenylmethane derivatives (see Fig. 2) are indeed linear (7) as predicted by eq. 3 suggesting that the association is micellar. The values of α derived from the slopes are in agreement with those from light scattering and, furthermore, ΔG_h^o values determined from the intercepts of such plots are in reasonable agreement with expected values derived from a consideration of the free energy associated with the transference of two phenyl rings from an aqueous to a non-aqueous environment.

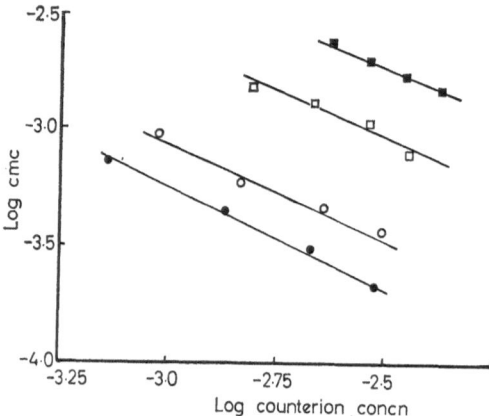

Fig. 2. Log cmc against counterion concentration for ○, bromodiphenhydramine hydrochloride; ●, chlorcyclizine hydrochloride; □, diphenylpyraline hydrochloride; and ■diphenhydramine hydrochloride. Concentrations are expressed as mole fractions (7).

There is strong evidence from equilibrium studies, therefore, that association of the diphenylmethane derivatives occurs by a micellar process. Relaxation techniques are currently being used to further investigate the mode of association of these systems.

TRICYCLIC COMPOUNDS

Typical drugs with tricyclic hydrophobic moieties are shown in Scheme 2.

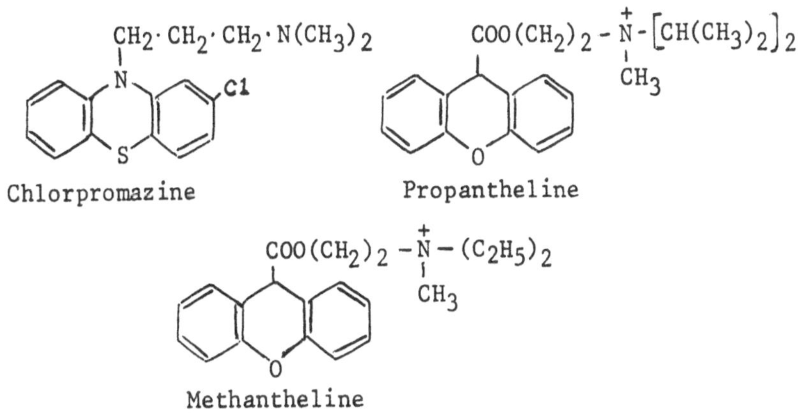

$$CH_2 \cdot CH_2 \cdot CH_2 \cdot N(CH_3)_2 \qquad COO(CH_2)_2 - \overset{+}{N} - \left[CH(CH_3)_2\right]_2$$

Chlorpromazine   Propantheline

$$COO(CH_2)_2 - \overset{+}{N} - (C_2H_5)_2$$

Methantheline

Scheme 2.   Structures of some typical tricyclic drugs.

Phenothiazine tranquillisers, of which chlorpromazine is an example, have been shown by a variety of techniques to undergo micellar association (8) (9). It is of interest however to note that nmr studies (10) (11) have suggested a stacked arrangement of molecules within the phenothiazine micelles. The association of the tricyclic antidepressants, for example amitriptyline, has similarly been shown to be micellar (12).

In contrast, the antiacetylcholine drugs, propantheline and methantheline, show a continuous increase in scattering intensity with increasing concentration with no apparent cmc (13) (see Fig. 3). From eq. 2, the sharpness of the cmc depends on the relative values of K_m, N and p. Combinations of low N and K_m can give rise to a gradual change of properties around the cmc region rather than a distinct inflection. However, it was not possible to fit the experimental scattering points using the simulation technique described above. The lines in Fig. 3, derived using this technique, are clearly a poor representation of the experimental data. The association pattern is thus nonmicellar and has been described using a stepwise association model. A complicating factor in the use of such a model for ionic systems is the nonideality which arises from the interaction of the charged aggregates. Adams and Williams (14) have proposed a rigorous method for the treatment of the weight-average molecular weight data of associating nonideal systems. There is, however, no simple way of using these equations unless the equality of values of the second virial coefficient for all species is assumed. An analytical treatment for ideal associating systems has been proposed by Steiner (15) and was applied to the

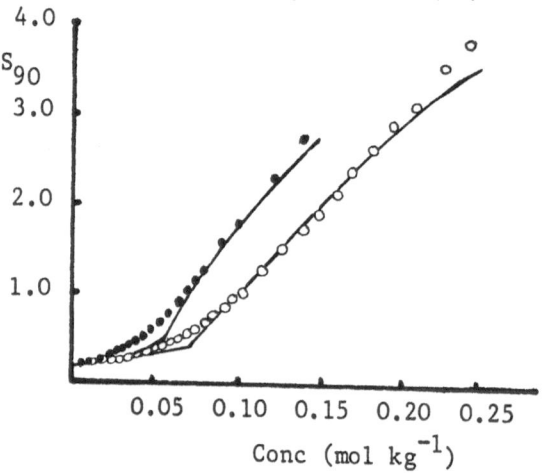

Fig. 3. Concentration dependence of the scattering ratio, S_{90}, for ●, propantheline bromide and ○, methantheline bromide. Continuous lines represent theoretical scattering as predicted by mass action equations (13).

experimental data for propantheline and methantheline. Integration of the light scattering data according to eq. 4 yields the weight fraction, x, of monomers as a function of the total solution concentration, c.

$$\ln x = \int_0^c \left[(M/M_w) - 1 \right] d \ln c \qquad (4)$$

M_w and M are the weight-average molecular weight of the aggregates and monomers respectively. Stepwise association constants, K_N, were calculated from

$$\left[(M_w/xM)-1 \right]/(xc/M) = 4K_2 + 9K_2K_3(xc/M) + \cdots\cdots$$
$$+ N^2 \left(\prod_{N=2}^{N} K_N \right) (xc/M)^{N-2} \qquad (5)$$

The average free energy change per monomer, ΔG_m, for the formation of the N-mer may be calculated from

$$\Delta G_m = -(RT/N) \ln \left(\prod_2^N K_N \right) \tag{6}$$

ΔG_m values for propantheline and methantheline were in reasonable agreement with values quoted for phenothiazine derivatives to which they bear some structural resemblance. This simple step-wise association model thus appears to adequately describe the light scattering data for these systems although the reason for the nonmicellar nature of the association is not yet clear.

Plots of surface tension of solutions of propantheline and methantheline against log concentration (16) were similar to those of micellar systems showing well-defined inflections normally interpreted as cmc's. This apparent discrepancy between the light scattering results which suggest nonmicellar association and surface tension results which suggest micellization may be resolved by consideration of the variation of monomer concentration with total solution concentration. Calculations using eq. 4 indicate that the monomer concentration in solutions of propantheline and methantheline reaches a limiting value, m_{mon}, at high concentrations. Surface tension is determined solely by monomer concentration and the inflections in the surface tension graphs of propantheline and methantheline are merely a consequence of a limiting monomer concentration in these systems rather than a micellar association process. Agreement between values of m_{mon} from the light scattering data and the inflections in the surface tension graphs is reasonable.

PYRIDINE DERIVATIVES

An interesting example of the apparent induction of non-micellar behaviour by alteration of the counterion associated with the drug is shown by antihistaminic drugs possessing a pyridine ring. Maleate salts of these compounds show nonmicellar light scattering plots whereas the light scattering behaviour of the hydrochloride salts is micellar (4) (17). Relaxation studies (18) on these compounds have indicated a single relaxation of weak amplitude for all compounds associated with Cl^- counterions (Fig. 4). In contrast, those compounds associated with maleate counterions showed an intense relaxation associated with more than one relaxation time. It was established that this intense relaxation arose from a proton transfer process involving the maleate counterions and the pyridine ring of the drug molecule. It could be eliminated at a pH of 1.8, when the spectrum was similar to that of the hydrochloride salts.

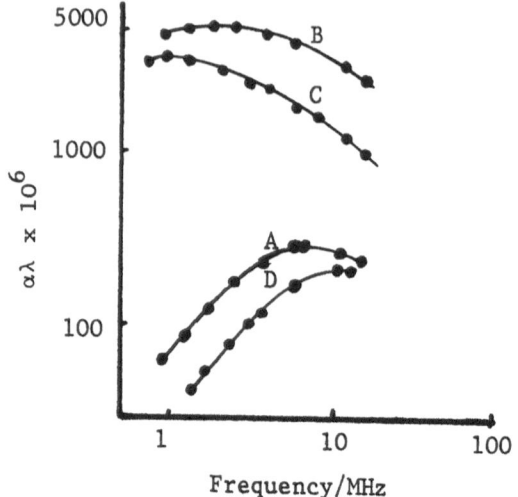

Fig. 4. Plots of relaxation spectra of (A) 0.1M tripelennamine
hydrochloride, (B) 0.1M chlorpheniramine maleate, pH 4.9,
(C) 0.1M pyridine + 0.05M sodium maleate, pH 5.0, and (D) 0.1M
chlorpheniramine maleate, pH 1.8 (18).

REFERENCES

(1) Florence, A.T.: 1968, Adv. Colloid Interface Sci. 2,
    pp. 115-149.
(2) Attwood, D.: 1972, J. Pharm. Pharmac. 24, pp. 751-752.
(3) Attwood, D.: 1976, ibid., 28, pp. 407-409.
(4) Attwood, D. and Udeala, O.K.: 1975, J. Phys. Chem. 79,
    pp. 889-892.
(5) Mukerjee, P.: 1974, J. Pharm. Sci. 63, pp. 972-981.
(6) Attwood, D. and Udeala, O.K.: 1974, J. Pharm. Pharmac. 26,
    pp. 854-860.
(7) Attwood, D. and Udeala, O.K.: 1974, ibid., 27, pp. 395-399.
(8) Scholtan, W.: 1955, Kolloid Z - Z. Polym., 142, pp. 84-104.
(9)Attwood, D., Florence, A.T. and Gillan, J.M.N.: 1974, J. Pharm.
    Sci., 63, pp. 988-993.
(10) Florence, A.T. and Parfitt, R.T.: 1970, J. Pharm. Pharmac.,
    22, pp. 121S-125S.
(11) Florence, A.T. and Parfitt, R.T.: 1971, J. Phys. Chem., 75,
    pp. 3554-3560.
(12) Attwood, D. and Gibson, J.: 1978, J. Pharm. Pharmac., 30,
    pp. 176-180.
(13) Attwood, D.: 1976, J. Phys. Chem., 80, pp. 1984-1987.
(14) Adams, E.T. and Williams, J.W.: 1964, J. Am. Chem. Soc.,
    86, pp. 3454-3461.
(15) Steiner, R.F.: 1952, Arch. Biochem. Biophys., 39, pp. 333-354.
(16) Attwood, D.; 1976, J. Pharm. Pharmac. 28 pp. 762-765.

(17) Attwood, D. and Udeala, O.K.: 1976, J. Pharm. Sci., 65, pp. 1053-1057.
(18) Gettins, J., Greenwood, R., Rassing, J.E. and Wyn-Jones, E.: 1976, J.C.S. Chem. Comm., pp. 1030-1031.

RELAXATION SPECTRA ASSOCIATED WITH MULTISTEP PROCESSES

J. E. Rassing

Department of Chemistry AD
Royal Danish School of Pharmacy, Copenhagen, Denmark

ABSTRACT

The present paper is concerned with the general problems involved in the interpretation of experimentally obtained relaxation spectra caused by multistep processes. Special emphasis is placed on the guidelines for 1) introducing approximations in order to reduce the number of independent and unknown parameters, 2) checking theoretically and experimentally the validity of these approximations, and 3) obtaining kinetic information from a measured ultrasonic relaxation spectrum caused by a multistep process. The topic is discussed in terms of the hydrogen bond polymerization of N-methyl acetamide the ultrasonic relaxation spectra of which have been thoroughly investigated by the author previously (1).

INTRODUCTION

In the field of ultrasonics the term "Relaxation spectrum" very often refers to a plot of a limited number of α/f^2 values (α being the sound absorption coefficient and f being the frequency of sound) vs f or log f where f covers 1-2 decades in the MHz region. If this plot shows a curvature almost S-shaped an ultrasonic relaxation is experimentally observed. If the plot shows a horizontal line no ultrasonic relaxation is observed. A given ultrasonic relaxation spectrum does not give direct information about the type of phenomenon that actually causes the observed effect. Since this information is necessary in order to extract

211

W. J. Gettins and E. Wyn-Jones (Eds.). Techniques and Applications of Fast Reactions in Solution. 211–224.
Copyright © 1979 by D. Reidel Publishing Company.

detailed information from the relaxation spectrum it must be obtained from other types of experiments or from theoretical consideration. When the phenomenon is known or suggested the further procedure in order to interpret the relaxation spectrum is to formulate a detailed reaction mechanism in terms of the molecular interactions that may account for the phenomenon on the molecular level. On the basis of this mechanism the theoretical expression for the ultrasonic relaxation spectrum is elaborated. The theoretically predicted relaxation spectrum is then one way or the other compared with the actually observed spectrum. Agreement between the predicted and the observed spectrum places evidence on the suggested phenomenon and the suggested mechanism. Usually this comparison furthermore gives rise to actual values of several rate constants, equilibrium constants, volume changes or enthalpy changes for the processes involved in the mechanism (2). If the mechanism in question is a multistep mechanism involving a large number of elementary reaction steps it becomes complicated to elaborate the theoretical expression for the relaxation spectrum. In addition the theoretical expression once derived very often contains a large number of independent and unknown parameters. Consequently a decent check of the suggested mechanism on the basis of the ultrasonic results turns out to be less convincing since any multi parameters equation is in agreement with a limited number of discrete experimental points.

The present paper deals with the problems involved when optimation of kinetic information and necessary approximations is desired in connection with the interpretation of ultrasonic relaxation spectra caused by a multistep process.

EQUILIBRIUM DESCRIPTION OF A MULTISTEP PROCESS

N-methyl acetamide, NMA, is a very interesting molecule because it is the smallest molecule that contains a peptide group. Consequently it might be a very convenient molecule to chose as a model compound for studies of the hydrogen bond entity of the peptide group. As a result a variety of different physico chemical investigations have been carried out on NMA dissolved in inert solvents. The general conclusion being that the amide group is planar, that NMA exists in trans form predominantly, and that this particular configuration together with a strong tendency to form hydrogen bonds result in a chain association of NMA molecules. The donor and acceptor sites of the hydrogen bonds are those of the peptide bond as illustrated below:

The following reaction scheme seems reasonable for the description of the stepwise chain association

$$N_1 + N_1 \rightleftarrows N_2$$

$$N_1 + N_2 \rightleftarrows N_3$$

. . .

. . . $\qquad\qquad$ (1)

$$N_1 + N_i \rightleftarrows N_{i+1}$$

. . .

. . .

$$N_1 + N_n \rightleftarrows N_{n+1}$$

The overall concentration of NMA calculated as monomer, C, is
given by

$$C = [N_1]^0 + \sum_{i=2}^{i=n+1} i \left[\prod_{j=1}^{j=i-1} K_j \right] ([N_1]^0)^i \qquad (2)$$

where n is the number of elementary reaction steps involved and
K_j is the equilibrium constant of reaction step number j. Eqn(2)
contains a large number of unknown and independent parameters,
K_j, and is consequently useless for any practical purpose. In
order to reduce the number on unknowns reasonable approximations
are needed. The approximation that gives the simplest description
is

$$K_1 = K_2 = K_3 = \cdots = K \qquad (3)$$

which reduces eqn. (2) to

$$C = \sum_{i=1}^{i=n+1} i \; K^{i-1} [N_1]^{oi} \qquad (4)$$

or

$$C = [N_1] \frac{1-(n+2)(K[N_1]^o)^{n+1} + (1+n)(K[N_1]^o)^{n+2}}{(1-K[N_1]^o)^2} \qquad (5)$$

Since it is unlikely theoretically to introduce a definite maximum size for a chain polymer and since the presence of large polymers is verified experimentally it seems reasonable to introduce that

$$n \to \infty \qquad (6)$$

which implicity involves that

$$K[N_1]^o < 1 \qquad (7)$$

The final expression for the overall concentration then becomes

$$C = \frac{[N_1]^o}{(1-K[N_1]^o)^2} \qquad (8)$$

It is possible to measure the free monomer concentration as a function of the overall concentration, and it turns out that eqn. (8) does not describe the data obtained over a wide concentration range. Consequently approximation (3) must be given up or revised. One possibility of introducing a slight revision is shown below

$$K_1 \neq K_2 = K_3 = \dots = K_n \qquad (9)$$

This approximation gives the following expression for the overall concentration

$$C = [N_1]^o \left[1 + \frac{K_1}{K_n} \sum_{i=2}^{i=n+1} i (K_n [N_1]^o)^{i-1} \right] \qquad (10)$$

If further more eqn. (6) is involved and

$$K_n [N_1]^o < 1 \qquad (11)$$

then eqn. (10) rewrites as

$$C = [N_1]^O \left[1 + K_1[N_1]^O \frac{2 - K_n[N_1]^O}{(1 - K_n[N_1]^O)^2} \right] \qquad (12)$$

Eqn. (12) describes the experimental data over a wide concentration range and by means of a fitting procedure actual values of K_1 and K_n are obtained. Table 1 gives the values obtained for NMA dissolved in different solvents measured by different techniques.

Solvent	$K_1 (M^{-1})$	$K_n (M^{-1})$	Temp. ($^{\circ}$C)	Method	Ref.
Carbon tetrachloride	1	38	25	IR	3
Carbon tetrachloride	1	30	35	IR	3
Carbon tetrachloride	1	21	50	IR	3
Carbon tetrachloride	9	61,6	20	Vapour pressure	4
Carbon tetrachloride	4.7(5.8)		25	IR	5
Benzene	6.7	14.3	25	Vapour pressure	6
Benzene	5.5	11.5	35	Vapour pressure	6
Benzene	4.2	8.9	49	Vapour pressure	6
$CDCl_3$	1.3	1.4	25	NMR	7
Dioxane	0.52(0.58)		25	IR	5

Table 1   Equilibrium constants for NMA dissolved in different solvents.

It appears that although there is some disagreement about the actual values of the equilibrium constants all the investigations show that $K_1 < K_n$. The fact that eqn. (12) describes the experimental data better than eqn. (8) is not surprising at all because eqn.(12) contains one more adjustable parameter. There are several possible approximations that result in an equation for the overall concentration with the same number of independent parameters as eqn.(12), for instance

$$K_i = a^i K \qquad a < 1 \qquad or \qquad K_i = \frac{K}{i} + K' \qquad (13)$$

The approximations given by eqns.(9) and (13) involve the same number of adjustable parameters and predict a decay in the distribution of matter on increasing polymer sizes. Furthermore they all agree reasonably well with the experimental results. The question then arises which type of approximation is the more reasonable to chose. This question can only be answered on the basis of theoretical considerations. The following statistical thermodynamical consideration may be used in order to chose the best approximation for the system in question.

In this treatment a certain number of possible orientations, p, with equal energy is ascribed to each monomer. It is then assumed that in a molecule linked by a hydrogen bond a particular one of the orientations is so favorable that it always occurs. If only rotational orientations are considered, the partition function of the monomer is given by

$$\psi_{N_1} = p\psi'_N \qquad (14)$$

where ψ'_N accounts for the other degrees of freedom. The partition function of the hydrogen bond linked polymer chain is given by

$$\psi_{N_i} = (\psi'_N)^i \exp\left(\frac{-(i-1)\Delta E}{kT}\right) \qquad (15)$$

where ΔE is the energy of association which is assumed to be independent of i. By neglecting the possible effects of difference in size and non random mixing of the polymer chains the following expression is obtained

$$K_1 = \frac{[N_2]}{[N_1]^2} = \frac{\psi_{N_2}}{(\psi_{N_1})^2} = \frac{1}{p^2} \exp\left(\frac{-\Delta E}{kT}\right) \qquad (16)$$

and

$$K_2 = \frac{[N_3]}{[N_1][N_2]} = \frac{\psi_{N_3}}{\psi_{N_1}\psi_{N_2}} = \frac{1}{p}\exp\left(\frac{-\Delta E}{kT}\right) \qquad (17)$$

and

$$K_3 = \frac{[N_4]}{[N_1][N_3]} = \frac{\psi_{N_4}}{\psi_{N_1}\psi_{N_3}} = \frac{1}{p}\exp\left(\frac{-E}{kT}\right) \qquad (18)$$

and

$$K_i = \frac{[N_{i+1}]}{[N_1][N_i]} = \frac{1}{p}\exp\left(\frac{-\Delta E}{kT}\right) \qquad (19)$$

Consequently the approximation given by eqn. (9) is predicted from this very simple statistical thermodynamical consideration. Furthermore it is predicted that $K_1 < K_n$ because the value of p must be larger than unity. This trend is actually observed as it appears from Table 1. Generally speaking the loss of entropy associated with the dimerization is larger than that associated with the further association because the dimerization causes two monomers to lose their orientations while further association i.e. addition of one monomer unit to an existing complex omly causes a loss for one monomer. On this basis the equilibrium description given by eqn. (12) seems to be the one to prefer. As mentioned above the approximations discussed in this part of the paper all lead to a decaying distribution function of the polymer concentration as a function of the degree of polymerization according to the situation that

$$[N_i]^\circ \to 0 \quad \text{for} \quad i \to \infty \qquad (20)$$

The distribution function has no extrema. Although this type of distribution function describes a vast majority of association phenomena such as hydrogen bond chain association and base stacking in several systems it does not apply to micellization or other association phenomena that are characterized by extrema in the distribution function.

In conclusion it may be stressed that the equilibrium description of a system that naturally is described by a multistep

scheme can be worked out with different approximations that reduce
the number of unknowns. In order to compare the ability of different approximations to describe the experimental data it is necessary to restrict the comparison to approximations giving the same
number of unknowns. Thus an equation with two unknowns describes
a given set of data better than an equation with a single unknown.
Very often it is possible to construct several different approximations that all cause an acceptable description of the data. The
choise between such approximations must be based on theoretical
considerations mainly and not on smaller fluctuations in the data
agreement.

KINETIC DESCRIPTION OF A MULTISTEP PROCESS

The kinetics of the hydrogen bond chain association is described
by n-1 differential equations n being the number of different
polymer sizes present. In terms of the previously discussed mechanism given by eqn.(1) the n-1 differential equations are the
following

$$[\dot{N}_1] = -k_1[N_1]^2 - [N_1]\sum_{i=1}^{i=n}k_i[N_i] + \sum_{i=1}^{i=n}k_{-i}[N_{i+1}] + k_{-1}[N_2]$$

$$[\dot{N}_2] = k_1[N_1]^2 - k_{-1}[N_2] - k_2[N_1][N_2] + k_{-2}[N_3] \tag{21}$$

$$\vdots$$

$$[\dot{N}_i] = k_{i-1}[N_1][N_{i-1}] - k_{-(i-1)}[N_i] - k_i[N_1][N_i] + k_{-i}[N_{i+1}]$$

The number of unknown and independent parameters is twice that
of the equilibrium description because each equilibrium constant
splits up in two rate constants. The approximation succesfully
applied in the equilibrium description thus only reduces the kinetic description to involve four independent parameters. Consequently additional approximations must be considered.

    The system in question polymerizes and therefore the kinetics
can not be followed by the traditional methods. The ultrasonic
technique, however, has proved usefull for the kinetic investigation (9-11). The ultrasonic relaxation spectrum indicates that
the kinetics can be described by a single relaxation time. This
might be surprising at a first glance because eqn.(21) predicts
n-1 relaxation times. The simplest way of approaching this situation is based on the point of view that the sound wave detects

a transition between two states of different energy. This transition between a non bonded state and a bonded state involves the formation of a hydrogen bond. Consequently the following two state mechanism may be suggested

$$Ac + Don \overset{k_1}{\underset{k_{-1}}{\rightleftarrows}} Bond \tag{22}$$

"Ac" and "Don" denote an acceptor site and a donor site for the hydrogen bond formation respectively. "Bond" denotes a hydrogen bond. The theoretical expression for the relaxation time derived for this single elementary reaction step is easily seen to be

$$\frac{1}{\tau} = k_1 ([Ac]^o + [Don]^o) + k_{-1} \tag{23}$$

The actual concentrations of "Ac", "Don" and "Bond" in the equilibrium situation can be calculated in terms of the equilibrium description discussed in the previous part of the present contribution. The monomer and each chain polymer contain one donor and one acceptor site. Consequently

$$[Ac]^o = [Don]^o = \sum_{i=1}^{i=\infty} [N_i]^o \tag{24}$$

and

$$[Bond]^o = \sum_{i=1}^{i=\infty} i [N_{i+1}]^o \tag{25}$$

The mechanism given by eqn.(22) implicitly involves that the rates of hydrogen bond formation and breaking are independent of the position from which the acceptor and donor sites are delivered and consequently are indepent of the degree of polymerization. The two state mechanism thus corresponds to the approximation given by eqn.(3). In terms of that approximation eqns.(24) and (25) rewrite as

$$[Ac]^o = [Don]^o = \frac{[N_1]^o}{1 - K[N_1]^o} \tag{26}$$

and

$$[\text{Bond}] = \frac{K([N_1]^{\circ})^2}{(1-K[N_1]^{\circ})^2} \qquad (27)$$

Combination of eqns. (23), (25) and (26) gives

$$\left(\frac{1}{\tau}\right)^2 = 4k_1k_{-1}C + k_{-1}^2 \qquad (28)$$

Eqn. (28) predicts linearity of $(1/\tau)^2$ vs the overall concentration of NMA. It is remarkable that the predicted concentration dependence of the relaxation time is identical with the one predicted from the far simpler monomer/dimer mechanism. Fig. 1 shows the concentration dependence of $(1/\tau)^2$ obtained from the ultrasonic experiments and it appears that the concentration dependence predicted by the two state mechanism agrees with the data.

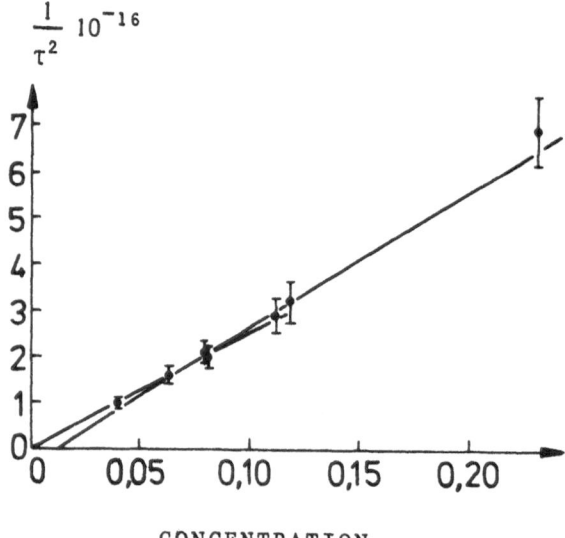

Fig. 1. The square of the reciprocal relaxation time vs the
overall concentration of NMA for NMA/CCl$_4$ mixtures.

From the data k_1 and k_{-1} are found to $2.0 \ 10^9 \ s^{-1}M^{-1}$ and $2.8 \ 10^7 \ s^{-1}$ respectively. The dynamic equilibrium constant defined as the ratio of the rate constants is 71 M^{-1}. This value of the equili-

brium constant corresponds surprisingly well with the value obtained from the equilibrium investigations in the concentration
range where the approximation given by eqn.(3) still applies.
When C is smaller than 0.1 M the IR-data indicate that
$K = 78 \pm 10$ % M^{-1}. Consequently it can be concluded that the ultrasonic relaxation spectrum is less sensitive to concentration
variations than the IR spectrum and therefore may be described by
the simpler approximation over a larger concentration range. It
seems interesting to mention that the two state type of approximation may give the clue to the understanding of why very complicated association systems very often cause surprisingly simple
relaxation spectra.

Rather than checking the results of the ultrasonic investigation with those of the IR investigation the latter might be
directly involved in the theoretical elaboration of the kinetic
equations. In that case it is necessary to consider the differential equations given by eqn.(21) in order to derive the equations for all the relaxation times involved. It can be shown that

$$\frac{1}{\tau_i} = k_{-n} |\lambda_i| \tag{29}$$

where λ_i is the eigenvalue number i of the coefficient matrix, $a_{i,j}$
and k_{-n} is the reverse rate constant in the mechanism given by
eqns.(11) and (3). If the first and last row and column of this
matrix are neglected the following triagonal matrix is obtained

$$\left\{ \begin{matrix} b & c & & & & \\ a & b & c & & & \\ & a & b & c & & \\ & & \cdot & \cdot & \cdot & \\ & & & \cdot & \cdot & \cdot \\ & & & & \cdot & \cdot & \cdot \\ & & & & & a & b \end{matrix} \right\} \tag{30}$$

where

$$a = -K[N_1]^o \tag{31}$$

$$b = K[N_1]^o + 1 \tag{32}$$

$$c = -1 \tag{33}$$

The eigenvalues for this matrix can be generated by means of the following equation

$$\lambda_r = 1+K[N_1]^0+2\sqrt{K[N_1]^0}\cos\frac{r\pi}{n-1} \tag{34}$$

where $r = 1,2,3,4,\ldots,n-1$

The distribution of eigenvalues and hence of relaxation times thus depends on the value of $K[N_1]^0$. In the limit of n going to infinity the maximum and minimum values of the eigenvalues are given by

$$\lambda_{max} = 1+K[N_1]^0+2\sqrt{K[N_1]^0} \tag{35}$$

and

$$\lambda_{min} = 1+K[N_1]^0-2\sqrt{K[N_1]^0} \tag{36}$$

In order for the average value to differ less than 10 % from the largest and the smallest value of the eigenvalues one of the following requirements must be fullfilled

$$\lambda_{av} = \begin{cases} K[N_1]^0 & \text{for } K[N_1]^0 \geq 400 \tag{37} \\ 1 & \text{for } K[N_1]^0 \leq 2.5 \times 10^{-3} \tag{38} \end{cases}$$

The inequality given by eqn.(37) is never fullfilled (see eqn.(7)) and fullfillment of the inequality given by eqn.(38) requires that the overall concentration is smaller than $4\ 10^{-5}$ M. The solution of the eigenvalue problem thus indicates a broad relaxation spectrum. The average relaxation time, however, is more precisely defined as

$$\left(\frac{1}{\tau}\right)_{av} = k_{-n}|\lambda_i|_{av} = k_{-n}\frac{\Sigma\beta_i\lambda_i}{\Sigma\beta_i} \tag{39}$$

where β_i are the weight factors to be considered for each eigenvalue. The weight factor or the amplitude parameter associated with each relaxation time can be elaborated in terms of the thermodynamic parameters describing the association. The theore-

tical consideration shows that the distribution of amplitude
parameters makes the multistep mechanism to agree with the experimental fact that a single relaxation time describes the ultrasonic experiments. One of the problems to get around is connected
with the fact that the number of elementary reaction steps is
infinitily large whereas the matrix to diagonalize numerically
must be finite. These problems are discussed in a previous paper
(12). The following kinetic information is now obtained: k_1, k_{-1},
k_n and k_{-n} equal 1.3 $10^{10}s^{-1}M^{-1}$, 2.6 10^9s^{-1}, 1.3 $10^{10}s^{-1}$ M^{-1} and
2.9 10^8s^{-1} respectively. It is important to notice that the de-
tailed analysis gives two different backward rate constants that
differ with a factor of 10 but both are larger than the backward
constant obtained from the two state treatment.

The interpration of the ultrasonic relaxation spectrum in
terms of the multistep mechanism is complicated but gives detailed
kinetic information. It requires, however, a basic knowledge of
the equilibrium distribution. In practice the ultrasonic technique
is often used to the investigate ion of systems for which the
equilibrium properties cannot be measured by other experimental
means and hence the detailed equilibrium description is not known.
In those situations it is still possible to obtain some information
about the rates involved by using the two state consideration (13).
It must be stated, that the rate constants obtained by this procedure are sort of mean values that may differ somewhat from the
individual true rate constants.

It has been demonstrated elegantly recently (14) that the
multistep scheme can be solved on closed form provided that the
approximation given by eqn.(3) applies. This solution which gets
around the eigenvalue problem predicts a single relaxation time
without taking   amplitude parameters into account. This new
consideration might threw new light on the evidence of the two
state consideration and this possibility is to be considered in
the very near future.

LOOKING AHEAD

Presumably the ultrasonic technique in the future is going to be
applied mainly to systems that cannot be kinetically investigated
by other experimental techniques. For instance systems that scatter the light and therefore makes spectrofotometric detection of
the ralaxation impossible or systems in which the electrical
conductivity is small and unchanged by the relaxation process.
Fast association phenomena is gels, emulsions, suspensions and
several systems within the field of the colloidal science may
contribute the future systems to be investigated by means of the
ultrasonic technique. Some of the systems may serve as models for
biological systems. If a very complicated mathematical treatment

is necessary in order to describe the model system a description
of the real system may seem impossible for the potential customers
of the ultrasonic results.
It is therefore important to present the simplest possible treatment leading to an understanding of the nature of the association
phenomenon as well as the sophisticated one leading to a higher
resolution of the kinetic details.

REFERENCES

1   J.Rassing, Advan.Mol.Relaxation Processes, 1 (1972) 55.

2   J.Rassing, in "Chemical and Biological Applications of
    Relaxation Spectrometry", ed. by E.Wyn-Jones, D.Reidel Publ.
    (1975) pp. 1-17.

3   H.Løvenstein,H.Lassen and Aa.Hvidt, ActaChem.Scand. 24 (1970)
    13.

4   R.D.Grigsby,Ph.D.Dissertation,Un. og Oklahoma,Norman,Okla. 1966.

5   I.M.Klotz and J.S.Franzen, J.Am.Chem.Soc. 84 (1962) 3461.

6   M.Davies and D.K.Thomas, J.Phys.Chem. 60 (1956) 767.

7   L.A.LaPlanche and M.T.Thompson, J.Phys.Chem. 69 (1965) 1482.

8   L.Sarolea-Mathot, Trans.Faraday Soc., 49 (1953) 8.

9   J.Rassing, Ber.Bunsenges.Physik.Chem. 75 (1971) 334.

10 J.Rassing and F.Garland, Acta Chem.Scand. 24 (1970) 2419.

11 J.Rassing, Acta Chem.Scand. 25 (1971) 1418.

12 J.Rassing,F.Garland and G.Atkinson, J.Phys.Chem. 75 (1971) 3182.

13 J.Rassing, J.Chem.Phys. 56 (1972) 52225.

14 M.Aizenman and T.A.Bak, to be published.

ULTRASONIC RELAXATION STUDIES OF THE KINETICS OF STACKING, CON-
FORMATIONAL TRANSITIONS AND PROTON TRANSFERS IN AQUEOUS SOLUTIONS
OF MOLECULES AND MACROMOLECULES OF BIOLOGICAL INTEREST

R. Zana, J. Lang, C. Tondre, J. Sturm, and S. Yiv

C.N.R.S., Centre de Recherches sur les Macromolécules,
Strasbourg, France.

ABSTRACT

A brief review of ultrasonic absorption studies of aqueous
solutions of biological compounds (amino-acids, bases, nucleosides, nucleotides, polypeptides, proteins and nucleic acids) in
relation with stacking, conformational transitions and proton
transfers is presented.

INTRODUCTION

Ultrasonic absorption methods have been extensively used for
the study of biological molecules (bases, nucleosides, nucleotides, amino-acids) and macromolecules (polypeptides, proteins,
nucleic acids). A survey of the literature reveals that these
studies have been carried out by workers coming from two widely
differing areas. On one side were physicochemists involved in
basic research, who used ultrasonic methods to gain kinetic informations on fast processes occuring in the time range between
1μs and 1ns (1). On the other side were workers who came from
the medical field and were interested in a rather practical
problem, namely the ultrasonic absorption of biological tissues
(2). The reason for this interest stems from the increasing use
of ultrasound in medical diagnosis and from the observation that
the attenuation of ultrasound in soft tissues is to a significant
extent occurring at a molecular level and determined by the molecules and macromolecules found in tissues (2).

Depending on the compound under investigation the results
have been essentially interpreted in terms of the following

225

W. J. Gettins and E. Wyn-Jones (Eds.). Techniques and Applications of Fast Reactions in Solution. 225–233.
Copyright © 1979 by D. Reidel Publishing Company.

processes : conformational equilibria of small molecules, polyion-
counterion interactions, stacking equilibria, conformational changes of biopolymers and proton-transfers. This brief review will
only deal with the last three processes. Indeed, the first one is
discussed in this Volume·by Prs Wyn-Jones and Hemmes and only
little work was done on polyion-counterion interactions since this
topic was last reviewed (3).

STACKING PROCESSES

Stacking usually involves molecules with a sufficiently large
aromatic moiety. In water these parts tend to stack. This interaction may provide a significant contribution to the stability of
secondary and tertiary structure of nucleic acids. Ultrasonic studies of stacked systems were undertaken in hope of gaining information on the kinetics of stacking processes as well as on the nature of stacking interactions.

So far the studies have only involved purine derivatives
(4-7). The results have been interpreted in terms of models which
are all derived from the so-called isodesmic association model,
which involves a series of association-dissociation equilibria
between a monomer A and an aggregate A_i of i monomers, all characterized by the same equilibrium constant.

The results relative to 6-methylpurine were interpreted in
terms of two models (4). The first one assumes that the dimerization constant is larger than the constants for the subsequent steps
(all taken equal) to account for the cooperativity of the system.
The second model, which gives a better account of the results,
assumes that the i-th association reaction has an equilibrium constant K/i where K is equal for all steps but dimerization. Both models are characterized by a spectrum of relaxation frequencies, in
agreement with the observation that the ultrasonic spectra are
slightly larger than for a single relaxation frequency process.

A similar observation was made in the study of N^6, N^9-dime-
thyladenine (5). However the difference.with spectra for a single
relaxation process was within the experimental error and was therefore neglected. Since the equilibrium studies showed no cooperativity, the ultrasonic results were interpreted in terms of the random isodesmic association model where all equilibria between any
two aggregates A_i and A_j to yield A_{i+j} are taken into account. The
same model was adopted in the recent and extremely accurate study
of stacking of N^6, N^6-dimethyladenosine (6). The reported relaxation spectra, determined by over 40 points, are definitely characteristic of a single frequency relaxation process. In an interesting theoretical development it is shown that the series of bimolecular reactions between any two stacked or unstacked species is

formally reduced to a single bimolecular reaction between two stacking surfaces in the calculation of the relaxation frequency and
amplitude. Note however that the expressions for the relaxation
frequency given in references (5) and (6) differ by a factor of 2.

Finally, a large part of the excess absorption of 5'-ATP at
around pH=3 was also attributed to stacking (7). The results could
be equally well accounted for in terms of dimerization or random
isodesmic association.

The main conclusion of these studies is that the stacking
association is close to being diffusion controlled (rate constants
of about 10^9 $M^{-1}s^{-1}$), in agreement with the results of studies conducted by means of T-jump for instance (8). Moreover, the stacking
interaction appears to be somewhat similar to the hydrophobic interaction. Indeed negative volume changes have been found for the
stacking reaction in references (5) and (6).

CONFORMATIONAL EQUILIBRIA OF BIOPOLYMERS

The ultrasonic studies dealing with the kinetics of conformational transitions of polypeptides, proteins and nucleic acids in
solution have been discussed or   reviewed in several papers (3,9-
11) to which the reader is referred to for work prior to 1974. The
ultrasonic studies performed after this date concern only polypeptides. It must be recalled that polypeptides are polyaminoacids
which can retain in non-polar solvents a helical conformation stabilized by H-bonds between the -C=O group of a residue with the
-NH- group of the third preceeding residue. In strongly polar solvents, or when residues become electrically charged, the helical
structure breaks down and polypeptides adopt a conformation similar
to a random coil. As polypeptides represent good models for proteins it was believed that the kinetics of the helix-coil transition would provide information on the kinetics of conformational
changes of proteins. This, however, did not prove to be the case.

Before looking at recent ultrasonic studies of helix-coil
transition we shall first briefly reexamine two papers (12,13)
often presented (14-16) as examples of ultrasonic studies of
helix-coil equilibrium in polypeptides. These papers deal with
polylysine in water (12) and polyornithine (PO) in water-methanol
mixtures (13), where the transition is induced by a pH change.
The observed excess absorption was attributed to the helix-coil
equilibrium and the results interpreted by means of Schwarz theory for this process (17a). This attribution was checked by performing comparative absorption measurements on aqueous solutions
of poly-L and DL-lysine (10). Identical absorption vs pH curves
were obtained for these two compounds, even though poly-DL-lysine
undergoes no transition in the pH range 3-13. Comparative

measurements were also performed on PO solutions in water and water + methanol mixtures. Again, the absorption vs pH curves were found (10) to be nearly coincident even though the maximum helical contents of PO in water and water + methanol differ by a factor 2.5. These results clearly indicate that the helix-coil equilibrium cannot be responsible for the largest part of the excess absorption of polylysine and polyornithine solutions. In fact this excess absorption has been shown to be due to proton transfer reactions (10, 18).

In 1975, a review on the kinetics of helix-coil equilibrium (19) concluded that polyglutamic acid (PGA) was probably the only polypeptide where the excess absorption is due to the helix-coil equilibrium. However, subsequent measurements on PGA (20) revealed that the excess absorption was time dependent and, moreover, that the equilibrium value of this absorption did not go through a maximum at the pH corresponding to mid-transition, in contradiction with previous observations (references (3) and (9) and references therein ; 17a, 21). Moreover, theoretical calculations were reported (22) where it was shown that internal proton transfers between helical and coiled regions of PGA in the transition range can quantitatively account for the excess absorption of PGA in the 10 kHz-1 MHz range, previously attributed to the helix-coil equilibrium (21). It should however be kept in mind that several adjustable parameters are involved in these calculations and render a critical evaluation of this work very difficult.

Although the studies reported in references (20) and (22) clearly suggest that the conclusions of ultrasonic studies of the kinetics of helix-coil transition are far from being definitive, the situation may not be as confusing as it may first look. Indeed, a recent ultrasonic study of poly-N^5-(3-hydroxypropyl)-L-glutamine (23), which is a water-soluble, non-ionic polypeptide, avoids most of the drawbacks usually encountered in studies of ionic polypeptides where the excess absorption may be due to proton transfers and counterion-polyion interactions (10,11), in addition to helix-coil equilibrium. This polymer was investigated in water-methanol mixtures where solvation equilibria may contribute to the absorption. Nevertheless, all the results could be quantitatively interpreted using the more recent theoretical treatment of the kinetics of helix-coil transition (17b), valid for polymers of chain length comparable to the cooperative length. Moreover, the kinetics of helix-coil transition of polypeptides has also been investigated by E-field jump (24-27) and T-jump (28,29). Two of the E-field jump studies (24,25) are not completely free from criticism concerning the attribution of the observed relaxation transients (22). However the other two E-field jump studies of PGA (26) and polylysine (27) as well as the T-jump investigations of polylysine labelled with a fluorescent dansyl group (28) and of PGA (29) using a fluorescence and an optical rotation detection, respectively, all yielded relaxation times very close to those obtained in ultrasonic studies of PGA in the 10 kHz - 1 MHz range (21). The excess ultrasonic absorption measured in these studies may therefore be reasonably attributed to a perturbation of the helix-coil equilibrium by ultrasonic waves.

Contradictory results have been reported for the helical content θ where the ultrasonic relaxation time or amplitude go through a maximum, when the polymer chain length becomes comparable to the cooperative length. The theory (17b) predicts that the maximum should then occur at $\theta < 0.5$. This prediction has been confirmed in references (23) and (28). On the contrary a maximum at $\theta > 0.5$ was found for short-chain PGA and attributed to fluctuations of the number of helix sequences (30,31). More experiments should be performed to check this point which is of considerable importance since it concerns the attribution of the excess absorption of PGA.

The rate constant k_F for the growth of helical segments has been found to be about 10^8 s^{-1} for ionic polypeptides (21,25-29) and 10^{10} s^{-1} for non-ionic polypeptides (see reference (23) and references therein). The slow rate found for ionic polypeptides has been attributed to a competition between solvent molecules and -NH- (or -C=O) groups for H-bonding to -C=O (or -NH-) groups (23, 28). This explanation is not fully convincing because all solvents or solvent mixtures used in helix-coil transition studies are usually hydrogen bonding agents (water, methanol, m-cresol, dichloroacetic acid, etc...).

PROTON TRANSFER PROCESSES

The various types of proton transfer reactions are given as reactions (1) to (3). As pointed out by Eigen (32) these reactions are kinetically coupled.

$$HX \rightleftharpoons H^+ + X^- \tag{1}$$

$$HX + OH^- \rightleftharpoons X^- + H_2O \tag{2}$$

$$HX + Y^- \rightleftharpoons X^- + HY \tag{3}$$

The characteristics of the excess ultrasonic absorption associated with reactions (1) to (3) are now well established (10,33, 34). In particular, this excess absorption goes through a maximum when the pH is modified.

Proton transfers constitute a common source of ultrasonic absorption by aqueous solutions of biological compounds. Indeed, the solubility of these compounds most often arises from chemical groups such as amino, carboxylic and phenolic groups which can be protonated or deprotonated. Since proton transfers involving these groups are usually fast (32) and accompanied by large volume changes, a sizeable excess absorption is often measured. Due to lack of space, this review will be restricted to some specific studies.

Hydrolysis-Protolysis

Amino-acids have been investigated both in the acid range (protolysis, k_{H^+}) and alkaline range (hydrolysis, k_{OH^-}). The reported values of the rate constants (34-37) are all close to the diffusion controlled limits. Nevertheless, significant differences exist between values obtained in different studies. The main reason for these discrepancies appears to lie in the different concentration ranges investigated by the various workers, and in some cases, in the small number of experimental points determining the relaxation spectra from which are extracted the rate constants. In reference (36) a coupling has been noted for some amino-acids between hydrolysis of the α-amino group and of the protonable side-chain group.

Excess ultrasonic absorptions due to protolysis and hydrolysis have also been observed for a variety of bases, nucleosides and nucleotides (38a). These proton transfers involve the secondary phosphoric acid function of nucleotides and/or functional groups or aromatic nitrogen atoms in bases and nucleosides. The latter are also observed with nucleic acids (38c).

The study of alkylamines and ethyleneimine oligomers (39) has revealed that the rate constant k_{OH^-} decreases only slightly upon increasing molecular weight of X^-. This decrease is smaller than that brought about by an increasing steric hindrance and/or hydrophobicity at the proton transfer site, as is indicated by a study of amines substituted by an increasing number of methyl and ethyl groups (40). In another study involving cyclic amines (41) it has been found that k_{OH^-} shows only very little dependence, if any, on the conformational properties of the investigated amines, which could undergo ring reversal and/or nitrogen inversion. The values of k_{OH^-} reported in references (39) to (41) are all close to the diffusion controlled limits.

Hydrolysis and protolysis have also been shown to be responsible for the absorption maxima observed in the alkaline range with proteins (9b) and polypeptides having a side chain amino group (10,18), and in the acid range with proteins. These maxima were first attributed to conformational equilibria (see paragraph 3). From the relaxation spectra for polylysine and polyornithine (10,18), k_{OH^-} values of about 10^{10} M^{-1} s^{-1} were obtained, i.e., only 2 to 3 times smaller than those for the side chain amino group of the corresponding α-amino-acids. Such a small difference was to be expected in view of the above results for alkylamines

and ethyleneimine oligomers. However, an abnormal dependence of
the relaxation frequency on pH was reported for polylysine (18)
and has not received a satisfactory explanation yet. The rate
constant k_{OH^-} obtained from the relaxation spectra of bovine serum
albumine has also been found to be close to that for small model
molecules (9b).

Proton exchange

The existence of bimolecular proton exchange was demonstrated in aqueous solutions of some amino-acids (36,42) and nucleotides (38b,43) through ultrasonic absorption measurements. These
exchanges occur between differently ionized forms of the same
amino-acid or nucleotide. With nucleotides the bimolecular exchange occurs between the secondary phosphoric acid function of the
phosphate moiety and a protonatable group of the base ring, and is
coupled with a fast stacking process, as well as with protolysis
reactions at each of the two protonatable sites (38b,43).

Evidence has been presented for intramolecular proton exchanges in solutions of various amino-acids (36,42). The corresponding
rate constants are of 10^7-10^8 s^{-1}. Intramolecular proton exchanges
between helical and coiled regions of polypeptides in the transition range, have been put forward (22) to explain the excess ultrasonic absorption of these compounds, heretofore attributed to
helix-coil equilibrium (9,17a,21). Finally similar processes involving various side chain protonatable groups of protein residues
may contribute to the absorption of protein solutions at neutral
pH, which has thus far received no definitive attribution (22,38b).

REFERENCES

1. "Chemical and Biological Applications of Relaxation Spectrometry", Wyn-Jones, E., Ed., D. Reidel Publ.Corp., Dordrecht
   Holland, 1975.
2. "Interactions of Ultrasound and Biological Tissues", Reid, J.,
   and Sikov, M., Eds, U.S.Dept. HEW Publication 73-8008, BRH/
   DBE, 1972.
3. Zana, R. : 1975, J.Macromol.Sci.-Revs.Macromol.Chem.C12, p.
   165.
4. Garland, F., and Patel, R. : 1974, J.Phys.Chem., 78, p.848 ;
   Garland, F., and Christian, S. : 1975, J.Phys.Chem. 79, p.1247.
5. Pörschke, D. and Eggers F. : 1972, Eur.J.Biochem. 26, p.490.
6. Heyn, M., Nicola, C., and Schwarz, G. : 1977, J.Phys.Chem.
   81, p. 1611.
7. Yiv, S., Lang, J., and Zana, R. : 1976, Stud.Biophys., 57,p.13.
8. Turner, D., Juan, R., Flynn, G., and Sutin, N. : 1974, Biophys.
   Chem. 2, p.385 ; Hammes, G., and Hubbard, C. : 1966, J.Phys.
   Chem., 70, p.1615 ; Inaoka, W., Harada, S., and Yasunaga, T. :

1978, Bull.Chem.Soc. Jpn. 51, p.1701 and references therein.

9. (a) Zana, R., reference (1), p.489 ; (b) Zana, R., Lang, J.,
   Tondre, C., and Sturm, J. : reference (2), p.21, and references
   therein.

10. Zana, R., and Tondre, C. : 1972, J.Phys.Chem. 76, p.1737.

11. Zana, R. : 1972, J.Am.Chem.Soc.,94, p.3646 and references thereii

12. Parker, R., Slutsky, L., and Applegate, K. : 1968, J.Phys.Chem.
    72, p. 3177.

13. Hammes, G., and Roberts, P. : 1969, J.Am.Chem.Soc.91, p. 1812.

14. Mathieson, A. : 1971, "Molecular Acoustics", Wiley ; p. 235.

15. Blandamer, M. : 1973, "Introduction to Chemical Ultrasonics",
    Academic Press, p. 88.

16. Bernasconi, C. : 1976, "Relaxation Kinetics", Academic Press,
    p. 264.

17. (a) Schwarz, G. : 1965, J.Mol.Biol. 11, p.64 ; (b) 1968, Biopo-
    lymers, 6, p.873 ; 1972, J.Theor.Biol. 36, p. 569.

18. Heywood , P., Rassing, J. and Wyn-Jones, E. : 1976,Adv.Mol.Relax.
    Proc., 8, p.95.

19. Zana, R. : 1975, Biopolymers, 14,p.2425 and references therein

20. Eggers, F., and Funck, T. : 1976, Stud.Biophys., 57, p.101.

21. Inoue, H. : 1972, J.Sci.Hiroshima Univ.A2, 34, p.37 ; Barksdale
    K., and Stuehr, J. : 1972, J.Am.Chem.Soc., 94, p.334.

22. Madsen, L., and Slutsky, L. : 1977, J.Phys.Chem.,81, p.2264.

23. Gruenewald, B., Nicola, C.,and Schwarz, G. : 1978, in press.

24. Lentz, D., Hutchins, J., and Eyring, E. : 1974, J.Phys.Chem.78,
    p.1021.

25. Yasunaga, T., Tsuji, Y., Sano, T.,and,Takenaka, H. : 1976, J.
    Am.Chem.Soc., 98, p.813.

26. Cummings, A., and Eyring, E. : 1975, Biopolymers, 14, p.2107.

27. Inoue, S., Sano, T., Yakabe, Y., Ushio, H. and Yasunaga, T. :
    Biopolymers, in press.

28. Bosterling, B., and Engel, J. : 1979, Biophys.Chem. in press

29. Yasunaga, T., Tsujii, Y., Sano, T., and Takenaka, H., ref.(1),
    p. 504.

30. Michels, B., and Cerf, R. : 1972, C.R.Acad.Sci.Paris, 274,p.1096

31. Cerf, R. : 1975, Adv.Chem.Phys. 33, p.73.

32. Eigen, M. : 1964, Ang.Chem.Int.Ed., 3, p.1.

33. Lang, J. : reference (1), p. 397.

34. Applegate, K., Slutsky, L., and Parker, R. : 1968, J.Am.Chem.
    Soc., 90, p.6909.

35. Hussey, M., and Edmonds, P. : 1971, J.Acoust.Soc.Am. 49,pp.1309
    and 1907.

36. Inoue, H. : 1970, J.Sci.Hiroshima Univ. A2, 34, p.17.

37. Brun, S., Rassing, J. and Wyn-Jones E. : 1973, Adv.Mol.Relax.
    Proc. 5, p. 313 ; Grimshaw, G., Heywood, P. and Wyn-Jones E.:
    1973, J.C.S. Faraday II, 69, p.756.

38. (a) Lang, J., Sturm, J., and Zana, R. : 1973, J.Phys.Chem.77,
    p. 2329 ; (b) 1974, J.Phys.Chem., 78, p.80 ; (c) Sturm, J., Lang
    J., and Zana, R. : 1971, Biopolymers, 10, p.2639.

39. Emara, M., Atkinson, G., and Baumgartner, E. : 1972, J.Phys.

Chem. 76, p.334 ; Nishikawa, S., Yasunaga, T., and Tatsumoto N. : 1973, Bull.Chem.Soc. Jpn 46, p.1657.

40. Eigen, M., Maass, G., and Schwarz, G. : 1971, Z.Phys.Chem., (N.F.) 74, p.319.
41. Grimshaw, D., and Wyn-Jones, E. : 1973, J.C.S. Faraday II, 69, p.168.
42. Maass, G. and Peters, F. : 1972, Angew.Chem. 11, p.428
43. Yiv, S., Lang, J., and Zana, R. : 1978 "Protons and Ions Involved in Fast Dynamic Phenomena", Elsevier, Amsterdam,p.33.

RAMAN LASER TEMPERATURE JUMP STUDY OF SOLVENT EFFECTS ON SIMPLE
STACKING SYSTEMS

T. G. Dewey and Douglas H. Turner

Department of Chemistry
University of Rochester
Rochester, New York 14627 U.S.A.

The kinetics of stacking have been investigated for three
systems in aqueous and mixed aqueous solutions. The three
systems, thionine, proflavin, and polyriboadenylic acid (poly A),
represent strong, moderate, and weak stacking, respectively.
Since all relaxations for these reactions are faster than 1 μsec.,
the Raman laser temperature jump technique was used (see Fig.1).[1,2]
The role of solvent in these reactions can be probed by using
cosolvents to perturb the aqueous environment. From an understanding of these effects insight into the forces dominating
stacking interactions can be gained.

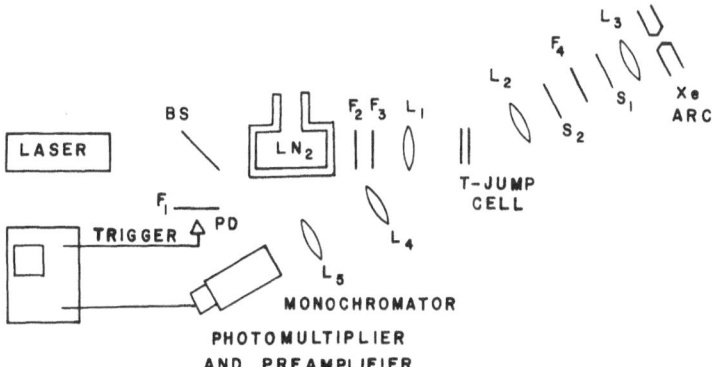

Fig. 1. Raman laser temperature-jump apparatus. BS is beam
splitter. F1-F4 are filters. L1-L5 are lenses. S1, S2 are
shutters. PD is photodiode.

W. J. Gettins and E. Wyn-Jones (Eds.). Techniques and Applications of Fast Reactions in Solution. 235–238.
Copyright © 1979 by D. Reidel Publishing Company.

For the dyes, thionine and proflavin, the reaction considered is the dimerization of the cationic species:

$$T + T \underset{k_{-1}}{\overset{k_1}{\rightleftharpoons}} T_2$$

For both dyes in aqueous solution (10 mM phosphate) the forward
rate is almost diffusion controlled, so that the stability of
the dimer is reflected in the reverse rate. For thionine and
proflavin the measured rate constants at 22°C are listed below.[3,4]
For thionine in ethanol and propanol the effect on the forward
rate is much too large to be attributed to changes in any of
the macroscopic properties of the solutions. To interpret these
results specific interactions at the molecular level must be
considered.

Table 1. Rate Constants for Thionine and Proflavin Association
          at 22°C. (Proflavin results in parentheses)

Solvent	$10^{-9}k_1$ $(M^{-1}s^{-1})$	$10^{-6}k_{-1}$ (s^{-1})
H_2O	2.4 (1.1)	0.9 (3.0)
1% EtOH	0.49	0.8
5% EtOH	0.20 (0.63)	1.4 (6.2)
10% EtOH	0.18 (0.48)	5.1 (9.5)
1% PrOH	0.36 (0.16)	1.0 (5.8)
10% Urea	0.61	1.7

In considering a molecular interpretation of the decrease
in the forward rate constant in aqueous alcohol solutions, two
simple mechanisms might be proposed. First, the alcohol can act
directly on the dye by strongly solvating it. This strong solvation can prevent a diffusion controlled encounter group from
collapsing to the dimer complex and, thereby, decrease the forward rate. On the other hand, the alcohol may act indirectly
by its effect on the water solvent structure. If ethanol acts
as a "structure maker", the encounter group may have difficulty
releasing its solvation sphere to a structured environment. The
rate of collapse to a dimer would then be decreased.

By examining the ethanol data more closely, it is seen that
these results are not entirely consistent with the water structure mechanism. First of all, most physical properties suggest
that the maximum structure in ethanol-water systems occurs around
10 mole %. The kinetic results show the forward rate constant
to have already plateaued at 5 mole %. Also an order of magnitude calculation shows that the rate of collapse of the diffusion
controlled encounter group to the dimer is $9 \cdot 10^7$ sec^{-1}. This
would imply that the solvent structure responsible for this rate
must be stable for about 10^{-8} sec., which seems to be an unrea-
sonably long time period. Furthermore, the effects observed in

10 mole % urea, a "structure breaking" solvent, are similar to those observed with ethanol and propanol. The slowing of the forward rate by a cosolvent which should have the opposite effect from ethanol argues against the water structure mechanism. Therefore, it appears that strong solvation of thionine by the non-aqueous cosolvent is a dominant effect competing with the stacking reaction.

The forces determining the solvation of the dyes are of some interest. If solvent-solvent interactions (e.g., hydrophobic effects) are important, then thionine and proflavin should behave similarly since they have essentially the same size and shape. Inspection of the rates in Table 1 indicates that, compared with thionine, the forward rate for proflavin dimerization is less sensitive to addition of ethanol, but more sensitive to addition of propanol. This suggests the solvation is dependent on dye-alcohol electronic interactions rather than solvent-solvent forces.

A comparison of the reverse rate of the two dyes in water shows the proflavin dimer to be about one-third as stable as the thionine one. The increase of the reverse rate with alcohol concentration may be interpreted as due to a cosolvent attack on the dimer. For proflavin the reverse rate is linearly related to the molarity of ethanol, suggesting that attack of one ethanol molecule is effective in breaking up the dimer. The reverse rate in thionine, on the other hand, is better fitted to the third power of the ethanol concentration. Thus the greater stability of the thionine dimer is reflected in the higher molecularity of alcohol attack.

Preliminary results have been obtained for poly A in mixed solvents. In poly A the stacking reaction is a unimolecular process. Rate constants could not be determined in earlier kinetic studies of poly A in aqueous, high salt solutions because of a wavelength dependence to the relaxation process.[5] However, with the laser temperature jump technique low ionic strength solutions can be investigated. At low ionic strength a single, wavelength independent, relaxation is observed for poly A in aqueous solution, and rate constants have been measured as a function of cosolvent.[6] The forward rates are two to three orders of magnitude slower than those measured for the dyes, even though the local concentration of adenine is very high. This indicates the stacking is conformationally controlled by barriers to rotation in the sugar-phosphate backbone.

References
(1) D. H. Turner, G. W. Flynn, N. Sutin, and J. V. Beitz, J. Am. Chem. Soc., 94, 1554 (1972).

(2) T. G. Dewey and Douglas H. Turner, Adv. Molec. Relaxation,
    13, 331 (1978).
(3) T. G. Dewey, Paulette S. Wilson, and Douglas H. Turner,
    J. Am. Chem. Soc. 100, 4550 (1978).
(4) T. G. Dewey, Dorothy Raymond, and Douglas H. Turner, in
    preparation.
(5) D. Pörschke, Eur. J. Biochem. 39, 117 (1973).
(6) T. G. Dewey and Douglas H. Turner, in preparation.

Acknowledgments.   This work was supported by NIH
                   Grant GM22939-03.

THE KINETICS OF MICELLE FORMATION

G. Platz

Institut für Physikalische Chemie, Universität Bayreuth
Bayreuth, West Germany

ABSTRACT

Anianssons equations for the fast monomer relaxation process in micellar systems are explained in an elementary way. The relations which must exist between distribution curves and rate constants for the stepwise aggregation processes in order to obtain one single relaxation time are discussed. Anianssons method is finally applied to the general case of a multistep process with stepwise monomer association.

INTRODUCTION

Two well separated relaxation times can always be observed by relaxation measurements on micellar systems. (1) The best model that can account for the two processes has been proposed by Aniansson and Wall. (2) For the derivation of the relaxation expressions, the micellar distribution curve was divided into three different regions: The monomers and oligomers, the nuclei in the distribution minimum and finally the proper micelles around the distribution maximum. If the equilibrium is suddenly perturbed, the reequilibration process between the proper micelles of different size can proceed rapidly. The number of the micelles is not changed by this process. Aniansson showed that the monomer relaxation leads exactly to a single relaxation time

W. J. Gettins and E. Wyn-Jones (Eds.). Techniques and Applications of Fast Reactions in Solution. 239–248.
Copyright © 1979 by D. Reidel Publishing Company.

at all concentrations, if the rate constants k_{s-1}^{+} and k_s^{-} for the aggregation step $A_{s-1}+A_1= A_s$ are a linear function of the aggregation number s, and if the micellar distribution curves obey special conditions. (2) For example, a gaussian distribution gives a single monomer relaxation time, if the k_s-values are independent of s. The general solution is eq. (1)

$$1/\tau_1 = \overline{k^{-}}/\sigma^2 + \overline{k^{+}} \cdot \sum_2 \overline{A_s}$$

$$= \overline{k^{-}}/\sigma^2 + \overline{k^{+}} \cdot (c_o-\overline{A_1})/\overline{n} \tag{1}$$

$\overline{k^{+}}$ and $\overline{k^{-}}$ are the mean values of the association and dissociation rate constants.

$$\overline{k^{\pm}} = \sum_2 k_s^{\pm}\overline{A_s}/ \sum_2 \overline{A_s} \tag{2}$$

$\sum_2 \overline{A_s}$ is the total concentration of the proper micelles. σ^2 is the variance of the micellar distribution:

$$\sigma^2 = \sum_2 s^2\overline{A_s}/\sum_2 \overline{A_s}-(\sum_2 s\overline{A_s}/\sum_2 \overline{A_s})^2 = \overline{n^2}-\overline{n}^2 \tag{3}$$

$\overline{n}$ and $\overline{n^2}$ are mean aggregation number and mean square of the aggregation number. c_o and A_1 are total and monomer concentration.

The formation or dissolution of whole micelles can only occur in a stepwise aggregation process through the distribution minimum. The concentrations of the particles in this region are usually several orders of magnitude smaller than the concentrations of the proper micelles and the process in which the number of the micelles is changed is therefore much slower than τ_1.

THE FUNDAMENTAL KINETIC EQUATIONS FOR THE MONOMER RELAXATION

The change of the monomer concentration by a single association step $A_{s-1} + A_1 = A_s$ is given by eq. (4)

$$dA_1/dt = -k^+_{s-1}A_{s-1}A_1 + k^-_s A_s \qquad (4)$$

In order to simplify the calculations, Aniansson introduced the relative deviations ξ_s from the equilibrium concentration $\overline{A_s}$ of the species A_s:

$$\xi_s = (A_s - \overline{A_s})/\overline{A_s} \qquad (5)$$

A_s is the actual concentration. By neglecting the term $\xi_{s-1}\xi_1$ due to the assumption $\xi_s \ll 1$, which is generally used for relaxation methods and with the equilibrium condition $k^+_{s-1}\overline{A_s} \cdot \overline{A_1} = k^-_s \overline{A_s}$, eq. (6) is obtained.

$$\overline{A_1}\,\dot{\xi}_1 = k^-_s \overline{A_s}(\xi_s - \xi_{s-1} - \xi_1) \qquad (6)$$

The total change of the monomer concentration is the sum of all steps. In order to include the first association step $2A_1 = A_2$ into eq. (7), the rate constants k^+_1 and k^-_2 must be in accord to $dA_1/dt = -k^+_1 A_1^2 + k^-_2 A_2$.

$$\overline{A_1}\dot{\xi}_1 = -\sum_2 k^-_s \overline{A_s} \cdot \xi_1 - \sum_2 k^-_s \overline{A_s}(\xi_{s-1} - \xi_s) \qquad (7)$$

The mass balance condition eq. (8) gives eq. (9).

$$A_1 + \sum_2 sA_s = \overline{A_1} + \sum_2 s\overline{A_s} \qquad (8)$$

$$\overline{A_1}\xi_1 + \sum_2 s\overline{A_s}\xi_s = 0 \qquad (9)$$

In micellar systems, the concentrations of the oligomers A_2, A_3... are so small that their contribution to the monomer relaxation is zero. The sum of the micelles does not change during the fast process:

$$\sum_2 \overline{A}_s \xi_s = 0 \tag{10}$$

The monomer relaxation leads to a single relaxation time, if the second term of eq. (7) is proportional to ξ_1:

$$-\sum_2 k_s \overline{A}_s (\xi_s - \xi_{s-1}) = p\xi_1 \tag{11}$$

In this case the monomer relaxation time τ_1 is given by eq. (12).

$$1/\tau_1 = \sum_2 k_s \overline{A}_s / \overline{A}_1 + p/\overline{A}_1 \tag{12}$$

SOLUTION FOR GAUSSIAN DISTRIBUTION CURVES AND k_s-VALUES INDEPENDENT OF THE AGGREGATION NUMBER

Gaussian distribution curves are commonly used to describe micellar systems. Therefore, the derivation of the equation for τ_1 will be shown for this case separately.

The $\overline{k}_s$-values are assumed to be constant:

$\overline{k}_s = \overline{k}$. Eq. (11) is transformed into eq. (13)

$$p\xi_1 = -\overline{k} \sum_2 \overline{A}_s (\xi_s - \xi_{s-1}) \tag{13}$$

Because the number of the micelles remains constant during the fast process (eq. (10)), eq. (13) is reduced to eq. (14).

$$p\xi_1 = \overline{k} \sum_2 \overline{A}_s \xi_{s-1} \tag{14}$$

The index for the summation may be changed:

$$p\xi_1 = k^- \sum_1 A_{s+1}\xi_s \qquad (15)$$

The concentrations of the oligomers A_2, A_3...
are so small, that their contribution to
eq. (15) is zero. Therefore the lower limit of
the summation index can be choosen as a number
u being greater one, but small enough to take
into account all micellar material. If the
distribution of the micelles is broad, it is
possible to use the approximation (16) with
s = 1. Eq. (15) leads to eq. (17).

$$\overline{A_{s+\Delta s}} = \overline{A_s} + \frac{d}{ds}\,\overline{A_s}\Delta s \qquad (16)$$

$$p\xi_1 = k^- \sum_u (\overline{A_s} + \frac{d\overline{A_s}}{ds})\,\xi_s \qquad (17)$$

Due to eq. (10), eq. (17) reduces to (18)

$$p\xi_1 = k^- \sum_u \frac{d\overline{A_s}}{ds}\,\xi_s \qquad (18)$$

The micellar distribution curve has a gaussian
form.

$$\overline{A_s} = A_o \exp\left(-\frac{(s-\bar{n})^2}{2\sigma^2}\right) \qquad (19)$$

The first derivative is given by eq. (20).

$$d\overline{A_s}/ds = -\frac{s-\bar{n}}{\sigma^2}\,\overline{A_s} \qquad (20)$$

Inserting eq. (20) into eq. (18) yields eq. (21).

$$p\xi_1 = -\frac{k^-}{\sigma^2} \sum_u s\overline{A_s}\,\xi_s + \frac{\bar{n}k^-}{\sigma^2} \sum_u \overline{A_s}\,\xi_s \qquad (21)$$

The second part on the right is zero due to
eq. (10). The first term can be expressed by
the mass balance condition eq. (9). This yields
eq. (22).

$$p\xi_1 = \frac{\bar{k}}{\sigma^2} \overline{A_1}\xi_1 \tag{22}$$

Eq. (22) inserted into eq. (12) gives the
expression (23) for the fast relaxation time.

$$1/\tau_1 = \bar{k}/\sigma^2 + \bar{k} \underset{2}{\Sigma} \overline{A_s}/\overline{A_1}$$

$$= \bar{k}/\sigma^2 + \bar{k} \cdot (c_o - \overline{A_1})/\bar{n} \tag{23}$$

RELATIONS BETWEEN DISTRIBUTION CURVES AND
RATE CONSTANTS TO OBTAIN A SINGLE RELAXATION
TIME FOR THE FAST MONOMER-RELAXATION

The monomer relaxation leads to one single
relaxation time, if eq. (11) is valid. Therefore
this equation must contain the relations between
k_s-values and distribution functions.

Eq. (11) is transformed into eq. (24).

$$p\xi_1 = \underset{2}{\Sigma} (\bar{k}_{s+1}\overline{A_{s+1}} - \bar{k}_s\overline{A_s})\xi_s + \bar{k}_2\overline{A_2}\xi_1 \tag{24}$$

During the fast process, the only relations between
the ξ_s and ξ_1 are the mass balance condition eq.
(9) and eq. (10), which means that the number of
the micelles remains constant. Therefore, eq. (11)
is satisfied, if eq. (25) is valid.

$$\underset{2}{\Sigma} (\bar{k}_{s+1} \overline{A_{s+1}} - \bar{k}_s\overline{A_s}) \; \xi_s = \underset{2}{\Sigma} (\alpha + \beta s)\overline{A_s}\xi_s =$$

$$-\beta\overline{A_1}\xi_1 \tag{25}$$

α and β must not depend on the ξ_s. The relaxation
time is given by eq. (26).

$$1/\tau_1 = \sum_2 k_s \overline{A_s}/\overline{A_1} - \beta + (k_2\overline{A_2}/\overline{A_1}) \qquad (26)$$

The term in brackets is zero for micellar systems. Eq. (26) must hold for any ξ_s-values. This is always and only the case, if each single part of the sum is zero. Therefore, eq. (27) is obtained.

$$(k_{s+1}^-\overline{A_{s+1}} - k_s^- \overline{A_s}) = (\alpha + \beta s)\overline{A_s} \qquad (27)$$

The constants α and β are estimated by the micellar distribution. This can be easily shown from eq. (27) if the following procedures are done:

$$\sum_2 (k_{s+1}^-\overline{A_{s+1}} - \sum_2 k_s^-\overline{A_s}) = \sum_2 \alpha\overline{A_s} + \sum_2 \beta s\overline{A_s} \qquad (28)$$

$$-k_2^-\overline{A_2} = 0 = \alpha\sum_2 \overline{A_s} + \bar{n}\beta \sum_2 \overline{A_s} \qquad (29)$$

$$\bar{n} = - \alpha/\beta \qquad (30)$$

If eq. (27) is multiplied with the aggregation number s and the same steps are carried out, eq. (31) is obtained.

$$-\sum_3 k_s^-\overline{A_s} = \alpha \sum_2 s\overline{A_s} + \beta \sum_2 s^2\overline{A_s} \qquad (31)$$

With $\bar{k}^-$ from eq. (2) and σ^2 from eq. (3), eq. (31) is transformed into eq. (32).

$$-\bar{k}^- = \alpha \bar{n} + \beta\overline{n^2} = \beta(\overline{n^2}-\bar{n}^2) = \beta\sigma^2 \qquad (32)$$

The result $\beta = -\bar{k}^-/\sigma^2$ inserted into eq. (26) gives the expression for the fast monomer relaxation eq. (1). α and β from eq. (27) may be replaced by eqs. (30) and (32). This yields eq. (33).

$$(k_{s+1}^-\overline{A_{s+1}} - k_s^-\overline{A_s}) = \bar{k}^-/\sigma^2 \cdot (\bar{n}-s)\overline{A_s} \qquad (33)$$

The equilibrium condition $k_s^+ \overline{A_s} \cdot \overline{A_1} = k_{s+1}^- \overline{A_{s+1}}$
leads to eq. (34), which shows that the
association rate constants must depend linearely
on s, if the dissociation rate constants do so.

$$k_s^+ \overline{A_1} - k_s^- = \overline{k^-}/\sigma^2 \cdot (\overline{n}-s) \tag{34}$$

Eq. (33) determines the form of the distribution
curves. The approximation eq. (16) leads to eq.
(35).

$$\frac{d}{ds} (k_s^- \overline{A_s}) = \frac{\overline{k^-}}{\sigma^2} (\overline{n} - s) \overline{A_s} \tag{35}$$

Now the distribution curves are obtained with
good accuracy by integration. The result for
dissociation rate constants not depending on
s is a gaussian distribution. (eq. (19)). If
the k_s-values change linearly with s (eq.(36)),
the distribution curves must be skew. A measure
for the skewness is the difference $n-n_{max}$ between
the mean aggregation number $\overline{n}$ and the aggregation
number n_{max} of the particles with the highest
concentration. $d/ds(\overline{A_s})$ is zero for $s = n_{max}$. From

$$k_s^- = k_o + sk \tag{36}$$

and eq. (35) follows:

$$(\overline{n} - n_{max}) = \sigma^2 \cdot k/\overline{k^-} \tag{37}$$

GENERAL MULTISTEP REACTIONS WITH STEPWISE MONOMER ASSOCIATION

In the general case the concentrations of the
dimers and trimers cannot be neclected. This
means that the number of the aggregates does
not remain constant during the fast process.
Therefore eq. (10) is not valid. The only
relation between ξ_1 and ξ_s is the mass balance
equation. In order to get a single relaxation
time, α must be zero in eq. (25) and eq. (27)

is transformed into eq. (38).

$$(k^-_{s+1}\overline{A_{s+1}} - k^-_s\,\overline{A_s}) = \beta s\overline{A_s} \tag{38}$$

β is determined by eqs. (39) and (40).

$$\sum_s (k^-_{s+1}\overline{A_{s+1}} - k^-_s\overline{A_s}) = \beta \sum_2 s\overline{A_s} \tag{39}$$

$$-k_2\overline{A_2} = \beta \sum_2 s\overline{A_s} = \beta(c_o - \overline{A_1}) \tag{40}$$

It can be easily shown that the k^-_s and the k^+_s values must be proportional to the aggregation number s in order to obtain a single relaxation time at all concentrations. The equilibrium constants and the concentrations are given by eq. (41) and (42).

$$K_s = k^+_{s-1}/k^-_s = \frac{(s-1)k^+}{sk^-} = \frac{s-1}{s}\,K \tag{41}$$

$$\overline{A_s} = \frac{2\overline{A_2}}{s}\,(K\overline{A_1})^{s-2} \tag{42}$$

The amount of monomers in the A_s decreases in a geometrical row and eq. (43) is valid.

$$\sum_2 s\overline{A_s} = 2\overline{A_s}(1 + K\overline{A_1} + (K\overline{A_1})^2 + \ldots) = 2\overline{A_2}/(1 - K\overline{A_1}) \tag{43}$$

Therefore

$$\beta = -k^-(1 - K\overline{A_1}) = -k^- + k^+\overline{A_1} \tag{44}$$

β into eq. (26) gives the relaxation expression eq. (45), which can be transformed into eq. (46).

$$1/\tau = \sum_2 sk^-\,\overline{A_s}/\overline{A_1} + 2k^-\overline{A_2}/\overline{A_1} + k^- - k^+\,\overline{A_1} \tag{45}$$

$$1/\tau = k^+ c_o + 2(k_1^+ - k^+)\overline{A_1^-} + k^-$$

(46)

Note: k_1^+ may be different from k^+

CONCLUSIONS

The form of the distribution curves for any investigated system may differ from the ideal behaviour, which leads to a single relaxation time. Nevertheless, in a lot of cases only one fast relaxation time is obtained for multistep processes. Deviations from the distribution curves discussed above lead to nonexponential relaxations. (2) It can be concluded that systems, which show a single fast monomer relaxation time within the experimental errors, have distribution curves not differing too much from the ideal form.

REFERENCES

1.) E.A.G. Aniansson, S.N. Wall, M. Almgren, J. Lang, C. Tondre, R. Zana, H. Hoffmann, I. Kielmann, W. Ulbricht, J. Phys. Chem., 80, 905 (1976).

2.) M. Almgren, E.A.G. Aniansson, K. Holmåker Chem. Phys. 19, 1 (1977).

A TREATMENT OF THE KINETICS OF MIXED MICELLES

Gunnar E.A. Aniansson

Physical Chemistry GU, University of Göteborg and
Chalmers University of Technology, S-412 96 Göteborg,
Sweden

The theory of step-wise micelle association and disinter-
gration has been extended to mixed micelles. The relaxation
process will again split into a fast and a slow one. During the
first one internal (pseudo-)equilibrium is established in the
micellar and monomer regions at a constant total number of
micelles and characterized by a number of relaxation times equal
to the number of components in the micelles. The slow process
will be characterized by a single relaxation time the value of
which is mainly determined by the properties at a saddle-shaped
narrow passage between the micellar and monomer regions. Closed
expressions for the relaxation times are deduced and their concentration dependence discussed.

A theory for the stepwise association and dissociation of
surfactant micelles was developed a few years ago [1]. Its application to a large quantity of experimental data has provided a consistent interpretation of these data and enabled the extraction
of basic kinetic and equilibrium parameters for these systems [2].
In the following extension to mixed micelles the concepts and
assumptions used are closely analogous to those of the previous
treatment to which the reader is referred for more details. For
simplicity the treatment is limited to two-component micelles but
the extension to larger number of components is quite straightforward.

The elementary steps are the following ones:

$$M_{r-1,s} + A \underset{k_1^-(r,s)}{\overset{k_1^+(r,s)}{\rightleftharpoons}} M_{r,s} \tag{1a}$$

W. J. Gettins and E. Wyn-Jones (Eds.), Techniques and Applications of Fast Reactions in Solution. 249–258.
Copyright © 1979 by D. Reidel Publishing Company.

and

$$M_{r,s-1} + B \underset{k_2^-(r,s)}{\overset{k_2^+(r,s)}{\rightleftharpoons}} M_{r,s} \tag{1b}$$

where A and B denote the two kinds of surfactant monomers and $M_{r,s}$ denotes a mixed micelle consisting of r A-components and s B-components ($M_{r,s}$ is then $A_r B_s$ in ordinary chemical notation).

Defining the relative deviations ξ, η, and μ, assumed to be small, from the equilibrium concentrations by

$$A = \bar{A}(1+\xi) \; ; \; B = \bar{B}(1+\eta) \; ; \; \text{and } M_{r,s} = \bar{M}_{r,s}(1+\mu_{r,s}) \tag{2}$$

where A, B and $M_{r,s}$ also denote the corresponding concentrations and the bar signifies the equilibrium value, the reaction flow components can be written

$$J_1(r,s) = k_1^+(r,s)AM_{r-1,s} - k_1^-(r,s)M_{r,s} =$$

$$-k_1^-(r,s)\bar{M}_{r,s}(\mu_{r,s} - \mu_{r-1,s} - \xi) \tag{3a}$$

and

$$J_2(r,s) = k_2^+(r,s)BM_{r,s-1} - k_2^-(r,s)M_{r,s} =$$

$$-k_2^-(r,s)\bar{M}_{r,s}(\mu_{r,s} - \mu_{r,s-1} - \eta) \tag{3b}$$

The time equation take the forms

$$\bar{M}_{r,s} \frac{d\mu_{r,s}}{dt} = -[J_1(r+1,s) - J_1(r,s)] - [J_2(r,s+1) - J_2(r,s)], \tag{4a}$$

$$\bar{A}\frac{d\xi}{dt} = -2J_1(2,0) - \sum_{r=3}^{\infty} J_1(r,0) - \sum_{r,s=1}^{\infty} J_1(r,s), \tag{4b}$$

$$\bar{B}\frac{d\mu}{dt} = -2J_2(0,2) - \sum_{s=3}^{\infty} J_2(0,s) - \sum_{r,s=1}^{\infty} J_2(r,s) \tag{4c}$$

Again the equations 3a, 3b, and 4a exhibit a close analogy to the equations of diffusion if differences are related to partial derivatived, $\mu_{r,s}$ to concentration, $M_{r,s}$ to the height of the vessel above the r-s-plane, and $k_1^-(r,s)$ and $k_2^-(r,s)$ to the diffusion coefficients in the two directions.

In normal cases $\bar{M}_{r,s}$ may be expected to be large in the micellar region around the mean aggregation numbers (n_1,n_2) and in the monomer region, see figure 1a. In a three dimensional diagram of $\bar{M}_{r,s}$ as a function of r and s the surface would consist of a rounded hill around (n_1,n_2) connected with the monomer region by

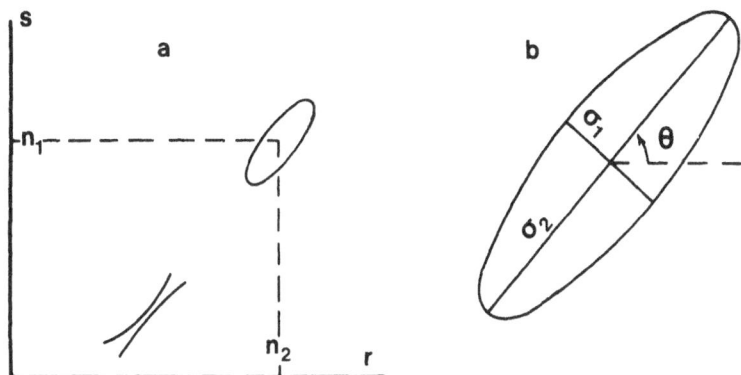

Figure 1. a) Sketch of a mixed micelle population around the mean
aggregation numbers n_1 and n_2 in the plane of the two aggregation
numbers r and s. The two arcs indicate the saddle shaped "narrow
passage" between the micellar and monomer regions. b) Illustration
of the main parameters determining a Gaussian micelle distribution.

a ridge with a saddleformed lowest part somewhere in between.
The ellipsoidal curve around (n_1, n_2) and the two arcs at the
saddle point in figure 1a can be considered as representative
parts of the niveau curve system in a topographical map of the
$\overline{M}_{r,s}$ surface.

The diffusion analogy again predicts two different relaxation processes: an initial fast establishment of pseudoequili-
brium in the micellar och monomer regions at a constant number of
micelles i.e. until

$$\mu_{r,s} - \mu_{r-1,s} \cong \xi \quad \text{and} \quad \mu_{r,s} - \mu_{r,s-1} \cong \eta \ , \tag{5}$$

followed by a slow one during which the number of micelles reaches
the equilibrium value by migration of aggregates in the ridge and
the $\mu_{r,s}$, ξ, and η go to zero.

THE FAST PROCESS

A solution can be found fairly easily of the kinetic equations
in ξ and η, of which the quantity measured in relaxation studies
is normally a linear function, in the case that the micellar
distribution is doubly Gaussian

$$\ln \overline{M}_{r,s} = \ln M_o + 2d(r-n_1)(s-n_2) - e(r-n_1)^2 - f(s-n_2)^2, \tag{6}$$

where M_o is the top value of $\overline{M}_{r,s}$, e, f>0 and ef>d^2, and $k_1^-(r,s)$ and $k_2^-(r,s)$ can be considered constant in the micellar region. The niveau curves of equation 6 are ellipses with the large main axis directed in the first and third quadrants of d>0 and in the second and fourth quadrants if d>0, figure 1b. 1/2e is the variance in aggregation number r at constant s, 1/2f is the variance of s at constant r, and d is measure of the correlation between r and s. Introducing new coordinats x and y along the large and small main axis, respectively, equation 6 takes the form

$$\ln \overline{M}(x,y)=\ln M_o-x^2/2\sigma_1^2-y^2/2\sigma_2^2 \tag{7a}$$

with

$$1/2\sigma_1^2=(e+f)/2-\sqrt{(e-f)^2/4+d^2} \tag{7b}$$

and

$$1/2\sigma_2^2=(e+f)/2+\sqrt{(e-f)^2/4+d^2} \tag{7c}$$

The angle θ between the larger main axis and the r-direction is given by

$$\text{tg } 2\theta=d/(f-e) \tag{7d}$$

To solve the equations for ξ and η one may start from equations 4b and c retaining only the double sums since the J:s are comparatively very small for low r and s values during the fast process. Replacing the sums by integrals and putting $\mu_{r,s}-\mu_{r-1,s}\cong$ $\partial\mu/\partial r$ and $\mu_{r,s}-\mu_{r,s-1}\cong\partial\mu/\partial s$ one obtains through partial integrations and use of the material balances (dimers, trimers, etc., seem not to occur in amounts that are of significance here).

$$\iint r\mu_{r,s}\overline{M}_{r,s}drds\cong-\overline{A}\,\xi \tag{8a}$$

and

$$\iint s\mu_{r,s}\overline{M}_{r,s}drds\cong-\overline{B}\,\eta \tag{8b}$$

the coupled equations

$$d\xi/dt=-\lambda_{11}\xi-\lambda_{12}\eta-\lambda_{10} \tag{9a}$$

and

$$d\eta/dt=-\lambda_{21}\xi-\lambda_{22}\eta-\lambda_{20}\,, \tag{9b}$$

where

$$\lambda_{11}=k_1(2e+C_m/\overline{A});\ \lambda_{12}=-k_12d\overline{B}/\overline{A};\ \lambda_{10}=k_12(en_1-dn_o)\delta C_m/\overline{A} \tag{10a}$$

$$\lambda_{21}=-k_2 2d\overline{A}/\overline{B}; \; \lambda_{22}=k_2(2f+C_m/\overline{B}); \; \lambda_{20}=k_2\cdot 2(fn_2-dn_1)\delta C_m/\overline{B} \quad (10b)$$

C_m denotes the equilibrium number of micelles and δC_m the excess number. The two relaxation times resulting from equations 9a and b are

$$1/\tau_1^f=(\lambda_{11}+\lambda_{22})/2+\sqrt{(\lambda_{11}-\lambda_{22})^2/4+\lambda_{12}\lambda_{21}} \quad (11a)$$

and

$$1/\tau_2^f=(\lambda_{11}+\lambda_{22})/2-\sqrt{(\lambda_{11}-\lambda_{22})^2/4+\lambda_{12}\lambda_{21}} \quad (11b)$$

Various combinations of the two relaxation times exist which facilitate the extraction of the unknown quantities from experimentally determined concentration dependencies of the relaxation times.

In the special case where one of the rate constants, for example k_2, is much larger than the other one there hold the simple relations

$$1/\tau_1^f\cong k_1[2e+C_m/\overline{A}-4d^2/(2f+C_m/\overline{B})] \quad (12a)$$

and

$$1/\tau_2^f=k_2(2f+C_m/\overline{B}) \quad (12b)$$

Results equivalent to those of equations 10 and 11 can also be obtained without replacing sums and differencies by integrals and derivatives in case that k_1^+, k_1^-, k_2^+ and k_2^- are linear functions of r and s. The details of the change of the micellar population can, as for one-component micelles, be obtained through expansion of $\mu_{r,s}$ in Hermite and other polynomials. This procedure yields results that include equations 10 and 11.

THE SLOW PROCESS

For the treatment of this process it is convenient to introduce

$$\mu_{r,s}'=\mu_{r,s}-r\xi-s\eta \quad (13)$$

Equations 3a and b then take the forms

$$J_1(r,s)=-k_1^-(r,s)\overline{M}_{r,s}(\mu_{r,s}'-\mu_{r-1,s}')\cong-k_1^-(r,s)\overline{M}_{r,s}\partial\mu'/\partial r \quad (14a)$$

and

$$J_2(r,s)=-k_2^-(r,s)\overline{M}_{r,s}(\mu_{r,s}'-\mu_{r,s-1}')\cong-k_2^-(r,s)\overline{M}_{r,s}\partial\mu'/\partial s \quad (14b)$$

Pseudoequilibria in the monomer and micellar regions then mean that μ' is independent of r and s in each of these regions. Since $\mu_{1,0}=\xi$ we have that in the monomer region

$$\mu'_{r,s}=0 \tag{15a}$$

and in the micellar region

$$\mu'_{r,s}=\mu'_m \tag{15b}$$

Between μ'_m and the total flow J through the lowest part of the ridge there holds the simple relation

$$\mu'_m=-RJ \tag{16a}$$

The "resistance" R has the general form

$$R=\frac{1}{wk_R M_R} , \tag{16b}$$

where M_R is the value of $\overline{M}_{r,s}$ at the saddle point of the lowest part of the ridge, l and w are measures of the length and width of this saddle region and k_R is given by

$$1/k_R=(\cos\theta')^2/k'_1+(\sin\theta')^2/k'_2 . \tag{16c}$$

θ' is the angle between the direction of the ridge in the saddle region and the r axis, $k'_1=k_1^-(r,s)$ in the same region and similarly for k'_2. Equation 16c was derived by replacing the saddle regions by a long and narrow parallelepipedal one with constant $\overline{M}_{r,s}$, $k_1^-(r,s)$, and $k_2^-(r,s)$. The results 16a and b can also be derived rigorously when $k_1^-(r,s)=k_2^-(r,s)=k_R$ in the saddle region and when the latter is of Gaussian form. l/w is then the ratio of the two corresponding "sigma parameters".

The derivation is thereafter closely analogous to that for onecomponent micelles and results in the following expression for the relaxation time

$$\frac{1}{\tau_s}=\frac{1}{RC_m}\frac{(\overline{A}+\overline{r^2}C_m)(\overline{B}+\overline{s^2}C_m)-(\overline{rs})^2C_m^2}{\left[(\overline{A}+(\overline{r^2}-n_1^2)C_m)(\overline{B}+(\overline{s^2}-n_2^2)C_m)\right]-\overline{(r-n_1)(s-n_2)}^2C_m^2} \tag{17}$$

The bar in $\overline{r_1^2}$ etc., signifies the average value of r_1^2 etc., in the micellar region.

CONCENTRATION DEPENDENCE

Space does not permit a comprehensive discussion[3] of the

concentration dependence of the expressions for the relaxation
times. The number of aspects to consider is of a higher order
than those for one-component systems. One may note, however, that
for a Gaussian micelle distribution the following quantities are
concentration independent:

$$\overline{r^2}-n_1^2=f/2(ef-d^2) \ , \tag{18a}$$

$$\overline{s^2}-n_2^2=e/2(ef-d^2) \ , \tag{18b}$$

and

$$\overline{rs}-n_1n_2=\overline{(r-n_1)(s-n_2)}=d/2(ef-d^2). \tag{18c}$$

For illustration, figures 2-4 display some numerical results
for the case when the two components are present in equal total
amounts. The micellar population chosen (at twice the cmc) was
that of figure 1a with $n_1=50$, $n_2=40$ and the saddle point at
$r=25$ and $s=10$. The correlation in the micellar distribution is
fairly pronounced and positive. Thus $\sigma_1=10$, $\sigma_2=3$ and $tg\theta=1.5$.
The saddle region is also Gaussian in shape, fairly long and
narrow, and its shift in position with total concentration was
taken into account. k_1 was taken $=1$ and $k_2=3$ (units are irrelevant
for the present purpose).

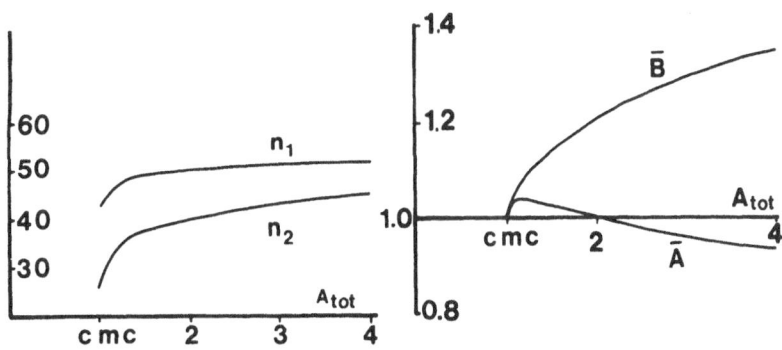

Figure 2. The dependence of the mean aggregation numbers n_1
and n_2 and the free monomer concentration $\overline{A}$ and $\overline{B}$ on the total
concentration $A_{tot}=B_{tot}$ for a particular case sketched in figure 1.

Figure 2 shows the variation of $\overline{A}$ and $\overline{B}$ with total concentration. After an early maximum slightly above the cmc it decreases
monotonously to values below that at the cmc. $\overline{B}$ increases mono-

tonously rather strongly at all concentration, the reason being
that the component B is bound to lesser extent than A to the
micelles. Both n_1 and n_2 increase with total concentration and
this effect is most pronounced just above the cmc. The larger
increase in n_2 than in n_1 is due to the opposite behaviour of
$\overline{B}$ and $\overline{A}$. Experimental determination of these four quantities in
mixed micelles would be most interesting.

C_m, not displayed, varies almost linearly with A_{tot} but a
downward bend is noticable.

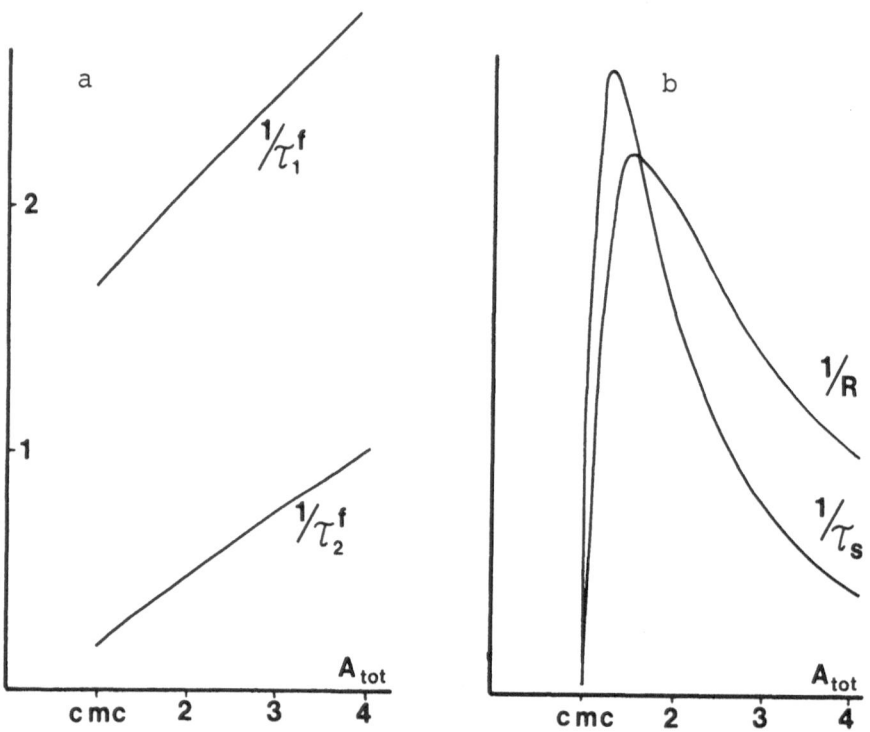

Figure 3. The reciprocals of, a) the two fast relaxation times
τ_1^f and τ_2^f, and, b) of the slow relaxation time τ_s and the
resistance R in the narrow passage as functions of A_{tot} for
the particular case.

The two fast relaxation times increase almost linearly with
total concentration, figure 3a, which is similar to the behaviour predicted for the one relaxation time in one-component
micelles. It is noteworthy that the values at the cmc are
surprisingly close to those predicted by equations 12a and b.

For C_m=0 these equations predict that the inverse of the longer relaxation time is approximately equal to the smaller rate constant over the total variance of r defined in equation (18a), whereas the shorter one is close to the larger rate constant over the variance, 1/2f, of s at constant r.

For the slower relaxation time the particular case chosen predicts an early sharp maximum of its reciprocal, see figure 3b. This is mainly caused by a similar behaviour of 1/R but the influence of the other factors of equation (17) is significant.

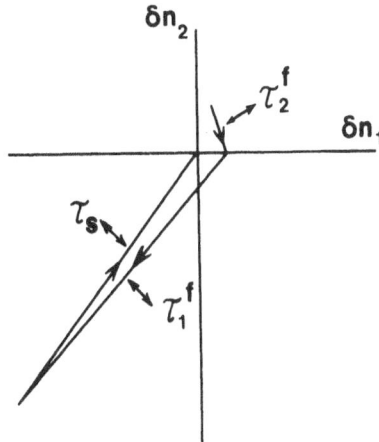

Figure 4. The shifts δn_1 and δn_2 in the two mean aggregation numbers during the three relaxation modes after a small concentration jump of the particular case.

The last figure shows the shifts δn_1 and δn_2 in the mean aggregation numbers during the three relaxation modes upon a small concentration jump at constant B_{tot}/A_{tot}. During the fastest relaxation mode the shift is small and at almost constant mean r value (n_1); c.f. the discussion of the apparent variance deduced above from the value of $1/\tau_2^f$ close to the cmc. The shifts during the following two modes are large and directed along the large main axis of the micellar distribution. The absolute magnitudes of these latter shifts are also much larger than the total shift during the whole process.

If the counterions of an ionized surfactant could be treated as a second component the treatment given above would be applicable to such surfactants. It is, however, probable that such a treatment would in general be too crude an approximation, but some features might still be essentially correctly predicted in this way.

REFERENCES

1. a. G. Aniansson and S. Wall, J.Phys.Chem. 78, 1024 (1974);
      79, 857 (1975).
   b. M. Almgren, G. Aniansson and K. Holmåker, Chem.Phys.,
      19, 1 (1977);
   c. G. Aniansson, Ber.Bunsenges.Phys.Chem. 82, 981 (1978).

2. a. G. Aniansson, S. Wall, M. Almgren, H. Hoffmann, I. Kielmann,
      W. Ulbricht, R. Zana, J. Lang and C. Tondre, J.Phys.Chem.
      80, 805 (1966);
   b. H. Hoffmann, Ber.Bunsenges.Phys.Chem. 82 989 (1978).

3.   G. Aniansson, to be published.

SOLUBILIZATION OF ACRIDINE ORANGE BY AQUEOUS TRITON-305

D. Y. Chao, G. Sumdani and Z. A. Schelly

Department of Chemistry
University of Texas at Arlington, TX 76019, USA

Solvent-jump relaxation kinetic, equilibrium spectral photometric and relaxation amplitude results are presented on aqueous Triton-305 at 5°, using acridine orange as indicator. The relaxation can be interpreted as the solubilization of the dye adsorbed on the micelle, at a rate determined by the dissolution of the micelle (Aniansson's slow process).

INTRODUCTION

The solvent-jump method (1) was used to investigate the relaxation kinetics of aqueous Triton-305 (octylphenyl polyoxyethylene, OPE_x, with x = 30) at 5° C. To facilitate the monitoring of the reaction, acridine orange (AO) was used as an indicator, since it was found (2) for nonionic detergents that the relaxation times characterizing the micellar solution are unaffected by the added dye, as long as the ratio r = (surfactant)/(dye) > 20 is satisfied.

EXPERIMENTAL

OPE_{30} was obtained from the Rohm and Haas Co. Its water content was found 29% by weight. It was used without purification in order to allow for direct comparison with previous studies (3,4). The average monomer molecular weight of 1526 and aggregation number n = 11 for OPE_{30} was adopted from the literature (4).

259

W. J. Gettins and E. Wyn-Jones (Eds.). Techniques and Applications of Fast Reactions in Solution. 259–264.
Copyright © 1979 by D. Reidel Publishing Company.

Acridine orange hydrochloride (AO), purchased from the Eastman Kodak Co., was purified by three recrystallizations from absolute methanol. The dye stock solution was stored in the dark and it had a natural pH≈7. All solutions were prepared using distilled deionized water.

The stopped-flow apparatus used in the solvent-jump experiments has been described previously (5).

ABSORPTION SPECTRA AND CMC

The spectral method (6) was used to determine the CMC = 6×10^{-4} M at 5° C of OPE_{30}-AO (at a constant AO concentration of 3.4×10^{-5} M)(7).

If an OPE_{30}-AO solution at a concentration above the CMC is diluted by a factor of two to a new concentration still above the CMC, the absorption maximum is slightly blue shifted (from 493.5 to 493 nm) and the main band decreases in intensity by more than a factor of two. This fact is utilized in the solvent-jump kinetic experiments, where the transient change of the absorption is monitored after dilution.

KINETIC STUDIES (7)

In the jump experiments, solutions of Triton-AO were mixed with an equal volume of water. Relaxations could be observed only if the surfactant concentrations were above the CMC after dilution. The change of absorbance with time was followed at the main absorption maximum (λ = 493.5 nm) of the solutions. Single relaxations in the lower millisecond range were observed in all experiments (Figure 1). The relaxation times τ and the relaxation signal amplitudes ΔA (Figure 2) were determined at 5° over a range of final total Triton concentration from 1.25 x CMC to 3.5 x CMC, containing a constant initial concentration of AO (3.4×10^{-5} M). No relaxation can be observed if the total dye concentration is kept constant before and after mixing. This, as well as the manner in which the transient change of the absorbance prior to relaxation takes place, suggest that a perturbation is needed which promotes the monomer–dimer equilibrium $2M \rightleftharpoons D$ of the free dye to shift rapidly to the left, releasing monomer AO that can subsequently be solubilized by the micelles. This latter slow process causes the observed exponential relaxation.

The kinetic results indicate (7) that the solubilization of the dye is taking place at a rate described by Aniansson's slow

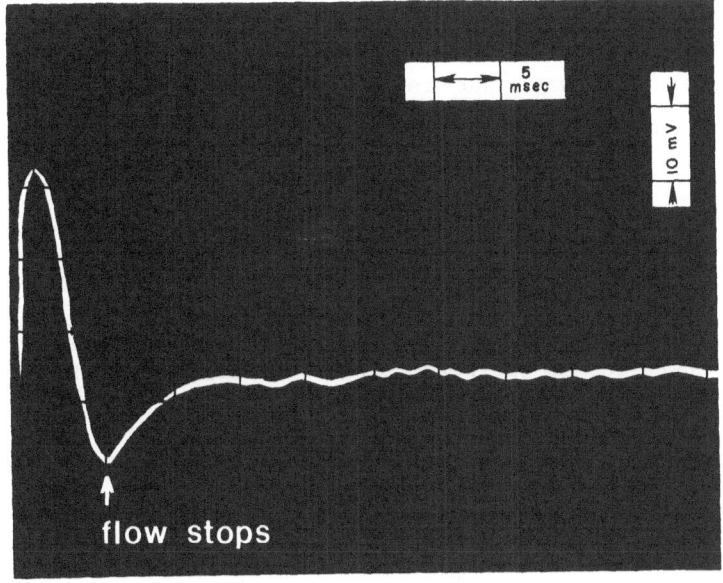

Figure 1. Typical solvent-jump relaxation curve (transmitted light intensity I $\underline{vs}$. time) of the OPE_{30}-AO system at 5°, using pretriggering. Monitoring is at λ = 493.5 nm.

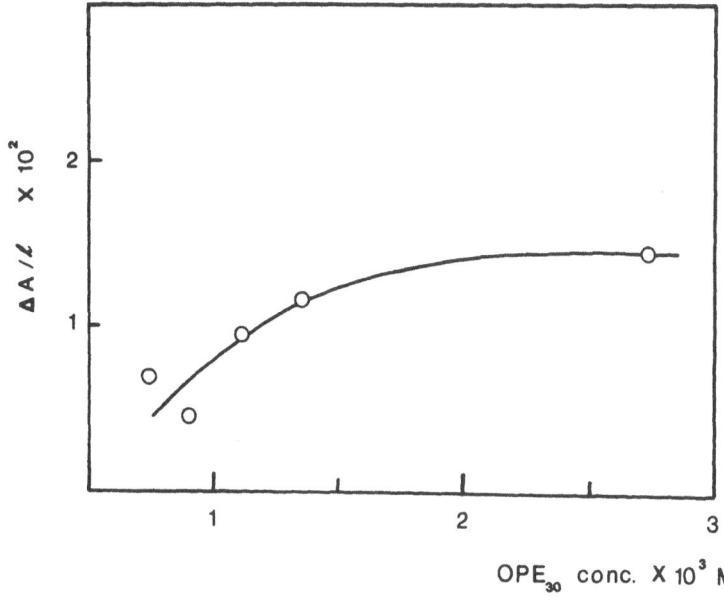

Figure 2. Relaxation signal amplitude per cm, $\Delta A/\ell$, $\underline{vs}$. the surfactant concentration for OPE_{30}-AO at 5°.

process (8). This is the case if the rate of surrounding the dye by the surfactant molecules is determined by the rate of replacement of surfactant monomers A_1 in the solution, where the A_1 concentration ($\sim$CMC) must be maintained. But this rate is the dissolution of micelles, characterized by τ_2 of Aniansson's theory (8).

RELAXATION AMPLITUDES (9)

The solubilization itself can occur in several different ways. Seven detailed models were considered, each of which includes the monomer-dimer equilibrium of the free AO:

1) Nucleation on monomer dye M.

2) Nucleation on monomer M and dimer D dye.

3) Micellization with monomer dye adsorption on the micelle A_n, followed by an intramolecular rearrangement of the adsorption complex A_nM', where the dye ends up inside the micelle (A_nM). Robinson et al. (10) have proposed a similar model for the solubilization of AO by anionic surfactants.

4) Nucleation on monomer dye, plus independent micellization simultaneously.

5) Micellization with monomer dye adsorption on the micelle.

6) Nucleation on adsorbed monomer dye, with simultaneous independent micellization. This mechanism is the same as 3) but the intramolecular rearrangement step $A_nM' \rightleftharpoons A_nM$ is replaced by a multistep sequence in which the monomer surfactant A_1 is deposited on M adsorbed on the "surface" of a micelle. During this process the other parts of the micelle are melting away, maintaining the size of the micelle as well as the A_1 concentration ($\sim$CMC) constant. Thus the nucleation steps

$$A_nM' \rightleftharpoons A_{n-1}M' + A_1 \qquad \text{melting away}$$

$$A_{n-1}M' + A_1 \rightleftharpoons A_nM'' \qquad \text{deposition}$$

$$\left.\right\} (n-1) \text{ times}$$

are repeated $(n-1)$ times, where in the final $(n-1)$th step A_nM'' becomes identical with A_nM of model 3). Since the A_1 concentration is maintained constant, it is reasonable to assume that the rate of deposition is controlled by the rate of melting.

7) Nucleation on adsorbed monomer M and dimer D dye, with simultaneous independent micellization. This mechanism is the same as 6) but with the additional D-adsorption equilibrium $A_n + D \rightleftharpoons A_n D'$ and the following multistep sequence

$$\left. \begin{array}{l} A_n D' \rightleftharpoons A_{n-1} D' + A_1 \\[2ex] A_{n-1} D' + A_1 \rightleftharpoons A_n D'' \end{array} \right\} \quad \text{(n-1 times)}$$

Although the distinguishing of the seven mechanisms based on rate measurements is practically impossible, analysis of the relaxation amplitudes can provide information needed in the decision making.

One can define the overall mean relaxation amplitude $\Delta \xi_o$ of an unrestricted number of coupled equilibria perturbed by solvent-jump (11). $\Delta \xi_o$ can be measured experimentally and calculated for the seven models as well(9). Comparison of the measured and calculated values can be used to decide in favor of a certain mechanism. The calculations support models 6) and 7). At the present time one cannot distinguish whether the nucleation takes place on the adsorbed monomer (model 6) or also on the adsorbed dimer dye (model 7). Due to the low dye concentration, however, model 6 is more likely.

Acknowledgement. This work was supported by the R. A. Welch Foundation and the Organized Research Fund of The University of Texas at Arlington.

REFERENCES

1. D. Y. Chao and Z. A. Schelly, J. Phys. Chem., 79, 2734 (1975). Also see article on the "Solvent-Jump Method" in this volume.

2. C. Tondre, J. Lang and R. Zana, J. Colloid Interface Sci., 52, 372 (1975).

3. J. Lang, J. J. Auborn and E. M. Eyring, J. Colloid Interface Sci., 41, 484 (1972).

4. J. Lang and E. M. Eyring, J. Polym. Sci., 10, Pt A-2, 89 (1972).

5. M. M. Wong and Z. A. Schelly, Rev. Sci. Instrum., 44, 1226 (1973).

6. M. L. Corrin and W. D. Harkins, J. Am. Chem. Soc., 69, 679 (1947).

7. D. Y. Chao, G. Sumdani and Z. A. Schelly, J. Phys. Chem., submitted.

8.   E. A. G. Aniansson and S. N. Wall, J. Phys. Chem., <u>78</u>, 1024
     (1974) and <u>79</u>, 857 (1975).
9.   Z. A. Schelly, D. Y. Chao and G. Sumdani in "Micellization,
     Solubilization and Microemulsions" Vol. 3, K. L. Mittal,
     ed., Plenum Press, New York, NY, 1979.
10. B. H. Robinson, N. C. White and C. Mateo, Adv. Mol. Relax.
     Processes, <u>7</u>, 321 (1975).
11. Z. A. Schelly and D. Y. Chao, Adv. Mol. Relax. Processes,
     1979, in press.

KINETIC STUDIES OF SYNTHETIC POLYMER/SURFACTANT COMPLEXES

D. M. Bloor, W. Knoche* and E. Wyn-Jones

Department of Chemistry, University of Salford, Salford M5 4WT, U.K.

*Max-Planck-Institut für Biophys. Chemie, Göttingen, West Germany

ABSTRACT

Using a pressure-jump apparatus the dissolution time for sodium dodecyl sulphate micelles has been measured in solutions which also contain polyvinyl pyrrolidone. The relaxation time associated with the association/dissolution of the micelles depends on the amount of polymer present and also its molecular weight. In addition the activation energy of the dissociation/association process decreases in the presence of the polymer. These data suggest that the micelles are incorporated along the polymer chain and a simple mechanism for the dissolution of micelles on the polymer is proposed.

INTRODUCTION

The interactions between surface active agent ions and synthetic polymers dominate their solution properties (1) as well as being extremely important for the study of fundamental phenomena. Although equilibrium measurements (2) on the bulk physical properties of these solutions have proved to be of particular value in identifying and developing an understanding of the factors influencing these interactions, the mechanistic details of the various processes taking place in solution can only be obtained through kinetic studies. Such studies have so far been neglected due mainly to the extremely short reaction times involved. We describe here relaxation studies using a pressure-jump

265

W. J. Gettins and E. Wyn-Jones (Eds.). Techniques and Applications of Fast Reactions in Solution. 205–270.
Copyright © 1979 by D. Reidel Publishing Company.

apparatus on polyvinyl pyrrolidine solutions containing sodium
dodecyl sulphate.

EXPERIMENTAL

These studies were undertaken using a pressure-jump apparatus over the time range $10-10^{-4}$ sec. incorporating a rapid data
capture and analysis system. The polymer, polyvinylpyrrolidone,
PVP, (Aldrich) was used without further purification. Sodium
dodecyl sulphate, SDS, (Henkel) was chosen as the surfactant
system, since a considerable wealth of equilibrium data were already available involving this surfactant. SDS was purified by
repeated recrystallisation from two solvents until the relaxation
time at the CMC agreed with literature values.

RESULTS AND DISCUSSION

It has already been established that the relaxation spectra
of micellisation in pure surfactants are characterised by two
relaxation times (3); a fast process in the range $10^{-6}-10^{-9}$
sec., attributed to a monomer/micelle exchange resulting in a
change in the aggregation number of the micelles following perturbation. The second much slower relaxation is observable in the
time range $10-10^{-4}$ sec. and is associated with the dissolution of
complete micelles. During this latter process, there is a change
in the actual number of micelles in solution, whilst the average
aggregation number of the remaining micelles remains effectively
constant. It is this slow relaxation process that can be studied
using the pressure-jump method.

The binding isotherm for the SDS/PVP system indicates three
distinct regions of binding as shown in Figure 1. Initially at
very low surfactant concentrations there is no detectable binding
(region I) until a concentration is reached at which time the
polymer begins to bind surfactant monomers (region II). It is
claimed that as the surfactant concentration increases, these
bound monomers may form clusters or premicellar aggregates and
eventually when the free monomer concentration becomes constant,
true micellar aggregates are formed (region III).

In the pressure-jump experiments, a single relaxation (10^{-4}
$> \tau < 10$ sec.) was always observed in solutions containing micelles
and the results are summarised below in Figure 2. This relaxation
is associated with the association/dissociation of micellar aggregates, and these data also show that the polymer affects the
relaxation time of this process. It should be noted that there
is a polymer molecular weight dependence as well.

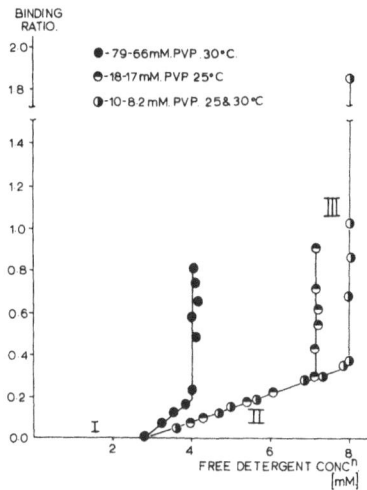

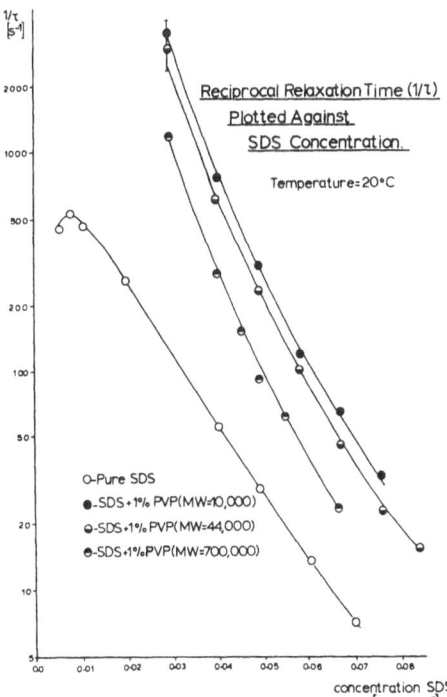

Figure 1   Plot of binding ratio against the free detergent concentration. The binding ratio is defined as the quantity
[conc^n of bound surfactant/total number of monomer units in the
polymer].

Figure 2

From an examination of the temperature dependance of the
relaxation time, the overall activation energy (Ea) of the
micelle association/dissociation process has been measured as a
function of the polymer concentration and is shown in Figure 3.

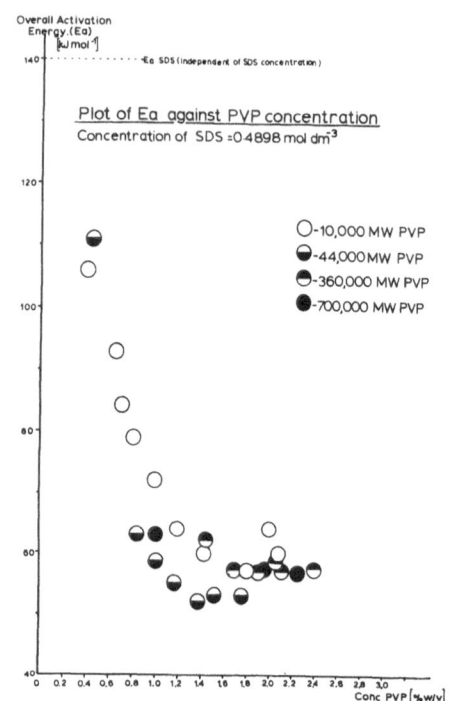

Figure 3

These results show a decrease in Ea with increasing PVP concentration to a limiting value of 57 kJ mol^{-1}. In a pure micellar
solution of SDS Ea has a constant value of 140 kJ mol^{-1}.

It is reasonable to argue that in polymer solutions containing surfactant micelles there will be free micelles present in
solution as well as micelles incorporated on the polymer as a
polymer/micellar complex. The evidence from the present work,
suggests that for corresponding solutions the dissolution time
of the free micellar aggregates will differ from that of those
present on the polymer. Experimentally, only one such relaxation
time was observed. This suggests that the majority of the
micelles are incorporated on the polymer and possibly that the
collision times between free micelles and the polymer/micelle
complex are at least an order of magnitude more rapid than the
observed slow relaxation time. These conclusions therefore lead
us to propose the following mechanism:

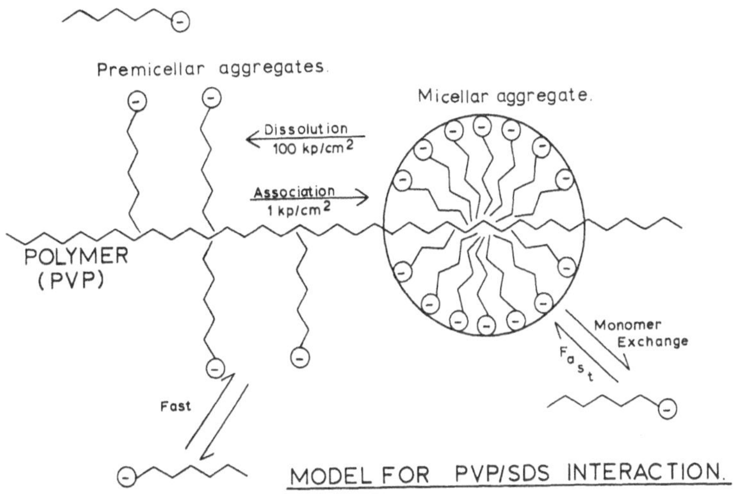

Figure 4

For the formation of micelles in these solutions, such a model would allow faster relaxation times than the pure surfactant because the dissociation-association process has been restricted to the one dimension along the polymer chain; these micellar aggregates are formed by diffusion of monomers along the polymer chain. In this model the polymer acts as a nucleus for surfactant aggregates by promoting clusters of monomers in the form of pre-micellar aggregates or sub-micelles.

The relationship between the relaxation time and the molecular weight of the polymer is more difficult to explain in terms of this model and is probably associated with a conformation change in the polymer. This is presently being investigated with a range of different polymers.

CONCLUSIONS

The data show that in micellar solutions containing polymer, the micellar aggregates are incorporated on the polymer chain which lends support to the "necklace" model invoked by Fishman et al. (4) and Shirahama et al. (5) from equilibrium studies.

ACKNOWLEDGEMENTS

We wish to thank the SRC for a European Short Visit Grant (E.W.-J.), a research grant and an additional award to one of us (D.M.B.) so that these measurements could be carried out at the

Max-Planck-Institute for Biophysical Chemistry, Göttingen, W. Germany. We also wish to thank Professor Strehlow for his hospitality in allowing D.M.B. to work in this laboratory.

REFERENCES

1. Breuer, M.M., Robb, I.D., Chem. and Ind., July 1972, 530.
2. Fishman, M.L., Eirich, F.R., J. Phys. Chem., 75, 1971, 3135.
3. Aniansson, E.A.G., Ber. Bunsenges. Phys. Chem., 82, 1978, 981.
4. Fishman, M.L., Eirich, F.R., J. Phys. Chem., 79, 1975, 2749.
5. Shirahama, K., Tsujii, K. and Takagi, T., J. Biochem., 75, 1974, 309.

ULTRASONIC INVESTIGATION OF THE SOLUBILIZATION OF AN ALCOHOL IN VISCOELASTIC MICELLE FORMING SYSTEMS

O. Bauer, S. Gravsholt*, J. Rassing and J. Thamsen

Department of Chemistry AD, Royal Danish School of Pharmacy, Copenhagen, Denmark

*Physical Chemical Institute, The Technical University, Lyngby, Denmark

ABSTRACT

The ultrasonic relaxation data on solutions of mixed micelles containing CTAB (Cetyl Trimethyl Ammonium Bromide), Sodium Salicylate and Pentanol have been compared with the data obtained for the system without the presence of Sodium Salicylate. The remarkable difference observed may be related to structural changes associated with the viscoelastic behavior that is caused by the presence of the latter component.

INTRODUCTION

Probably the most significant property of micelle forming systems, from a technical and biological point of view, is the ability of dissolving substances that are insoluble in pure water. The mechanism for the solubilization may be formulated in terms of the pseudo-phase separation model for micellization according to which the solute is distributed among the micelle phase and the water phase obeying the general equilibrium distribution law. The equilibrium is dynamic in the way that a continual exchange of solute molecules takes place between the micelles and the bulk water. It has been shown previously (1) that the exchange process involving the alcohols propan-1-ol to hexan-1-ol and CTAB micelles is characterized by a single relaxation time in the ultrasonic frequency range 0.7 - 105 MHz. The dynamics of this process was described in terms of the kinetic principles previously

W. J. Gettins and E. Wyn-Jones (Eds.). Techniques and Applications of Fast Reactions in Solution. 271–274.
Copyright © 1979 by D. Reidel Publishing Company.

applied to the interpretation of micelle/monomer interactions in
micelle forming systems (2). The Treatment has recently been extended in a way that allows calculations of the rate constants
for the dynamic exchange of solute molecules between micelles and
bulk solution (3). Furthermore it has been shown that this kinetic
treatment predicts an equilibrium condition that alternative experimental methods verify.

It has been discovered that addition of relatively small amounts
of different chemicals makes the CTAB-system viscoelastic (4).
The viscoelasticity has a vast influence on the mechanical and
optical properties of the system but the mode of this influence
is not clearly resolved. The current discussion concerning the
molecular mechanism responsible for the viscoelasticity involves
suggestions about changing micelle shapes, micelle sizes, and the
existance of different types of suddenly dominating forces for
instance electrostatic forces. Here we report our ultrasonic data obtained from the viscoelastic system that is created by adding Sodium Salicylate to CTAB dissolved in water.

EXPERIMENTAL

The ultrasonic absorption data for all solutions studied were
measured over the frequency range 0.7-100 MHz with a resonator (5)
and a conventional pulse technique. The CTAB was obtained from
MERCK in the grade "zür analyse". The viscoelastic behavior was
detected by a visual method and by the light scattering observed
when the system placed between crossed nicols was exposed to
mechanical stress. The characteristic trend in the observed data
appears from Table 1.

TABLE 1

Typical values of α/f^2 (α denotes the sound absorption coefficient and f denotes the frequency) at certain selected frequencies for CTAB (0.1 M), Sodium Salicylate (0.05 M) and different concentrations of Pentanol. The values given in the brackets are obtained for the systems with no Sodium Salicylate present.

	α/f^2 10^{17} cm^{-1}s^2			
[Pentanol]	f=0.72	f=1.2	f=3.2	f=9.9 MHz
0.1	371 (280)	310 (265)	262 (204)	90.1 (91.0)
0.2	741 (483)	618 (474)	439 (370)	123 (148)
0.3	2664 (575)	1715 (560)	676 (500)	140 (166)
0.4	3742 (780)	2304 (800)	824 (740)	183 (236)

DISCUSSION

The ultrasonic relaxation observed in CTAB/Pentanol/water-systems has previously been interpreted in terms of a perturbation of the equilibrium between alcohol molecules in the water phase and micelle phase respectively. Generally speaking the sound wave detects the alcohol molecules jumping in and out of the CTAB-micelles. The interesting question is whether or not this process is affected by additives that make the system viscoelastic. It is important to notice that the ultrasonic absorption data at different frequencies for CTAB in pure water are pretty close to those obtained from CTAB made viscoelastic by adding Sodium Salicylate. Thus the ultrasonic wave hardly detects the change that causes the viscoelasticity.

From Table 1 it appears, however, that the addition of Sodium Salicylate (0.05 M) affects the values of α/f^2 to a very large extent at lower frequencies for the system in question. Qualitatively it may be argued that the relaxation time for the alcohol jumping in and out of the micelles is relatively independent of

the presence of Sodium Salicylate whereas the amplitude parameter
as indicated by the large values of α/f^2 at lower frequencies
increases considerably. It is shown that this change is not directly correlated with the observed variations in the viscosity.
The results may be explained by either a shift in the equilibrium
constant for the distribution of alcohol between the two phases
and/or by variation in the volume change accompanying the solubilization process. Thus the observed effect may be understood
in terms of a shift in the micelle size and/or shape including
changes in the packing of monomers on the micelle surface. It is
important to mention that the system containing 0.4 M Pentanol
forms liquid crystals and that the system containing 0.2 M Pentanol as an exception does not show any viscoelastic behavior at
all. At the present stage of the work it seems reasonable to
conclude that the observed effect in the ultrasonic parameters
is related to structural changes in the micelles and not specifically to the viscoelastic behavior. Further investigation to
threw light on this hypotheses is in preparation.

ACKNOWLEDGEMENT

The authors want to express their gratitude to The Danish Science
Research Council for providing funds for the resonance gear.

REFERENCES

1) D. Hall, P.L. Jobling, E. Wyn-Jones, and J. Rassing
    J.C.S.Faraday II, 73 (1977) 1582.

2) J. Rassing, P.J. Sams, and E. Wyn-Jones, J.C.S.Faraday II
    70 (1974) 1247.

3) D. Hall, P.L. Jobling, J. Rassing, and E. Wyn-Jones,
    J.C.S.Faraday II, 74 (1978) 1957.

4) S. Gravsholt, J.Colloid.Sci. 57 (1976) 575.

5) F. Eggers, Acustica, 19 (1968) 323.

REACTION KINETICS IN AQUEOUS MICELLAR SYSTEMS: THE MECHANISM OF THE REACTION BETWEEN Ni^{2+}(aq) AND NEUTRAL BIDENTATE LIGANDS OF DIFFERING HYDROPHOBICITIES IN THE PRESENCE OF SODIUM DODECYL-SULPHATE MICELLES.

V. C. Reinsborough* and B. H. Robinson

Chemical Laboratory, University of Kent at Canterbury, U.K.

ABSTRACT

The kinetics of the reaction between Ni^{2+}(aq) and the ligands 2,2'-bipyridyl and 4,4'-dimethyl-2,2'-bipyridyl have been investigated in micellar SDS solutions. A maximum rate enhancement is found close to the critical micelle concentration. Since the ligand partitions significantly between the micellar pseudo-phase and the aqueous region, the increase in reaction rate observed is related to the hydrophobicity of the ligand. A quantitative expression for the variation in rate with micelle concentration is derived, which includes terms allowing for partitioning of both Ni^{2+} and the ligand between the micellar and the aqueous phases.

A detailed analysis of the results suggests that when the ligands are associated with the micelle they are entirely available for reaction at the micelle surface. The kinetic results also indicate that the ligand is not involved in the rate-limiting step for reaction on the micelle surface. The rate-limiting step is, in fact, associated with release of a water molecule from the solvation shell of the metal ion prior to complexation with the ligand. The rate constant for this process is similar to that measured in aqueous solution in the absence of micelles.

---

* On sabbatical leave from:
  The Department of Chemistry,
  Mount Allison University,
  Sackville, New Brunswick, Canada.

W. J. Gettins and E. Wyn-Jones (Eds.). Techniques and Applications of Fast Reactions in Solution. 275–281.
Copyright © 1979 by D. Reidel Publishing Company.

The number of papers dealing with the kinetics of chemical reactions in the presence of micelles has greatly increased since 1960. For most reactions, only modest rate changes are observed in the presence of micelles but for some reactions rates are increased by factors approaching 10^6. Results reported in the literature up to 1974 have been extensively reviewed (1).

Until recently, a sound theoretical basis for the mechanistic interpretation of these kinetic measurements has been lacking. It is now recognised that several distinct factors can influence reactivity in micellar solutions. These include (i) partitioning coefficients for the reactants between the micelle and aqueous solution, (ii) kinetic effects due to a change in the environment in which the reaction takes place (either through a medium effect or direct involvement of the surfactant head-groups in the reaction) and (iii) the influence of the micelle surface potential on the pK_a's etc. of the reactants. These factors have recently been incorporated into a general kinetic theory by Berezin et al. (2).

In this paper, the physico-chemical approach outlined above will be applied to the analysis of metal-ligand complex formation reactions occurring in the presence of charged micelles. A kinetic study should provide some insight into the following relevant questions:

(i)   Where does the reaction predominantly occur?

(ii) How rapidly do the reactants (which are an ion and a molecule in our experiments) diffuse on/in micelles?

(iii) How rapidly do the reactants exchange between micelles?

(iv) Are the reactants in the correct configuration for reaction when associated with the micelle?

(v)   Can any change in rate which is observed be explained solely in terms of the concentrative effect of the micelles on the reactants (i.e. in terms of partitioning coefficient effects) or is there some additional specific influence of the micellar environment on the kinetics?

From previous results (3) obtained for the reaction using Ni^{2+}(aq) and the related coloured ligand pyridine-2-azo-p-dimethylaniline (PADA), it was concluded that in the presence of excess micellar SDS, both Ni^{2+}(aq) and PADA were strongly adsorbed at the micelle surface where reaction occurs. The rate of reaction was markedly pH-dependent due to the protonation of the ligand, and a shift in the pK_a of ~2 units was observed on interaction of the ligand with the micelle.

The main aim of the present study is, therefore, to investigate the effect of varying the hydrophobicity of the ligand on the kinetics and mechanism of the reaction.

EXPERIMENTAL METHODS

The stopped-flow method with UV detection was used to follow the kinetics. To prevent protonation of the ligands on the micelle surface the pH of the solutions was controlled within the region 7.8 → 8.1 using low concentrations ($<10^{-3}$ mol dm^{-3}) of Na_2HPO_4/NaH_2PO_4 buffers. The solubilities of the ligands in water and SDS solutions were measured by preparing saturated solutions, followed by analysis using UV spectrophotometry.

RESULTS

The observed rate constants (k_{obs}) for complexation in micellar solutions were found to be in the order PADA > Dimbipy > Bipy. In all cases maximum values were determined in the region of the cmc. In this region the reaction rate (compared with that measured in aqueous solution for the same Ni^{2+} concentration) was enhanced by a factor of ~100 for Bipy, at least a factor of 100 with Dimbipy, and up to a factor of 10^3 for PADA.

From Table 1, it can be seen that the order of ligand solubilities in aqueous solution is PADA < Dimbipy << Bipy, so that Bipy is considerably more hydrophilic than the other two ligands. From solubility measurements in 0.1 mol dm^{-3} SDS, a solubility enhancement factor (SHF) can be determined. The order of values obtained is PADA >> Dimbipy >> Bipy.

Ligand	Solubility in Water (S_o) /mol dm^{-3}	Solubility in 0.1 mol dm^{-3} SDS (S') /(mol dm^{-3})	SHF $(=S'/S_o)$	K_L /(dm^3 mol^{-1})
PADA	1.5×10^{-4}	4.3×10^{-2}	285	2800
Dimbipy	4.1×10^{-4}	1.0×10^{-2}	24.5	250
Bipy	4.0×10^{-2}	1.1×10^{-1}	2.7	40

Table 1. Solubility of Ligands in Water and in 0.1 mol dm^{-3} SDS at 291K.

The increase in solubility of the ligands in SDS solution
is due to them being associated with the micelle in some way.
The order of free energies of transfer of the ligands to the
micelle is reflected in the SHF values. (Detailed kinetic and
solubility measurements are given in the full paper (4).

DISCUSSION

Reactions between Ni^{2+}(aq) and bidentate ligands in aqueous
solution generally take place according to the Eigen-Wilkins
mechanism.   (See the paper by Greenwood, Robinson and White
(5) for a fuller discussion.)   Formation of the monodentate
complex is rate-limiting.   Since the observed rates are
essentially independent of the ligand, the rate-determining
step is loss of a solvating water molecule from the metal ion.

In the presence of SDS, the reactants will be adsorbed onto
the micelle surface to an extent depending on the hydrophobicity
of the ligand.   The Ni^{2+}(aq) ion will be located in the Stern
layer and held there by coulombic attraction to the micelle
surface.   However, ion migration over the surface should be
rapid.   The micellar binding site for the ligand is more
uncertain.   Because of the polar groups on the ligand it might
be expected that it would be located close to the micelle-water
interface rather than in the hydrophobic core of the micelle.
(There is evidence (6) that even non-polar molecules like
benzene are preferentially located in the interface region).

For the 1:1 complexation reaction

$$Ni^{2+}(aq) + L \rightleftharpoons Ni^{2+}L \qquad\qquad (A)$$

if reaction   occurs on the micelle surface, the rate of complex
formation will depend on (i) the nickel ion surface concentration
rather than its bulk concentration value (in mol dm^{-3} of
solution) and (ii) the accessibility of the ligand at the micelle-
water interface.   If both reactants are totally adsorbed at
the micelle-water interface, then for total concentrations of
$[Ni^{2+}]_T \gg [L]_T$ and for rapid mobility of metal ion and/or
ligand on/between micelles, we have:

$$k_{obs} = k_1^M [Ni^{2+}]_T / \{\{[SDS]_T - cmc\}A\, N_{AV}\} + k_{-1}^M \qquad (1)$$

It should be noted that the surface concentration of Ni^{2+},
described as $[Ni^{2+}]_S$, can be simply expressed as:

$$[Ni^{2+}]_S = [Ni^{2+}]_T / \{\{[SDS]_T - cmc\}A\, N_{AV}\} \quad mol\ m^{-2} \qquad (2)$$

where A is the surface area per SDS head group.   The rate constants k_1^M and k_{-1}^M then represent rate constants appropriate to a surface reaction and have the dimensions $m^2 \, mol^{-1} \, s^{-1}$ and s^{-1} respectively.

Although the mean aggregation number of SDS micelles may be concentration dependent, the head group of a micellar surfactant molecule is always present at the micelle-water interface, so that A is not likely to be significantly concentration dependent unless the packing in the micelles changes.   Hence a plot of k_{obs} vs $[Ni^{2+}]_T/\{[SDS]_T-cmc\}$ should be linear if reaction takes place exclusively on the micelle surface.   Such a linear plot is found for the reaction between Ni^{2+} and PADA (for concentrations > 2 cmc), from which a slope of 3000 s^{-1} is obtained (3).   When such a plot is attempted for the ligands Bipy and Dimbipy a non-linear plot is obtained, the deviations being most pronounced for Bipy.   Such kinetic measurements indicate that the ligand is not confined solely to a micellar binding site.

When the reactants partition significantly between the micellar and aqueous regions, the simple equilibrium scheme (A) requires expansion as in scheme (B) below:

$$
\begin{array}{ccccc}
Ni^{2+}{}_{aq} & + & L_{aq} & \underset{k_{-1}}{\overset{k_1}{\rightleftharpoons}} & NiL^{2+}{}_{aq} \\[2ex]
\Big\updownarrow K_{Ni} & & \Big\updownarrow K_L & & \Big\updownarrow K_C \\[2ex]
Ni^{2+}{}_M & + & L_M & \underset{k_{-1}^M}{\overset{k_1^M}{\rightleftharpoons}} & NiL^{2+}{}_M
\end{array}
\qquad (B)
$$

K_{Ni}, K_L and K_C represent equilibrium constants for partitioning of reacting components between the micellar and aqueous phases, as indicated in (B).

We can define K_{Ni} etc. by expressions of the type:

$$
K_{Ni} = \frac{\text{surface concentration of nickel}}{\text{aqueous concentration of nickel}} =
$$

$$
[Ni^{2+}]_M / \{[Ni^{2+}]_{aq}\{[SDS]_T-cmc\}\}
$$

$$(3)$$

$[Ni^{2+}]_M$ is the micelle-bound nickel ion concentration expressed as mol per dm^3 of solution, so that K_{Ni} has units of $dm^3 \, mol^{-1}$.

The value of K_{Ni} can be obtained by application of the coupled indicator reaction (Ni^{2+} and $Murexide^-$) as indicated elsewhere in this publication (7). The value of K_L can be obtained from solubility data since $K_L = [L]_M/\{[L]aq([SDS]_T-cmc)\}$ which is approximately equal to $(S-S_o)/(S_o([SDS]_T-cmc))$ where S and S_o represent the solubilities in mol dm^{-3} of the ligand in SDS solution and pure water respectively. The derived values of K_L at 291K are as shown in Table 1.

Equation (4) below can be derived when metal ion and ligand partitioning is rapid compared with the complexation reaction, and for $[Ni^{2+}]_T \gg [L]_T$.

$$k_{obs} = \frac{(k_1^M/(AN_{AV}))[Ni^{2+}]_T}{([SDS]_T-cmc)(1+K_{Ni}^{-1}([SDS]_T-cmc)^{-1})(1+K_L^{-1}([SDS]_T-cmc)^{-1})}$$
$$+ k_{-1}^M \qquad (4)$$

For PADA, K_{Ni}^{-1} and K_L^{-1} are much less than $([SDS]_T-cmc)$ for $[SDS]_T > 2cmc$, so that the simpler equation (1) can be used. However, for both Bipy and Dimbipy, the term $K_L^{-1}([SDS]_T-cmc)^{-1}$ contributes significantly to eqn. (4). After due allowance has been made for partitioning effects, we obtain values of $k_1^M/(AN_{AV})$ of 2800 s^{-1} for Dimbipy and 3300 s^{-1} for Bipy at 298K. These values are very similar to that found previously for PADA ($k_1^M/(AN_{AV})$ = 3000 s^{-1} at 298K). That these values are more or less independent of the nature of the ligand suggests that the ligand is not involved in the rate-determining step of the reaction, and further, that all three ligands are entirely available for reaction at the micelle surface. Taking a value for A of 60×10^{-20} m^2, we obtain an average value of k_1^M of 1.1 $(\pm0.1)\times10^9$ m^2 mol^{-1} s^{-1} for the rate constant for reaction on the micelle surface for all three ligands. From this data it is possible to derive a value of $k_{ex}^M \sim 10^4$ s^{-1} for the rate constant to water loss from the metal ion on the micelle surface (4).

Acknowledgements: We thank Mount Allison University, Sackville, New Brunswick, for a sabbatical grant (to VCR) and the SRC for providing financial support for equipment used in this investigation.

REFERENCES

(1) J.H. Fendler and E.J. Fendler, "Catalysis in Micellar
     and Macromolecular Systems", Academic Press, New York,
     1975.

(2) I.V. Berezin, K. Martinek and A.K. Yatsimirski, Russ. Chem.
     Revs. (Usp. Khim.), 1973, 42, pp. 787.

(3) A.D. James and B.H. Robinson, J. Chem. Soc., Faraday Trans.
     I, 1978, 74, pp. 10.

(4) V. Reinsborough and B.H. Robinson, J. Chem. Soc., Faraday
     Trans. I, to be published, 1979.

(5) R.C. Greenwood, B.H. Robinson and N.C. White, this
     publication.

(6) P. Mukerjee, Ber. Bunsenges. Phys. Chem., 1978, 82, pp. 931.

(7) M. Fischer, W. Knoche, B.H. Robinson, J. MacLagan Wedderburn
     and N.C. White, this publication.

KINETICS AND MECHANISM OF METAL-COMPLEX FORMATION IN AQUEOUS MICELLAR AND OIL INVERSE-MICELLAR SYSTEMS

P. D. I. Fletcher and B. H. Robinson

Chemical Laboratory, University of Kent at Canterbury, U.K.

ABSTRACT

The kinetics of the reaction between Ni^{2+}(aq) and the bidentate ligand pyridine-2-azo-phenol has been studied in different colloidal systems, namely, the aqueous micellar system formed by sodium dodecyl sulphate and an inverse-micellar system based on aerosol-OT, water and heptane. The reaction occurs at the surfactant-water interface in both colloidal systems and a kinetic treatment appropriate to a surface reaction is derived. The rates, mechanism and energetics are found to be similar for the reaction in both media. Activation energies are close to that measured in bulk water indicating that the rate-determining step in the reaction is loss of a water molecule from the nickel ion in all the media investigated.

The kinetics and mechanisms of ligand-substitution reactions involving labile metal ions in aqueous solution are now well understood (1). It is therefore of interest to investigate how such reactions are affected when the reaction medium is changed either by addition of charged micelles to an aqueous solution or by dispersing the reactants in an inverse-micellar (water-in-oil microemulsion) system.

MECHANISM OF THE REACTION IN WATER

We have studied the reaction between Ni^{2+}(aq) and the bidentate ligand pyridine-2-azo-p-phenol (p-PAP) (I) to form the metal-ligand complex (II).

W. J. Gettins and E. Wyn-Jones (Eds.). Techniques and Applications of Fast Reactions in Solution. 283–290.
Copyright © 1979 by D. Reidel Publishing Company.

$$\text{Ni}(H_2O)_6^{2+} + \text{(pyridyl-azo-phenol, I)} \xrightleftharpoons[\text{Fast}]{K_{os}} [(H_2O)_5\text{Ni}(H_2O)p\text{-PAP}]^{2+}$$

outer-sphere complex

I

II

In aqueous solution the reaction proceeds via a fast pre-equilibrium involving an outer-sphere complex. The rate of complex formation is dominated by the rate of loss of a water molecule from the first co-ordination shell of the metal ion. (A relatively fast ring-closure step to form the bidentate complex from the monodentate complex is omitted in the kinetic scheme for clarity.)

For initial concentrations $[\text{Ni}^{2+}]_T \gg [p\text{-PAP}]_T$ and $K_{os} \ll [\text{Ni}^{2+}]_T^{-1}$ the observed rate constant for complex formation (k_{obs}) is given by equation 1:

$$k_{obs} = K_{os}\, k_{ex}\, [\text{Ni}^{2+}]_T + k_b \qquad \ldots\ldots(1)$$

K_{os} is the equilibrium constant for outer-sphere complex formation, k_{ex} is the first order rate constant for water loss from the nickel aquo-ion and k_b is the first-order rate constant for the decomposition of the complex.

THE NATURE OF THE REACTION IN THE DISPERSED MICELLAR SYSTEMS

The reaction was studied both in aqueous micellar solutions and in oil inverse-micellar solutions in order to investigate the effect of charged interfaces on metal ion reactivity.

The anionic surfactant sodium dodecyl sulphate was used to form the aqueous micellar system. In order to ensure that no protonated or de-protonated species were participating in the reaction the pH behaviour of the system was studied spectro-photometrically in both water and micellar solutions. The approximate derived pK_a values in both bulk water and micellar SDS are shown below.

$$PAP^- \xrightleftharpoons{8.1} PAP \xrightleftharpoons{\sim 3} PAPH^+ \left.\right] \begin{array}{l} \text{bulk} \\ \text{water} \end{array}$$

$$NiPAP^+ \xrightleftharpoons{6.5} NiPAP^{2+} \xrightleftharpoons{<3} PAPH^+$$

$$PAP^- \xrightleftharpoons{9.5} PAP \xrightleftharpoons{4.8} PAPH^+ \left.\right] \begin{array}{l} \text{micellar} \\ \text{SDS} \end{array}$$

$$NiPAP^+ \xrightleftharpoons{\sim 9.5} NiPAP^{2+} \xrightleftharpoons{3} PAPH^+$$

The pK_a values are increased by 2-3 pH units in the micellar SDS due primarily to a micelle surface potential effect.   This is discussed in more detail in references (2) and (3).   It can be seen that the optimum pH to follow the kinetics of the reaction $Ni^{2+} + PAP \rightleftharpoons NiPAP^{2+}$ is approximately pH 5 in bulk water and approximately pH 7 in micellar SDS solution.   By working at these pH values the subsequent kinetic analysis is greatly simplified.

The inverse micellar system used was that of water dispersed in heptane using the surfactant sodium bis-ethylhexyl sulphosuccinate (AOT) as a dispersant.

AOT

The inverse-micellar system $AOT/H_2O/Heptane$ has been characterised by a variety of physical methods including light scattering (4), ultracentrifugation (5), and viscosity measurements (4).   The water phase is dispersed as micro-droplets in the organic phase; the structure being shown on the following page.

The water droplets are spherical and the size has been found to depend primarily on the ratio (R) of water concentration to AOT concentration.   Typically, as the R value was varied from 5 → 20 the diameter of the droplet water core increased from 20 → 50 Å and the number of water molecules contained in a droplet increased from 250 → 2500.

It was found the pK_a's involved in the inverse micelle reaction could not readily be measured spectrophotometrically

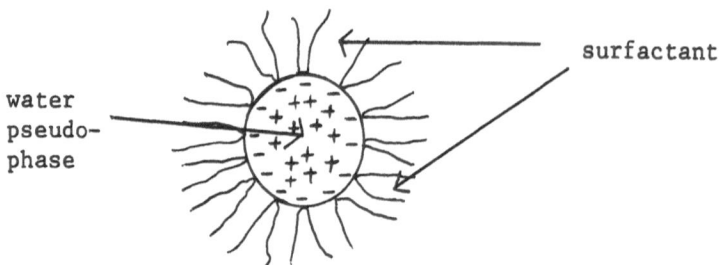

Structure of an AOT-stabilised inverse micelle.

since the inverse micelles were unstable on addition of high
concentrations of acid or base.   Furthermore, in weakly acidic
or basic solutions there are problems in defining the pH of the
aqueous droplets.   However, the spectra were unaffected by the
addition of small amounts of acid or base, and, by analogy with
the spectra obtained in bulk aqueous solution, it was concluded
that the reaction being studied was Ni^{2+} + PAP $\rightleftharpoons$ $NiPAP^{2+}$.

LOCATION OF THE REACTION

     The reaction can proceed in three possible locations:
1.   The bulk dispersion medium; i.e. water in the aqueous micellar
system or heptane in the inverse-micellar system.
2.   The micellar core which for the aqueous micelle is hydrocarbon and for the inverse-micelle is water.
3.   In the surfactant-water interface region.

     Since partitioning between the phases can, in principle,
occur, the reaction could, however, be occurring simultaneously
in different environments.

     We believe the reaction in both the aqueous micellar and
inverse-micellar systems takes place at the surfactant-water
interface for the following reasons:

1.   The doubly-charged Ni^{2+} aquo ion has been shown to be
located at the SDS surfactant-water interface for aqueous micelles
using a spectrophotometric probe method (6) (7).   This would be
expected since a doubly-charged cation should be strongly
attracted by coulombic forces to a negatively-charged surface.

2.   p-PAP is insoluble in heptane, only very sparingly soluble
in water but readily soluble in SDS aqueous solutions and AOT
solutions in heptane.   From solubility measurements in various
aqueous solutions of SDS it was possible to show that more than
99% of the p-PAP was associated with the surfactant at concentrations greater than twice the cmc of SDS.   (The cmc of SDS in
pure water is ~8 x 10^{-3} mol dm^{-3} at 298.2K.)

3. The rate enhancement observed suggests that a bulk reaction is unlikely.

4. The apparent pK_a values were all shifted in the presence of SDS micelles. This suggests that the p-PAP was experiencing a pH different from the bulk aqueous value.

5. The kinetic analysis is consistent with the site of reaction being located at the surfactant-water interface region.

KINETIC RESULTS AND DISCUSSION

The kinetics were measured using a small volume stopped-flow instrument incorporating optical detection.

In the inverse-micellar system the pseudo-first-order rate constant (k_{obs}) for $[Ni^{2+}] > 10$ [p-PAP] was measured for a range of AOT concentrations (0.035 mol $dm^{-3} \rightarrow 0.11$ mol dm^{-3}) and R values (= $[H_2O]/[AOT]$) ranging from ~5 $\rightarrow$ 20. A plot of k_{obs} vs $[Ni^{2+}]_T/[AOT]$ gives a single straight line independent of the water concentration except at high R values. Since the concentration of Ni^{2+} at the surfactant-water interface is inversely proportional to the AOT concentration, this result is consistent with the proposed reaction scheme. (At high water concentrations there is some indication of partitioning of the Ni^{2+} into the core of the aqueous droplet. Hence it is possible from these kinetic measurements to obtain information on the distribution of Ni^{2+} ions within the aqueous droplet.)

The aqueous micellar system yields analagous results. The kinetic measurements were carried out over a range of SDS concentrations (0.011 mol $dm^{-3} \rightarrow 0.101$ mol dm^{-3}). In this case the surface concentration of metal ion is proportional to $[Ni^{2+}]_T/([SDS]-cmc)$. A plot of k_{obs} vs $[Ni^{2+}]_T/([SDS]-cmc)$ again yielded a straight line plot up to a percentage surface coverage by Ni^{2+} of ~2.5%. (The cmc was corrected for the effect of lowering of the cmc on addition of metal ion.)

For the bulk aqueous reaction the observed pseudo-first-order rate constant is given by equation (2)

$$k_{obs} = K_{os}^b \, k_{ex}^b \, [Ni^{2+}]_T + k_b^b \qquad \dots.(2)$$

where the superscript b refers to reaction in the bulk medium.

For reaction at the charged surfactant-water interface a similar equation (equation (3)) holds. The superscript s denotes that the parameter refers to reaction at the surface (i.e. surfactant-water interface region).

$$k_{obs} = K^S_{os}\, k^S_{ex}\, [Ni^{2+}]_s + k^S_b \qquad \dots (3)$$

It should be noted that K^S_{os} is an equilibrium constant involving surface concentrations. From the definition below it can be seen that it has units of $mol^{-1}\, m^2$. $[Ni^{2+}]_s$ has units of $mol\, m^{-2}$.

$$K^S_{os} = \frac{[\text{outer-sphere complex}]_s}{[Ni^{2+}]_s\, [p\text{-PAP}]_s}$$

In the SDS micellar solutions the surface concentration of Ni^{2+} ($= [Ni^{2+}]_s$) can be expressed as the total Ni^{2+} concentration ($[Ni^{2+}]_T$) divided by the total area of surfactant-water interface and is given in equation (4)

$$[Ni^{2+}]_s = \frac{[Ni^{2+}]_T}{([SDS]_T - cmc)\, A\, N_{AV}} \qquad \dots (4)$$

where A = the area of one surfactant molecule headgroup in m^2
N_{AV} = Avogadro's number
cmc = critical micelle concentration.

In the reversed-micellar system a similar equation can be derived (equation 5), assuming the concentration of free AOT in the heptane is negligible.

$$[Ni^{2+}]_s = \frac{[Ni^{2+}]_T}{[AOT]_T\, A\, N_{AV}} \qquad \dots (5)$$

A surface rate constant for metal-complex formation (in units s^{-1}), k_{surf}, can be defined for these systems as follows.

$$k_{surf} = \frac{K^S_{os}\, k^S_{ex}}{A N_{AV}}\ s^{-1}$$

Values for the reaction between Ni^{2+} and p-PAP in both systems are given in the table. The values in both systems are comparable since the values of A for both SDS and AOT have been shown to be ~60 $\overset{\circ}{A}^2$.

Activation energies were measured for both types of micellar system assuming that negligible partitioning of the reactants away from the surface occurred (8).

The surface rate constants and the enthalpies of activation for both systems are shown below and are compared with data obtained previously for the system Ni^{2+} and pyridine azo dimethylaniline (PADA) in aqueous SDS solution. The table refers to 298.2K.

	$k_{surf}/(s^{-1})$	$\Delta H^{\ddagger}_{surf}/kJ\ mol^{-1}$
Ni/PAP/SDS	1700	46
Ni/PAP/AOT	3600	50
Ni/PADA/SDS	3000	52

We conclude:

(a) The rate enhancement effects observed are due to the reactants being concentrated at the surfactant-water interface.

(b) The rate and energetics of these reactions in aqueous micellar solutions are not very dependent on the ligand used.

(c) Reaction at the micellar and inverse-micellar surfactant-water interfaces occurs at similar rates.

(d) We can further infer that the rate-determining step in metal-ligand substitution in all systems studied is loss of water from the metal ion. This is apparently unaffected by any perturbation of water structure due to the interface.

(e) Communication of reactants between micelles and availability at the surface is rapid compared with the rate of water loss from $Ni^{2+}(aq)$.

REFERENCES

1   Hewkin, D.J., and Prince, R.H.: Coordination Chem. Revs., 1970, 5, pp. 45-73.

2   James, A.D., and Robinson, B.H.: J. Chem. Soc. Faraday Trans. I, 1978, 74, pp. 10.

3   Mukerjee, P., and Banerjee, K.: J. Phys. Chem., 1964, 68, pp. 3567.

4   Day, R.A., Robinson, B.H., Clarke, J.H.R., and Doherty, J.V.: J. Chem. Soc. Faraday Trans. I, 1979, 75, pp. 132.

5   Robinson, B.H., Steytler, D.C., and Tack, R.D.:   J. Chem.
     Soc. Faraday Trans. I, 1979.   To be published.

6   Fischer, M., Knoche, W., Robinson, B.H., and Maclagan
     Wedderburn, J.H.:   J. Chem. Soc., Faraday Trans. I, 1979,
     75, pp. 119.

7   Holzwarth, J., Knoche, W., and Robinson, B.H.:   Ber.
     Bunsenges. Phys. Chem., 1978, 82, pp. 1001.

8   Reinsborough, V., and Robinson, B.H: this publication.

APPLICATION OF A METAL ION/MUREXIDE INDICATOR SYSTEM TO THE STUDY OF MICELLAR PHENOMENA

Brian H. Robinson, John H. Maclagan Wedderburn,
Neal C. White, Michael Fischer* and Wilhelm Knoche*

University of Kent, Canterbury, U.K.

*Max-Planck-Institut für biophysikalische Chemie,
Göttingen, West Germany

Introduction

The chromophoric anion murexide (Mu^-) shown in fig. 1 is used to determine metal ion concentrations in aqueous solution.

The reaction of a divalent metal ion M^{2+} with Mu^- ($M^{2+} + Mu^- \rightleftharpoons M Mu^+$) can also be conveniently used as an indicator system for the study of various aspects of micellisation, in particular the determination of critical micelle concentrations (cmc) and rates of micelle formation of anionic surfactants. Before this system can be successfully employed for the study of micellar phenomena, it is necessary to have a clear understanding of the indicator reaction itself in aqueous solution in the absence of surfactant.

This indicator reaction is also of interest for several other reasons:
(i) It is associated with a large volume change (for Ni^{2+}, $\Delta v^o = 23\ cm^3 mol^{-1}$ [1]) so that it is a very suitable test system for evaluating the performance of the pressure-jump technique with optical detection [2].
(ii) both rates and equilibrium concentrations for the reaction are strongly pH-dependent in the region of pH 4.0 to 7.5 where no hydrolysis or protonation of the reactants occurs. Hence the reaction must be more complicated than the simple reaction scheme (fig. 1) indicates.

W. J. Gettins and E. Wyn-Jones (Eds.). Techniques and Applications of Fast Reactions in Solution. 291–299.
Copyright © 1979 by D. Reidel Publishing Company.

(iii) The reaction involves charged species so that activity coefficients need to be introduced when the reaction is studied at
varied ionic strength. Since reaction rates can be obtained to
high precision, it is possible to test the validity of the activity corrections when fast reactions are being studied.

In the course of this work the stopped-flow and pressure-jump
techniques have been used, both equipped with optical detection.
In the case of the pressure-jump measurements, all solutions were
carefully degassed before use and data were analysed on-line after
being processed by an analog to digital converter.

The Ni^{2+}/Mu^{-} Reaction in Aqueous Solution

A suitable metal ion for study with the above techniques is
the aquo-nickel ion Ni^{2+}. Since murexide decomposes rapidly below
pH 4 and Ni^{2+} hydrolyses above pH 7.5, we confined our
measurements to the pH range 4.0 to 7.5. We found the reaction is
strongly pH dependent in this region so that the complex must be
involved in a protonation reaction.

A reaction scheme consistent with our thermodynamic and
kinetic data is

$$Ni^{2+} + Mu^{-} \underset{k_{-1}}{\overset{k_1}{\rightleftarrows}} NiMu^{+} \rightleftharpoons (NiMu)_{dep} + H^{+} \qquad (1)$$

$$(+) \qquad (-) \qquad\qquad (1) \qquad\qquad (2) \qquad (H)$$

$(NiMu)_{dep}$ represents a deprotonated $NiMu^{+}$ species in which a
proton is lost either from the ligand or from a water molecule
solvating the nickel ion in complex (1). The deprotonation reaction is fast compared with metal-complex formation.

Equilibrium Analysis

For our thermodynamic and kinetic analysis we employ the
Davies equation for activity coefficients (γ_i):

$$\log \gamma_i = -0.5 \ z_i^2 \ [I^{0.5}/(1 + I^{0.5}) - 0.3 \ I] \qquad (2)$$

in which γ_i depends only on the ionic charge z_i of species i and
the ionic strength (I). Since $\gamma_- = \gamma_H = \gamma_1 = \gamma_I$, $\gamma_+ = \gamma_{II}$, and $\gamma_2 = 1$
the equilibrium is described by

$$K_1 = k_1/k_{-1} = c_1/(c_+ c_- \gamma_{II}) \qquad (3)$$

and

$$K_H = c_2 c_H/c_1$$

At low pH values (pH < pK_H) the equilibrium is determined mainly by K_1. Measurements at pH > pK_H yield values of $K_1 \cdot (1 + K_H c_H^{-1})$. A typical set of spectrophotometric measurements at pH = 6 is shown in fig 2. Data obtained over the pH range 4.0 to 7.3 were

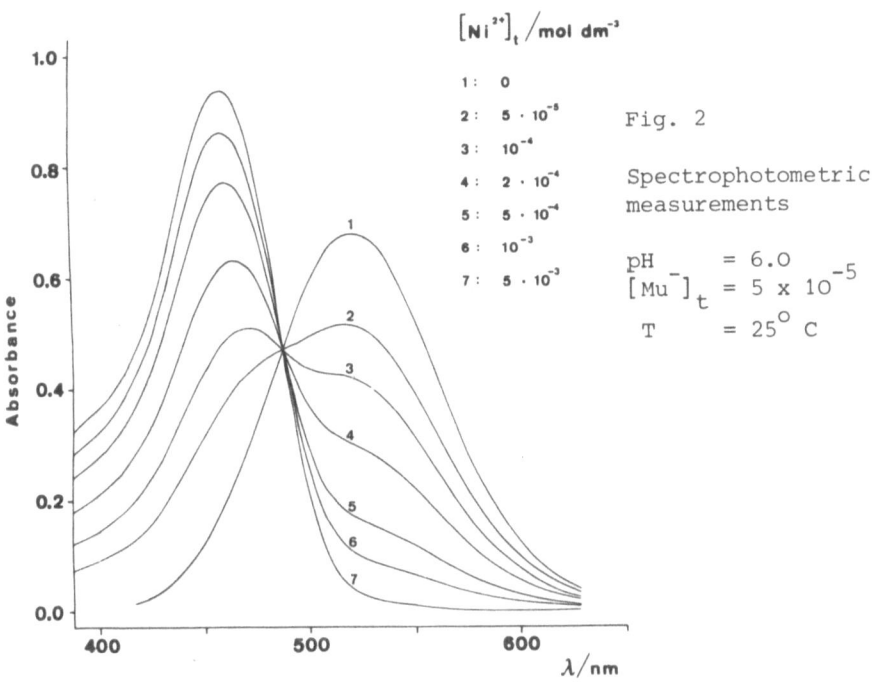

$[Ni^{2+}]_t / mol\ dm^{-3}$

1: 0

2: $5 \cdot 10^{-5}$

3: 10^{-4}

4: $2 \cdot 10^{-4}$

5: $5 \cdot 10^{-4}$

6: 10^{-3}

7: $5 \cdot 10^{-3}$

Fig. 2

Spectrophotometric measurements

pH $\quad = 6.0$

$[Mu^-]_t = 5 \times 10^{-5}$

T $\quad = 25^\circ$ C

analysed by computer to determine K_1 and K_H (See reference 3 for details of the calculation) with the result that

$$K_1 = (3.8 \pm 0.2) \cdot 10^3 dm^3 mol^{-1}$$

$$K_H = (1.8 \pm 0.3) \cdot 10^{-6} mol\ dm^{-3}$$

Results were independent of the wavelength used for the evaluation.

Kinetic Analysis

The kinetic measurements showed a single relaxation effect in the time range 100 µs to 5 s which was dependent on concentration and pH. Since it can be assumed that the protonation reaction is fast compared with metal complex formation we obtain the following kinetic expression working under buffered conditions:

$$1/\tau = k_1 \gamma_{II} (c_+ + c_- \beta) + k_{-1} (1 + K_H/c_H)^{-1} \qquad (5)$$

in which

$$\beta = 1 + d \ln \gamma_{II}/d \ln c_+ \tag{6}$$

Under pseudo-first-order conditions (i.e. $c_+ \gg c_-$), we obtain the simplified equation

$$1/\tau = k_1 \gamma_{II}[Ni^{2+}]_t + k_{-1}(1 + K_H/c_H)^{-1} \tag{7}$$

$[Ni^{2+}]_t$ is the weighed-in concentration of nickel ion. Therefore at constant ionic strength (i.e. γ_{II} constant) a plot of $1/\tau$ vs. $[Ni^{2+}]_t$ will yield a straight line with slope $k_1 \gamma_{II}$ (independent of pH) and intercept $k_{-1}(1 + K_H/c_H)^{-1}$ which strongly depends on pH. When the ionic strength is not kept constant by addition of inert electrolyte, a plot of $1/\tau$ vs. $[Ni^{2+}]_t$ will be curved as shown in fig. 3. This figure gives an indication of the magnitude

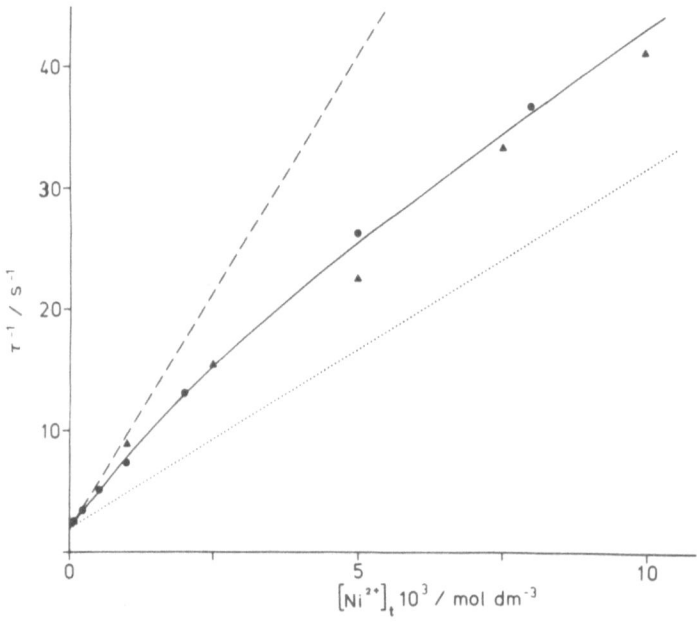

Fig. 3   Plot of $1/\tau$ vs. $[Ni^{2+}]_t$ at pH = 5.0 and 25° C.
Pressure-jump data ●; Stopped-flow data ▲ ;
── Calculated line using equ. (5) and (6);
--- Calculated line assuming $\gamma_{II} = 1$ (I = 0);
··· Calculated line assuming $\gamma_{II} = 0.38$ (I = 0.1)

of the activity coefficient corrections involved. Significant deviations from ideal behaviour ($\gamma_{II} = 1$, dashed line in fig 3) are observed even at nickel concentrations of about 10^{-3} mol dm^{-3}. The continuous line in fig 3 is calculated using eqn. (5) and (6) together with the values of K_1 and K_H quoted above. A value of k_1 of $7.8 \cdot 10^3$ dm^3 mol^{-1} s^{-1} was used, which was obtained from a best fit to all the kinetic data in the pH range 4.0 to 7.3 . (Further

details of the derivations, experimental data and computations
are given in ref. [3]). The mean value of k_1 obtained by pressure-
jump measurements was $(8.05 \pm 0.45) 10^3 dm^3 mol^{-1} s^{-1}$ and that by
stopped-flow measurements was $(7.5 \pm 0.6) 10^3 dm^3 mol^{-1} s^{-1}$. Thus
both methods give good agreement with an average value of $k_1 =$
$(7.8 \pm 0.4) 10^3 dm^3 mol^{-1} s^{-1}$.

Discussion

    According to the theory of Eigen and Tamm, the mechanism of
ligand exchange is given by the following scheme:

$$Ni^{2+} + Mu^- \underset{fast}{\overset{Kos}{\rightleftharpoons}} Ni^{2+}(H_2O)Mu^- \underset{}{\overset{k_{ex}}{\rightleftharpoons}} NiMu^+ \qquad (8)$$

Since the equilibrium constant for the outer-sphere complex
formation $K_{os} \ll 1/[Ni^{2+}]_t$,

$$k_1 = K_{os} \cdot k_{ex} \qquad (9)$$

K_{os} can be estimated by the Fuoss equation [4]. Assuming a formal
charge of -1 on the murexide anion, K_{os} is calculated to be
$\sim 6\ dm^3 mol^{-1}$, and when the effective charge is zero,
$K_{os} \sim 0.3\ dm^3 mol^{-1}$. Since k_{ex} is known to be $\sim 10^4 s^{-1}$, our data
indicate a value for K_{os} of $\sim 0.8\ dm^3 mol^{-1}$ calculated from eqn. 9.
This indicates that the kinetic results are consistent with the
Eigen mechanism, but with considerable charge delocalisation in
the murexide anion.

Application of the Indicator Reaction to the Study of Micellar Phenomena

Equilibrium Measurements

    We have used the $Ni^{2+} + Mu^-$ reaction as a simple spectro-
photometric probe method to monitor micelle formation involving
the anionic surfactant sodium dodecylsulphate (SDS). Fig. 4 shows
spectra obtained as the concentration of SDS is increased at
constant nickel and murexide concentrations. The spectra clearly
indicate that the extent of complex formation decreases as the
concentration of micelles increases.

    An explanation for the spectral changes in fig 4 is as follows:
When micelles form in solution, nickel ions are strongly attracted
into the region of the micelle surface. On the other hand, the
murexide anion is repelled away from the negatively charged
micelle surface and is located solely in the bulk aqueous region.
Therefore, as anionic micelles form in solution, Ni^{2+} ions and
Mu^- ions are effectively separated resulting in less complex
formation. Furthermore it is reasonable to assume that, over the
SDS concentration range studied, the binding of $NiMu^+$ to the
micelle surface may be ignored, since the complex is hydrophilic

Fig. 4

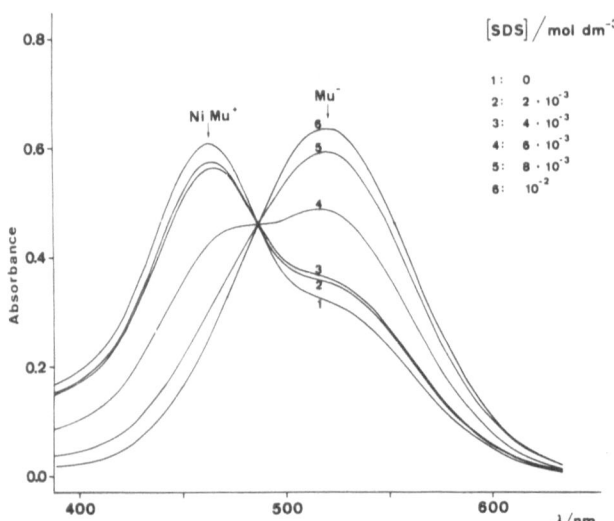

Spectromohotometric measurements for the $Ni^{2+} + Mu^-$ system in the presence of SDS
$[Ni^{2+}]_t =$
$2 \cdot 10^{-4}$ mol dm^{-3}
$[Mu^-]_t =$
$5 \cdot 10^{-5}$ mol dm^{-3}
pH = 6.0
T = 25° C

and only singly-charged. Therefore the reaction scheme 1 has to be modified to

$$Ni^{2+}_B + Mu^-_B \rightleftharpoons NiMu^+_B \rightleftharpoons NiMu_B + H^+ \qquad (10)$$
$$\downarrow$$
$$Ni^{2+}_S$$

(subscript B denotes the bulk aqueous region and subscript S the surface region of the micelle)

Qualitatively it can be seen from fig. 4 that there is a marked change in the spectrum at concentrations of SDS between 4 and $6 \cdot 10^{-3}$ mol dm^{-3} for $[Ni^{2+}]_t = 2 \cdot 10^{-4}$ mol dm^{-3}. The change in the spectra below [SDS] = $4 \cdot 10^{-3}$ mol dm^{-3} is caused by a change in γ_{II} due to the presence of SDS in monomer form.

For a quantitative discussion we define an apparent equilibrium constant K_{app} as

$$K_{app} = \frac{[NiMu^+]_B}{([Ni^{2+}]_B + [Ni^{2+}]_S)[Mu^-]_B \gamma_{II}} \qquad (11)$$

which is related to the equilibrium constant K_1 measured in the absence of micelles by

$$K_{app}/K_1 = [Ni^{2+}]_B/([Ni^{2+}]_B + [Ni^{2+}]_S) \qquad (12)$$

i.e. K_{app}/K_1 = fraction of Ni^{2+} free in aqueous solution
A plot of this fraction vs [SDS] is shown in fig 5. The onset of

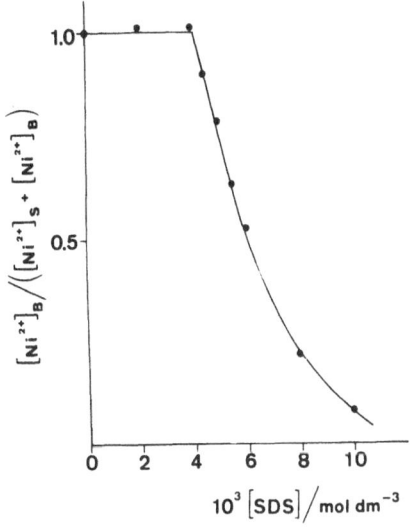

Fig. 5

Fraction of Ni^{2+} free in
aqueous solution vs [SDS]
Experimental conditions
as in fig. 3

micelle formation is readily detected. Furthermore it can be seen
from the figure that over 90 % of the Ni^{2+} is bound at the surface
for [SDS] $\sim 10^{-2}$ mol dm^{-3}. This indicates strong binding of Ni^{2+} to
the mice le surface.

This binding may be described by a binding constant K_{Ni} defined by the equilibrium

$$Ni^{2+}_{B} + \begin{matrix} \text{micellar} \\ \text{binding site} \end{matrix} \underset{\overset{\longleftarrow}{}}{\overset{K_{Ni}}{\longrightarrow}} Ni^{2+}_{S} \qquad (13)$$

Since the concentration of micellar binding sites is proportional
to the concentration of surfactant in micellar form, i.e.([SDS] –
cmc), we can define K_{Ni} as

$$K_{Ni} = \frac{[Ni^{2+}]_S}{[Ni^{2+}]_B([SDS]-cmc)} \qquad (14)$$

Our results indicate a value of K_{Ni} of $\sim 10^3$dm^{3}mol^{-1}. This value
will depend primarily on the surface potential of the micelle
since Ni^{2+} is bound to the micelle solely by electrostatic
attraction. The surface potential may vary with SDS concentration
[5] just above the cmc so that K_{Ni} is not constant over the whole
range. This will be discussed in a subsequent paper [6].

It is clear from fig. 5 that this simple spectrophotometric
method gives a very sensitive indication of the first appearance
of micelle surface in the solution. The effect of the small amount
of Ni^{2+} present is to lower the cmc significantly. However, the

study of micellisation in the presence of metal ions is particularly relevant to the study of micellar catalysis of reactions
such as metal-ligand exchange and electron transfer processes.

Kinetic Measurements

The kinetics of Ni^{2+}/Mu^- complex formation has also been
studied in the presence of SDS and sodium tetradecylsulphate
(STS). In a previous study of the kinetics of the reaction between
Ni^{2+} and the hydrophobic ligand pyridine-2-azo-p-dimethyl-aniline
the rate of complex formation was dramatically promoted in the
region of the cmc of SDS [7,8]. In this case the reacting species
are concentrated in the region of the micelle surface. In contrast,
the rate of complex formation between Ni^{2+} and Mu^- in solutions
containing anionic micelles decreases as the concentration of
micelles increases. This is because the reactants are separated,
and reaction can only occur in the bulk aqueous phase. Figure 6

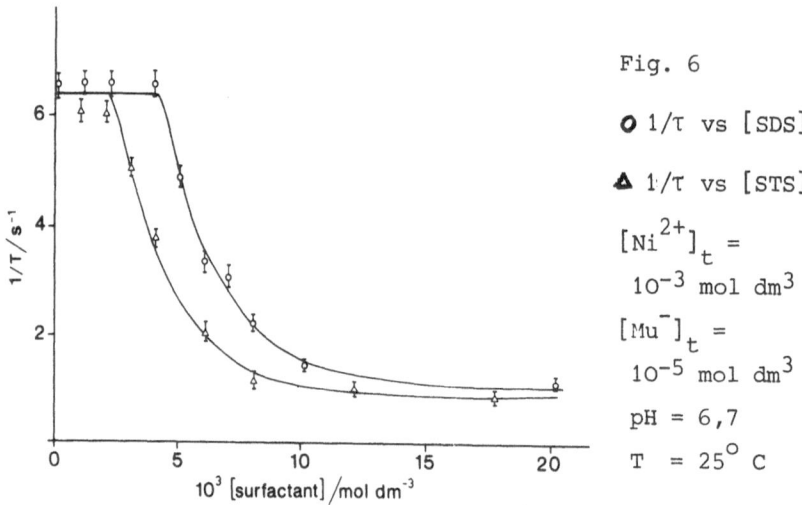

Fig. 6

O $1/\tau$ vs [SDS]

▲ $1/\tau$ vs [STS]

$[Ni^{2+}]_t =$
10^{-3} mol dm^3

$[Mu^-]_t =$
10^{-5} mol dm^3

pH = 6,7

T = 25° C

shows stopped-flow data obtained for the reaction between Ni^{2+}
and Mu^- in the presence of SDS and STS

Referring to scheme (10), under the condition $c_+ \gg c_-$ the
relaxation time is described by a modified version of eqn. (7),
which takes into account fast partitioning of Ni^{2+} between the
micelle surface and the bulk aqueous phase:

$$1/\tau = \frac{k_1 \gamma_{II} [Ni^{2+}]_t}{1 + K_{Ni}([\overline{SDS}] - cmc)} + k_{-1}(1 + K_H/c_H)^{-1} \qquad (15)$$

This equation can be rearranged to give

$$\frac{k_1 \gamma_{II} [Ni^{2+}]_t}{1/\tau - k_{-1}(1 + K_H/c_H)^{-1}} - 1 = K_{Ni} ([SDS]-cmc) + 1 \qquad (16)$$

from which K_{Ni} is again readily obtained [6], using values of k_1, k_{-1}, and K_H determined in the absence of surfactant.

Fig. 6 shows the high sensitivity of the kinetic measurements in indicating the formation of micelles in SDS and STS solutions. This approach can readily be extended to studies of the effect of additive species (e.g. dodecanol, NaCl) on micelle formation and the binding of metal ions to negatively charged surfaces.

Using a fast indicator system such as Zn^{2+}/Mu^- (which has a response time of a few microseconds) instead of Ni^{2+}/Mu^-, it is possible to follow spectrophotometrically the rate of micelle formation involving anionic surfactants.

Acknowledgement:

We thank the Science Research Council and NATO for financial support.

References

[1] A. Jost, Ber. Bunsenges. Phys. Chem., 69, 617 (1963).
[2] W. Knoche and G. Wiese, Rev. Sci. Instr., 47, 209 (1976).
[3] M. Fischer, W. Knoche, B.H. Robinson, and J.H. Maclagan Wedderburn, J. Chem. Soc. Faraday I, 75, 119 (1979).
[4] R.M. Fuoss, J. Amer. Chem. Soc., 80, 5059 (1958).
[5] A.A. Bhalekar and J.B.F.N. Engberts, J. Amer. Chem. Soc., 100, 5914 (1978).
[6] M. Fischer, W. Knoche, B.H. Robinson, and J.H. Maclagan Wedderburn, to be published.
[7] A.D. James and B.H. Robinson, J. Chem. Soc. Faraday I, 74, 10, 1978.
[8] J. Holzwarth, W. Knoche, and B.H. Robinson, Ber. Bunsenges. Phys. Chem., 82, 1001 (1978).

THE APPLICATION OF FAST-REACTION METHODS TO REACTIONS OF TRANSITION-METAL IONS AND COMPLEXES

E.F. Caldin

University of Kent, Canterbury, Kent. U.K.

The reactions of metal ions and metal complexes form a recognisable subject within what is still called "inorganic" chemistry although it has been greatly influenced by the use of ligands (1, 2). Transition-metal chemistry is a major part of this field and has provided much of the impetus towards mechanistic studies in it (3-6). Fast-reaction methods have made possible the investigation of the reactions of transition-metal ions, expecially those of the series V^{2+}, Cr^{2+}, Mn^{2+}, Fe^{2+}, Co^{2+}, Ni^{2+}, Cu^{2+}, Zn^{2+} (7-11). These studies have contributed greatly to the general picture, so much so that the classical reactions which are convenient for spectrophotometry on the conventional time-scale can no longer be regarded as typical.

Among the fast-reaction methods, the stopped-flow technique is the most widely applicable, and has been much used, especially to investigate the reactions of nickel (II) and cobalt (II); but it cannot be applied to the faster reactions such as those of manganese (II) or copper (II). These have been studied by relaxation methods, either single-pulse (temperature-jump, pressure-jump, or flash) or periodic-disturbance (ultrasonic absorption); these methods have the advantage of being able to measure high rates, but the disadvantage that an equilibrium with a suitable equilibrium constant and enthalpy change must be set up. For the special case of solvent exchange at metal cations, where the entering and leaving ligand are the same, the main bulk of the results has been obtained by n.m.r. methods (measurements of relaxation times, directly or by line-broadening) which give uniquely detailed information. Effects of high pressure (giving values of activation volumes) have been studied by adaptations of the temperature-jump, flash, stopped-flow and n.m.r. methods (12).

W. J. Gettins and E. Wyn-Jones (Eds.). Techniques and Applications of Fast Reactions in Solution. 301–311.
Copyright © 1979 by D. Reidel Publishing Company.

The main classes of reactions in transition-metal chemistry which have been studied by fast-reaction methods are as follows: (a) electron-transfer reactions, for which see the article in this volume by J.F. Holzwarth (b) ligand substitution, which includes many reactions of biological interest, and (c) solvent exchange, which though thermodynamically simpler than other ligand substitutions is kinetically more complex, because the solvent is simultaneously the incoming group, the leaving group and the medium.

Ligand substitution and solvent exchange at a divalent cation M^{2+} may be represented in a preliminary way by equations 1 and 2, in which L represents the ligand, S the solvent, and MS_6 the solvated metal ion; the charges are omitted.

$$L + MS_6 \rightleftharpoons LMS_5 + S \tag{1}$$

$$S + MS_6 \rightleftharpoons MS_6 + S \tag{2}$$

SOLVENT EXCHANGE AT TRANSITION-METAL CATIONS

The rate of solvent exchange at divalent metal cations in the first transition-metal series has been measured for various solvents by n.m.r. methods (13-16). The proton n.m.r. signal from a water molecule in bulk water is due to transitions of the ^1H nucleus between two energy levels, and the nuclei have a definite average lifetime in the upper level. When a water molecule approaches close to a paramagnetic ion such as Ni^{2+}, its ^1H nuclei comes into the strong field associated with unpaired electrons, which can effect the transition. The time spent by such a nucleus in the upper level is therefore shortened, and the lifetime for ^1H nuclei in the solution becomes uncertain. In accordance with the principle of indeterminancy, the energy in the upper state becomes correspondingly uncertain; this uncertainty is reflected in the frequency of absorption, and the signal is therefore broadened. There are several other factors that affect the line-width, but fortunately they have different temperature-dependencies, and from an Arrhenius-type plot it is possible to locate with confidence the temperature-range in which the line-width is controlled by the rate of solvent exchange at the cation. This is more easily done if measurements are made on the nucleus ^{17}O, for solvents bonding through oxygen, or ^{14}N for exchange of acetonitrile and other solvents bonding through nitrogen (16).

For the divalent ions of the first transition-metal series, in aqueous solution, the rate constant and activation enthalpies determined in this way run parallel with the crystal-field stabilisation energies of the ions; this is in accordance with the view that the interaction of the solvent with the cation is the important factor, and that it influences enthalpy rather than entropy changes (Table 1, Fig 1). (The charge density at the ion, which is proportional to the reciprocal radius, will no doubt have an influence, but the crystal-field effect

	Cr^{2+}	Mn^{2+}	Fe^{2+}	Co^{2+}	Ni^{2+}	Cu^{2+}	Zn^{2+}	V^{2+}
$\log_{10}k_s$	>10	>7.6	6.5	6.4	4.5	>10	7.5	2.0
$\Delta H^{\ddagger}$	3	8	8	10	11			
CFSE	-6	0	+1	+2	+5	-6	0	+7

Table 1
Solvent exchange at transition-metal ions in aqueous solution. Rate constants, activation enthalpies, and crystal-field stabilisation energies.

k_s = rate constant for solvent exchange (s^{-1}) at 25°C.
$\Delta H^{\ddagger}$ = activation enthalpy for solvent exchange (kcal mol^{-1})
CFSE = crystal-field stabilisation energy (kcal mol^{-1}); values from A.L. Companion, J. Physical Chem., 1969, 73, 739 (assuming $\rho = 1$)
Kinetic data mainly rounded values from refs 32, 33.

is much larger.) For a given cation in different solvents, the rate constant varies over and about two orders of magnitude (Table 2), the rate constants for the various cations lying generally in the same order.

The interpretation of these variations is not yet entirely clear. There is no satisfactory correlation with the bulk properties of the solvent, such as dielectric constant or polarisability; nor would such a correlation be expected, since the solvent here takes part in the reaction as a nucleophilic entering and leaving group, as well as a solvating medium. Factors that might be expected to be important are the ion-solvent interaction energies, the nucleophilicity and steric requirements of the solvent molecule, the difference of solvation between the reactants and the transition state, and (for some cations) the possibility of a change of coordination number. No one of these factors appears to dominate, and a fuller interpretation must await more

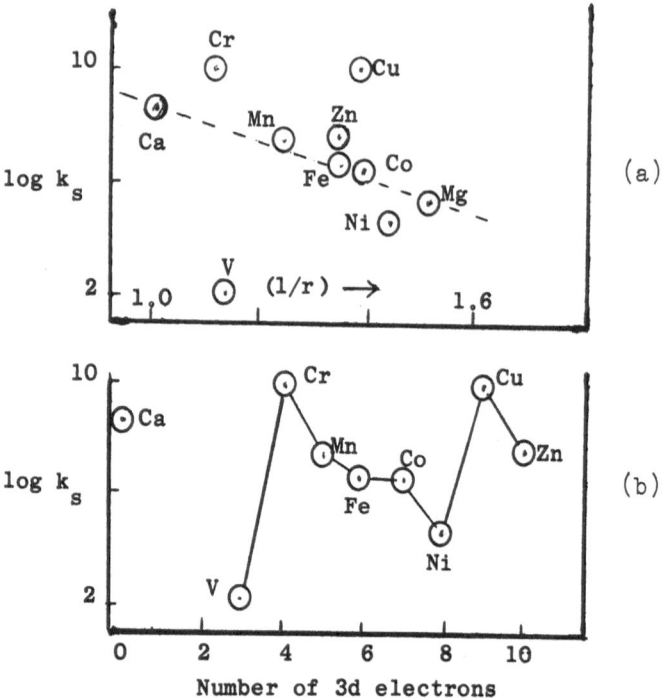

Fig. 1. Rate constants (k_s/s^{-1}) for water exchange at
divalent metal ions: (a) as function of reciprocal
ionic radius ($r^{-1}/Å^{-1}$); (b) as function of the number
of 3d electrons on M^{2+}. The dashed line in (a) has been
drawn through the points for the two $3d^0$ cations Mg^{2+}
and Ca^{2+}. (After D.N. Hague, ref. 6).

Table 2

	NH_3	Water	DMF	DMSO	MeCN	MeOH
Ni^{2+}	5.0	4.5	3.9	3.6	3.5	3.0
Co^{2+}	6.8	6.1	5.4	5.2	5.2	4.2

Solvent exchange rates for Ni (II) and Co (II) ions
in various solvents.

Value of $\log_{10}k_s$ at $25°C$ (k_s in s^{-1})

quantitative data. Determinations of enthalpies and entropies of solution of the cations, by calorimetric and solubility methods, and of activation volumes of the reactions by n.m.r. measurements under high pressure (17), should provide useful information. In the meantime, we can make helpful comparisons with ligand-substitution reactions, as follows.

LIGAND SUBSTITUTION AT DIVALENT TRANSITION-METAL CATIONS

Solvent exchange and ligand substitution are alike in involving the breaking of an ion-solvent bond, but differ as regards the bond-forming processes; thus a comparison of their kinetics should be profitable. Moreover in ligand substitution the incoming ligand and its concentration can be varied (as they cannot in solvent exchange), so that far more information can be obtained.

Aqueous Solutions

The rate law for ligand substitution at the divalent metal ions of the first transition series, under suitable conditions of concentration, corresponds to the simple scheme $M + L \rightleftharpoons ML$, where M = metal ion, L = ligand, and charges are omitted. The simplest results are obtained in aqueous solution; when the cation M is varied, the forward rate constant k_L runs roughly parallel to that for solvent exchange, and when the ligand L is varied the rate constant is related to the product of the charges on M and L. These results suggest that the formation of the transition state involves both the entering of a ligand molecule L and the departure of a solvent molecule S from the first coordination shell of the solvated metal cation MS_6^{2+}, the energy required being largely due to the loss of S. The mechanism is thus of the interchange-dissociative type (I_d). There must also be a preliminary formation of a weak outer-sphere complex, at least for negatively charged ligands. Thus we formulate the reaction scheme for monodentate ligands as in equation 3;

$$L + MS_6 \underset{\phantom{K_{12}}}{\overset{K_{12}}{\rightleftharpoons}} (L, MS_6) \underset{k_{32}}{\overset{k_{23}}{\rightleftharpoons}} L..MS_5..S \rightarrow LMS_5 + S \qquad (3)$$

Under the simplest conditions, this gives for the observed forward rate constant, k_L, the expression $k_L = k_{12}k_{23}$.

This mechanism was put forward by Eigen and Wilkins (18, 19), who also assumed (a) that k_{23} could be identified with the solvent-exchange rate constant k_S determined by n.m.r. methods, i.e. that the rate of solvent loss is unaffected by the proximity of the ligand; and (b) that K_{12} is controlled by electrostatic

forces, determined by the product of the charges on metal ion
and ligand and the distance of closest approach. (For uncharged
ligands, the electrostatic value K_{ES} reduces to a form identical
with that for random encounters). The resulting expression
(equation 4) gives values of the forward rate constant for ligand

$$k_L = K_{12}k_{23} = K_{ES}k_s \qquad (4)$$

substitution in aqueous solution which agree with the observed
values within an order of magnitude. This is true not only for
monodentate ligands such as pyridine or Cl^-, but for bidentate
ligands such as bipyridyl or o-phenanthroline, where the
complete scheme must include a final ring-closure step, as in
equation 5; in aqueous solution this step appears to be fast
relative to the preceding rate-limiting step.

$$LL + MS_6 \xrightleftharpoons{K_{12}} (LL,MS_6) \xrightleftharpoons[k_{32}]{k_{23}} LL..MS_5-S \longrightarrow LLMS_5 + S \longrightarrow \genfrac{}{}{0pt}{}{L}{|MS_5} + S$$

$$(5)$$

Activation volumes have been determined for the reactions
of Ni^{2+}, Co^{2+}, Cu^{2+} and Zn^{2+} with various ligands (NH_3, pada,
glycinate, murexide), by temperature-jump experiments at high
pressure. They are positive, as expected for a dissociative
mechanism, and are in general agreement with the Eigen-Wilkins
model (20-22). (For the reactions of the trivalent ion Fe^{3+} with
Cl^- or SCN^-, the activation volumes are negative, as for an
associative mechanism (23-25)).

A modification of the theory is required when the metal
cation is already bound to a molecule of a different ligand:
$ML_1 + L_2 \longrightarrow ML_1L_2$ (26-28). The temperature-jump method was
used by Hague to investigate the reactions of several charged
and neutral ligands with various complexed Ni(II), Co(II), Mn(II),
Zn(II) and Mg(II) species in aqueous solution. The reaction of
a monodentate ligand L_2^- with Mg(II) complexed with the negatively-
charged bidentate ligand ATP, for example, would be:-

$$Mg(ATP)(H_2O)_4^{2-} + L_2^- \longrightarrow Mg(ATP)(L_2)(H_2O)_3^{3-} + H_2O \qquad (6)$$

If K_{12} were controlled by the product of the charge on L_2 and
the overall charge on ML_1, and equation (4) applied, we should
expect a large decrease in k_L as the charge on the ML_1 species
is changed from +2 to -3. This is not observed. It appears that
the overall charge on this species is less important than the
local charge density on the metal ion.

Non-aqueous Solvents

In non-aqueous solvents, matters are less simple.
Monodentate ligands give rates agreeing quite well with the
Eigen-Wilkins scheme, and hence values of the ratio $k_L/K_{ES}k_s$
around unity, as equation 4 predicts. Bidentate ligands,
however, exhibit much more specificity from one ligand to another,
and larger variations from solvent to solvent, giving values of
$k_L/K_{ES}k_s$ deviating from unity in both directions and varying
by orders of magnitude (Table 3 and Fig 2).

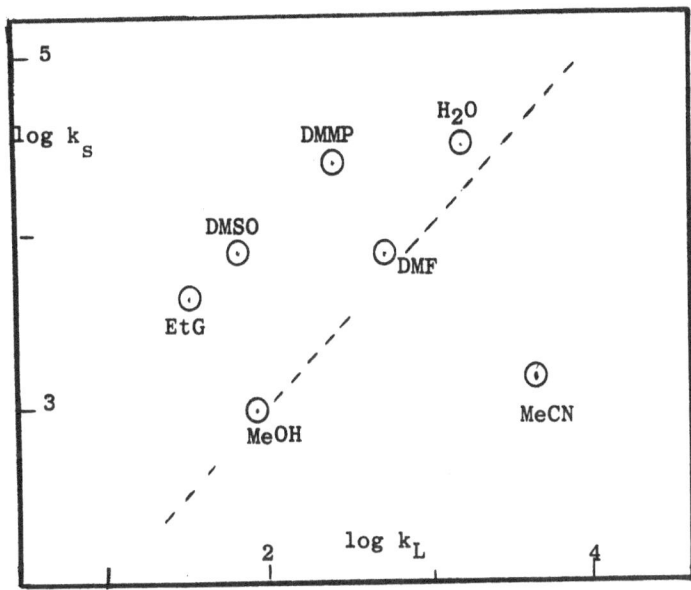

Fig. 2. Solvent exchange and ligand-substitution
at Ni^{2+} in various solvents. Plot of log k_s for
solvent exchange against log k_L for ligand substitution
with bipyridyl at 25°C.

These deviations from equation 4 for bidentate ligands in
non-aqueous solvents might conceivably have their origin in any
one of the three steps of the reaction (equation 5). The third
step has been considered and rejected as unlikely by Coetzee,
who has contributed much to this field including excellent
reviews (29-31). The second step was considered by Bennetto and
Caldin (32, 33); the possibility that k_{23} is influenced by the
solvent-solvent interactions ("solvent structure") was
considered as an explanation of the observed correlations between
$k_L/K_{ES}k_s$ and some physical properties of solvents related to the

Solvent	Ligand	log k_L	log k_s	log R_1	Comments
Water	water	—	4.5	0	R_1 varies < 10-fold in aqueous solution for various ligands.
	pyridine	3.6	4.5	-0.3	
	bipy	3.2	4.5	-0.7	
	phen	3.5	4.5	-0.4	
	SCN⁻	4.5	4.5	-0.7	
	oxalate²⁻	4.9	4.5	-0.7	
MeCN	pyridine	2.9	3.4	-0.4	R_1 varies < 10-fold for monodentate ligands in various solvents.
water		3.6	4.5	-0.3	
DMSO		3.4	3.9	+0.1	
MeCN	SCN⁻	5.0	3.4	-0.1	
water		4.5	4.5	-0.7	
DMSO		4.9	3.9	0.0	
MeCN	bipy	3.61	3.4	+0.3	R_1 varies 100-fold for bidentate ligands in various solvents
water		3.2	4.5	-0.7	
DMSO		1.84	3.9	-1.7	
MeCN	phen	4.70	3.4	+1.4	
water		3.5	4.5	-0.4	
DMSO		2.6	3.9	-0.9	

Table 3

Kinetics of ligand substitution and solvent exchange at nickel (II) ion in various solvents

k_L = rate constant ($M^{-1}s^{-1}$) at 25°C for forward reaction M + L → ML.

k_s = rate constant (s^{-1}) at 25°C for solvent exchange.

R_1 = $(4/4)\frac{k_L/K_{12}k_s}{}$ (as in ref. 29).

K_{12} = $10^3(4\pi N/3) a_1^3$ for neutral ligands, where a_1 is the distance of closest approach of the centre of the solvent cation to the reaction site on the ligand. For charged ligands this expression is multiplied by a term representing the electrostatic interaction.

Data mainly from Coetzee, refs. 29-31.

strength of their structures; a general picture of the effects of the incoming ligand molecule L on the local solvent structure was developed, in the line of Frank and Wen's model of aqueous ionic solutions (34) and its extension to non-aqueous solvents by Cox, Parker et al. (35). Monodentate ligands, however, do not show these large deviations. Partly for this reason, Coetzee emphasises the first step; he concludes from his survey that the deviations arise mostly in the outer sphere, because the solvent influences both the stability of the outer-sphere complex and the orientation of the ligand in the outer sphere. The deviations will then depend on a combination of factors: the hydrogen-bonding donor strength of the solvent, the strength of the cation-solvent interaction, and the steric requirements in the inner sphere. These factors may reinforce or oppose one another, and so produce the observed variety of behaviour in different solvents(31). The solvent structure is also influenced by such factors, however, and it is not easy to distinguish their effects on outer-sphere complexation from those on solvent structure.

The role of the solvent should become clearer when we know more about its effects on the transition state and the reactants separately; so far there have been few studies of the thermodynamics of transfer of ligands and metal ions from one solvent to another, which would permit the kind of 'dissection' of the rate parameters that has been very useful in other fields, such as (for example) the kinetics of reactions involving anions (35). At all events it is clear that water is not here a typical solvent, but an anomalous one, as in other kinetic and thermodynamic contexts, and that the role of the solvent is more complex and subtle than the results in water and would suggest.

OTHER REACTIONS

The reader is referred elsewhere for discussions of substitution reactions with macrocyclic ligands (2-6), and in micelles (36); like the effects of bound ligands, these reactions are of importance in reactions of biological interest. Complex-formation reactions involving a change of covalency, such as reactions of square-planar nickel (II) complexes with nucleophiles, are also omitted (37). All these reactions offer interesting applications of fast-reaction methods.

REFERENCES

1. F. Basolo and R.G. Pearson, "Mechanisms of Inorganic
    Reactions", 2nd edn, Wiley, New York, 1967.
2. J. Burgess, "Metal Ions in Solution", Ellis and Horwood,
    Chichester, 1978 .
3. R.G. Wilkins, "The Study of Kinetics and Mechanism of
    Reactions of Transition Metal Complexes", Allyn and Bacon,
    Boston, 1974.
4. D.N. Hague, "Fast Reactions", Wiley, London 1971, Chap 3.
5. D.N. Hague, in "Inorganic Reaction Mechanisms", ed. J.
    Burgess and others (Chem. Soc. Sp. Per. Reports), vol. 1,
    pp. 209, 240 (1971); vol. 2, pp. 196, 225 (1972); vol. 3,
    pp. 261, 329 (1973); vol. 4, pp. 209, 249 (1976); vol. 5,
    pp. 240, 277 (1977).
6. D.N. Hague, in "Chemical Relaxation in Molecular Biology",
    ed. I. Pecht and R. Rigler, Springer-Verlag, Berlin, 1977,
    pp. 84-106.
7. A. McAuley and J. Hill, Quart. Rev. Chem.Soc., 1969, 23, 18.
8. D.J. Hewkin and R.H. Prince, Coord. Chem. Rev., 1970, 5, 45.
9. R.G. Wilkins, Acc. Chem. Res., 1970, 3, 408.
10. K. Kustin and J. Swinehart, Progr. Inorg. Chem., 1970, 13, 107.
11. T.W. Swaddle, Coord. Chem. Rev., 1974, 14, 217.
12. For references, see E.F. Caldin, this volume   ; see
    also ref. 17, and E.F. Caldin and R.C. Greenwood, this
    volume.
13. T.R. Stengle and C.H. Langford, Coord. Chem. Rev., 1967, 2, 349.
14. J.P. Hunt, Coord. Chem. Rev., 1971, 7, 1.
15. S.F. Lincoln, "Progr. Reaction Kinetics", ed. K.R. Jennings
    and R.B. Cundall, 1977, 9, 1.
16. S.F. Lincoln and R.J. West, Austr. J. Chem., 1973, 26, 255.
17. A.E. Merbach and H. Vanni, Helv. Chim. Acta, 1977, 60, 1124;
    W.L. Earl, F.K. Meyer and A.E. Merbach, Inorg. Chim Acta.,
    1977, 25, L91.
18. M. Eigen, Z. Elektrochem., 1960, 64, 155; Pure Appl. Chem.,
    1963, 6, 97.
19. M. Eigen and R.G. Wilkins, "Mechanisms of Inorganic Reactions",
    Advances in Chemistry Series, 1965, 49, 55.
20. E.F. Caldin, M.W. Grant and B.B. Hasinoff, J.C.S. Faraday
    Trans. I, 1972, 68, 2247.
21. M.W. Grant, J.C.S. Faraday I, 1973, 69, 560.
22. A. Jost, Ber. Bunsenges. Phys. Chem., 1975, 79, 850.
23. B.B. Hasinoff, Can. J. Chem., 1976, 54, 182.
24. A. Jost, Ber. Bunsenges.Phys. Chem., 1976, 80, 316.
25. K.R. Brower, J. Amer. Chem. Soc., 1968, 90, 5401.
26. D.N. Hague et al., Trans. Faraday Soc., 1971, 67, 786, 3069;
    J.C.S. Faraday I, 1972, 68, 37, 932, 2259; J.C.S. Dalton,
    1974, 254; and refs 5 and 6.
27. D.W. Margerum et al., J. Amer. Chem. Soc., 1967, 89, 1088;
    1970, 92, 1875.

28. G.G. Hammes and J.I. Steinfeld, J. Amer. Chem. Soc.,
    1962, 84, 4639.
29. J.F. Coetzee in "Solute-Solvent Interactions", ed. J.F.
    Coetzee and C.D. Ritchie, vol. 2, Marcel Dekker, New York,
    1976, Chap 14.
30. J.F. Coetzee, Pure Appl. Chem., 1977, 49, 27.
31. J.F. Coetzee in "Protons and Ions Involved in Fast Dynamic
    Processes", ed. P. Laszlo, Elsevier, Amsterdam, 1978,
    pp. 77-90.
32. H.P. Bennetto and E.F. Caldin, J. Chem. Soc., A, 1971,
    2191, 2198.
33. E.F. Caldin and H.P. Bennetto, J. Solution Chem., 1973,
    2, 217.
34. H.S. Frank and W.Y. Wen, Disc. Faraday Soc., 1957, 24, 133.
35. B.G. Cox, G.R. Hedwig, A.J. Parker and D.W. Watts, Austral.
    J. Chem., 1974, 27, 477.
36. B.H. Robinson, this volume.
37. M. Cusumano, J.C.S. Dalton, 1976, 2133, 2137; K.J. Ivin,
    R. Jamison and J.J. McGarvey, J. Amer. Chem. Soc., 1972,
    94, 1763.

KINETIC STUDIES OF THE COMPLEX FORMATION OF NICKEL (II) IONS
WITH PYRIDINE-2-AZO-p-DIMETHYLANILINE (PADA) IN NON-AQUEOUS
SOLVENTS AT PRESSURES FROM 1 BAR TO 2 KBAR.

R. C. Greenwood and E. F. Caldin

University of Kent, Canterbury, Kent.

PHOTOCHEMICAL-RELAXATION RATE MEASUREMENTS AT HIGH PRESSURES

High-pressure kinetic measurements in several solvents have
been carried out on the ligand-substitution reaction between
Ni^{2+} ions and the bidentate ligand pyridine-2-azo-p-dimethylaniline,
PADA (1). The laser photochemical relaxation method described
elsewhere in this volume (2) uses a 3-μs-pulse of laser light
to dissociate the Ni^{2+}-PADA complex. The excited-state species
produced returns rapidly to the ground state and the subsequent
recombination reaction is monitored by a conventional
spectrophotometric detection system. The laser flash is used
in conjunction with a high-pressure system described previously
(3), which uses a hydraulic pump to exert pressures of up to
3 kbar upon the sample solution. The forward rate constant for
ligand substitution (k_L) is determined at various pressures, and
$\Delta V_L^{\ddagger}$ is obtained from a plot of log k_L against pressure.

THE MECHANISM OF LIGAND SUBSTITUTION REACTIONS OF Ni(II)

The study of the effect of solvent upon the rate of
complexation has revealed details of the reaction mechanism, and
it is hoped that a study of the solvent-dependence of activation
volumes may permit further elucidation of metal-ligand, metal-
solvent and solvent-solvent interactions. According to the
Eigen-Wilkins mechanism for ligand substitution reactions (4)(5),
the first step is formation of an outer-sphere complex,
characterised by an equilibrium constant K_{12}. Subsequently the
ligand enters the first coordination sphere and the forward rate
constant for this step is identified with k_s for exchange of

W. J. Gettins and E. Wyn-Jones (Eds.). Techniques and Applications of Fast Reactions in Solution. 313–314.
Copyright © 1979 by D. Reidel Publishing Company.

solvent molecules, as measured by n.m.r. methods. Finally if
the ligand is bidentate there is a ring-closure step. In the
simplest case, when metal ion is in large excess, the overall
forward rate constant predicted by this model is given by
$k_L = K_{12}k_s$. This equation fits the observations quite well
in aqueous solution, but in non-aqueous solvents large deviations
occur with bidentate ligands.

In considering solvent effects, a useful parameter is the
ratio of the rate constants for ligand substitution and solvent
exchange, k_L/k_s, which for a bidentate ligand is not constant
as predicted but may vary from solvent to solvent as much as two
orders of magnitude (5). There are no obvious correlations
between solvent properties and the values of k_s or k_L alone,
but their ratio k_L/k_s shows a good correlation with (for example)
the enthalpy of evaporation, ΔH_{evap}, which can be considered
as the energy required to produce a hole of molecular size in
the solvent. Similarly there are no obvious correlations between
solvent properties and the activation enthalpies $\Delta H_s^{\ddagger}$ and $\Delta H_L^{\ddagger}$,
but their difference $\Delta\Delta H^{\ddagger} = \Delta H_L^{\ddagger} - \Delta H_s^{\ddagger}$ again shows a good
correlation with ΔH_{evap}. We may expect therefore that the
difference of activation volume $\Delta\Delta V^{\ddagger} = \Delta V_L^{\ddagger} - \Delta V_s^{\ddagger}$ will permit
further understanding of the role of solvent in the mechanism
of ligand-substitution reactions. (The value of $\Delta V_L^{\ddagger}$ in aqueous
solution has been found to be generally independent of the
nature of the ligand L, and is thought to arise mainly from the
stretching of the metal-ligand bond (1b, 6).) Values of $\Delta V_L^{\ddagger}$
have been measured in water, DMSO and DMF; the variations are
not large. Measurements in other solvents are in progress.

REFERENCES

(1) (a) Bennetto H.P. and Sabet-Imani Z., J.C.S. Faraday I,
          1974, 71, 1143.
     (b) Caldin E.F., Grant M.W. and Hasinoff B.B., J.C.S.
          Faraday I, 1972, 68, 2247.
     (c) Robinson B.H., and White N.C., J.C.S. Faraday I,
          1978, 74, 2625.
(2) Greenwood R.C., Robinson B.H. and White N.C., this volume.
(3) Caldin E.F., Grant M.W., Hasinoff B.B. and Tregloan P.A.,
     J. Physics E., Sci. Instrum., 1973, 3, 73.
(4) Eigen M. and Wilkins R.G., in Mechanisms of Inorganic
     Reactions (Advances in Chemistry Series, no. 49), 1965,
     p. 55.
(5) For a brief account see E.F. Caldin, this volume.
(6) M.W. Grant, J.C.S. Faraday Trans. I, 1973, 69, 560.

KINETICS OF THE REACTION OF IMIDAZOLES WITH IRON(III) PORPHYRIN CHLORIDES.

D.A. Sweigart and W. Fiske

Department of Chemistry, Swarthmore College, Swarthmore, PA 19081

The reaction of five-coordinate high-spin iron(III) porphyrin chlorides, Fe(Por)Cl, with imidazoles, RIm, to produce six-coordinate low-spin complexes according to reaction (1) has recently received attention [1-7]. The monoimidazole

$$Fe(Por)Cl + 2\ RIm \longrightarrow Fe(Por)(RIm)_2^+,Cl^- \tag{1}$$

adduct, Fe(Por)(RIm)Cl, is generally not seen in thermodynamic studies because the formation constant for the addition of the second imidazole (K_2) is greater than that for the first addition (K_1). Herein we report a kinetic study of reaction (1), the porphyrins being tetraphenylporphyrin (TPP) and protoporphyrin IX dimethyl ester (PPIXDME). With N-alkyl imidazoles (N-R imidazoles) the unstable six-coordinate intermediates Fe(Por)(RIm)Cl were detected and their optical spectra and formation constants measured. We also report that Fe(Por)Cl reacts with N-R and N-H imidazoles by different mechanisms, the difference being due to the hydrogen bonding capability possessed by N-H but not N-R imidazoles.

Reaction (1) was followed at 25°C by the stopped-flow method in acetone as solvent and under pseudo first-order conditions with the nucleophile in large excess. The metalloporphyrin concentration was about 5×10^{-5} $\underline{M}$. The results are given in the Table and a typical kinetic plot in Figure 1.

The results show that N-R imidazoles react more slowly and follow a different rate law than N-H imidazoles. With N-R imidazoles, we propose the following mechanism:

W. J. Gettins and E. Wyn-Jones (Eds.). Techniques and Applications of Fast Reactions in Solution. 315–320.
Copyright © 1979 by D. Reidel Publishing Company.

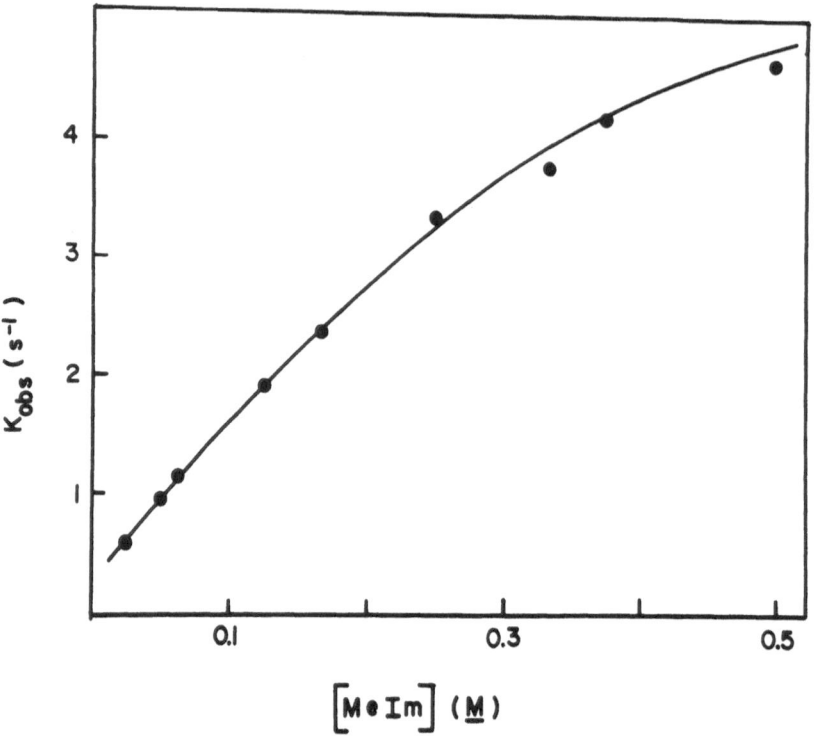

Figure 1. Rate data for the reaction of Fe(PPIXDME)Cl
with N-MeIm.

TABLE. Kinetic Data for Reaction (1)

	RIm	k_{obs}/s^{-1}
Fe(TPP)Cl	Im	$6.1 \times 10^3 [Im]^2$
	2-MeIm	$15 \times 10^3 [RIm]^2 + 8.1$
	4-PhIm	$1.0 \times 10^3 [RIm]^2 + 0.7$
	N-PrIm	$41[RIm]/(1+8.3[RIm])$
Fe(PPIXDME)Cl	Im	$7.8 \times 10^3 [Im]^2$
	N-MeIm	$21[RIm]/(1+2.6[RIm])$

$$Fe(Por)Cl + RIm \underset{k_1}{\overset{K}{\rightleftharpoons}} Fe(Por)(RIm)Cl$$

$$Fe(Por)(RIm)Cl \underset{k_2}{\overset{}{\rightleftharpoons}} Fe(Por)(RIm)^+,Cl^-$$

$$Fe(Por)(RIm)^+,Cl^- + RIm \underset{k_4}{\overset{k_3}{\rightleftharpoons}} Fe(Por)(RIm)_2^+,Cl^-$$

Using the steady state approximation for the five-coordinate intermediate, and assuming that $k_3[RIm] \gg k_2$, k_{obs} is calculated according to equation (2).

$$k_{obs} = \frac{k_1 K[RIm]}{(1+K[RIm])} + \frac{k_2 k_4}{k_2+k_3[RIm]} \qquad (2)$$

The second term in eqn. (2) is due to the reverse reaction, and except for very low values of [RIm] is not important. Under these conditions, a plot of $1/k_{obs}$ vs. $1/[RIm]$ should be linear with intercept and slope $1/k_1$ and $1/k_1K$, respectively. Such plots are, in fact, linear, yielding results discussed below.

Compelling evidence for the above mechanism and in particular for the formation of Fe(Por)(RIm)Cl in a preequilibrium step is given by the data in Figure 2 which show that for Fe(PPIXDME)Cl and N-MeIm the calculated absorbance change $(A_\infty-A_0)$ and the observed change (ΔA_{obs}) in general differed and at some wavelengths were of opposite sign. Furthermore, the discrepancy between ΔA_{obs} and $A_\infty-A_0$ was itself a function of the N-MeIm concentration. (Analogous results were obtained for Fe(TPP)Cl and N-PrIm [7].) For the particular value of [RIm] indicated, Figure 2 shows that upon mixing the reactants, the absorbance changed from A_0 to $A_\infty-\Delta A_{obs}$ at a rate too rapid to measure. This then decayed to products, A_∞, in the rate determining step. By fixing the porphyrin concentration and wavelength and varying [RIm], a plot of equation (3) was used [8] to calculate K from the slope. The values of K extracted from

$$A_\infty-\Delta A_{obs} = \frac{1}{K}\frac{(A_\infty-A_0-\Delta A_{obs})}{[RIm]} + A_{INT} \qquad (3)$$

the curvature in the rate constant plots (e.g., Figure 2) and those obtained from eqn. (3) are compared below and seen to be in excellent agreement:

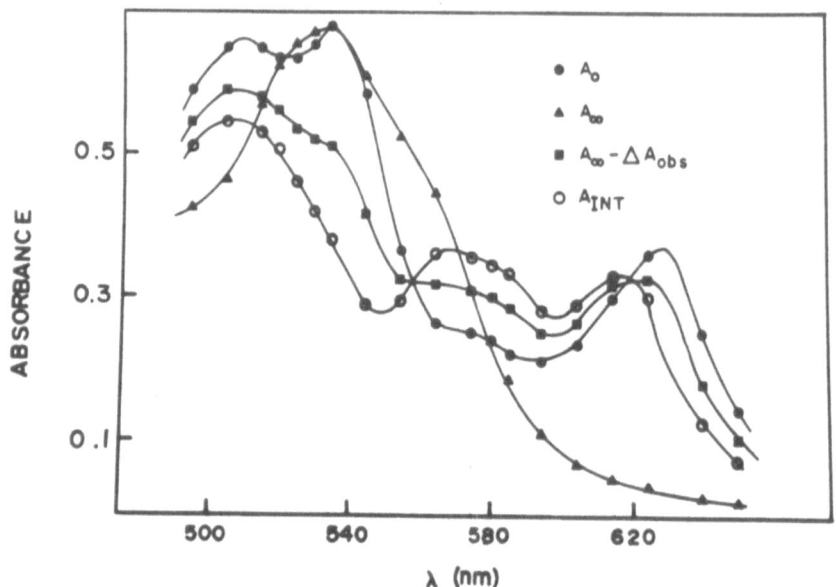

Figure 2.   Absorbance data for Fe(PPIXDME) and N-MeIm
(0.46 $\underline{M}$) at 25°C. A_0 is Fe(PPIXDME)Cl; A_∞ is
Fe(PPIXDME)(N-MeIm)$_2^+$,Cl$^-$; A_{INT} is Fe(PPIXDME)(N-MeIm)Cl;
$A_\infty-\Delta A_{obs}$ is the spectrum immediately after mixing (< 5 msec) and
corresponds to an equilibrium mixture of Fe(PPIXDME)Cl and
Fe(PPIXDME)(N-MeIm)Cl.

Complex	k_1/s^{-1}	K(kinetic)/M^{-1}	K(static)/M^{-1}
Fe(TPP)(N-PrIm)Cl	5.0 ± 0.3	8.3 ± 0.8	9.5 ± 0.5
Fe(PPIXDME)(N-MeIm)Cl	8.1 ± 0.7	2.6 ± 0.3	2.9 ± 0.2

The intercept in Eqn. (3), A_{INT}, is the absorbance of the intermediate, and this can be calculated as a function of wavelength
according to eqn. (4), wherein $x = K[RIm]/(1+K[RIm])$.   The

$$A_{INT} = \frac{A_\infty-\Delta A_{obs}-(1-x)A_0}{x} \qquad (4)$$

resulting optical spectrum of Fe(PPIXDME)(N-MeIm)Cl was calculated using a value of 2.9 $\underline{M}^{-1}$ for K and is shown as A_{INT} in
Figure 2.   Although this complex decays to the final products
$t_{1/2} \approx 0.1$s at 25°C, at $\approx$ -50°C it persists long enough to be
seen visually as a green species that gradually changes to red

$Fe(PPIXDME)(N-MeIm)_2^+$. The spectra in Figure 2 suggest [9] that $Fe(PPIXDME)(N-MeIm)Cl$ is high-spin. We are attempting to obtain the esr spectra of the $Fe(Por)(RIm)Cl$ intermediates.

The species formed in the preequilibrium step in reaction (1) with N-R imidazoles could be formulated as the five-coordinate $Fe(Por)(RIm)^+,Cl^-$. In this case, however, we are unable to formulate a reasonable mechanism that explains the kinetic behavior, in particular the fact that a limiting rate is reached at high nucleophile concentration.

With N-H imidazoles as the nucleophile, no intermediates were detected, possibly because the speed of reaction (1) precluded the use of sufficiently high concentrations of RIm.

N-H imidazoles differ from N-R imidazoles in that the former are well known for their hydrogen bonding ability. We propose that N-H imidazoles react much more rapidly than N-R ones and obey a different rate law because they assist the removal of the coordinated chloride _via_ hydrogen bonding. N-R imidazoles can not hydrogen bond and the chloride dissociation (k_1) is the rate limiting step. In particular, we propose the following mechanism:

$$Fe(Por)Cl + RIm \overset{K}{\underset{}{\rightleftharpoons}} Fe(Por)(RIm)Cl$$

$$Fe(Por)(RIm)Cl + RIm \overset{k_1}{\underset{k_2}{\rightleftharpoons}} Fe(Por)(RIm)_2^+,Cl^-$$

$$k_{obs} = \frac{k_1 K[RIm]^2}{1+K[RIm]} + k_2$$

In contrast to the fundamental difference we find between N-H and N-alkyl imidazoles, Pasternack _et al._ [5] report that Im and N-MeIm react with $FeTPP(DMSO)_2^+$ according to the same rate law (second order in nucleophile). The behavior we find should occur only when the leaving group can hydrogen bond, as in $Fe(Por)Cl$ or $Fe(Por)F$.

ACKNOWLEDGMENT

Part of this work was supported by a Camille and Henry Dreyfus Teacher-Scholar grant to D.A.S.

REFERENCES

1. Coyle, C.L., Rafson, P.A., Abbott, E.H. Inorg. Chem.,
   1973, 12, 2007.
2. Ciaccio, P.R., Ellis, J.V., Munson, M.E., Kedderis, G.L.,
   McConnille, F.X., Duclos, J.M. J. Inorg. Nucl. Chem., 1976,
   38, 1885.
3. Walker, F.A., Lo, M.W., Ree, M.T. J. Am. Chem. Soc., 1976,
   98, 5552.
4. Satterlee, J.D., LaMar, G.N., Bold, T.J. J. Am. Chem. Soc.,
   1977, 99, 1088.
5. Pasternack, R.F., Gillies, B.S., Stahlbush, J.R. J. Am.
   Chem. Soc., 1978, 100, 2613.
6. Momenteau, M. Biochim. Biophys. Acta, 1973, 304, 814.
7. Burdige, D., Sweigart, D.A. Inorg. Chim. Acta, 1978, 28,
   L131.
8. Rossotti, J.F.C., Rossotti, H. The Determination of Stability Constants, McGraw-Hill: New York, N.Y., 1961; p.
   274.
9. Smith, D.W., Williams, R.J.P. Struc. Bonding, 1970, 7, 1.

THE APPLICATION OF A DYE-LASER PHOTOCHEMICAL RELAXATION
TECHNIQUE TO THE STUDY OF METAL-LIGAND SUBSTITUTION REACTIONS
IN AQUEOUS AND MICELLAR SOLUTIONS

R. C. Greenwood, B. H. Robinson and N. C. White

Chemical Laboratory, University of Kent at Canterbury,
U.K.

ABSTRACT

The dye-laser photochemical relaxation method perturbs a
reaction by photodissociation of one of the reacting species,
which has several advantages over more conventional relaxation
methods such as the temperature and pressure-jump methods. In
order to assess the usefulness of the technique in the study of
the kinetics of metal-ligand complexation reactions, a system
was chosen which had previously been studied by both temperature-
jump and stopped-flow methods, namely, bidentate complex
formation involving Ni^{2+}(aq) and PADA. Two relaxation effects
are produced by photochemical perturbation of this system, the
faster one being identified with a ring-closure reaction, and the
slower one with the recombination of Ni^{2+}(aq) and PADA. For the
latter process, excellent agreement is found between the kinetic
data from this and the previous studies. The effect of micelles
(of the anionic surfactant sodium dodecyl sulphate) on the
kinetics of these reactions was also investigated.

THE DYE-LASER PHOTOCHEMICAL RELAXATION TECHNIQUE

The photochemical relaxation method employed in this work
initiates a chemical reaction by the direct absorption of light
into one of the reacting species (1). In contrast, the
temperature and pressure-jump techniques perturb the position of
equilibrium by changing the conditions of the reaction medium (2).
The photochemical dissociation method thus has several inherent
advantages over the temperature-jump and related relaxation
methods:-

W. J. Gettins and E. Wyn-Jones (Eds.), Techniques and Applications of Fast Reactions in Solution. 321–327.
Copyright © 1979 by D. Reidel Publishing Company.

(1)   The relaxation amplitude produced by temperature or pressure-jump methods depends upon the position of equilibrium. Since the photochemical method directly perturbs one of the reacting species the extent of the induced reaction depends only upon the quantum yield for photodissociation (3).

(2)   Thermodynamic constraints (e.g. the requirement of a non-zero $\Delta H°$ for temperature-jump studies), are eliminated.

(3)   If light is absorbed by a specific chemical species in a complex reaction system, then the technique can be used to generate selected intermediates.   This may facilitate the elucidation of the details of a reaction mechanism.

In this work a 3µs pulse from a Rhodamine 6-G dye-laser was used to photochemically dissociate metal ion complexes, kinetic results being obtained from the subsequent recombination reactions.   The ligand used to test the applicability of the approach was pyridine-2-azo-p-dimethylaniline (PADA) since the kinetics of its complexation have been studied in aqueous solution with $Ni^{2+}(aq)$ (4).   The influence of micelles formed by the anionic surfactant sodium dodecyl sulphate (SDS) on this reaction was also investigated.   The arrangement of the apparatus used is shown in figure (1).

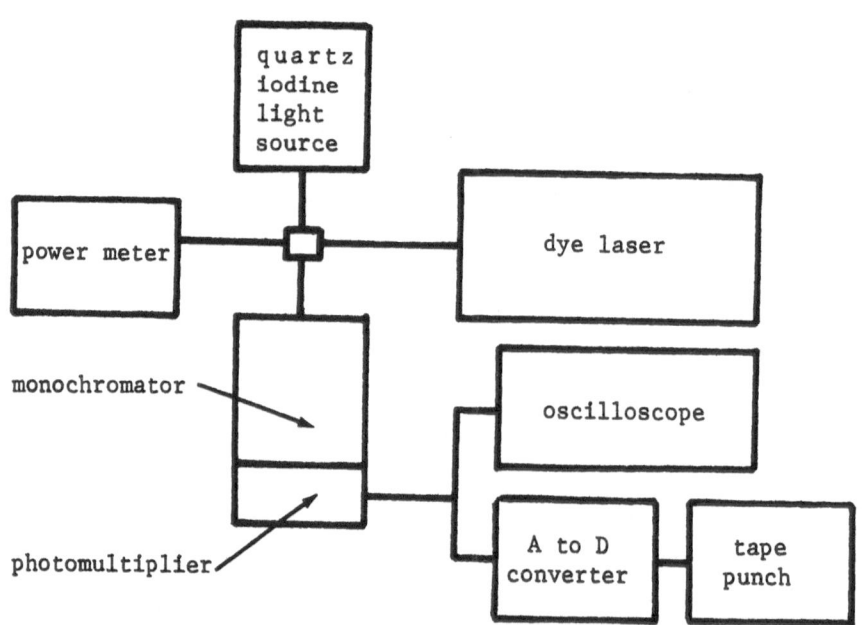

Figure 1.   Block diagram of the laser apparatus with spectrophotometric detection.

THE MECHANISM OF THE COMPLEXATION REACTION

It has been shown in previous studies by several fast reaction techniques that the complexation reaction between a metal ion and a bidentate ligand involves several discrete steps (5). As shown in figure 2, step 1 represents rapid (diffusion controlled) formation of an outer-sphere complex and step 2 the formation of a monodentate complex for which the forward rate constant can be identified with that for water exchange (k_{ex}) measured by n.m.r. methods. Since PADA is a bidentate ligand, a further ring closure step (3) takes place in which the final product is formed. The rate of ring closure is generally assumed to be rapid compared with the rate of dissociation of the monodentate complex (i.e. $k_{RC} \gg k_{-2}$).

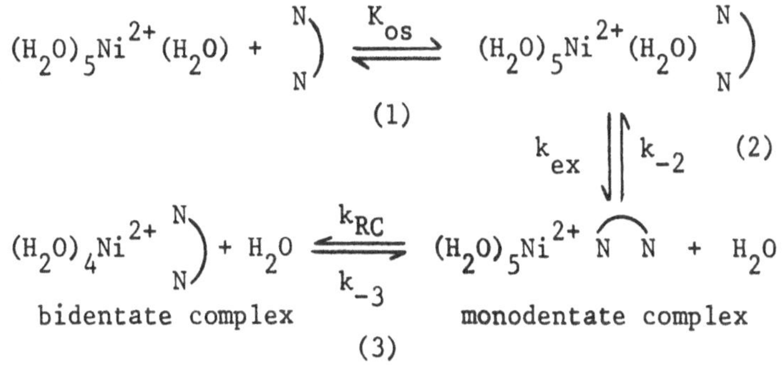

$$(H_2O)_5Ni^{2+}(H_2O) + \left.\begin{matrix} N \\ N \end{matrix}\right) \xrightleftharpoons{K_{os}} (H_2O)_5Ni^{2+}(H_2O)\left.\begin{matrix} N \\ \\ N \end{matrix}\right) \quad (1)$$

$$k_{ex} \big\Updownarrow k_{-2} \quad (2)$$

$$(H_2O)_4Ni^{2+}\left.\begin{matrix} N \\ N \end{matrix}\right) + H_2O \underset{k_{-3}}{\overset{k_{RC}}{\rightleftharpoons}} (H_2O)_5Ni^{2+}\,\overset{\frown}{N\ \ N} + H_2O$$

bidentate complex   monodentate complex

$$(3)$$

Figure 2. Reaction scheme for the overall complexation of $Ni^{2+}(aq)$ by PADA.

Providing pseudo-first order conditions are employed, i.e. the initial concentration of aquo-nickel $[Ni^{2+}]_T \gg [PADA]_T$ and $K_{os} \ll [Ni^{2+}]_T^{-1}$ then for the overall complexation reaction;

$$k_{obs} = \frac{K_{os}k_{ex}k_{RC}[Ni^{2+}]_T}{k_{-2} + k_{RC}} + \frac{k_{-2}\,k_{-3}}{k_{-2} + k_{RC}} \quad (1)$$

In our experiment the dye-laser pulse can induce the photo-chemical dissociation of the bidentate metal complex (6). In figure 3 S represents a solvent (water) molecule, L-L a bidentate ligand (PADA) and M the metal ion (Ni^{2+}). The excited state species $S_4M\overset{L}{\underset{L}{)}}$ produced by the laser pulse can be deactivated by several processes. These are: radiationless decay (thermal deactivation), fluorescence, and chemical reaction i.e. photosubstitution pathways A and B. If after perturbation the ground state species $S_5M\,\overset{\frown}{L\ \ L}$ and $S_5\,MS\,\overset{\frown}{L\ \ L}$ are obtained rapidly then their subsequent reactions can be monitored.

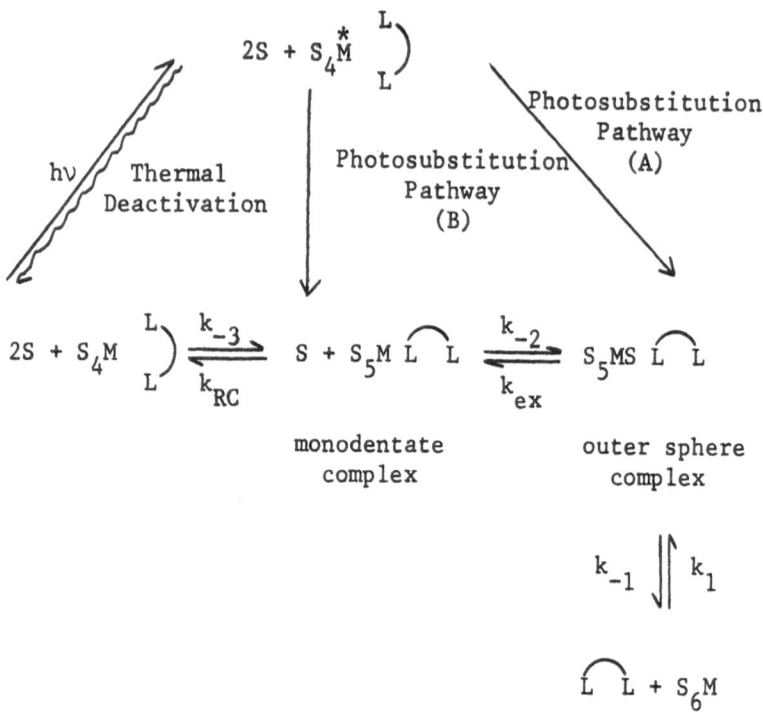

Figure 3. Reaction scheme for the photodissociation
of a metal complex.

RESULTS AND DISCUSSION

     The photochemical perturbation of the system Ni^{2+} and PADA
produces two relaxation effects in both aqueous and micellar
media characterised by widely differing relaxation times. The
faster process is independent of the concentration of both
Ni^{2+} and PADA but is dependent on temperature. In bulk water
at 298.2 K the fast (F) relaxation time τ_F has a value of
35($\pm$4) μs, which gives a first order rate constant of 2.9($\pm$0.3) x
10^4 s^{-1}. This fast process is identified with ring closure
(step 3 of figure 2) following photochemical generation of the
monodentate complex. The activation energy $(E_A)_F$ for this
process is found to be 40 $\pm$ 4 k J mol^{-1}.

     The slower (S) process (τ_S ~ 10-100 ms) gives values of
$(k_{obs})_S$ which are linearly dependent upon the Ni^{2+} concentration
but independent of PADA under pseudo-first-order conditions
(i.e. $[Ni^{2+}]_T \gg [PADA]_T$). The data can be represented by the
rate equation;

$$\tau_S^{-1} = (k_{obs})_S = k_f [Ni^{2+}]_T + k_b$$

$$\text{where } k_f = \frac{K_{os} k_{ex} k_{RC}}{k_{-2} + k_{RC}} \quad \text{and} \quad k_b \frac{k_{-2} \, k_{-3}}{k_{-2} + k_{RC}}$$

A linear plot of $(k_{obs})_S$ versus $[Ni^{2+}]_T$ is obtained, from which the results in Table (1) are derived. Excellent agreement is found between the values of k_f and k_b obtained from these plots and those found previously by other techniques (4). The activation energy for the overall forward reaction $(E_a)_S$ is found to be 57($\pm$5) k J mol^{-1}.

THE INFLUENCE OF MICELLAR SDS ON THE KINETICS OF COMPLEXATION OF Ni^{2+} BY PADA

Previous stopped-flow studies (7) (8) have shown that micelles formed by SDS have a pronounced effect on the kinetics of complex formation between Ni^{2+} and PADA. It has been shown that the reaction occurs at the negatively-charged surface of the micelle, both PADA and Ni^{2+} being adsorbed on this surface. Using the laser perturbation technique, under conditions such that the reaction has gone essentially to completion, two relaxation processes are observed in micellar SDS solutions which are similar to those found in aqueous solutions in the absence of SDS. The fast relaxation process is again independent of Ni^{2+} and PADA concentration. The fast (F) relaxation time τ_F^* (where the superscript * indicates reaction in micellar solution) is found to be 100($\pm$10) µs at 298.2 K, corresponding to a first-order rate constant of 10^4 s^{-1}. This process is identified with the rate constant for the ring-closure reaction on the micelle surface.

The slow (S) relaxation times τ_S^* gives values of $(k_{obs}^*)_S$ which are linearly dependent on $[Ni^{2+}]_T$ and also show a dramatic dependence on [SDS], especially in the region of the critical micelle concentration (c.m.c.). The rate is markedly enhanced (by a factor of 10^3), compared with the bulk water value, when the concentration of SDS exceeds the c.m.c., in agreement with previous measurements (7) (8). Close to but above the c.m.c. $(k_{obs}^*)_S$ is approximately constant but at higher micelle concentrations it falls gradually to values closer to that measured in the absence of micelles $(k_{obs})_S$. Above the c.m.c. the reaction is occurring exclusively on the micellar surface and a simple model, presented previously (7), suggests that the observed rate constant $(k_{obs}^*)_S$ is inversely proportional to the concentration of micellar binding sites at the surface, and first-order in excess reagent, i.e. $(k_{obs}^*)_S = k_f^* [Ni^{2+}]_T/$ $([SDS]_T - c.m.c.) + k_b^*$. The plot of $(k_{obs}^*)_S$ versus $[Ni^{2+}]_T/$ $([SDS]_T - c.m.c.)$ is found to be linear up to values of

$[Ni^{2+}]_T/([SDS]_T-c.m.c.)$ of ~ 0.05.   From the slope a value of 3.0×10^3 s^{-1} is derived, in good agreement with that obtained previously.

	L	SF	L	SF
$k_f/10^3 dm^3 mol^{-1}s^{-1}$	1.4	1.3		
$k_f^*/10^3 s^{-1}$			3.0	3.0
k_b/s^{-1}	0.1	0.1		
$k_{RC}/10^4 s^{-1}$	2.9			
$k_{RC}^*/10^4 s^{-1}$			1.0	
$(E_a)_S/kJmol^{-1}$	57±5	55±4		
$(E_a^*)_S/kJmol^{-1}$			55±4	52±4
$(E_a)_F/kJmol^{-1}$	40±4			
$(E_a^*)_F/kJmol^{-1}$			40±4	

Table 1.   Summary of the kinetic data, L = laser photodissociation method, SF = stopped flow method, the superscript * indicates a measurement obtained in micellar solution.

The dye-laser photodissociation apparatus has also been linked to a high pressure device (9).   By measurement of reaction rates at different pressures up to 3K bar it is possible in principle to determine the volume of activation $\Delta V_f^{\ddagger}$ for the reaction in any solvent medium.   It is found that for the complexation of Ni^{2+}(aq) by PADA in water the value of $\Delta V_f^{\ddagger}$ agrees with that determined previously by the laser temperature-jump method (10).

REFERENCES

(1) (a) Ivin, K.J., Jameson, R. and McGarvey, J.J., J. Amer. Chem. Soc., 1972, 94, 1763.
     (b) Young, R.C., Keene, F.R. and Meyer, T.J., J. Amer. Chem. Soc., 1977, 99, 2468.
     (c) Goodall, D.M. and Greenhow, R.C., Chem. Phys. Letters, 1971, 9, 583.
(2) Caldin, E.F., Chem. in Britain, 1975, 11, 4.
(3) Creutz, C. and Sutin, N., 1973, 95, 7177.

(4) Cobb, M.A. and Hague, D.N., J.C.S. Faraday I, 1972, 68, 932.
(5) Eigen, M. and Wilkins, R.G., in "Mechanism of Inorganic
     Reactions", ed. Gould, R.F., Adv. Chem. Series No. 49,
     Amer. Chem. Soc., Washington D.C., p. 65.
(6) "Concepts of Inorganic Photochemistry", eds. Adamson, A.W.
     and Fleischauer, P.D., Wiley Interscience, New York, 1975.
(7) James, A.D. and Robinson, B.H., J.C.S. Faraday I, 1978,
     74, 10.
(8) Holzwarth, J., Knoche, W. and Robinson, B.H., Ber.
     Bunsenges. Phys. Chem., 1978, 82, 1001.
(9) Caldin, E.F., Grant, M.W., Hasinoff, B.B. and Tregloan, P.A.,
     J. Phys. E., 1973, 3, 73.
(10) Caldin, E.F., Grant, M.W., Hasinoff, B.B., J.C.S. Faraday I,
     1974, 71, 1143.

THE KINETICS OF THE FORMATION OF THE BeSO$_4$ COMPLEX IN H$_2$O-DMSO MIXTURES

D. H. Devia and H. Strehlow

Max-Planck-Institut für biophysikalische Chemie
D-3400 Göttingen, West Germany

The work being reported in this paper is part of a continuing study on metal-complex formation using BeSO$_4$ as a test electrolyte. In pure aqueous solution the reaction has been extensively studied (1) and is shown to follow the Eigen-Tamm mechanism:

$$Be^{2+} + SO_4^{2-} \xrightarrow{\quad K_o \quad} \underset{\substack{\text{outer sphere} \\ \text{complex}}}{Be(H_2O)SO_4} \rightleftharpoons BeSO_4$$

As the process free ions ⟶ outer sphere complex is fast, the reaction in pressure-jump experiments is characterised by a single relaxation time.

The work was then extended to H$_2$O-formamide mixtures (2). For the slow process two relaxation times, separated by a factor of about 3, were observed. To account for this the Eigen-Tamm scheme has to be extended as shown in figure 1. The triangles represent co-ordination tetrahedra and W and F are the two solvents. In the slow process (double arrows) the labels W and F refer to interchange between W and SO$_4^{2-}$ and F and SO$_4^{2-}$ respectively.

In H$_2$O-DMSO mixtures, owing to the greater difference in properties of the two components in this solvent pair, further separation of the various processes shown in figure 1 was hoped for. Up to five different relaxation times, separated by factors ranging between 3 and 10, have been observed in H$_2$O-DMSO mixtures. In figure 2 the relaxation times measured in 0.1 mol l^{-1} BeSO$_4$ solutions at different solvent compositions are displayed.

329

W. J. Gettins and E. Wyn-Jones (Eds.), Techniques and Applications of Fast Reactions in Solution, 329–332.
Copyright © 1979 by D. Reidel Publishing Company.

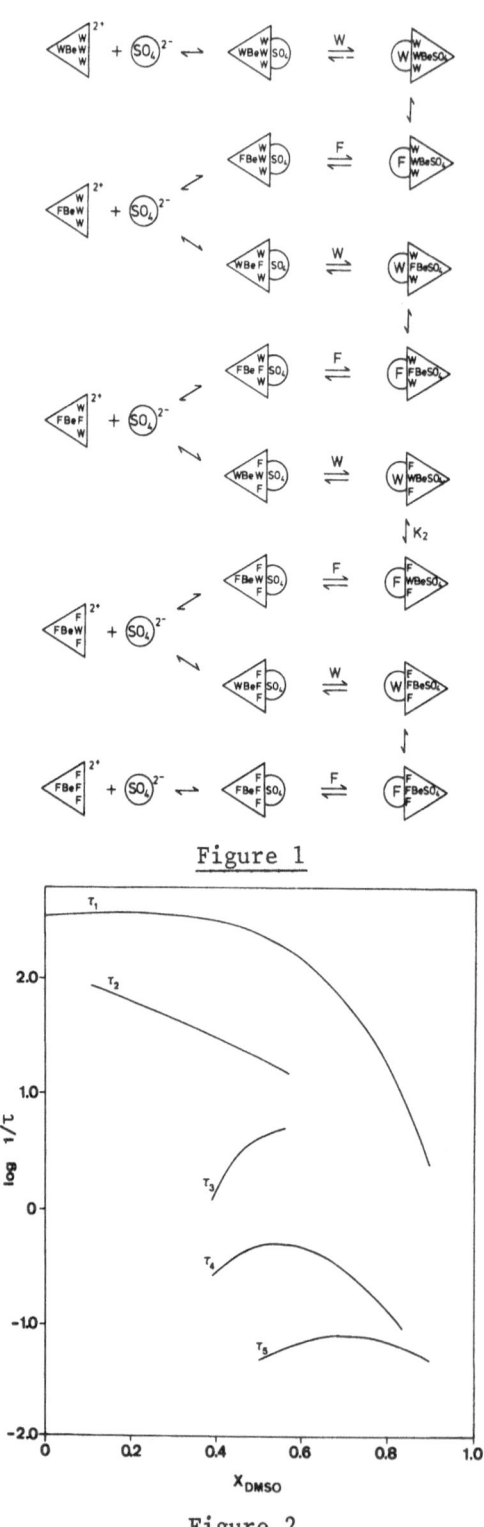

Figure 1

Figure 2

From these results and from conductivity measurements it was found that at $x_{DMSO} > 0.35$, $Be(SO_4)_2^{2-}$ is also a significant species. This leads to the reaction scheme:

$$
\begin{array}{ccccccc}
& & BeSO_4 & & & & BeSO_4 \\
& & + & & K_o & & + \\
Be^{2+} & + & SO_4^{2-} & \rightleftharpoons & & & BeLSO_4 \\
& & \big\updownarrow{}^{k_{14}} \, k_{41} & & & & \big\updownarrow{}^{k_{23}} \, k_{32} \\
& & & & k_{34} & & \\
Be^{2+} & + & Be(SO_4)_2^{2-} & \rightleftharpoons & & & 2BeSO_4 \\
& & & & k_{43} & &
\end{array}
$$

This scheme has to be superimposed on the scheme displayed in figure 1, that is the free Be^{2+} ions are present as the solvated species $Be(H_2O)_i(DMSO)_{4-i}^{2+}$, which react with differing lability.

When the shortest relaxation time is called τ_1 and the largest τ_5, then the following assignments can be made:

τ_1 corresponds to an interchange of a H_2O molecule in the inner sphere of Be^{2+} for SO_4^{2-}

τ_2 corresponds to an interchange of a DMSO molecule for SO_4^{2-} in $Be(H_2O)_3DMSO^{2+}$

τ_3 corresponds to an interchange of a DMSO molecule for SO_4^{2-} in $Be(H_2O)_2(DMSO)_2^{2+}$

τ_4 corresponds to the formation of the $Be(SO_4)_2^{2-}$ complex

τ_5 corresponds to an interchange of a DMSO molecule for SO_4^{2-} in $Be(H_2O)(DMSO)_3^{2+}$.

The species $Be(H_2O)_i(DMSO)_{4-i}^{2+}$ have been characterised by nmr measureements (3).

This work is to be reported in full in a forthcoming publication in the Berichte der Bunsengesellschaft.

References

1.   W. Knoche, C. A. Firth, and D. Hess, Adv. Molec. Relax. Proc., 6, 1 (1974);

H. Strehlow and W. Knoche, Ber. Bunsenges., $\underline{73}$, 427 (1969).

2.   R. Lachmann, I. Wagner, D. H. Davia, and H. Strehlow, Ber. Bunsenges., $\underline{82}$, 492 (1978).

3.   H. H. Fuldner, D. H. Devia, and H. Strehlow, Ber. Bunsenges., $\underline{82}$, 499 (1978).

VIBRATIONAL AND ULTRASONIC RELAXATION SPECTRA OF SOME SODIUM SALTS IN DIMETHYLACETAMIDE

D.E. Irish, S.Y. Tang, H. Talts and S.Petrucci

Depts. of Chemistry University of Waterloo, Ontario, Canada and Polytechnic Institute of New York, N.Y., U.S.A.

ABSTRACT

Raman bands originating from solvated SCN^- anions and from Na^+SCN^- ion-pairs in solutions of NaSCN in N, N-dimethylacetamide (DMA) have been detected. Integrated Raman band intensities have been measured and the distribution of the concentrations of free and bound ligand and hence the stability constant have been obtained. Ultrasonic relaxation spectra at t=25°C, in the frequency range 3 to 350 MHz, show a single Debye relaxation for NaSCN, $NaNO_3$ and $NaNO_2$ solutions. These kinetic data have been interpreted in terms of the Eigen theory by employing the appropriate information derived from the Raman spectral study.

ACKNOWLEDGEMENT

The authors are grateful to the NATO organization and to the National Research Council of Canada for grants which made this collaboration possible.

INTRODUCTION

Polyatomic anions (free and associated with cations in solution), can be used as probes, through their vibrational spectra, to identify the species present in solution and determine their relative concentrations (1). Ultrasonic relaxation techniques provide information about the kinetics of the formation of the same

W. J. Gettins and E. Wyn-Jones (Eds.). Techniques and Applications of Fast Reactions in Solution. 333–344.
Copyright © 1979 by D. Reidel Publishing Company.

species (2). On a time spectrum, the vibrational spectra identify species with a lifetime as short as 10^{-13} secs, whilst ultrasonic methods are generally used for trans- formations relaxing in the range 10^{-9} to 10^{-6} seconds. Correlation of the results of the two methods is of utmost use because concomitant structural and dynamic information about the same systems is thereby obtained.

To this end it was decided to study vibrational and ultrasonic relaxation spectra of solutions of NaSCN in the solvent DMA. The CN stretching frequency of the SCN$^-$ anion has been used for the Raman study.

In order to study the effect of changing the ligand on the ultrasonic relaxation spectra, solutions of $NaNO_3$ and $NaNO_2$ in the same solvent DMA were investigated.

EXPERIMENTAL

DMA (Aldrich 99+%) was redistilled twice in an all glass apparatus under vacuo. NaSCN (Baker analyzed) $NaNO_3$ (Fisher certified), $NaNO_2$ and NH_4SCN (Fisher certified) and NH_4NO_3 (Baker analyzed) were dried to constant weight in an oven at 120°C and atmospheric pressure. Solutions were prepared by weight, and the ones used for the Raman work were filtered through Millipore FHL-PO 1300 filters directly into pyrex glass capil- laries. The solutions used for the Raman work con- tained both the electrolyte of interest and a second salt to provide internal intensity standards.

The 514.5 nm line of a Spectra Physics argon ion laser was used as the exciting source for the Raman spectra. A Jarrel-Ash spectrometer was used to record the spectra (3). Digitized band profiles were resolved into component bands by computer analysis (4).

The equipment and procedure for the pulsed ultrasonic measurements have been described previously (5). Crystal displacements in the liquid were always below $R^2/2\lambda$ (with R the radius of the emitting crystal and λ the wavelength of sound) so as to remain within the Fresnel zone. The attenuation coefficients, α, and corresponding frequencies, f, were fitted to the Debye single relaxation function: (2)

$$\alpha/f^2 = [A/(1+(f/f_R)^2)] + B. \qquad (I)$$

RESULTS AND DISCUSSION

I- Vibrational spectra. The linear thiocyanate anion
generates a three-line vibrational spectrum, nominally
as follows: ν_1(CN) 2059 cm^{-1}; ν_2 (degenerate bend)
480 cm^{-1}; ν_3 (C-S) 735 cm^{-1}. The formation of either
N- or S-bonded complexes gives rise to spectral changes
that are well documented (6). For N-bonded coordination ν_1 increases, ν_3 increases and ν_2 is largely
unchanged. For S-bonded coordination ν_1 increases
but ν_3 decreases. ν_2 also decreases and splits due
to removal of degeneracy for nonlinear-S-bonded complexes. In the systems described below SCN$^-$ is
believed to be N-bonded to the cation as in the studies
of Chabanel et al. (7) Representative spectra for
NaSCN in DMA are reported in fig. 1 together with the
Gaussian-Lorentzian band components used to analyze

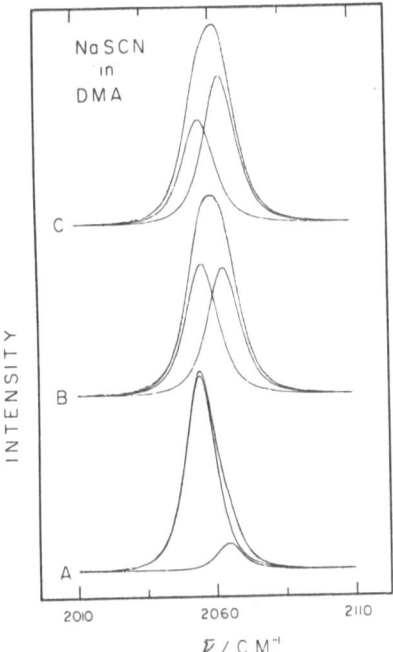

Fig. 1-The CN stretching region of the digitized Raman Spectra of NaNCS/DMA solutions: A)0.101M;
B)1.11M; C)1.54 M

the spectra as discussed below. The two bands are
centered at 2058 and 2068 cm^{-1}; the former is assigned
to the free SCN$^-$ ion and the latter to the Na$^+$SCN$^-$
contact ion-pair. Notice that the 2068 cm^{-1} band
increases at the expense of the 2058 cm^{-1} band as the

concentration of salt increases, in accord with the
mass law. Also attempts to interpret the spectra by
a single Gaussian-Lorentzian product function (8)

$$ I = I_0 \left[\exp \left(-\frac{(\bar{\nu}-\nu_0)^2}{2\theta^2} \right) \right] \left[1 + \frac{(\bar{\nu}-\bar{\nu}_0)^2}{\theta^2} \right]^{-1} $$

(II)

(with ν_0 the center frequency (cm^{-1}) of the band),

fail if one keeps the band's full width at half peak
height constant at $\sim$ 12 cm^{-1} over all the concentration range studied (as done in this work). (For
LiSCN, in the same solvent, a larger separation,
2059 cm^{-1} and 2075 cm^{-1}, occurs with appearance of
two separate bands). Both the 480 and 735 cm^{-1} bands
are obscured by intense lines of the solvent.

In order to deduce the formation constant for the
process, the Raman intensities of the NaSCN/DMA solutions were quantitatively analyzed and used to calculate the average ligand number $\bar{n}=C_B/C_T$, where C_B is
the concentration of bound SCN$^-$ and C_T is the total
concentration of Na$^+$. In order to calculate C_F, it
was necessary to measure the molar scattering coefficient J_F of the free SCN$^-$ ion, where the intensity
$I_F=J_FC_F$. The quantity I_F is the relative integrated
intensity ratio $I_F=(A_{2058}/A_{931})C_{ClO_4}$, namely it is
the ratio of the area of the 2058 "free" band SCN$^-$,
relative to that of the ClO_4^- ion at 931 cm^{-1} (symmetrical stretch) normalized to 1 molar ClO_4^- concentration.

In order to determine $J_F(SCN^-/ClO_4^-)$, solutions of
NH$_4$SCN containing a small amount of Bu$_4$N ClO$_4$ were
prepared and their Raman intensities determined.
These solutions generate a single CN band centered at
2059 cm^{-1} and it was assumed, therefore, that the
stoichiometric salt concentration was equal to the
"free" SCN$^-$ concentration. Relative integrated intensity ratios $I_F=(A_{2059}/A_{931})$ C_{ClO_4} were plotted
against C_F (fig. 2) and linearity was observed. The
internal intensity standard (ClO_4^-) compensates for
differences in refractive index and eliminates cell
corrections. The slope of the curve, calculated by
least squares analysis, was $J_F=1.71$. The concentration of the "free" SCN in the NaSCN solutions were
then calculated from the I_F/J_F ratios and C_B was
obtained from C_T-C_F. Some samples containing more
than the stoichiometric amount of Na$^+$ were prepared

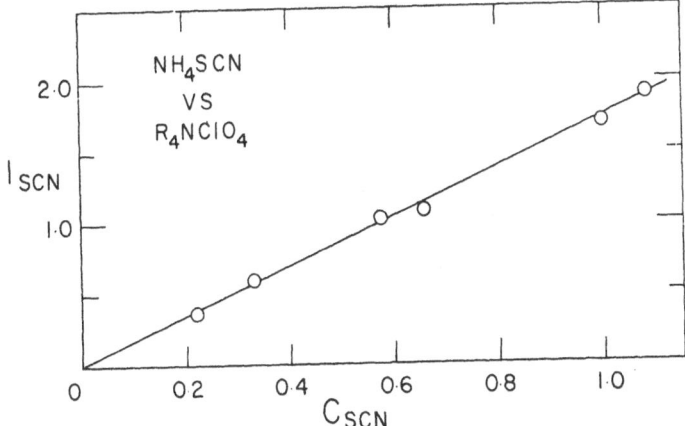

Fig. 2-Relative intensity of the CN stretching band with respect to the 931 cm^{-1} band of the ClO_4^- ion, vs the total concentration of the SCN^-

by dissolving small amounts of $NaClO_4$ (as the internal standard) in the $NaSCN/DMA$ solutions. The data for $\bar{n}$ from both $NaSCN/Bu_4NClO_4/DMA$ and $NaSCN/NaClO_4/DMA$ solutions fall on the same curve, (fig. 3) within

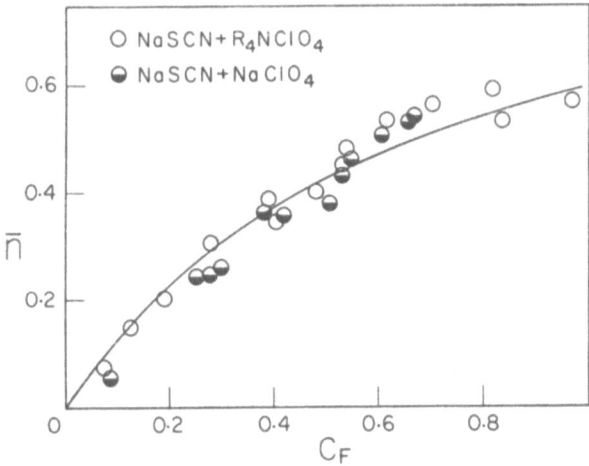

Fig. 3-Observed average ligand number $\bar{n}$ vs. the concentration of "free" SCN^- for $NaSCN/DMA$ solutions. The solid line has been calculated for $K_1 = 1.49$ M^{-1}.

experimental error.   Clearly, only mononuclear species
are formed (1b).   The solid line in fig. 3 was cal-
culated from the function $\bar{n} = \beta_1 C_F / 1 + \beta_1 CF$ by nonlinear
least squares using a single value of $\beta_1 = K_1 = 1.49 \pm 0.01$
M^{-1}.   Therefore only the species NaNCS in addition to
the free ions exists in the concentration range studied
$(0.1 < C_0 < 1.6 \text{ M})$.   Notice also that β_1 does not need to
be changed in numerical value.   As β_1 is the concentration quotient :

$$\beta_1 = K_1 = \frac{[\text{NaNCS}]}{[\text{Na}^+] \ [\text{SCN}^-]}$$

it implies that over the concentration range investi-
gated the activity coefficients ratio $(\gamma \text{NaNCS} / \gamma \pm^2)$
remains practically constant as discussed below.

II-Ultrasonic Relaxation. The ultrasonic results for
NaSCN, NaNO$_3$, and NaNO$_2$ DMA solutions are reported in
Table I in the form of f_R, A and B parameters (Eq I).
Fig. 4 reports representative functions for the NaNO$_2$
solutions at 25°C.   The solid lines are the computed
Debye functions (2).

Table I.   Ultrasonic Relaxation parameters f_R, A and
B for the electrolytes investigated in DMA at 25°C.

System	C(M)	f_R (MHz)	$A \times 10^{17}$ $(cm^{-1}s^2)$	$B \times 10^{17}$ $(cm^{-1}s^2)$
Solvent	--	--	--	26.0±0.7
NaClO$_4$	0.10$_1$	--	--	27.0±0.4
NaSCN	0.69	177	50.4	33.6
	0.48	156	41.0	31.5
	0.27$_4$	139	24.4	30.0
	0.18$_7$	137	13.9	28.5
	0.10	125	9.2	27.0
Na NO$_3$	0.60	90	98	30
	0.40	75	88	32
	0.19	60	74	28
	0.10	48	65	27
	0.05	40	42	28
NaNO$_2$	0.15	20	182	28
	0.10	16	173	27
	0.05	12.5	153	27
	0.04$_6$	12.5	160	27

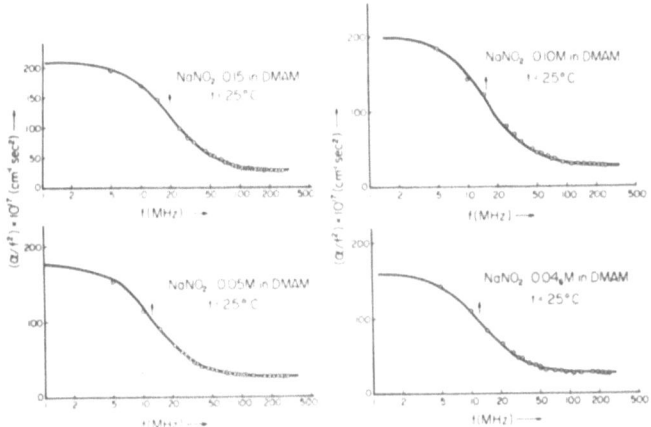

Fig. 4- α/f^2 vs the frequency f for NaNO$_2$ in DMA

The ultrasonic relaxation data were analyzed as follows: consistent with the interpretation of the Raman data, a kinetic process of the type :

$$Na^+ + L^- \underset{k_R}{\overset{k_f}{\rightleftharpoons}} NaL, \qquad (L = ligand),$$

is assumed to occur in the solutions.

According to Eigen (9), following the treatment of Stuehr and Yeager (10), the relaxation time of this overall process, τ, should correlate with the ionic concentration by the equation :

$$\tau^{-1} = k_f \ \sigma \ C_o \ \frac{\gamma \pm^2}{\gamma_{NaL}} \ [2 + \frac{\partial \ln(\gamma \pm^2 / \gamma \ NaL)}{\partial \ln \sigma}] + k_r \qquad (III)$$

In the above $\tau^{-1} = 2\pi f_R$; σ is the degree of dissociation.

Nixon and Plane (11) drew attention to the constancy of the concentration quotients K=[MeL$^{(n-1)+}$]/[Me^{n+}][L$^-$], obtained from Raman intensities in rather concentrated solutions and this constancy has been demonstrated many times since for metal ion complexes (1b).

The ionic medium seems to become "saturated" and the ionic strength no longer appears to be an important

variable for the β's. Consistently we have pointed
out the constancy of $\beta_1=K_1=1.49\pm0.01$ M^{-1} for NaSCN in
DMA. As a consequence the term $\gamma\pm^2/\gamma_{NaL}$ can be taken
as practically constant in eq. III and then:

$$\tau^{-1} = 2k_f\sigma\ C_o(\gamma\pm^2/\gamma_{NaL}) + k_r \qquad (IV)$$

By plotting τ^{-1} against σC_o a slope of $2k_f(\gamma\pm^2/\gamma_{NaL})$
and an intercept of k_r can be determined. The
ratio of 1/2 slope over the intercept gives the concentration quotient K_1

$$\frac{k_f\gamma\pm^2}{k_r\gamma_{NaL}} = K_1 = \frac{[NaL]}{[Na^+][L^-]} = \frac{1-\sigma}{\sigma^2C_o} \qquad (V)$$

The slopes and intercepts were obtained by iterative
linear regression, assuming $\sigma=1$ as a starting point,
evaluating K_1 and hence a second approximation for σ
and recycling until convergence was achieved.

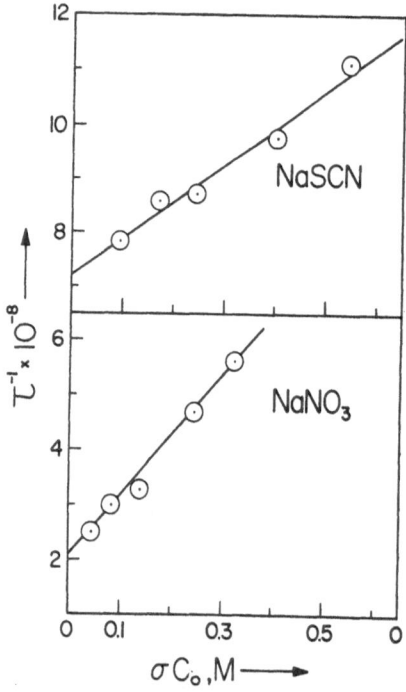

Fig. 5- Plots of τ^1 vs. $\sigma\ C_o$ for NaSCN/DMA and
NaNO₃/DMA solutions at 25°C.

Fig. 5 illustrates the final results for NaSCN and NaNO$_3$. The results for $k_f(\gamma\pm^2/\gamma_{NaL})$, k_r and k_1 are reported in Table II. It may be seen that the quantities $k_f(\gamma\pm^2/\gamma_{NaL})$ for the three salts are of the same order of magnitude; the maximum difference is of the order of 40% of the average value $(4.8\pm1)10^8 M^{-1}S^{-1}$. This is consistent with the expectation of an Id mechanism of solvent substitution around Na$^+$ (12).

Table II. Rate constants and stoichiometric equilibrium constants for the Na$^+$ salts investigates in DMA at 25°C

Electrolyte	$k_f(\gamma\pm^2/\gamma_{NaL})$ $(M^{-1}S^{-1})$	k_r (S^{-1})	K_1 (M^{-1})
NaSCN	3.5×10^8 *	7.2×10^8 *	0.48
NaNO$_3$	5.4×10^8	2.1×10^8	2.6
NaNO$_2$	5.5×10^8	4.1×10^7	13.4

*By using $K_1 = K_{Raman} = 1.49$ M^{-1} one calculates for NaSCN in DMA

$$k_f \left(\frac{\gamma\pm}{\gamma_{NaNCS}} \right) = 4.6 \times 10^8 M^{-1}S^{-1}$$

and $k_r = 7.0 \times 10^8$ S^{-1}.

On the other hand there is a factor of 17 between the extreme k_r values. The value of $K_1 = 0.48 M^{-1}$ for NaSCN compares favorably with the value $K_1 = 1.49$ M^{-1} (obtained from the Raman spectra) in view of the precision obtained from kinetic data.

The isoentropic change of volume ΔV_s for the complexation of Na$^+$ by SCN$^-$ was obtained as follows. The equation (2,9)

$$\mu_{max} = \frac{\pi}{2\beta_s} \frac{(\Delta V_s)^2}{RT} \left\{ \frac{1}{[Na^+]} + \frac{1}{[SCN^-]} + \frac{1}{[NaSCN]} + \frac{d\ln\gamma\pm^2}{d[Na^+]} + \frac{d\ln\gamma_{NaNCS}}{d[NaCS]} \right\}^{-1}$$

becomes, by introducing the degree of dissociation σ :

$$\mu_{max} = \frac{\pi}{2\beta} \frac{(\Delta V_s)^2}{RT} \left\{ \frac{2}{\sigma} + \frac{1}{1-\sigma} + \frac{d(\ln\gamma\pm^2/\gamma_{NaSCN})}{d\sigma} \right\}^{-1} C_o.$$

With the assumption that $(\gamma\pm^2/\gamma_{NaSCN})$ is approximately constant one obtains

$$\mu_{max} = \frac{\pi}{2\beta_s} \frac{(\Delta V_s)^2}{RT} \frac{\sigma(1-\sigma)}{2-\sigma} \; C_o . \qquad\qquad (VI)$$

$\beta_s = \dfrac{1}{\rho u^2}$ is the adiabatic compressibility, ρ the density and u, the sound velocity.

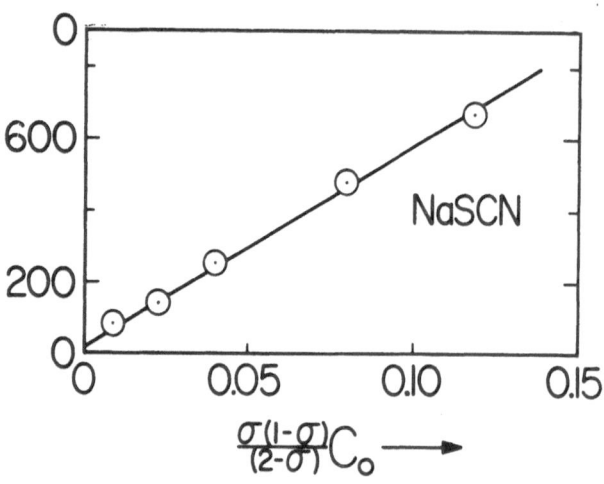

Fig. 6- Plot of μ_{max} vs. $[\sigma(1-\sigma)/(2-\sigma)] \; C_o$ for NaSCN/DMA solutions, $\mu_{max}=(A/2)uf_R$. The ordinate is multiplied by 10^5.

Fig. 6 reports a plot of μ_{max} vs. $[\sigma(1-\sigma)/(2-\sigma)] \; C_o$ for NaSCN in DMA. The slope was obtained by linear regression: $r^2=0.997$; $I=18$; slope$=5.65 \times 10^{-2}$ (r is the correlation coefficient, I the intercept).

By expressing C_o in mole/cm^3 $\Delta V_s=6.5$ cm^3/mole, a value of the same order of magnitude as corresponding ones determined for other ionic conplexation processes (13).

It should be noted that the above treatment for the calculation of K_1 from the Raman intensities precludes accounting for the possible presence of sizeable concentrations of outer-sphere ion-pairs. With the hypothesis of the existence of the latter species and calling $Co=C_1+C_2+C_3$ with C_1 the concentration of free ions, C_2 the concentration of outer-sphere and C_3 the concentration of contact species, it follows that $C_b=C_3$ and $C_f=C_1+C_2$ where C_b and C_f are the "bound" and "free" species concentrations as determined by Raman spectroscopy. These definitions are consistent with the fact that Raman spectroscopy provides signals characteristic of two classes of anion - viz. the contact species and the rest now designated solvent-separated and free species. The Raman β_1 would then be $K_1 = C_3/(C_1+C_2)^2$, instead of the Eigen $(C_2+C_3/C_1^2$ value. However, presence of a sizeable amount of outer-sphere species would result in an additional ultrasonic relaxation given by (14)

$$\mu_{mI} = \frac{\pi}{2\beta} \frac{(\Delta V_s)_I^2}{RT} C_0 \frac{\sigma}{\sigma_{12}} \frac{\sigma_{12}(1-\sigma_{12})/(2-\sigma_{12})}{1+\frac{\sigma_{12}(1-\sigma_{12})}{2-\sigma_{12}} (\frac{d\ln\gamma\pm^2}{d\sigma_{12}})} \quad (VII)$$

where $\sigma_{12} = \dfrac{C_1}{C_0-C_3} = \dfrac{C_1}{C_1+C_2}$, and $\sigma = \dfrac{C_1}{C_0}$.

When $C_2 \to 0$, σ_{12} tends to unity and $\mu_{mI} \to 0$. In the present work only one relaxation process of the Debye type has been observed and attributed to the formation of contact species. This is evident from the fact that k_R depends on the anion and $k_R < k_{-D}$ (diffusion) $\cong 4 \times 10^9 S^{-1}$ as calculable from the Eigen equation (15)

$$k_{-D} = \frac{2kT}{\pi \eta a^3} \frac{b}{e^b-1} \quad ; b = \frac{|Z_+Z_-|e^2}{a\epsilon kT}$$

With $a = 5 \times 10^{-8}$ cm the assumed parameter. Given that, in general (16) $\Delta V_I > \Delta V_{II}$, the absence of an observable relaxation process associated with the outer-sphere ion-pair formation leads to the conclusion that $C_3 >> C_2$ for the system investigated in this work.

REFERENCES

1a.   Irish, D.E. in "Ionic Interactions" Vol. II, Petrucci, S. Ed., Academic Press NY 1971, Chapter 9.
  b.   Irish, D.E. and Brooker, M.H. in "Advances in Infrared and Raman Spectroscopy" Vol. 2, Clark R.J.H. and Hester R.E. Eds. Heyden, London 1976, Chapter 6.
2.   Petrucci, S. in "Ionic Interactions, Vol. II, Petrucci, S. Ed. Academic Press NY 1971, Chapter 7.

3.   Bulmer, J.T., Irish, D.E., Grossman, F.W., Herriot, G., Tseng, M., and Weerheim, A.J., Appl. Spectr. 29, 506 (1975).

4.   Davis, A.R., Irish, D.E., Rodin, R.B. and Weerheim, A.J., Appl. Spectr. 26, 384 (1972).

5.   Petrucci, S., J. Phys. Chem. 71, 1174 (1967); Darbari, G.S., Richelson, M.R., and Petrucci, S., J. Chem. Phys. 53, 859 (1970).

6.   Gans, P., "Vibrating Molecules", Chapman and Hall, London 1971, page 192.
Burmeister, J.L., Coord. Chem. Revs. 1, 205 (1966).

7.   Paoli, D., Lucon, M., and Chabanel, M., Spectrachimica Acta, part A Vol. 34, Issue 11, 1087 (1978).

8.   Ref. 4; σ is the variance $\theta = \Delta\nu/1.46$, where $\Delta\nu$ is the width of the band at $I_0/2$.

9.   Eigen, M. and DeMaeyer, L., in "Rates and Mechanisms of Reactions" part II, Weissberger, A., Ed. Intersci. 2nd. ed. 1963 Chapter XVIII.

10.   Stuehr, J., and Yeager, E., in "Physical Acoustics", Mason, W.P., Ed. Acad. Press NY 1965 Vol. II part A, p. 386.

11.   Nixon, J., and Plane, R.A., J. Am. Chem. Soc., 84, 4445 (1962).

12.   Langford, C.H., and Gray, H.B., "Ligand Subst. Processes", W.A. Benjamin, NY 1966.

13.   Fisher, I.H., J. Phys. Chem. 66, 1607 (1962).

14.   Tamm, K., in "Dispersion and absorption of Sound by Molecular Processes", Sette, D. Ed., Acad. Press, NY 1963, page 192.

15.   Eigen, M., Z. physik. Chem. N.F., 1, 176 (1954).

16.   Hemmes, P., J. Phys. Chem. 76, 895 (1972).

MICROWAVE DIELECTRIC RELAXATION OF LITHIUM SALTS IN MEDIA OF LOW PERMITTIVITY AT t=25°C

D.Saar ,J.Brauner ,H.Farber and S.Petrucci

Depts. of Chemistry and Electrical Engineering
Polytechnic Institute of New York, Brooklyn
and Farmingdale Campuses, New York, U.S.A.

ABSTRACT

This paper reviews data(taken in our laboratory)of
complex permittivities of solutions of lithium salts
at U.H.F. and microwave frequencies. The solvents
used were tetrahydrofuran, dimethyl- and diethylcarbonate.

Parametric fitting of the data to the sum of a Debye
relaxation for the solvent, and to a Cole-Davidson
distribution (LiClO$_4$, NaClO$_4$ in THF) and to a Debye
function (other systems) for the solute, is reported.
The solute relaxation is interpreted as due to rotational relaxation of the dipolar species present in
solution. Structural information from IR-Raman spectra
and electrical conductance data from the literature
are used to substantiate the above interpretation.

INTRODUCTION

This paper represents a condensed and, at times,
slightly altered review of a research program in
existence at our laboratory since 1973. Because most
of the material has been published elsewhere, details
are referred to the original papers.

Electrolyte solutions show, at times, additional dielectric relaxations with respect to the solvent one,
at microwave frequencies. The problem we tried to
attack was the characterization of the molecular nature
of the solute dielectric relaxation. Whenever feasible,
it was decided to use alternate structural information
from other methods on the same systems, in order to

W. J. Gettins and E. Wyn-Jones (Eds.). Techniques and Applications of Fast Reactions in Solution. 345–353.
Copyright © 1979 by D. Reidel Publishing Company.

substantiate the interpretation. In aqueous 2:2
electrolytes Pottel (1) had reported for the concentration c≈1M (1 mole/dm^3) a narrow spectrum of relaxation times in addition to the one of water. He
attributed this spectrum to the rotational relaxation
of two solvated forms of outer-sphere ion-pair complexes, namely Me(H$_2$O)$_2$SO$_4$ and Me(H$_2$O)SO$_4$. More recently (2) this work was extended at lower concentrations (0.1-0.2M). Only a single Debye relaxation was
observable for the solutes. This was attributed to the
rotational relaxation of one of the two solvated species,
the one being probably the predominant form in this
concentration range.

2:1 and 1:2 electrolytes, which had been classified (3)
basically as nonassociated, did not show a solute
relaxation (2). Similarly a review of dielectric data
for 1:1 electrolytes in water (1), methanol (2),
DMSO (3), where ionic association is negligible, shows
that no additional dielectric relaxation, due to the
solute, is present.

On the contrary, in acetone, both Bu$_4$NBr (4) and
LiNO$_3$ (5) show an additional dielectric relaxation with
respect to the solvent. Bu$_4$NBr has been reported to be
partially associated in acetone (6) at 25°C.

It seems difficult, in view of these observations, to
escape the conclusion that the observed solute relaxation is due to the presence of dipole ion-pairs and
to dismiss alternate interpretation of the solute relaxation as due to the relaxation of the transport of
ionic charges. In fact, should this be the case, it
is not apparent why one cannot observe the solute
dielectric relaxation when the electrolyte is completely dissociated, therefore having the largest
possible number of electrical carriers per unit volume.
The interpretation becomes more challenging for electrolytes in media of permittivity lower than 15; other
complex ionic species possessing permanent dipole
moments may cause the appearance of additional contributions to the relaxation spectrum of the solute.
One would expect the spectrum to become more complex
and expressable by more than two relaxation regions in
these media. It was with the above observations and
ideas in mind that we engaged in the research program
whose progress up to the present time is shown below.

EXPERIMENTAL PART

Three methods are currently in use for microwave work up to f ∿ 90 GHz :

a) Waveguides containing resonant cavities to be filled by the liquid under study. This method is generally limited to liquids having very low dielectric loss (or tan $\delta = \frac{\varepsilon''}{\varepsilon'} << 1$, with ε'' and ε' the coefficients of the imaginary and real part of the complex permittivity $\varepsilon^* = \varepsilon' - J\varepsilon''$). Exceptions to the above statement may be made if one succeeds in filling only a small portion of the cavity by the lossy material.

b) Standing guided waves with various instrumental options (slotted lines or directional couplers) with audiomodulated waves and crystal detectors. These methods are usable up to tan $\delta \leq 0.3 - 0.4$.*

c) Travelling guided waves and microwave bridges (balance methods (8) whose limitations are frequency f≤40 GHz but that can work up to tan $\delta \leq$ 2-3.

Because of the planned work with electrolytes in media of low permittivity and consequently because of the relatively low electrical conductances, we chose the standing wave methods. In particular reflectometers with directional couplers sampling the returning wave (with components from the air-liquid and liquid-reflector interface) were used (9). The frequency range was 2-4 GHz for co-axial lines and 10, 16, 35, 55 and 66 GHz for waveguide instruments. For the latter ones, four reflectometers corresponding to the 8-12 GHz band, 12-18 GHz band, 26-40 GHz band and 50-70 GHz band were used. For the frequency range 0.3-1.5 GHz a co-axial General Radio admittance bridge was employed. Details of these instrumentations and of their use have been already described (9),(10).

Three solvents, tetrahydrofuran- dimethyl- and diethyl-carbonate ranging in permittivity from 7.4 to ∿ 3 were used.

In THF a number of electrolytes as $LiClO_4$, $NaClO_4$, Bu_4NNO_3, $LiNO_3$, NaPicrate at concentrations ≤0.1M were used.

In dimethylcarbonate $LiClO_4$, LiSCN and LiBr ∿ 0.1M were investigated whereas in diethylcarbonate LiSCN and $LiClO_4$ ∿ 0.1M were studied.
*With the exception of the Von Hippel method: "Dielectrics and Waves" M.I.T. press.

RESULTS

For $LiClO_4$ in THF a Cole-Davidson distribution function (11) for the solute with a distribution parameter $\beta<1$, and a Debye function for the solvent seem to describe well the complex dielectric permittivity ε^* in accord with the function

$$\varepsilon^* = \varepsilon_{\infty 2} + \frac{\varepsilon_0 - \varepsilon_{\infty 1}}{(1+J\omega\tau_1)\beta} + \frac{\varepsilon_{\infty 1} - \varepsilon_{\infty 2}}{1+J\omega\tau_2} \tag{I}$$

In the above ε_0 is the static permittivity of the solution, $\varepsilon_{\infty 1}$, the static permittivity of the solvent in the solution $\varepsilon_{\infty 2}$ the residual value of the permittivity at frequencies far above the relaxation of the molecular orientation of both solute and solvent. $\tau_1 (=2\pi f_{R1})$ and $\tau_2 (=2\pi f_{R2})$ are the two corresponding relaxation times.

For $LiClO_4$ 0.05M in THF $\beta=0.8$ (9). The appearance of a Cole-Davidson distribution is confirmed by literature data (12) at c=0.25 M, 0.4 M and 0.6 M and t = 30°C. The distribution parameter β becomes lower in value by increasing the electrolyte concentration. The same effect is found by adding benzene to THF. For $C(LiClO_4)=$ 0.8 M and a molar ratio $(THF/LiClO_4)=3$, $\beta=0.5$.

For $NaClO_4$ 0.05 M and 0.1M in THF (13) also a Cole-Davidson distribution with parameters $\beta=0.9$ and $\beta = 0.8$ respectively describes the solute relaxation.

For Bu_4NNO_3 0.05M, $LiNO_3$ 0.1M and NaPicrate 0.05M a single Debye relaxation, within experimental error, seems to describe the data for the solute dielectric relaxation. The same holds true for LiSCN, $LiClO_4$ and LiBr 0.1M in dimethylcarbonate and for LiSCN and $LiClO_4$ 0.1M in diethylcarbonate. For all these cases eq. I with $\beta=1$ and the appropriate parameters ε_0, $\varepsilon_{\infty 1}$, $\varepsilon_{\infty 2}$, τ_1 and τ_2 describe the total dielectric spectrum investigated. However, for dimethyl- and diethylcarbonate literature data (14) for the pure solvents indicate contributions of non-Debye nature at frequencies above 35 GHz in the millimeter and submillimeter range. These additional effects, probably associated to nonrotational motion of the solvent, have been neglected here.

DISCUSSION

Results from other methods of investigation as electrical conductance and Raman spectra and literature information from IR spectra (15) have been used to interpret the dielectric data.

For $LiClO_4$ (9a)and $NaClO_4$ (13) in THF,conductance data reveal the presence of non conducting pairs as the preponderant species in the concentration range 0.01-0.1M. Free ions are practically absent in this concentration range and small amounts of conducting triple ions are formed. The latter species reveal their presence by the eventual appearance of a minimum in the conductance-concentration function.

Hence the dielectric relaxation process cannot possibly be assigned to the relaxation of the transport of charges; their presence either as free ions or as triplets accounts only to a minute minority of the total solute population around $c \approx 0.1M$.

A discussion of the statistical distribution of the population of the ion-pairs (9a) with respect to the interionic distance, based on the Bjerrum probability ratio and a continuum for the solvents reads:

$$\frac{P(r)}{P(a)} = \frac{4\pi L \times 10^{-3} c r^2 (\exp\left(\frac{|Z_+Z_-|e^2}{r\epsilon kT}\right))dr}{4\pi L \times 10^{-3} c a^2 e^b dr} = \frac{r^2 \exp\frac{|Z_+Z_-|e^2}{r\epsilon kT}}{a^2 e^b}$$

$$(II)$$

In the above a is the contact distance between the ions takes as 3×10^{-10} meters; $\epsilon \approx \epsilon = 7.4$, the solvent permittivity, neglecting the increase of permittivity due to the solute. The probability ratio is shown in Table I. It results that the majority of ion-pairs in THF are contact species. Therefore arguments trying to assign the increase of the polarization of the medium (hence the origin of the dielectric relaxation) to the inversion of position of long range solvent-separated pairs (within the Bjerrum radius $q \approx 38 \times 10^{-10}$ meters in THF) are without context.

Table I

Bjerrum probability ratios as function of the interionic distance r for pairwise interaction between ions

$r \times 10^{10}$	$|Z_+Z_-|e^2/r\epsilon kT$	$P(r)/P(a)$
(meters)	--	--
a=3.0	25.224	1

4.0	18.918	3.24×10^{-3}
5.0	15.134	1.15×10^{-4}
10.0	7.567	2.38×10^{-7}
$q=37.84$	2	1.30×10^{-8}

Arguments questioning the lifetime of the ion-pairs, or, in other words, questioning whether an ion-pair can be considered a permanent dipole upon passage of the electromagnetic wave, could be raised.

The shortest possible lifetime of an ion-pair in THF may be judged from the diffusion decomposition rate constant of an ion-pair in accord with Eigen (16,17)

$$k_{-D} = \frac{2kT}{\pi a^3 \eta} \quad \frac{b}{e^b-1} = 5.93 \times 10^4 \, sec^{-1}$$

In the above $a=3 \times 10^{-10}$ meters whereas the viscosity of the solvent THF at 75°C (9) $\eta=0.004583$ poise. The time constant for the decomposition of an ion-pair is then $t=k_{-D}^{-1}=1.69 \times 10^{-5}$ seconds in THF. This time is far longer than the period of the electromagnetic wave $T=\frac{1}{f} \approx 10^{-9}$ sec in the frequency range corresponding to the relaxation times of the solutes in THF.

One could try to explain the observed dielectric relaxation for the ionic solutes invoking the presence of a network of ions forming a kind of loose crystal aggregates. The observed relaxation would then correspond to the relaxation of the short range inversion of the position of ions of contrary signs, without being able to assign an individual partner to each ion. These arguments, possibly reasonable for ions of spherical distribution of charge at high concentrations, close to a solvated fused salts range, do not apply to our measurements, as shown below, both on theoretical and experimental grounds.

The average distance between pairs is 2r with r the radius of the sphere containing an ion-pair. The volume containing an ion-pair (assuming random distribution of pairs) is $V=\frac{4}{3}\pi r^3$, where $V=1/c \times 10^{-3}$L. Numerically at $c \approx 0.1M$ $V= 1.66 \times 10^{-26} m^3$/molecule and $r \approx 15.8 \times 10^{-10}$ meters, namely $2r \approx 31.7 \times 10^{-10}$ meters whereas the interionic distances calculated from apparent dipole moments (18,19) range from 2 to 6×10^{-10} meters. Hence the distance between the pairs if far larger than the interionic distance at the concentration investigated.

Evidence of specificity of the nature of the interionic

bands and of the contact nature of pairs for LiSCN in dimethylcarbonate comes from Raman spectra (10) and IR spectra as well (15).

Experimentally the Raman spectrum of 0.24M LiSCN in dimethyl carbonate (10),corresponding to the anti-symmetric stretching of the SCN anion,shows only two bands. These have been assigned to the Li^+-N-bonded pair, LiNCS ($\tilde{\nu}=2067$ cm^{-1}) and to the aggregate (LiNCS)$_2$ ($\tilde{\nu}=2052$ cm^{-1}). No spectroscopically "free" SCN$^-$ ion exists, namely no solvent separated species is detect-able due to the absence of the CN stretching band at $\sim$2060 cm^{-1}. The presence of only two bands has been confirmed quantitatively by band analysis of the spectrum by two Gaussian-Lorentzian functions of the type:

$$I = I_{oi} \exp\left(-\frac{(\bar{\nu}-\bar{\nu}_i^0)^2}{2\sigma^2}\right) \left(\frac{(\bar{\nu}-\bar{\nu}_i^0)^2}{\sigma^2}\right)^{-1} \qquad (III)$$

where i=1,2; I_{oi} is the intensity of the band at the wavenumber ν_i^0, σ is the variance $\sigma = \frac{\Delta\nu1/2}{1.46}$ with $\Delta\nu1/2$ the band width at half intensity $Io/2$. For this analysis $\Delta\nu_{1/2}\approx12$cm^{-1}which is charac-teristic of the "free" SCN$^-$ at $\sim$2060 cm^{-1} (20).

Final evidence of the correlation between the observed dielectric relaxation and the ion-pairs comes from the linearity of the relaxation strength $\varepsilon_0-\varepsilon_{\infty1}$,with the concentration of ion-pair c_p for both LiClO$_4$ in THF (9) and LiBr, LiSCN and LiClO$_4$ 0.1M in dimethyl-carbonate (10).

In accordance with the Böttcher equation (22)

$$(\varepsilon_0-\varepsilon_{\infty1})=\frac{10^3 L}{\bar{\varepsilon}_0(1-\alpha f)^2} \frac{\mu^2}{3kT} c_p \frac{3\varepsilon_0}{2\varepsilon_0+1} \qquad (IV)$$

$$\bar{\varepsilon}_0 =(1/36\pi)10^{-9} \text{ Farad m}^{-1}$$

plots of $(\varepsilon_0-\varepsilon_{\infty1})$ (for the same electrolyte)or plots

of $[(\varepsilon_0-\varepsilon_{\infty1})/\mu^2$ $\underline{\text{vs}}$ $c_p \frac{2\varepsilon_0}{2\varepsilon_0+1}$ (for different elec-

trolytes) should give a straight line passing through the origin if one neglects the polarizability-reaction field factor term $(1-\alpha f)^2$. For LiClO$_4$ in THF and LiBr, LiSCN and LiClO$_4$ 0.1M in dimethyl carbonate this is roughly the case, a linear correlation between $(\varepsilon_0-\varepsilon_{\infty1})$and $\varepsilon_0-\varepsilon_{\infty}/\mu^2$ respectively being observed (9,10) with respect to the ion-pair concentration term.

There is, however, the additional question of the contribution of the presence of other species as triplets
and quadrupoles to the dielectric strength if these
species possess a non zero dipole moment. This is an
aspect of research still in full development in our
laboratory, therefore only preliminary information is
available.

In diethylcarbonate, it appears that quadrupoles or
dimers as $(LiNCS)_2$ and $(LiClO_4)_2$ have a negligible
dipole moment. This is evident from the absence of an
additional dielectric relaxation and it is confirmed
by IR spectra appeared in the literature (22) for LiSCN.
Both the CN and CS bands frequencies decrease in wavenumber ν_0. This has been assigned (22) as due to the
formation of a symmetrical structure of zero dipole
moment

$$S \equiv C - N \overset{\diagup Li \diagdown}{\underset{\diagdown Li \diagup}{}} N - C \equiv S$$

However it seems, from IR spectra (22),that in THF
LiSCN shows a band at ~ 2087 cm^{-1} which has been
assigned to a S-bonded species probably $L_i^+ \overset{+}{SCN} L_i^+$
which is the precursor of the formation of nonlinear
chains (eventually showing splitting of the degenerate
bending frequency $\nu_0 = 480$ cm^{-1} of the SCN anion). Such
structural evidence seems to appear at relatively high
concentrations (22) (c>>0.1M). If these chains exist
also for $LiClO_4$ in THF, the appearance of a Cole-
Davidson distribution, especially evident (12) at
c>0.1M, might be due to the presence of polar species
other than neutral pairs rather than to the rotation-
collision mechanism proposed in our initial research
(9) to explain the data.

It seems apparent, from the above, that symbiosis of
dielectric relaxation with structural information may
allign microwave dielectric relaxation with the existing techniques for the dynamic study of electrolyte
solutions.

REFERENCES

1. R. Pottel in "Chemical Physics of Ionic Solutions"
    B.E. Conway and R.G. Barrados Ed. J. Wiley, N.Y.
    1966 page 581
2. E.A.S. Cavell and S. Petrucci J. Chem. Soc., Farad.
    Trans. II 74 1019 (1978)

3. C.W. Davies "Ion Association", Butterworths, London 1962

4. E.A.S. Cavell Trans Far. Soc. 61, 1578 (1965)

5. J. Barthel, H. Behret and F. Smithals Ber. Bunsenges. phys. chem. 75, 305 (1971)

6. M.B. Reynolds and C.A. Kraus J. Am. Chem. Soc. 70, 1709 (1948)

7. W.E.Vaughan, in "Dielectric Properties and Molecular Behaviour" N. Hill et al. Ed. Nostrand-Reinhold, London 1969

8. E.A.S. Cavell J. Sci. Instr. 44, 401 (1967); J.C.S. Faraday II, 70, 78 (1974)

9. H. Farber and S. Petrucci J. Phys. Chem. 79, 1221 (1975)

9a P. Jagodzinski and S. Petrucci J. Phys. Chem. 78, 917 (1974)

10. D. Saar, J. Brauner, H. Farber, S. Petrucci J.Phys. Chem. 82, 545 (1978)

11. N. Hill in "Dielectric Properties and Molecular Behaviour" N.Hill et al Ed. Nostrand-Reinhold, London 1969

12. J.P. Badiali, H. Cachet, A. Cyrot and J.C. Lestrade J.C.S. Faraday II 69, 1339 (1973)

13. H. Farber and S. Petrucci, J. Phys. Chem. 80, 327 (1976)

14. M.W. Evans, M.N. Afsar, G.J. Davies, C. Menard and J. Goulon, Chem. Phys. Letter 52, 388 (1977)

15. M. Chabanel, C. Menard and M. Guiheneuf C.R. Acad Sci. Paris 272, 253 (1971)

16. M. Eigen Z. Physik. Chem. (Frankfurt)1, 176 (1974)

17. S. Petrucci J. Phys. Chem. 71, 1174 (1967)

18. M. Chabanel, D. Menard J. Phys. Chem. 82, 1943 (1978)

19. D. Saar, J. Brauner, H. Farber, S. Petrucci J. Phys. Chem. 82, 1943 (1978)

20. D.E. Irish, S.Y.Tang, H. Talts and S. Petrucci, these proceedings

21. C.F. Böttcher, "Theory of Electrical Polarization" Elsevier Amsterdam 1973

22. J. Vaes, M. Chabanel and M.L. Martin, J. Phys. Chem. 82, 2420 (1978) and references quoted therein.

CROWN ETHER REACTION KINETICS

Edward M. Eyring, Michael M. Farrow, Licesio J. Rodriquez, Lindsay B. Lloyd, Ronald P. Rohrbach, and Evan L. Allred

Department of Chemistry, University of Utah, Salt Lake City, Utah 84112, U.S.A. and Office Products Division, IBM Corporation, Boulder, Colorado 80302, U.S.A.

ABSTRACT. The macrocycles 18-crown-6 and 15-crown-5 in aqueous solution undergo at least one rapid conformational change detectable by ultrasound. A correlation between ring size and rigidity of such macrocycles and the rapidity of the conformational change is suggested. A linear inverse correlation exists between the rate constant for formation of crown ether-alkali and alkaline earth metal cation complexes and the charge density of the cation. 2,6-dimethylylbenzoic acid-18-crown-5 does not significantly accelerate cleavage of glycine p-nitrophenyl ester in aqueous solution whereas crown ethers with functionalized side arms do catalyze this type of reaction. Ultrasonic absorption kinetic data are reported for aqueous amino acid-crown ether complexation reactions that indicate the complexation step is unlikely to be rate limiting in any such crown ether catalyzed process in water-like solvents.

It was C.J. Pedersen (1) who coined the name crown ether in 1967 to describe molecules such as 1,4,7,10,13,16-hexaoxacyclooctadecane

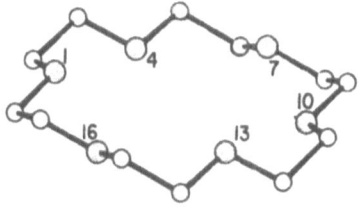

This particular macrocycle with eighteen atoms in the ring, six of which are oxygens, is called 18-crown-6. The crown ethers have commanded wide attention because of their capacity for

W. J. Gettins and E. Wyn-Jones (Eds.). Techniques and Applications of Fast Reactions in Solution. 355–361.
Copyright © 1979 by D. Reidel Publishing Company.

selectively binding cations. Among their many applications the
crown ethers were at first viewed (2) only as models of naturally
occurring antibiotic facilitators (e.g. valinomycin) of ion transport through biological membranes. However, the crowns have
lately assumed potential commerical importance as ion selective
carriers through organic liquid membranes designed to separate
different (radioactive) isotopes of the same element as well as
ions of different elements dissolved in concentrated aqueous
media (3,4). Extensively modified crown ethers have also been
studied for their potential utility as homogeneous catalysts
(5-7). Our own interest in the kinetic properties of crown
ethers grew out of these two general applications of crown ethers
in other laboratories.

A Joule heating temperature jump (T-jump) relaxation method
study (8) of the kinetics of complexation of monovalent cations
in methanol by dibenzo-30-crown-10 particularly intrigued us.
Chock had found the rate of complexation of several monovalent
cations (Na^+,K^+,NH_4^+,etc.) to be almost diffusion controlled and
essentially too fast for precise determination by T-jump equipment then available to him. He also noted an even faster relaxation process that was completely inaccessible. This latter relaxation process Chock ascribed to a conformational change of the
dibenzo-30-crown-10 between two ligand conformers one of which
is more suitable for complexing the cation. Such an inference
is entirely consistent with known, rapid conformational equilibria in solutions of valinomycin (2), for example.

Valinomycin and dibenzo-30-crown-10 are both large enough
macrocycles to envelope a monovalent cation with geometries
(9,10) for the backbones of the macrocycles in the complex that
have been likened to a deformed bracelet and to the seam on a
tennis ball. Smaller crown ethers such as 18-crown-6 can only
girdle (11) a monovalent cation, so it is natural to wonder
whether a rapid conformational preequilibrium involving so small
a crown must precede complexation of the cation. The absence
(8) of any systematic trends of complexation rate constants
versus cation charge density for dibenzo-30-crown-10 reacting
with various monovalent cations in methanol also piqued our
curiosity and gave rise to a series of ultrasonic absorption
kinetic studies of 18-crown-6 (12-15) and 15-crown-5 (14,16)
interacting with a variety of monovalent and divalent cations
in aqueous solution. Several conclusions emerge from this work:

    1) The small 15-crown-5 and 18-crown-6 rings in aqueous
        solution undergo a rapid conformational change [analo-
        gous to the one inferred by Chock (8) for dibenzo-30-
        crown-10] with the dominant conformer being the one that
        complexes the cations.

    2) A plot of $\log_{10} k_r$, where k_r is the complexation rate

constant, versus (ionic charge/ionic radius) for the
ions $Na^+, K^+, Rb^+, Cs^+, NH_4^+, Sr^{2+}$, and Ba^{2+} and the ligands
15-crown-5 and 18-crown-6 in water gives rise to a very
satisfactory straight line of negative slope (14).

3) The cations $Ag^+, T\ell^+, Pb^{2+}$ and Hg^{2+} react more rapidly
   with either ligand in water than do the alkali metal
   (and alkaline earth metal) cations of corresponding
   charge density.

4) All the complexation rate constants, k_r, for these cations lie within a factor of ten of an average value of
   $\sim 2 \times 10^8 M^{-1} sec^{-1}$ that is in turn two powers of ten smaller than the diffusion controlled limiting value.

5) Variations in stability constants of 18-crown-6 complexing various cations in water arise principally from
   significant differences in rate constants for dissociation of the complex ions rather than from k_r.

6) The conformational relaxation times in Table I suggest
   that a correlation exists between macrocycle rigidity
   and the speed at which the ring undergoes the conformational change.

Table I. Comparison of Concentration Independent, Ultrasonic
Relaxations in Several Dissolved Ring Compounds

	f_r, MHz	$\tau^{-1} = k_{12} + k_{21}$, sec^{-1}	Ref
18-crown-6 (aq)	100.7 ± 3.3	6.3×10^8	(15)
15-crown-5 (aq)	22.9 ± 0.4	1.4×10^8	(16)
cyclohexaamylose (aq)	12.3 ± 0.5	7.7×10^7	(17)
dinactin (methanolic)	28	1.8×10^8	(18)
4-methyl-1,3-dioxan (in p-xylene)	8	5.0×10^7	(19)

Note that $2\pi f_r = \tau^{-1}$ and that $k_{12} = 6.2 \pm 0.2 \times 10^8$ sec^{-1}, see
ref. (15), for the particular case of 18-crown-6 (aq).

The correlation mentioned in conclusion 6) is clearly only
a very rough one. For instance, dinactin has three carbon atoms
between each pair of ether oxygen atoms in its thirty-two atom
ring quite unlike crown ethers that have two methylene groups between neighboring ether oxygen atoms. One also must ignore the
presence of four methyl and four ethyl side chains as well as
four furan rings in asserting for purposes of this comparison
that dinactin approximates (an impossible) 32-crown-8. On the

other hand, the bottomless woven basket structure of cyclohex-
amylose (19) does give rise to a more rigid structure for this
ring than one would anticipate for any of the crown or crown-
like single chain macrocycles. Ultrasonic techniques should be
used to test our speculation that 21-crown-7, 24-crown-8, and
dibenzo-30-crown-10 all undergo conformational changes at frequencies exceeding 22.9 MHz whereas the smaller 12-crown-4 ring
changes shape with a frequency under 22.9 MHz.

The notion of relative ring flexibility is not easy to
make quantitative from theory. However, quantum mechanical calculations of the energies associated with different conformations
of such molecules are possible (20,21). The results of such calculations could possibly be correlated with kinetically derived
thermodynamic parameters for the conformational change (15).
At worst, such an investigation would reveal the importance of
solvation effects on the conformational equilibria.

Crown ethers have been modified to render them more effective
homogeneous (5-7,22) and heterogeneous catalysts (23,24). Our
one preliminary effort in this area was a kinetic study (25) of
the accelerated cleavage of glycine p-nitrophenyl ester (I) in
the presence of 2,6-dimethylylbenzoate-18-crown-5 (II) in water.

I   II

The ring of (II) (26,27) should capture and hold the substrate
ester (I) while the carboxyl group facilitates the hydrolysis
reaction. Extensive examination has shown far less acceleration
of the reaction than expected. Thus, for example, an 18-crown-
6 to which two thioalkyl sidearms have been added (6) appears
to be a much more effective catalyst for ester solvolysis.

The one other useful insight derived from this study (25)
of (II) is that the rate of complexation of the ester by the crown
ether is probably much too rapid to be rate limiting in any such
catalyzed hydrolysis. This we infer from the very rapid rates
of complexation of simple amino acids by 18-crown-6 and 15-crown-
5 in water shown in Table II.

Table II. Laser Debye-Sears Ultrasonic Rate and Kinetic Equilibrium Constants for Aqueous Amino Acid-Crown Ether Complexation at 25°C.

Amino acid	$k_f, 10^7 M^{-1} s^{-1}$	$k_r, 10^7 s^{-1}$	K, M^{-1}
	18-Crown-6		
Glycine	8.4 ± 1.0	12.2 ± 1.0	0.7 ± 0.1
α-Alanine	6.1 ± 0.7	5.4 ± 0.8	1.1 ± 0.3
β-Alanine	6.6 ± 0.5	7.2 ± 0.5	0.9 ± 0.1
γ-Aminobutyric acid	5.1 ± 0.2	8.3 ± 0.2	0.6 ± 0.1
Threonine	3.8 ± 1.3	6.0 ± 1.3	0.6 ± 0.4
	15-Crown-5		
Glycine	5.6 ± 0.8	<2	>2

As workers succeed in anchoring crown ether catalysts on surfaces, it will be interesting to see the extent to which catalyzed reaction rates are further enhanced by what is known as a "reduction in dimensionality"(28). The essence of this idea is that a substrate only needs to hunt for an "active site" in two dimensions once it sticks on the catalytic surface. This is in contrast to a longer three dimensional search for the catalyst in the case of homogeneous solutions.

ACKNOWLEDGEMENT. This work was sponsored by a contract from the Office of Naval Research and by Grant AFOSR 77-3255 from the Directorate of Chemical Sciences, Air Force Office of Scientific Research.

REFERENCES

1. C.J. Pedersen, J. Am. Chem. Soc., 89, 7017 (1967); in
Synthetic Multidentate Macrocyclic Compounds, R.M. Izatt and J.J.
Christensen, eds., Academic Press, New York, NY, 1978, Chapter 1.

2. E. Grell, Th. Funck, and F. Eggers, in *Membranes*, Vol.
III, G. Eisenman, ed., Dekker, New York, NY, 1974, Chapter 1.

3. R.A. Schwind, T.J. Gilligan, and E.L. Cussler, in *Synthetic
Multidentate Macrocyclic Compounds*, R.M. Izatt and J.J. Christensen, eds., Academic Press, New York, NY, 1978, Chapter 6.

4. J.J Christensen, J.D. Lamb, S.R. Izatt, S.E. Starr, G.C.
Weed, M.S. Astin, B.D Stitt, and R.M. Izatt, J. Am. Chem. Sec.,
100, 3219 (1978).

5. J.H. Fendler and E.J. Fendler, *Catalysis in Micellar and
Macromolecular Systems*, Academic Press, New York, NY, 1975, Chapter II.

6. T. Matsui and K. Koga, Tetrahedron Lett., 1978, 1115.

7. R.D. Gandour, D.A. Walker, A. Nayak, and G.R. Newkome, J.
Am. Chem. Soc., 100, 3608 (1978).

8. P.B Chock, Proc. Natl. Acad. Sci. (U.S.), 69, 1939 (1972).

9. K. Neupert-Laves and M. Dobler, Helv. Chim. Acta, 58, 432
(1975).

10. M.A. Bush and M.R. Truter, J. Chem. Soc. Perkin Trans., 2,
345 (1972).

11. M. Dobler, J.D. Dunitz, and P. Seiler, Acta Crystallogr.,
B30, 2741 (1974).

12. G.W. Liesegang, M.M. Farrow, N. Purdie, and E.M. Eyring, J.
Am. Chem. Soc., 98, 6905 (1976).

13. G.W. Liesegang, M.M. Farrow, F.A. Vazquez, N. Purdie, and
E.M. Eyring, J. Am. Chem. Soc., 99, 3240 (1977).

14. L.J. Rodriguez, G.W. Liesegang, M.M. Farrow, N. Purdie,
and E.M. Eyring, J. Phys. Chem, 82, 647 (1978).

15. G.W. Liesegang, M.M. Farrow, L.J. Rodriguez, R.K. Burnham,
E.M. Eyring and N. Purdie, Int. J. Chem. Kin., 10, 471 (1978).

16. L.J. Rodriguez, G.W. Liesegang, R.D. White, M.M. Farrow, N. Purdie and E.M. Eyring, J. Phys. Chem., 81, 2118 (1977).

17. R.P. Rohrbach, L.J. Rodriguez, E.M. Eyring, and J.F. Wojcik, J. Phys. Chem., 81, 944 (1977).

18. P.B. Chock, F. Eggers, M. Eigen, and R. Winkler, Biophys. Chem., 6, 239 (1977).

19. P.C. Hamblin, R.F.M. White, and E. Wyn-Jones, Chem. Commun., 1968, 1058.

20. A. Pullman, C. Giessner-Prettre, and Y.V. Kruglyak, Chem. Phys. Lett., 35, 156 (1975).

21. W.F. Richards, *Quantum Pharmacology*, Butterworth, Inc., Boston, Mass., 1977, Chapter 13.

22. L. Wong and J. Smid. J. Am. Chem. Soc., 99, 5637 (1977).

23. E. Blasius, Z. Anal. Chem., 284, 337 (1977).

24. M. Tomei, O. Abe, M. Ikeda, K. Kihara, and H. Kakiuchi, Tetrahedron Lett., 1978, 3031.

25. R.P. Rohrbach, G.D. Lyon, L.J. Rodriguez, E.L. Allred, and E.M. Eyring, ONR Tech. Report 13, Univ. of Utah (1977).

26. M. Newcomb and D.J. Cram, J. Am. Chem. Soc., 97, 1257 (1975).

27. I. Goldberg, Acta Cryst., B32, 41 (1976).

28. P.H. Richter and M. Eigen, Biophys. Chem, 2, 225 (1974).

ESTIMATION OF THE RATIO OF SOLVENT SEPARATED TO CONTACT ION PAIRS BY HIGH FIELD CONDUCTIVITY MEASUREMENTS

Paul Hemmes[1] and John J. McGarvey[2]

[1] Department of Chemistry, Rutgers University, Newark, NJ 07102 U.S.A.   [2] Department of Chemistry, Queen's University of Belfast, Belfast BT9 5AG U.K.

Abstract: High field conductimetric measurements of the ratio of contact to solvent separated ion pairs have been made on the system $ZnSO_4$ in methanol ($K = 5.8 \times 10^4$). The method used has quite general applicability for such determination and is one of the few such generally useful methods.

The chemical significance of ion pairs subspecies (solvent separated pairs, contact pairs, etc.) has been well documented. (1) In only a handful of cases can one quantitatively estimate the ratio of contact pairs to solvent separated ones using spectrophotometry. Relaxation kinetics has been used for this purpose (2) but suffers the limitation that the dynamics of the ion pair formation must lie in the time domain of the experimental technique or else one obtains only composite constants (i.e. products of equilibrium and rate constants).

We have recently developed a technique of considerable generality which will enable the thermodynamics of the processes I and II to be determined

$$M^{+n} + L^{-n} \underset{}{\overset{I}{\rightleftharpoons}} M,L \underset{}{\overset{II}{\rightleftharpoons}} ML$$

M and L are free ions
M,L the solvent separated pair
ML the contact pair

The principle is basically that used by Patterson many years ago for a high field conductance study of aqueous CO_2 solutions. What was done was to take the weak electrolyte and subject it to a short pulse of high electric field. In such a solution the second Wien effect

W. J. Gettins and E. Wyn-Jones (Eds.). Techniques and Applications of Fast Reactions in Solution. 363–365.
Copyright © 1979 by D. Reidel Publishing Company.

causes an increase in the degree of dissociation of the electrolyte.
Quantitatively

$$\frac{\Delta \Lambda}{\Lambda} \quad \frac{1 - \alpha}{2 - \alpha} \quad \frac{9.64}{DT^2} \quad E$$

where E is the field strength (V/cm)
α the zero field degree of dissociation

Other terms have their usual significance. If, however, the dynamics
of the process II are such that during a field pulse of a few microseconds this process cannot respond, then the only change in conductance observed will be due to the equilibration of process I in the
field. Hence the electrolyte behaves as if its association constant
were K_0 = where

$$K_0 = \frac{(M,L)}{(M)(L)} = \frac{1 - \alpha}{\alpha^2 \gamma^{\pm 2} C}$$

The total association constant K_Σ can be measured by conventional conductance studies which both processes are given sufficient
time to equilibrate. Then

$$K_\Sigma = \frac{(M,L) + (ML)}{(M)(L)} = K_0(1 + K_1)$$

where $K_1 = \frac{(ML)}{(M,L)}$

Standard conductance measurements gave $K_\Sigma = 7.5 \times 10^6 \, M^{-1}$

By means of a high field conductance technique described
previously (4) the value of $K_0 = 1.3 \times 10^3 \, M^{-1}$ was obtained. Combination of these values gives

$$K_1 = \frac{K_\Sigma}{K_0} -1 = 5.8 \times 10^3.$$ Thus $ZnSO_4$ in methanol is very

largely in the form of contact ion pairs. Similar conclusions were
reached about the nature of the ion pairs in Zn meta benzene
disulfonate in methanol. (5) It is interesting to note that Petrucci
estimated K_0^5 using the Fuoss equation (6) with a = 1.35 nm and
obtained 979 M^{-1} for a 2:2 electrolyte in methanol. This should be
compared with our value of $K_0 = 1.3 \times 10^3$ from direct measurement.

Our results are only semiquantitative for a number of reasons
including the fact that the oscilloscopic measurement of $\Delta \Lambda / \Lambda$ is not
of very high accuracy. Furthermore our conductance study (7) shows

that a small number of triple ions (e.g. $Zn(SO_4)_2^{-2}$ exist in these solutions. These contribute to Λ but not appreciably to $\Delta\Lambda$ (8). Nevertheless we feel that this method is highly promising for the study of very weak electrolytes.

Acknowledgments: The authors wish to thank the Scientific Affairs Division of NATO for a travel grant which made this study possible.

References

(1)   M. Swarc in Ions and Ion Pairs in "Organic Reactions", Vol 2, M. Swarc, ed., J. Wiley and Sons, New York, 1974

(2)   S. Petrucci in "Ionic Interactions", S. Petrucci, ed., Academic Press, New York, 1971

(3)   D. Berg and A. Patterson: 1953, J. Am. Chem. Soc., 75, pp. 5197-5200

(4)   H. Hirohara, K. J. Ivin, and J. J. McGarvey: 1974, J. Am. Chem. Soc., 96, p. 3311

(5)   R. Lovas, G. Macri and S. Petrucci: 1970, ibid, 92, pp. 6502-06

(6)   R. M. Fuoss: 1958, J. Am. Chem. Soc., 80, pp. 5059-61

(7)   J. Curry and P. Hemmes: unpublished results

(8)   P. Hemmes: unpublished calculations

DYNAMIC ASPECTS OF IONIC AGGREGATION IN LOW POLAR MEDIA

J. Everaert and A. Persoons

Laboratory of Chemical and Biological Dynamics
Dept. of Chemistry, University of Leuven, Leuven, Belgium

ABSTRACT

From the pressure dependence of the dissociation process of the tetrabutylammoniumpicrate (TBAP) ion-pair in mixtures of benzene-chlorobenzene ($2.27 \leqslant D \leqslant 3.87$) the volume change of the dissociation (ΔV) and the activation volumina for ionic dissociation, respectively ionic recombination ($\Delta V_d^{\ddagger}$, resp. $\Delta V_r^{\ddagger}$) were obtained. The volume change of ionic dissociation was evaluated from conductance data at different pressures while the activation volumina were obtained from the pressure-dependence of the relaxation time of the dissociation equilibrium. Relaxation measurements were performed with the electric field modulation technique. All results obtained are consistent with a simple theoretical treatment of the ionic association-dissociation processes based on the sphere-in-continuum model.

INTRODUCTION

Important information on the dynamics of chemical processes is obtained from the study of their dependence on hydrostatic pressure. The effect of hydrostatic pressure is related to volume-changes and provides structural information as opposed to the effect of temperature which depends upon the energy changes in the chemical process. Pressure studies have become therefore an essential tool for a structural investigation of the transition state (1).
In these studies it is important to realize that the effect of pressure on rates and equilibria of reactions in solution is partly an intrinsic property of the process, but depends also

W. J. Gettins and E. Wyn-Jones (Eds.), Techniques and Applications of Fast Reactions in Solution. 367–372.
Copyright © 1979 by D. Reidel Publishing Company.

upon the pressure-dependent changes in the properties of the
reaction medium. From early experimental investigations (2) the
importance of electrostriction of the solvent in ionisation processes was quickly recognized and later theoretical developments
(3,4) were firmly based on electrostriction phenomena.
The ionisation phenomena in aqueous solutions can hardly be described by the above mentioned theories due to the complexities
of the pressure-dependence of the solvent-structure. However in
less structured media, e.g. low polar aprotic solvents, a quantitative description of pressure effects on equilibria and rates
of ion-pairing processes should be feasible. More specifically,
with the solvent considered as a continuum the permittivity and
viscosity of which increase with pressure, we may predict an
enhancement of dissociation of the ion-pair. Due to the higher
viscosity the mobility of the ions is decreased and this means,
from the viewpoint of dynamics, a decrease in recombination rate
of the free ions with increasing pressure.
Some pressure effects in ion-pair equilibria in low polar media
were studied by Fuoss and coworkers using conductance measurements (5,6). In our laboratory the kinetics aspects of ion-
pairing processes were investigated with chemical relaxation
techniques (7).

In this communication we report some results of pressure
effects on rates and equilibria for the dissociation of tetrabutylammoniumpicrate (TBAP) in benzene-chlorobenzene mixtures.

EXPERIMENTAL

Conductimetric and kinetic measurements were performed over a
pressure range up to 700 atm. The sample cell is a field modulation cell modified for pressure work (7). The conductimetric
cell constant is determined from dielectric calibration the constancy of which is checked after each pressure-run by remeasu-
ring at 1 atm ; deviations for all measurements reported were
always below 1 %. The necessary corrections for concentration
due to compressibility were performed.
Changes in cell-constant at high pressure are irrelevant for the
kinetic measurements performed by electric field modulation techniques (8) the results of which are independent of cell geometry.
The density of the benzene-chlorobenzene mixtures was calculated,
assuming ideality, from the compressibility of the pure solvents.
Measurements of the dielectric constant in the sample cell at
different pressures yielded the change of dielectric constant
with pressure $(\partial D/\partial P)_T$, which is the important parameter for the
description of electrostriction. The viscosity at different
pressures was extrapolated from the pressure dependence for the
pure solvents and the viscosity, at 1 atm, of the mixtures.

The data for the dielectric constant and the viscosity for the different solvent-mixtures at 1 atm are summarized in Table I.

Table I : Physical parameters of the solvent medium and TBAP ion-pair at T = 298 K.

vol % C_6H_5Cl	D	η cP	a Å	$10^5\ \partial\ln D/\partial P$ atm^{-1}	$10^4\ \partial\ln\eta/\partial P$ atm^{-1}
0	2.27	0.602	6.22	6.21	7.87
11	2.62	0.617	6.14	6.83	7.71
16	2.78	0.625	6.03	6.98	7.60
30	3.22	0.645	5.95	7.36	7.37
40	3.54	0.661	5.88	7.66	7.17
50	3.87	0.678	5.78	7.91	6.98

RESULTS AND DISCUSSION

Electrolyte solutions in low polar media are characterized by a very complex conductance behaviour mainly due to the long range of Coulombic interaction. However, at low electrolyte concentration, a simple ion-pair dissociation into free ions is the predominant process. This situation is experimentally come out by the reciprocal relation between equivalent conductance and the square root of electrolyte concentration. In relaxation experiments this simple behaviour is likewise characterized by a linear dependence of the reciprocal relaxation time of the ionisation equilibrium on the square root of total concentration. It is therefore relatively easy to delimit the experimental conditions within which the simple ion-pairing process is present and not obscured by the progressive emergence of higher ionic aggregates, e.g. triple ions, quadrupoles ...
Conductance and chemical relaxation were measured for TBAP ion-pairs in benzene-chlorobenzene mixtures, with composition as given in Table I, over a large concentration and pressure range. At low concentration and/or higher dielectric constant the simple ion-pair dissociation is the main ionisation process. From the conductimetric data the ion-pair dissociation constant K_D was obtained at different pressures in all solvent-mixtures. From the concentration dependence of the chemical relaxation time the dissociation k_d and recombination rate constant k_r for the simple ion-pairing equilibrium were obtained ; as indicated this was only possible in the concentration domain for which the reciprocal relaxation time is a linear function of the square root of electrolyte concentration.
The functional dependence of K_D, k_d and k_r upon pressure yielded

the volume change of dissociation ΔV and the activation volumina $\Delta V_d^{\ddagger}$ and $\Delta V_r^{\ddagger}$. These experimental values are given in Table II. The activation volumina for the solutions at the lower polarity are lacking because in this region the relaxation time is predominantly determined by triple-ion formation even at the lowest TBAP concentration measurable with the field modulation technique. If the ionisation equilibrium is treated as the association-dissociation of hard, charged, spheres subjected to Brownian motion in a continuous medium of dielectric constant D and viscosity η it is possible to give theoretical expressions for K_D, k_r and k_d (in the usual molarity units) :

$$\text{Fuoss} \qquad K_D = \frac{3000}{4\pi N_o a^3} \exp - \frac{e_o^2}{DkT\underline{a}} \qquad\qquad [1] \quad (9)$$

$$\text{Debye} \qquad k_r = \frac{8e_o^2 N_o}{3000} \frac{1}{D\eta\underline{a}} \qquad\qquad [2] \quad (10)$$

$$\text{Eigen} \qquad k_d = \frac{2e_o^2}{\pi} \frac{1}{D\eta\underline{a}^4} \exp - \frac{e_o^2}{DkT\underline{a}} \qquad\qquad [3] \quad (11)$$

The important parameters in these equations are the dielectric constant D and the viscosity η, characterizing the macroscopic properties of the medium, and the interionic distance $\underline{a}$ which is a structural property of the ion-pair. The distance parameter $\underline{a}$, calculated from the dissociation constant K_D with the Fuoss equation, was shown to be pressure-independent. The values of $\underline{a}$ for the different solvent mixtures are given in Table I.
We can therefore derive relatively simple expressions for the volume changes from the foregoing equations :

$$\Delta V = - RT \frac{e_o^2}{DkT\underline{a}} \frac{\partial \ln D}{\partial P} \qquad\qquad [4]$$

$$\Delta V_r^{\ddagger} = RT \left(\frac{\partial \ln D}{\partial P} + \frac{\partial \ln \eta}{\partial P} \right) \qquad\qquad [5]$$

$$\Delta V_d^{\ddagger} = RT \left[\left(1 - \frac{e_o^2}{DkT\underline{a}} \right) \frac{\partial \ln D}{\partial P} + \frac{\partial \ln \eta}{\partial P} \right] \qquad\qquad [6]$$

Table II : Experimental and theoretical values of the
volume change and activation volumina for the
dissociation of TBAP in benzene-chlorobenzene
at T = 298 K (volumes in ml.mol^{-1})

D	ΔV	ΔV (eq4)	$\Delta V_r^{\ddagger}$	$\Delta V_2^{\ddagger}$ (eq5)	$\Delta V_d^{\ddagger}$	$\Delta V_d^{\ddagger}$ (eq6)
2.27	-61.6 ± 5.1	-62.0	-	21.4	-	-40.7
2.62	-54.5 ± 3.3	-60.3	-	21.2	-	-38.9
2.78	-57.3 ± 1.0	-59.1	-	21.0	-	-37.9
3.22	-51.8 ± 1.3	-54.6	19.6 ± 0.6	20.5	-32.3 ± 1.3	-33.9
3.54	-48.5 ± 0.8	-52.2	18.0 ± 0.7	20	-30.8 ± 1.3	-32.0
3.87	-50.8 ± 1.0	-50.2	20.5 ± 0.9	19.6	-30.6 ± 0.5	-30.4

The theoretical values calculated from eq. 4-6 are also given in
Table II. The excellent agreement between theory and experimental data identifies the ion-pairing equilibrium in the media considered as composed of simple diffusion-controlled processes.
All relevant dynamic properties of these processes may be derived
from a knowledge of the macroscopic properties of the medium on
one hand and the structural parameter a on the other hand. From
the numerical calculation it is also established that the main
contribution to $\Delta V_r^{\ddagger}$ is the viscosity dependence of pressure while
$\Delta V_d^{\ddagger}$ is mainly determined from the change of dielectric constant
with pressure. This result is intuitively clear since the recombination process depends on the mobility of the ions in the viscous medium while the dissociation is mainly opposed by the
strong Coulomb attraction.

This investigation has clearly shown that the dissociation-
recombination processes of ionic species in low polar media can
be treated successfully as a problem of diffusional motion of
hard spheres in a continuous medium. The structural properties
of the ionic aggregate and the macroscopic properties of the
solvent-medium are sufficient to evaluate quantitatively the dynamic properties of ionic phenomena. Eventual discrepancies between experimental data and theoretical values are therefore indicative for specific solute-solvent interactions.

ACKNOWLEDGEMENTS

Financial support from the Belgian Science Foundation (F.K.F.O.
grant N° 2.0051.77N), the Belgian government (Programmatie van
het Wetenschapsbeleid, conventie N° 76/81-II4) and the Catholic

University Leuven (Derde Cyclus OT/III/19) is acknowledged.
J. Everaert is an aspirant of the N.F.W.O.

REFERENCES

(1) e.g. T. Asano, W.J. Le Noble, Chem. Rev., 78, 407 (1978)
(2) I. Fanjung, Z. Phys. Chem., 14, 673 (1894)
(3) P. Drude, W. Nernst, Z. Phys. Chem., 15, 79 (1894)
(4) M. Born, Z. Phys., 1, 45 (1920)
(5) J.F. Skinner, R.M. Fuoss, J. Phys. Chem., 69, 1437 (1965)
(6) J.F. Skinner, E.L. Cussler, R.M. Fuoss, J. Phys. Chem., 71,
     4455 (1967)
(7) F. Nauwelaers, Ph. D. thesis (Leuven, 1974)
(8) A. Persoons, J. Phys. Chem., 78, 1210 (1974)
     F. Nauwelaers, L. Hellemans and A. Persoons, J. Phys. Chem.,
     80, 767 (1976)
     A. Persoons and L. Hellemans, Bioph. J., 24, 119 (1978)
(9) R.M. Fuoss, J. Am. Chem. Soc., 80, 5059 (1958)
(10) P. Debye, Trans. Electrochem. Soc., 82, 265 (1942)
(11) M. Eigen, Z. Phys. Chem. (Frankfurt-am-Main), 1, 176 (1959).

EXCHANGE KINETICS OF COMPLEXES OF DIAMAGNETIC CATIONS WITH NONAMETHYLIMIDODIPHOSPHORAMIDE, STUDIED BY N.M.R.

P.R.Rubini, L.Rodehüser and J-J.Delpuech

Laboratoire de Chimie Physique Organique,
Equipe de Recherche Associée au CNRS,
Université de Nancy I, C.O.140,
F-54037 Nancy Cedex, France

ABSTRACT

Complexes of the bidentate ligand nonamethylimido-diphosphoramide (NIPA) with the diamagnetic cations Mg^{2+}, Ca^{2+}, Sr^{2+} and Li^+ have been studied and the kinetic parameters of the ligand exchange reaction in dipolar aprotic solvents have been determined. The kinetic stabilities of the complexes are several orders of magnitude higher than those of monodentate organophosphorous ligands (trimethylphosphate, hexamethylphosphorotriamide) and are comparable to those of ionic chelating ligands. Mechanisms for two types of ligand exchange reactions are discussed.

INTRODUCTION

The neutral ligand nonamethylimidodiphosphoramide (NIPA) (1) forms remarquably stable complexes with several diamagnetic cations such as Mg^{2+}, Ca^{2+}, Sr^{2+}, and even with the alcaline ion Li^+.

$$\underline{1}$$

W. J. Gettins and E. Wyn-Jones (Eds.). Techniques and Applications of Fast Reactions in Solution. 373–378.
Copyright © 1979 by D. Reidel Publishing Company.

1 might be considered as a dimerized hexamethylphosphorotriamide (HMPA), complexes of which having
been the objects of some of our earlier studies[1]. In
complex formation however NIPA acts as a bidentate ligand so that the results for exchange kinetics were
expected to differ significantly from those obtained
with HMPA.

A technique especially well suited for the investigation of symmetric exchange reactions in solution
is the dynamic nuclear magnetic resonance method. It
is based on the change in the nuclear transversal relaxation time T_2 of either the central cation or a
ligand nucleus, with varying rates of exchange. The
mechanism leads to a broadening of the respective n.m.r.
signals when the exchange rates increase (e.g. on raising the temperature) until coalescence of the lines
occurs.

EXPERIMENTAL

The complexes of a great number of metal cations
with NIPA can be prepared in the solid state[2]. Coordination numbers and structural data of many compounds
were determined by de Bolster and Groeneveld[3]. Following their procedure we prepared the complexes
$[Mg(NIPA)_3]^{2+}$, $[Ca(NIPA)_3]^{2+}$, $[Sr(NIPA)_3]^{2+}$ and
$[Li(NIPA)_2]^+$ in the form of their perchlorate salts.

The kinetic studies were carried out in nitrome-
thane (Mg-, Ca- complexes) or in mixtures of nitrome-
thane with dichloromethane 1 : 2 by volume (Sr-, Li-
complexes), solvents in which perchlorates are highly
dissociated[4].

The nuclei observed were ^{31}P decoupled from proton and ^1H. For simplification of the proton n.m.r.
spectrum of NIPA the phosphorus was decoupled too, leading to a simple two line spectrum, the ratio of the
signals being 8 : 1. The $^{31}P\{^1H\}$ spectra show one peak
for the bound and one for the free NIPA the latter
being shifted to higher magnetic fieldstrength.

RESULTS AND DISCUSSION

Reaction rates were obtained by total line-shape
analysis of the spectra, the equations used for the
simulation being Bloch's equations modified for exchange phenomena[5]. The experimental and theoretical spectra
for the Mg-NIPA system are shown in Figure 1.

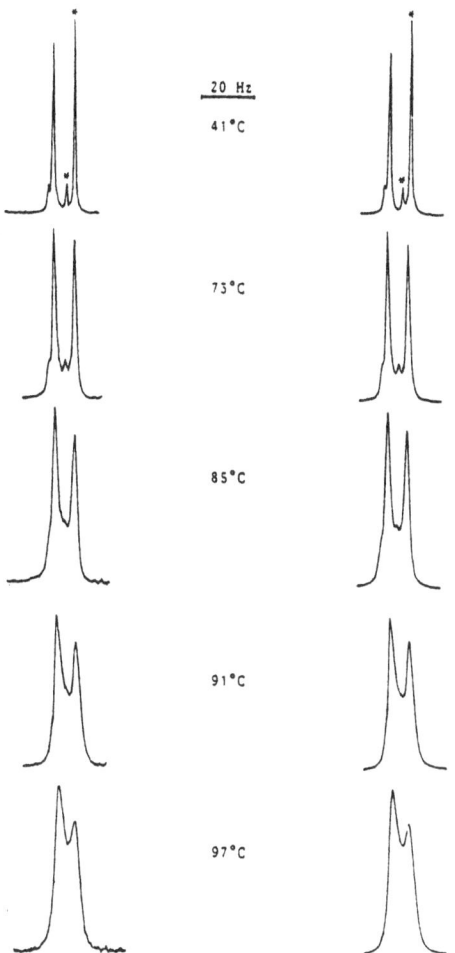

Fig.1 : Experimental (left side) and calculated spectra
        for the Mg-NIPA system. Concentrations are
        0.102 M of complex salt and 0.302 M of free
        ligand. (The signals of free NIPA are marked
        by an asterisk).

    In the case of the Mg- and Ca-complexes the tem-
perature range between the beginning of [1]H-line broadening and coalescence is 40 to 90°C and -20 to 30°C,
respectively. For Sr and Li compounds, coalescence of
the ^{31}P signals occurs at about -60 and -30°C, respectively.
    The exchange rates were found to be independent
of the concentration of free ligand for the Mg-complex
and proportional to it in the case of the three other

systems.

The rates and thermodynamic activation parameters for the overall reaction

$$M(NIPA)_m^{n+} + NIPA^x \xrightarrow{k} M(NIPA)_{m-1}(NIPA)^{x\ n+} + NIPA$$

are summarized in the following table.

TABLE I. Kinetic and thermodynamic parameters for the exchange of NIPA on complexes of several diamagnetic cations at 25°C.

Complex[a]	k_{298}	$\Delta H^{\neq}$ [kJ/mole]	$-\Delta S^{\neq}$ [J.K^{-1}.mole^{-1}]
Mg(NIPA)$_3$ $^{2+}$	$4.3.10^{-2}$ s^{-1}	66.3	− 50
Ca(NIPA)$_3$ $^{2+}$	$3.2.10^{1}$ M^{-1}.s^{-1}	30.7	−113
Sr(NIPA)$_3$ $^{2+}$	$3.7.10^{4}$ M^{-1}.s^{-1}	23.4	− 79
Li(NIPA)$_2$ $^{+}$	$1.4.10^{4}$ M^{-1}.s^{-1}	33.0	− 54

a) As perchlorate salt

For the reaction of the Mg-complex, we considered mechanisms I and II to be in agreement with the first order reaction type.

I

k_{obs} being k_1

II

with $k_{obs} = (k_1/k_{-1}).k_2 = K.k_2$

assuming that the equilibrium is established rapidly compared to the second step.

In the case of the other reactions for which the exchange rate is proportional to the concentration of free ligand, the order of reaction may be explained by mechanism III.

III

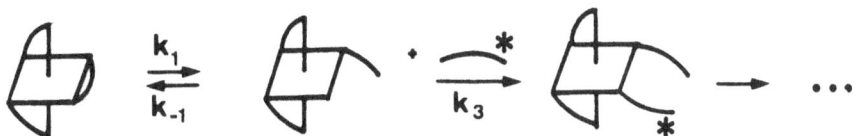

so that $k_{obs} = k_1 k_3 [L] / (k_{-1} + k_3 [L])$.

If the preceeding equilibrium is very rapid, $k_3 [L] << k_{-1}$ and therefore $k_{obs} = K . k_3 [L]$, with $K = k_1 / k_{-1}$ and $[L] = [NIPA]_{free}$.

Similar mechanisms implying at least two steps with partial detachment of a ligand molecule have been proposed in the literature for other bidentate ligands[6,7].

CONCLUSIONS

The kinetic stabilities of metal ion complexes with neutral organophosphorous compounds as ligands are largely enhanced if monodentate ligands are substituted by the bidentate compound nonamethylimidodiphosphoramide. This is due partly to the chelating effect of the molecule but also to the greater ease of electron delocalization in the complex and a change in the ligand substitution reaction mechanism.

The stability of the complexes of NIPA may be of practical interest, e.g. for the extraction and separation of cations.

REFERENCES

1. a) J-J.Delpuech, M.R.Khaddar, A.A.Péguy and P.R. Rubini, J.Amer.Chem.Soc., 97, 3373 (1975)
   b) L.Rodehüser, P.R.Rubini, and J-J.Delpuech, Inorg. Chem., 16, 2837 (1977)
2. K.P.Lannert and M.D.Joesten, Inorg.Chem., 7, 2048 (1968)

3. M.W.G. De Bolster and W.L. Groeneveld, *Rec.Trav.Chim. Pays-Bas*, <u>91</u>, 171 (1972)
4. J.E.Prue and P.J.Sherrington, *Trans.Faraday Soc.*, <u>57</u>, 1795 (1961)
5. a) R.Kubo, *J.Phys.Soc.Japn.*, <u>9</u>, 935 (1954)
   b) R.A.Sack, *Mol.Phys.*, <u>1</u>, 163 (1958)
   c) J.Jen, *J.Magn.Res.*, <u>30</u>, 111 (1978)
6. R.G.Pearson and R.D.Lanier, *J.Amer.Chem.Soc.*, <u>86</u>, 765 (1964)
7. A. Vasilescu, *Rev.Roum.Chim.*, <u>20</u>, 951 (1975)

POLYSACCHARIDE CONFORMATION AND INTERACTIONS
IN SOLUTIONS AND GELS

Edwin R. Morris

Unilever Research, Colworth Laboratory,
Sharnbrook, Bedford MK44 1LQ, England.

Polysaccharide physical properties may be traced to various levels
of molecular organisation. Individual sugar rings in the covalent
sequence, or primary structure, are essentially rigid, and chain
geometry (secondary structure) is determined by their relative
orientations. Chain flexibility is restricted by non-bonded
interactions between adjacent residues, thus leading to extended
coil dimensions, and high solution viscosity. In the solid state
most polysaccharides adopt ordered tertiary structures such as
nested ribbons or double helices. These may persist in solution,
or as the fundamental structural unit in gels, and frequently show
co-operative order-disorder behaviour. Tertiary structures may
further associate to build up higher levels of organisation
(quaternary structure). Changes in structural organisation may
be monitored chiroptically, rheologically, calorimetrically, or
by changes in molecular weight or n.m.r. relaxation behaviour, all
of which may, in principle, be used as probes for kinetic studies.

INTRODUCTION

Polysaccharides are naturally occurring carbohydrate polymers
which represent the principal solid component of most plants,
and are also of major importance in many animal and bacterial
tissues. In addition to their familiar role as energy reserves
the primary biological function of many polysaccharides is in
development of structure. This natural ability is widely
exploited industrially to create texture in, for example, food
and cosmetics. The molecular origin of bulk textural properties
may be traced to various levels of polysaccharide organisation.

W. J. Gettins and E. Wyn-Jones (Eds.). Techniques and Applications of Fast Reactions in Solution. 379–388.
Copyright © 1979 by D. Reidel Publishing Company.

PRIMARY STRUCTURE

The component sugars of polysaccharides occur in the polymer
chain as rings, as shown opposite. Substituents may lie either
equatorially (e) in the plane of the ring, or axially (a) above
or below the ring. Linkage of adjacent sugars may be either α
(axial) or β (equatorial), as shown. Branching is introduced
by disubstitution of a single residue. Polysaccharide primary
structures frequently show a simple repeating sequence, as
outlined below.

AGAROSE

IOTA
CARRAGEENAN

Alginate is a linear block co-polymer of 4-linked α-L-guluronate
and β-D-mannuronate residues, arranged in homopolymeric blocks
of both types, and in regions which approximate to an alternating
structure (1). Plant galactomannans have a linear backbone
of β-1,4 linked D-mannose, substituted to different degrees at
position 6 by α-D-galactose residues. In one member of this
family, locust bean gum (LBG), galactose substituents are
clustered together in 'hairy' regions, interspersed by 'smooth'
unsubstituted mannan blocks (2). The bacterial polysaccharide
xanthan is based on the β-1,4 linked D-glucose structure of
cellulose (see opposite), substituted on every second backbone
residue by a charged trisaccharide sidechain (3,4).

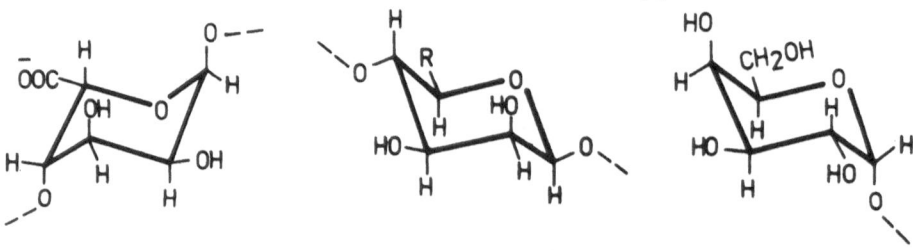

1,4 linked α-   1,4 linked β-D-mannuronate   α-D-galactose
L-guluronate   (R=COO⁻) or mannose (R=CH₂OH)

D - SUGARS   L - SUGARS

SECONDARY STRUCTURE

The geometry of the individual sugar rings in a polysaccharide
chain is essentially rigid. Overall conformation is therefore
determined by the relative orientations of component sugars, as
defined by the rotational angles ϕ and ψ, (see below). The
range of possible values of ϕ and ψ is severely restricted by
steric clashes and other non-bonded interactions between adjacent
residues, with consequent chain stiffening, and extended coil
dimensions in solution, hence leading to the extremely high
viscosity of polysaccharide solutions at relatively low
concentrations. Restriction is greatest for axial inter-
residue linkages, and for sugar rings with bulky equatorial
substituents (hydroxy groups or larger) on positions adjacent
to the inter-residue bond (5,6,7).

In addition to the effects of restricted rotation about
inter-residue bonds, overall coil dimensions depend significantly
on the bonding pattern of individual component sugars (7).
Polysaccharide chains with a 'trans' bonding arrangement across
each sugar ring (see below) are designated 'type A' (5), and
show extended coil dimensions, while 'cis' bonding ('type B')
introduces a 'U-bend', and favours more compact conformations.

CELLULOSE (TYPE A)
'TRANS' BONDING

AMYLOSE (TYPE B)
'CIS' BONDING

TERTIARY STRUCTURES

In the solid state most polysaccharides adopt sterically regular crystal-like conformations which can be characterised by x-ray fibre diffraction (8). These structures correspond to fixed values of the rotational angles ϕ and ψ between adjacent residues rather than a statistical distribution of orientations as for random coils in solution. Type A polysaccharides, such as cellulose or galactomannans pack together (5) in extended ribbon-like solid state conformations, while chains which include type B residues may form hollow helices, which often occur as compact multi-stranded structures such as the double helices of amylose, agarose, and carrageenan (9,10,11). In the solid state xanthan adopts an ordered 5-fold conformation, stabilised by non-bonded interactions between main-chain residues and the trisaccharide sidechains, which pack along the backbone (12). In favourable cases such ordered tertiary structures may persist under conditions of extensive hydration, either in solution, or as the fundamental structural units in polysaccharide gel networks (13).

In general ordered conformations are promoted by favourable non-covalent interactions, inflexible secondary structure, and efficiency of packing, and inhibited by loss of conformational entropy, energy of hydration, intermolecular electrostatic repulsion, structural irregularities, and branching. The balance of these opposing drives is often delicate, and may be tipped by relatively small perturbations. For example, thermally induced order-disorder transitions, which may or may not be accompanied by a gel-sol transition, have been observed for a number of polysaccharide systems (8,13,14), and show the sharp temperature profile characteristic of a co-operative process.

In addition to such thermal transitions, adoption of ordered conformations may be promoted by changes in ionic environment. At the simplest level of understanding, compact ordered conformations of polyelectrolytes may be stabilised under conditions of high ionic strength due to reduction of mutual electrostatic repulsions by charge screening. More directly, ordered conformations of charged polysaccharides may be stabilised by incorporation of counterions within the tertiary structure. Thus the primary event in the gelation of alginate is dimerisation of poly-L-guluronate sequences, with specific interchain chelation of calcium, or other divalent cations of appropriate size (15,16). Under hydrated conditions intermolecular associations of polysaccharides are stable only above a minimum critical chainlength necessary for co-operativity, typically in the range 15-20 residues (17,18). Thus, as shown schematically opposite, the traditional concept of a polysaccharide gel network as involving 'point cross-linking' of disordered chains is superceded by a 'junction-zone' model, with extensive regions

of ordered tertiary structure. Junctions are terminated by
occurence in the primary structure of residues incompatible
with the ordered conformation (13). Thus for carrageenan,
double helix formation is interrupted by the presence of 4-linked
α-D-galactose residues in the unbridged chair conformation,
rather than in the helix-compatible 3,6-anhydro ring form.
Similarly poly-L-guluronate dimerisation in alginate is terminated
by chain sequences involving D-mannuronate. Such junction
delimiting structural features are important for the formation
of gel networks, rather than insoluble precipitates, by allowing
each chain to participate in ordered associations with several
different partners.

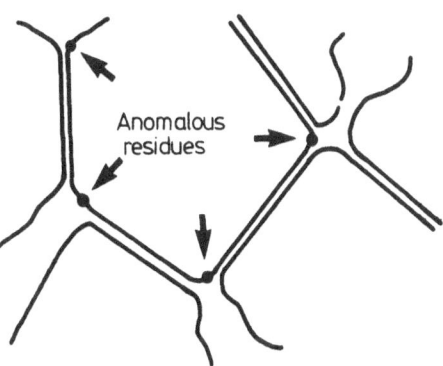

'Point-crosslinking' model   'Junction-zone' model

JUNCTION ZONE TYPES

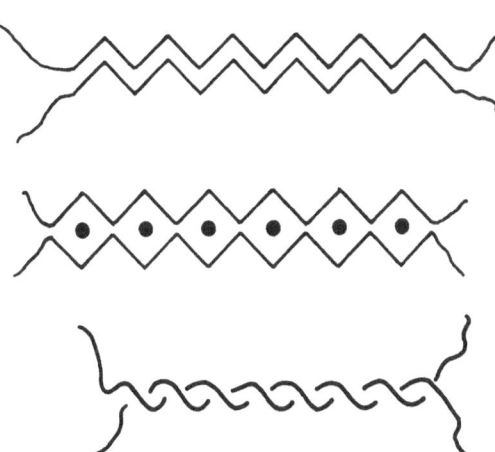

NESTED RIBBONS
    e.g. cellulose
            galactomannans

CATION 'EGG BOXES'
    e.g. alginate
            pectin

DOUBLE HELICES
    e.g. amylose
            carrageenan
            agarose

QUATERNARY STRUCTURE

While adoption of ordered tertiary structure by flexible polymer
chains involves considerable loss of conformational entropy,
subsequent aggregation of such rigid molecular assemblies is
entropically far less unfavourable, and indeed is to be expected
(19) in the absence of over-riding repulsions. Thus for
carrageenan and agar double helices, aggregation increases with
decreasing sulphate content (20), being most evident for uncharged
agarose helices. Association of charged tertiary structures
may be promoted by specific counterions whose radius and charge
are suitable for incorporation in the aggregate. Thus further
association of poly-L-guluronate dimers is promoted by calcium
ions (16), while aggregation of carrageenan helices is specifically
favoured by the presence of potassium ions (14). In some cases
such quaternary structure may be necessary for gelation. Thus
the primary event in the gelation of iota carrageenan is formation
of clusters or 'domains' of approximately 10 chains held together
by double helix junctions. Development of a cohesive gel
network, however, requires association of these domains by helix-
helix aggregation, promoted specifically by potassium ions (21).

Such interactions may also be important for the industrially
valuable solution properties of polysaccharides, particularly in
particle suspension and emulsion stabilisation. Rigid rod-like
molecules such as xanthan show extreme pseudoplasticity in
solution (23), behaving almost like true gels under conditions
of low shear, but thinning dramatically with increasing shear
rate. This behaviour is consistent with weak but extensive
intermolecular association under zero or low shear conditions,
to develop a tenuous gel-like network. On destruction of this
network by larger deformation, the ability of molecular rods to
align in the direction of flow presumably contributes to
subsequent further shear thinning.

Quaternary structure formation may also involve association
of ordered conformations of unlike polysaccharides. The most
fully characterised mixed junctions are those of agarose,
carrageenan and xanthan helices with unsubstituted backbone
regions of galactomannans and related 'type A' polysaccharides (24).

Xanthan Random Coil   Solution Order   Mixed Gel with
                                                 Locust Bean Gum

PERTURBATIONS AND PROBES

Changes in structural organisation of polysaccharide systems
may, as discussed above, be induced by changes in temperature,
ionic strength, pH, concentration of specific cations, and
water activity. In principle, therefore, these may all be used
as perturbations for kinetic studies. The response of the
system to such perturbations may be monitored by a range of
physical techniques. So far few kinetic measurements have been
made on polysaccharide systems, although ultrasonic studies of
agarose and carrageenan gels (25,26) and preliminary results on
kinetics of the iota carrageenan coil-helix transition (27) have
been reported. The scope and suitability of specific probes of
polysaccharide conformation and interactions, however, have been
established by extensive equillibrium studies, as outlined below.

For polysaccharides with an accessible u.v. chromophore,
circular dichroism (c.d.) provides a sensitive index of
structure, conformation, and interactions. This is illustrated
by the very different c.d. spectra shown (28,29) by the three
block types present in alginate, and, as shown below, by the
large spectral changes on chelation of calcium ions on specific
binding sites around the carboxy chromophore (14,15).

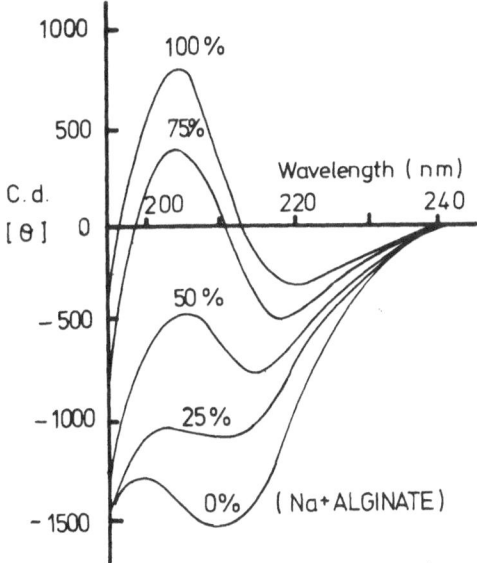

Recent studies in the vacuum ultraviolet spectral region (30,31)
suggest that all polysaccharides show two c.d. bands at around
155 and 175 nm, which are sensitive to chain conformation.
This is illustrated below for the thermally induced order-
disorder transition of agarose, as monitored by changes in
intensity of the higher wavelength band.   The pronounced
hysteresis is ascribed (31) to quaternary association (aggregation)
stabilising the double helix at temperatures substantially above
those at which it would form spontaneously in isolation.

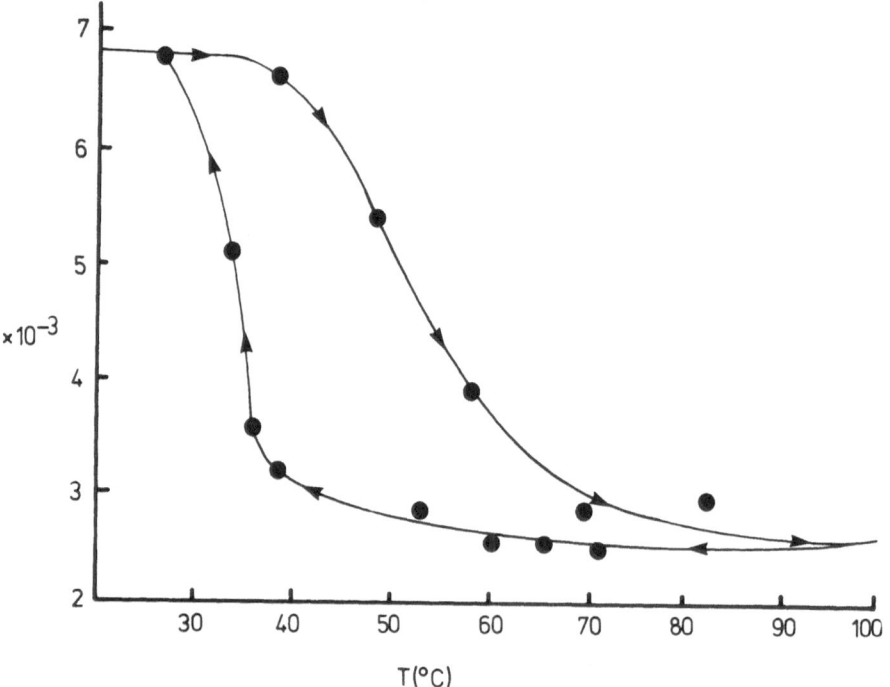

These bands are inaccessible to normal c.d. equipment,
but their effects can be monitored indirectly by optical rotation
studies at higher wavelengths.   Indeed a simple, semi-empirical
correlation has been established between optical rotation and
linkage angles ϕ and ψ between adjacent residues (32).   Chain
conformation may also be monitored by n.m.r. relaxation studies.
Relaxation behaviour can be measured directly as the time constant
for exponential decay of magnetisation, or indirectly from high
resolution linewidth.   Thus flexible polymers show broad but
discernable n.m.r. resonances, while for rigid conformations
such as the carrageenan double helix, linewidth is so great that
all peaks are flattened into the baseline.   Loss of detectable
signal in high resolution n.m.r. can therefore be used (33) as an
index of conformational ordering.   Order-disorder transitions
in solution may also be detected by changes in viscosity (η).

These three approaches are illustrated below for the thermally
induced order-disorder transition of xanthan (22,24,34).

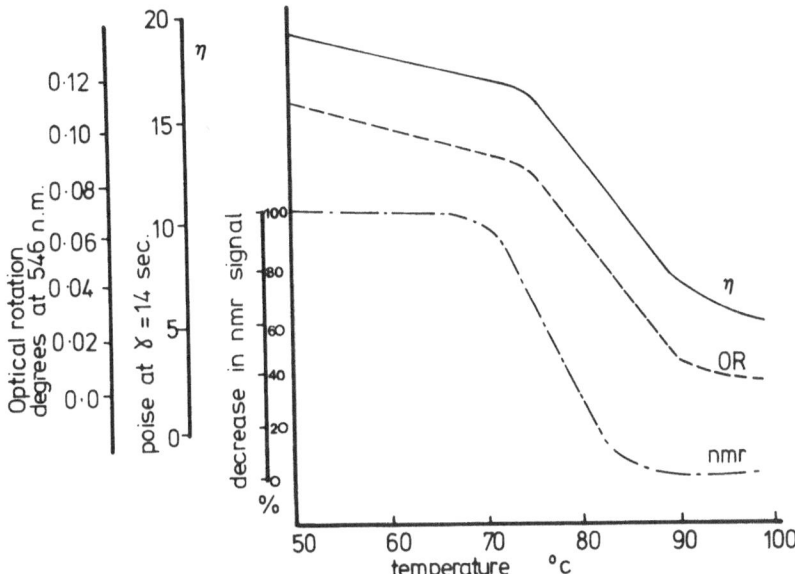

Under appropriate conditions of ionic environment, membrane
osmometry and light scattering show the expected doubling of
molecular weight on double helix formation by structurally regular
iota carrageenan chain segments, which display the same order-
disorder behaviour in solution as the intact polymer, but without
gelation (21). Order-disorder transitions may also be monitored
by differential scanning calorimetry, and by cation binding in
systems such as alginate and pectin, where specific cations
are incorporated within the ordered tertiary structure (15).

REFERENCES

1) Haug, A., Larsen, B., and Smidsrød, O. : 1966, Acta Chem.
   Scand. 20, pp. 183-190.
2) Baker, C.W. and Whistler, R.L. : 1975, Carbohyd. Res. 45,
   pp. 237-243.
3) Jansson, P.E., Kenne, L., and Lindberg, B. : 1975, Carbohyd.
   Res. 45, pp. 275-282.
4) Melton, L.D., Mindt, L., Rees, D.A., and Sanderson, G.R. :
   1976, Carbohyd. Res. 46, pp. 245-257.
5) Rees, D.A. and Scott, W.E. : 1971, J. Chem. Soc. B, pp.469-479.
6) Rees, D.A. : 1969, Advan. Carbohyd. Chem. Biochem. 24,
   pp. 267-332.

7) Morris, E.R., Rees, D.A., Welsh, E.J., Dunfield, L.G., and Whittington, S.G. : 1978, J. Chem. Soc. Perkin II, pp. 793-800.

8) Rees, D.A. : 1977, 'Polysaccharide Shapes', Chapman Hall, London.

9) Wu, H-C.H. and Sarko, A. : 1978, Carbohyd. Res. 61, pp. 7-40.

10) Arnott, S., Scott, W.E., Rees, D.A., and McNab, C.G.A. : 1974, J. Mol. Biol. 90, pp. 253-267.

11) Arnott, S., Fulmer, A., Scott, W.E., Dea, I.C.M., Moorhouse, R., and Rees, D.A. : 1974, J. Mol. Biol. 90, pp. 269-284.

12) Moorhouse, R., Walkinshaw, M.D., and Arnott, S. : 1977, ACS Symposium Series 45, pp. 90-102.

13) Rees, D.A. and Welsh, E.J. : 1977, Angew. Chem. Int. Ed. Engl. 16, pp. 214-224.

14) Morris, E.R., Rees, D.A., Thom, D., and Welsh, E.J. : 1977, J. Supramolec. Struct. 6, pp. 259-274.

15) Grant, G.T., Morris, E.R., Rees, D.A., Smith, P.J.C., and Thom, D. : 1973 FEBS Letts.,32, pp. 195-198.

16) Morris, E.R., Rees, D.A., Thom, D. and Boyd, J. : 1978, Carbohyd. Res. 66, pp. 145-154.

17) Kohn, R. and Luknar, O. : 1977, Collec. Czechoslov. Chem. Commun. 42, pp. 731-744.

18) Whistler, R.L. : 1973, Advan. Chem. 117, pp. 242-255.

19) Flory, P.J. : 1956, Proc. Roy. Soc. ser. A 234, pp. 50-73.

20) Rees, D.A. : 1972, Chem. Ind. (London), pp. 630-636.

21) Morris, E.R., Rees, D.A. and Robinson, G.R. : 1979, J. Mol. Biol., submitted.

22) Morris, E.R., Rees, D.A., Young, G., Walkinshaw, M.D., and Darke, A. : 1977, J. Mol. Biol. 110, pp. 1-16.

23) Jeanes, A. : 1974, J. Polym. Sci. Symp. No. 45, pp. 209-227.

24) Dea, I.C.M., Morris, E.R., Rees, D.A., Welsh, E.J., Barnes, H.A., and Price, J. : 1977, Carbohyd. Res. 57, pp. 249-272.

25) Gettins, W.J., Jobling, T.L., and Wyn-Jones, E. : 1978, J. Chem. Soc. Faraday II 74, pp. 1246-1252.

26) Pass, G., Phillips, G.O., Wedlock, D.J., and Wyn-Jones, E. : 1978, Macromolecules 11, pp. 433-435.

27) Norton, I.T., Goodall, D.M., Morris, E.R., and Rees, D.A. : 1978, Chem. Commun., pp. 515-516.

28) Morris, E.R. and Sanderson, G.R. : 1972, 'New Techniques in Biophysics and Cell Biology', Pain, R.H. and Smith, B.J. eds., pp. 113-147, Wiley, London.

29) Morris, E.R., Rees, D.A., Sanderson, G.R., and Thom, D. : 1975, J. Chem. Soc. Perkin II, pp. 1418-1425.

30) Balcerski, J.S., Pysh, E.S., Chen, G.C., and Yang, J.T. : 1975, J. Amer. Chem. Soc. 97, pp. 6274-6275.

31) Liang, J., Stevens, E.S., Morris, E.R. and Rees, D.A. : 1978, Biopolymers, in press.

32) Rees, D.A. : 1971, J. Chem. Soc. B, pp. 469-479.

33) Bryce, T.A., McKinnon, A.A., Morris, E.R., Rees, D.A., and Thom, D. : 1974, Faraday Discuss. Chem. Soc. 57, pp. 221-229.

34) Rees, D.A. : 1972, Biochem. J. 126, pp. 257-273.

BIOLOGICAL APPLICATION:
KINETICS ON THE INTERACTION BETWEEN PROFLAVINE AND DNA

T. Yasunaga and H. Ushio

Department of Chemistry, Faculty of Science,
Hiroshima University, Hiroshima 730, Japan

ABSTRACT.   Temperature-jump relaxation method has been employed
to study the kinetics on the interaction between proflavine and
native or denatured DNA.   Two relaxations could be observed in
the time range, 10^{-5}-10^{-3} sec, in both cases and the relaxation
phenomena have been interpreted by considering two binding sites
in DNA.   The rate and equilibrium constants have been calculated
for the elementary steps.   These obtained values show that the
proflavine bound to phosphate group is intercalated between two
consecutive bases.

1. INTRODUCTION

   The interaction of acridine dyes with nucleic acid, which
is of considerable biological interest, also displays cooperative
nature[1-3].   Although many equilibrium and kinetic studies have
been carried out on these systems[4,5], most of them have not
taken the cooperativity into account.
   In proflavine-DNA complex, two principal modes of binding
are known to exist[4], which differ in stability.   One is the inside binding which is attributed to Van der Waals type interactions between a dye and the bases of DNA.   The other is the
outside binding which is attributed to the electrostatic
interaction between a dye and a phosphate group.
   For the so-called strong binding intercalation model, two
models have been proposed.   In the complete intercalation model,
the dye molecule is totally enclosed inside the double helix
between two consecutive base pairs[6].   In the partial intercalation model, the dye lies between successive nucleotide bases on
the same polynucleotide chain and the positive charge of the dye

389

W. J. Gettins and E. Wyn-Jones (Eds.). Techniques and Applications of Fast Reactions in Solution. 389–403.
Copyright © 1979 by D. Reidel Publishing Company.

remains in the neighborhood of phosphate group[7]. The grounds for these models are based on the comparison of the equilibrium properties of the binding of dye between native and denatured DNA systems.

For the weak binding, it is merely reported that the dye binds itself to the phosphategroup of DNA[5]. However, the recombination rate constant is small by about one or two order of magnitude compared with that obtained in poly(α,L-glutamic acid)-dye system (10^9 M^{-1} sec^{-1})[8]. Although this fact is of interest concerning the binding mode, nothing has been discussed about the value of this rate constant. Moreover, it has been reported that this binding process includes a positive cooperativity[9] and the binding behavior differs between native and denatured DNA[7].

The purpose of the present work is to elucidate the above questions after consideration of cooperativity by the temperature-jump method.

2. EXPERIMENTAL

Materials.   Calf thymus DNA was purchased as a sodium salt from the Sigma Chemical Company (type V). Denatured DNA was obtained by heating the native DNA solution for 15 min. at 100°C and then cooling rapidly in ice to room temperature. The degree of denaturation was checked up by measuring the amount of hyperchromicity at 260 nm and its average percentage was more than 31%.

Proflavine was purchased as a hemisulfate form from Wako Pure Chemical Industries, Ltd. and purified by the procedure of Gupta et al.[10]

Method.   Spectrophotometric measurements were carried out with the Union Giken SM-401 spectrophotometer.

Kinetic studies were performed using a Joule-heating temperature-jump apparatus. Temperature jumps of about 4-5°C were normally applied. Heating times were 5 μsec in the 0.2 M KNO$_3$ solution. This orientational effect was canceled out by choosing the suitable direction of the polarizer which is inserted between the cell and the monochrometer. Kinetic experiments were carried out at 20°C with a thermostated cell and the others at room temperature. The relaxation was followed spectrophotometrically at 430 nm and the spectrophotometric data were analyzed at 444 nm.

3. THEORETICAL BASIS

Proflavine.   The equilibrium constant for dimerization of dye, K_d, in solution can be represented as

$$\{(\varepsilon_M - \varepsilon_T)/C_T\}^{\frac{1}{2}} = \{2K_d/(\varepsilon_M - \varepsilon_D)\}^{\frac{1}{2}}\{(\varepsilon_M - \varepsilon_D) - (\varepsilon_M - \varepsilon_T)\} \tag{1}$$

where ε and C are the extinction coefficient and concentration, respectively, and the subscripts T, M and D are the total, the monomer and the dimer of dye, respectively[11].

DNA-Dye system.   Choice of convenient expressions for describing binding equilibria and kinetics depends on the model adopted for the nature of the binding site. Two kinds of expressions were proposed: one is the Scatchard description and the other is the neighbor-exclusion model[12]. In the present work, the former expression was adopted in order to make easy the comparison of the data between native and denatured DNA.

The extinction coefficient of bound dye, ε_B, is determined by the following relation[5]:

$$\frac{1}{(\varepsilon_F - \varepsilon_1)} = \frac{1}{(\varepsilon_F - \varepsilon_B)(C_p{}^o - C_T)K(0)} + \frac{1}{(\varepsilon_F - \varepsilon_B)} \tag{2}$$

where ε_F is extinction coefficient of free dye; $C_p{}^o$ is total concentration of DNA; and $K(0)$ is defined to be the value of $K(r)(= C_{ap}/[C_p{}^o - C_{ap})C_F])$ at $C_p{}^o \ll C_T$. The concentration of an apparent bound dye, C_{ap}, can be calculated by Eq. (3).

$$C_{ap} = C_T(\varepsilon_F - \varepsilon_T)/(\varepsilon_F - \varepsilon_B) \tag{3}$$

For the system including two binding sites, the Scatchard equation is

$$r_{ap}/C_F = (r_I + r_{II})/C_F = K_I(g_I - r_I) + K_{II}(g_{II} - r_{II}) \tag{4}$$

where C_F is free dye concentration; C_i is bound dye concentration of complex i; r_i is ratio of bound dye of complex i to concentration of DNA; g_i is number of binding site of complex i; K_i is binding constant of complex i. For the system including one binding site, Eq. 4 can be written as

$$r_{ap}/C_F = K_{ap}(g_{ap} - r_{ap}) \tag{5}$$

where K_{ap} is apparent binding constant. The fraction of occupied binding sites for complex i, θ_i, can be easily related to the quantity r_i

$$\theta_i = r_i/g_i \tag{6}$$

Then the cooperative parameter q_i and binding constant K_i are determined from the following relation[13].

$$F(\theta_i)\sqrt{C_F} = \sqrt{q_i/K_i} - \sqrt{q_i K_i} C_F \tag{7}$$

where $F(\theta_i) = (1-2\theta_i)/\{\theta_i(1-\theta_i)\}$.

Kinetics.   For the following two step reaction having two binding sites,

$$\text{proflavine} \underset{k_{-1}}{\overset{k_1}{\rightleftharpoons}} \text{complex I} \underset{k_{-2}}{\overset{k_2}{\rightleftharpoons}} \text{complex II} \tag{8}$$

$$\text{site I} \qquad\qquad\qquad \text{site II}$$

the relaxation equations are expressed on the assumption that the relaxation time of the first step is much faster than that of the second.

$$\tau_f^{-1} = k_1\{X(1)\} + k_{-1} \tag{9}$$

$$\tau_s^{-1} = k_2\{X(II)\} + k_{-2} \tag{10}$$

with

$$X(I) = C_F + C_P^{\circ}(g_I - r_I)$$

$$X(II) = r_I C_P^{\circ} + C_P^{\circ}(g_{II} - r_{II})\frac{K_1^{-1} + C_F}{K_1^{-1} + C_F + C_P^{\circ}(g_I - r_I)}$$

On the other hand, when the strong cooperativity is present in the first process of Eq. 10 , relaxation equation can be written[14]

$$\tau_f^{-1} = 2k_1/\sqrt{q_I} \cdot \sqrt{r_I(g_I - r_I)} \cdot C_P^{\circ} \tag{11}$$

4. RESULTS

The value of ε_M was chosen so as to obtain the best linearity for the plots of $\{(\varepsilon_M - \varepsilon_T)/C_T\}^{\frac{1}{2}}$ vs. $(\varepsilon_M - \varepsilon_T)$ in Eq. 1 with $\varepsilon_M = 4.14 \times 10^4$ M^{-1} cm^{-1}. The value of K_d was determined from two intercepts of this straight line and turned out to be 8.3×10^2 M^{-1} which is in good agreement with that obtained by Schwarz et al.[11] Using this K_d value, the concentration of the dimeric dye was estimated to be negligible amount in the usual concentration (about 40 μM).

DNA-proflavine.   There is no difference in the wavelengths of
the extinction maxima between the two kinds of DNA-proflavine
systems. However, the wavelength of the isosbestic point in the
native DNA-proflavine system differs slightly from that in the
denatured DNA-proflavine system. This fact may suggest that the
forms of the complex are somewhat different between the two
systems. Therefore the values of ε_B and $K(0)$ were determined
using Eq. 2 to be 2.7×10^4 M^{-1} cm^{-1} and 2.5×10^4 M^{-1} for native
DNA system and 2.8×10^4 M^{-1} cm^{-1} and 2.8×10^4 M^{-1} for denatured
DNA system.
    The binding curves and the scatchard plots for these systems
are shown in Fig. 1. As can be seen in Fig. 1A, the values of
r_{ap} in both systems are at low C_F region, but slightly different
at the high C_F region. In this figure, a discontinuity of the
curves can be observed near $C_F = 5$ μM. This discontinuity can
also be observed more explicitly near $r_{ap} = 0.05$ in Fig. 1B.
From these findings, the Scatchard plots were divided into two
components which were defined as complex I ($0.1 < r_{ap} < 0.2$) and
complex II ($r_{ap} < 0.1$). The obtained values of g_{II} turned out to
be equal in the two kinds of DNA-proflavine systems but those of
g_I to be different. These facts are consistent with the binding
behaviors and suggest that the difference in the binding curves
in the two systems is caused by that in the nature of the
complex I.

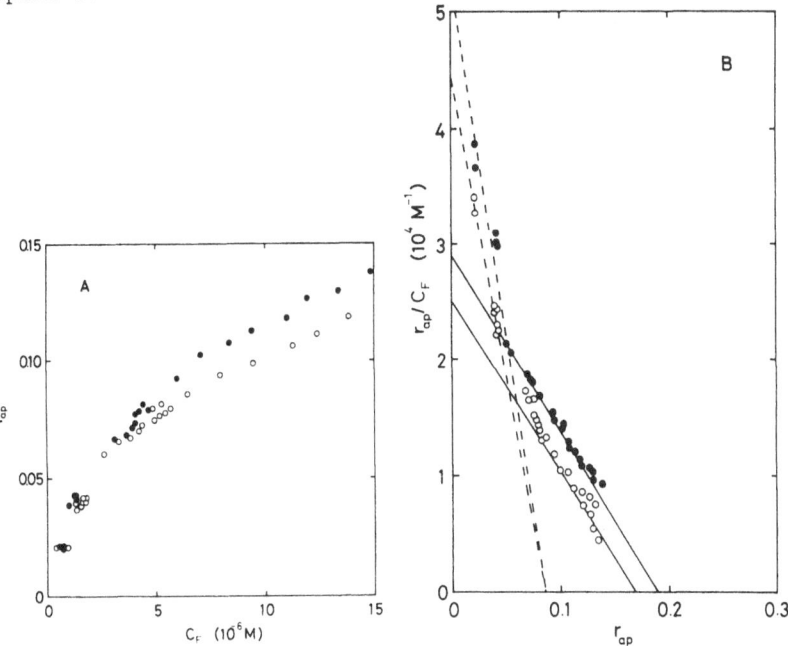

Fig. 1. Binding curves (A) and Scatchard plots (B): (o) native
DNA; (•) denatured DNA.

Table 1.   Equilibrium Properties for the Interaction of
Proflavine with DNA at pH 6.9[a].

	ε_B $10^4 M^{-1} cm^{-1}$	$K(0)$ $10^4 M^{-1}$	K_{ap} $10^5 M^{-1}$	K_I $10^4 M^{-1}$	K_{II} $10^5 M^{-1}$	K'_{II} $10^5 M^{-1}$
native DNA	2.7	2.5	1.6	5.0	5.2	1.5
denatured DNA	2.8	2.8	1.6	6.7	6.8	1.2

	g_{ap}	g_I	g_{II}	q_{ap}	q_I	q_{II}
native DNA	0.17	0.085	0.085	1	2.9	0.81
denatured DNA	0.19	0.105	0.085	1	5.9	0.82

[a] The solvent is 0.2M KNO_3-1 mM phosphate buffer.

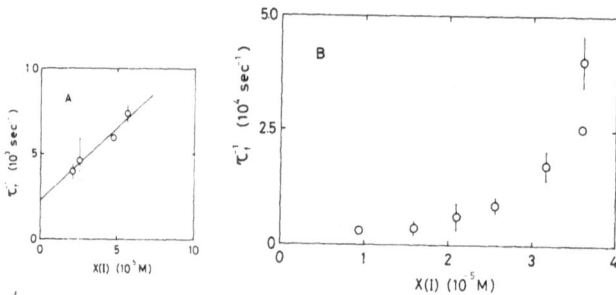

Fig. 2.   Plots of τ_f^{-1} vs. $X(I)$ according to Eq. 9 : (A) native
DNA at constant polymer to dye ratio of about 12; (B) denatured
DNA at constant polymer to dye ratio of about 6.

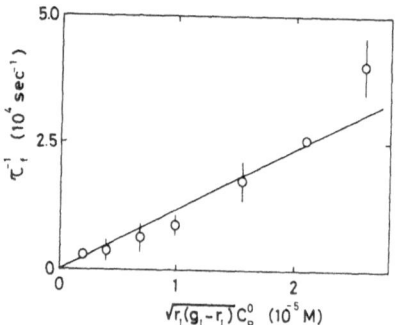

Fig. 3.   Plots of τ_f^{-1} vs. $\sqrt{r_I(g_I-r_I)}C_p^{o}$ according to Eq. 11.
Experimental condition is the same as Fig. 2B.

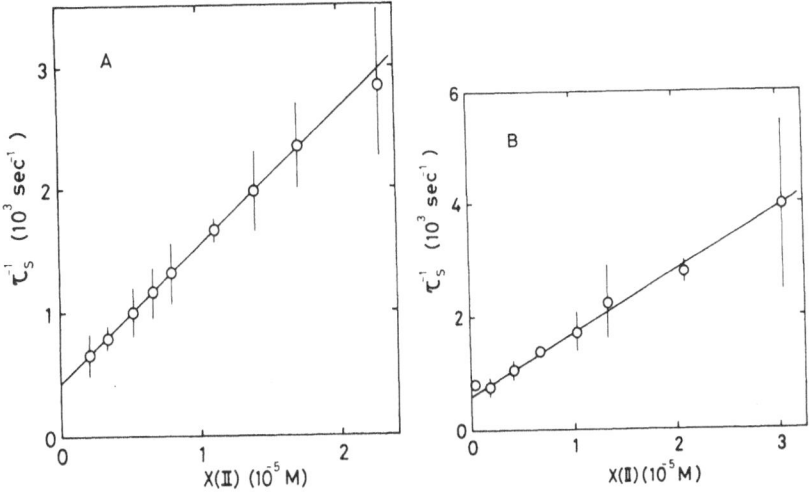

Fig. 4. Plots of τ_s^{-1} vs. X(II) according to Eq.10. Experimental conditions are the same as Fig. 2.

Table 2.   Kinetic Parameters for the Interaction of Proflavine with DNA at pH 6.9 and 20°C[a].

	k_1	k_{-1}	K_1	k_2	k_{-2}	K_2
	$10^7 M^{-1} sec^{-1}$	$10^3 sec^{-1}$	$10^4 M^{-1}$	$10^8 M^{-1} sec^{-1}$	$10^2 M^{-1} sec^{-1}$	$10^5 M^{-1}$
native DNA	8.4	2.3	3.7	1.1	4.3	2.6
denatured DNA	150	22		1.1	6.0	1.8

[a] The solvent is 0.2M KNO_3–1 mM phosphate buffer.

Using these values of g_{ap}, g_I and g_{II}, the binding constants K_{ap}, K_I and K_{II} and the cooperative parameters q_{ap}, q_I and q_{II} were determined. The value of K_{ap} obtained in the native DNA system is in good agreement with those in the literature which was obtained by the single binding curve[2]. All these obtained parameters are compiled in Table 1.

Kinetics.   In the observed relaxation curves, there are two well separated relaxation times, usually differing about tenfold in time scale. In the native DNA-proflavine system, however, the amplitude for fast process was so small that the faster relaxation time could not be measured with high precision.
    Obtained relaxation times were examined according to the reaction scheme reported by Li and Crothers[5]. As a result, it was found that this model cannot explain the experimental results in the denatured DNA system.

Fig. 2 shows the plots of the reciprocal relaxation times for the fast process vs. X(I) according to Eq. 9. As can be seen in this figure, a good linearity was obtained in the native DNA system but not in the denatured DNA system. Therefore, on the basis of the cooperativity, the reciprocal relaxation times in the denatured DNA system were plotted vs. $r_I(g_I-r_I)C_p^o$ in Fig. 3 which shows a straight line traversing the origin.

For the slow process, the relaxation times were fitted to the mechanism of Eq. 8 in the both systems, which are shown in Fig. 4. The values of binding constants, K_1 and K_2 ($=k_2/k_{-2}$), obtained kinetically were in good agreement with those obtained spectrophotometrically. (As the definition of K_{II} is different from that of K_2, K'_{II} ($=K_{II}/K_IB_I$) was used for comparison.) This fact confirms that the reaction mechanism of Eq. 8 is reasonable. All obtained kinetic constants are compiled in Table 2.

5. DISCUSSION

Complex I.   The obtained value of the rate constant, k_1, is about one order of magnitude greater than the usual one. This difference is caused by that in the number of binding site which is 0.085 in the present work and 1.0 in the usual one. From the fact that the kinetic data in the denatured DNA system could not be explained by the usual reaction scheme, it can be said that the present reaction scheme is more reasonable than the usual one and the value of k_1 in the native DNA system is about 10^8 M^{-1} sec^{-1}.

Complex II.   The complex II can be considered to be the same species as the usual intercalative complex. This is confirmed by the following similarities to the results reported hitherto[4]:
1. The binding constant was larger than that of complex I.
2. Negative cooperativity was observed.
3. The region of r_{ap}, where the complex II exists predominantly, was the same.

As a detailed feature of the intercalative complex, the following two models have been proposed;
1. complete intercalation model
2. partial intercalation model.

From the facts that the two relaxation processes were observed in both systems and that the values of k_2 were equal between two systems, it is confirmed that the model 2 is reasonable. In the denatured DNA system, the complete intercalation cannot occur on account of the difference in the structure. Therefore the above facts lead to the hypothesis that the model 1 occurs in the native DNA and model 2 occurs in the denatured DNA. In this case, the difference in the values of k_2 can be expected because of the difference in the flexibility of the backbone. However, present values of k_2 deny the above hypothesis and show that the complex II in each DNA system is the same.

Present reaction scheme shows that the complex I exists in the complex II as it is. This fact also supports the model 2 because the partial intercalated dye interacts with both base and phosphate group. Furthermore, the fact that the values of g_{II} were the same in two systems suggests the identity of complex II in the native DNA system and that in the denatured DNA system. From these facts, it can be concluded that the complex II is identically the same between two systems and its detailed mode is a partial intercalation.

REFERENCES

1. A. R. Peacocke and J. N. H. Skerrett, Trans. Faraday Soc., 52, 26 (1956).
2. R. W. Armstrong, T. Kurucsev, and U. P. Strauss, J. Amer. Chem. Soc., 92, 3174 (1970).
3. W. Bauer and J. Vinograd, J. Mol. Biol., 47, 419 (1970).
4. A. Blake and A. R. Peacocke, Biopolymers, 6, 1225 (1968).
5. H. J. Li and D. M. Crothers, J. Mol. Biol., 39, 461 (1969).
6. L. S. Lerman, ibid., 3, 18 (1961).
7. D. S. Drummond, V. F. W. Simpson-Gildmeister, and A. R. Peacocke, Biopolymers, 3, 135 (1965).
8. H. Ushio, T. Yasunaga, T. Sano, and Y. Tsuji, ibid., 15, 187 (1976).
9. D. F. Bradley and M. K. Wolf, Proc. Natl. Acad. Sci. U. S., 45, 944 (1959).
10. V. S. Gupta, S. C. Kraft, and J. S. Samuelson, J. Chromatography, 26, 158 (1967).
11. G. Schwarz, S. Klose, and W. Balthasar, Eur. J. Biochem., 12, 454 (1970).
12. D. E. V. Schmechel and D. M. Crothers, Biopolymers, 10, 465 (1971).
13. G. Schwarz, Eur. J. Biochem., 12, 442 (1970).
14. G. Schwarz and W. Balthasar, ibid., 12, 461 (1970).

INTERACTION BETWEEN POLY(STYRENE SULFONIC ACID) AND A "BIFUNCTIONAL" CATIONIC DYE

Vincenzo Vitagliano, Ornella Ortona, and Roberto Sartorio

Istituto Chimico, Università di Napoli, Naples Italy, and Istituto di Chimica, Università di Salerno, Salerno Italy

The spectrophotometric behaviour of solutions of 1-1'penta= methylene bis (3,6 - bis(dimethylamino) acridinium), D ,was recently discussed[1]. D is a "Bifunctional" or "dimer" dye obtained by con= necting the 10 - N nitrogens of two acridine orange (AO) molecules through a pentamethylene chain.

The D spectrophotometric behaviour was compared with that of the parent AO dye[2]: D showed a very strong intramolecular associ= ation in aqueous solution exhibiting an absorption spectrum in the visible region similar to that of the dimer AO , with a maximum at 473 nm (see fig.1). Neither temperature effects in the range 0 - 50 °C nor concentration effects were observed on the absorption spectrum shape.

On the other hand an alcoholic medium destroys the D intramolecular stacking and causes the absorption spectrum to be= come similar to that of monomeric AO, with a maximum at 495 nm.

D binds stoichiometrically to the SO_3^- groups of polystyrene sulfonic acid (PSS). The binding process was followed by titrating D with PSS in aqueous solution; a change on the absorption spectrum was observed and was attributed to the formation of intermolecular associated species among dye molecules crowded along the polyanion chain.

In the presence of a SO_3^- groups excess the dilution of D along the polymer chain was observed with the reappearance of the dye absorption spectrum shown in water.

The titration process exhibits a very sharp end - point which allows to titrate D by using PSS (see fig.2).

Dilution of the dye along the polyanion chain also promotes a gradual breaking of the intramolecular stacking. This effect was revealed by the increasing absorption at 495-500 nm; in the presence of a very high excess of SO_3^- groups the absorption spectrum of

W. J. Gettins and E. Wyn-Jones (Eds.). Techniques and Applications of Fast Reactions in Solution. 399–403.
Copyright © 1979 by D. Reidel Publishing Company.

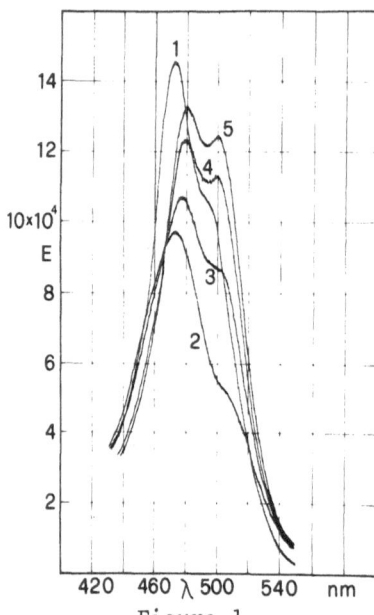

Figure 1
Equivalent Optical Densities
of D in Aqueous Solution (1)
and in the presence of vari‾
ous amounts of PSS: 2,3,4,5‾
P/D = 4, 340, 860, and 1420
respectively.

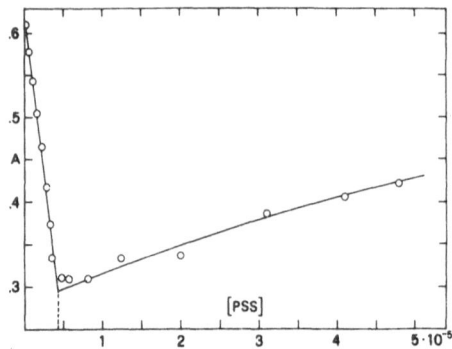

Figure 2
Titration of D with PSS. A is
the optical absorption of the
dye at 473 nm as a function of
SO_3^- equivalent concentration.
The dye concentration is constant.

unstacked bound D with a maximum at
500 nm appeared (see fig.1).

The breaking of intramolecul=
ar association was attributed to some
specific interaction between the PSS
benzene rings and the dye aromatic
systems.

From the analysis of all
experimental results it was concluded
that the interaction between D and PSS can be interpreted by a model
very similar to that proposed for the system AO-PSS [3,4].

The D - PSS interaction equilibrium shows a slight thermal
effect, therefore thermal relaxation measurements (T-jump) on this
system were carried out in order to compare the results with those
obtained for the AO - PSS system in the past [5].

Two sets of T-jump runs were taken at 20°C (final temper=
ature): the first one was taken at constant dye concentration
(D = 7 x 10^{-6} eq./1)(*) and varying PSS concentration (P) in the
P/D ratio range 10 to 1000. The second set was taken at constant
P/D ratio (P/D = 22) for various dilutions of an original stock
solution. The experimental results are shown in figure 3 A and B.

The peculiarity of the behaviour of the D - PSS system
as compared to the AO - PSS one is the presence of two main kinetic
processes having mean reciprocal relaxation times differing by a
factor of 100.

Although the knowledge of the D - PSS system is still at
an initial stage, a tentative interpretation of the two processes

----

(*) The dye concentration is the "equivalent concentration" with
    respect to the PSS sulfonic groups, as explained in our
    previous paper [1].

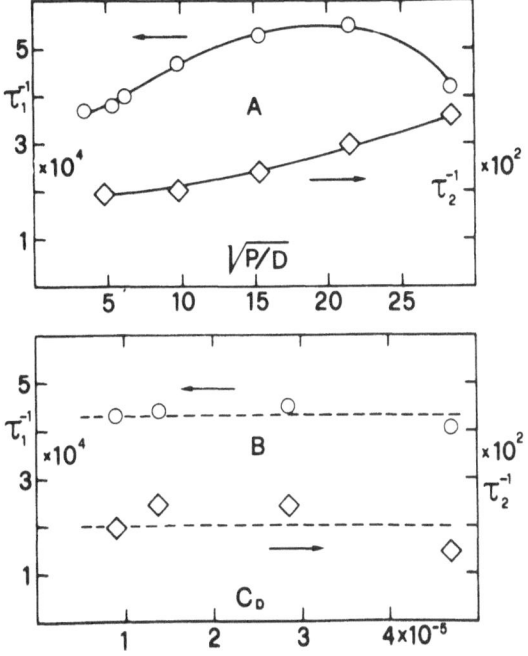

Figure 3

Interaction of D with PSS: Reciprocal mean relaxation times of the faster and slower kinetics: A, data at constant dye concentration (7×10^{-6} eq./1) and varying P/D ratio. B, data at constant P/D ratio (P/D = 22) and varying total dye concentration.

can be given on the basis of the present experimental results.

Both kinetics are almost independent of the total dye concentration at constant P/D (see fig.3 B). For this reason we must exclude the dye binding to the polyelectrolyte from the relax= ation processes:

$$D_{bound} \rightleftarrows D_{free} \qquad (1)$$

such a process is known to be a function of dye concentration[6] and it was found to be present in the system AO – PSS at lower P / D values[5].

We must assume that both kinetiks involve processes appearing inside the volume of each polyanion molecule; for this reason they are not affected by changes on the "bulk" concentration of either D or PSS, depending only on the ratio between the available binding sites concentration and the dye concentration, namely on P/D.

The faster kinetics has mean relaxation times of the same order of magnitude as those found for the system AO – PSS, $\tau_1^{-1} = 3 - 4 \times 10^4$ sec^{-1}. Furthermore the τ_1^{-1} values plotted as a function of P/D go through a maximum as predicted by Schwarz[7] and found for the systems AO – PSS [5] and proflavine – polyacrylic acid[8].

For these reasons we may attribute the faster kinetics to the temperature change of the dye distribution along the macro=ion chain as done for the system AO - PSS. The process can be described by the following phenomenological equation:

$$uaa \; + \; uuu \; \underset{k_{1-}}{\overset{k_{1+}}{\rightleftarrows}} \; uau \; + \; uua \qquad (2)$$

or:

$$(aau) + (uuu) \; \rightleftarrows \; (uau) + (auu) \qquad (2 \, a)$$

where u is an available free binding site on the polymer chain and a is a site occupied by a dye molecule.

Schwarz suggested that the reaction might proceed through a transient nucleation process such as [7]:

$$xaau \; \rightleftarrows \; xauu$$
$$\updownarrow \qquad\qquad\qquad\qquad (3)$$
$$D \; + \; uuu \; \rightleftarrows \; uau$$

Actually the free dye, D, should be considered as a molecule within the domain of the macroion.

The slower kinetics, not found for the system AO - PSS, has τ_2^{-1} values regularly increasing by increasing P/D and we think it reasonable to attribute this process to the relaxation of the intra molecular dye association:

$$D_s \; \underset{k_{2-}}{\overset{k_{2+}}{\rightleftarrows}} \; D_u \qquad (4)$$

(where s and u stand for stacked and unstacked).

If we assume a monomolecular kinetics for the process (4) we may easily justify the experimental increase of τ_2^{-1} brought about by an increase in P/D. In fact the relaxation equation for process (4) is:

$$\frac{d \, [D_u]}{d \, t} = - \, (k_{2+} + k_{2-}) \Big([D_u] - [D_u]_{eq} \Big) \qquad (5)$$

$$= - \, \tau_2^{-1} \Delta \, [D_u] \qquad (5 \, a)$$

By increasing the concentration of free SO_3^- groups inside the domain of each macroion (i.e. by increasing P/D) the equilibrium (4) shifts to the right and the equilibrium constant (k_{2+} / k_{2-}) increases. As a consequence τ_2^{-1} increases, as found experimentally.

ACKNOWLEDGMENT

The research was carried on with the financial support of the Italian C.N.R.

REFERENCES

(1) V.Vitagliano,O.Ortona,and M.Parrilli, J.Phys.Chem., $\underline{82}$,   (1978)

(2) V.Vitagliano, "Interaction between Cationic Dyes and
    Polyelectrolytes" in "Chemical and Biological Applications of
    Relaxation Spectrometry", E.Wyn-Jones, Ed., Reidel Publishing
    Dordrecht, Holland (1975) pp 437–466

(3) V.Vitagliano,L.Costantino,and A.Zagari, J.Phys.Chem., $\underline{77}$, 204
    (1973)

(4) V.Vitagliano,L.Costantino,and R.Sartorio, J.Phys.Chem., $\underline{80}$,
    959 (1976)

(5) V.Vitagliano, J.Phys.Chem., $\underline{77}$, 1922 (1973)

(6) G.Schwarz, Eur.J.Biochem., $\underline{12}$, 442 (1970)

(7) G.Schwarz, Ber.Bunsenges Phys.Chem., $\underline{76}$, 373 (1972)

(8) G.Schwarz and S.Klose, Eur.J.Biochem., $\underline{29}$, 249 (1972)

TEMPERATURE JUMP STUDIES ON ACRIDINE ORANGE - CARBOXYMETHYL-CELLULOSE COMPLEXES

A. Dawson, D. J. Wedlock*, E. Wyn-Jones & G. O. Phillips*

Department of Chemistry & Applied Chemistry, University of Salford, Salford M5 4WT

*North East Wales Institute, Deeside, Clwyd.

ABSTRACT. A kinetic study was made of the reaction between various carboxymethylcellulose (CMC) preparations and the dye acridine orange (AO). Appropriate spectral measurements have also been made. It has already been demonstrated by Hammes and Hubbard (1,2) that the monomer-dimer reaction of acridine orange is essentially diffusion controlled, dye stacking occurring in <10^{-10} sec. The relaxation effects observable in the CMC - AO system can be approximately described by a two relaxation process. Measurement of relaxation times were made at ambient pH on fully neutralised CMC in the sodium form, under conditions of low ionic strength (10^{-3}M) using sodium chloride as the inert electrolyte. These conditions were purposely chosen as a medium that would discourage dimer and oligomerisation i.e. dye aggregation, normally associated with the dye under conditions of high ionic strength (3).

The CMC-AO system was investigated for the effect of; molecular weight, degree of substitution (d.s.) in carboxymethyl groups, ratio of anionic site (S) concentration to cationic dye (D) concentration and also concentration of dye, on the dye-polyanion relaxation phenomenon.

EXPERIMENTAL

The temperature jump equipment comprised a Hartley Measurements, joule heating temperature jump assembly with both oscilloscope output and a Varian trace superposition noise averaging facility for increasing signal to noise ratio, also equipped with an analogue to digital (paper tape) converter for data storage. The relaxation effect was monitored by fast kinetic spectrophoto-

W. J. Gettins and E. Wyn-Jones (Eds.). Techniques and Applications of Fast Reactions in Solution. 405–413.
Copyright © 1979 by D. Reidel Publishing Company.

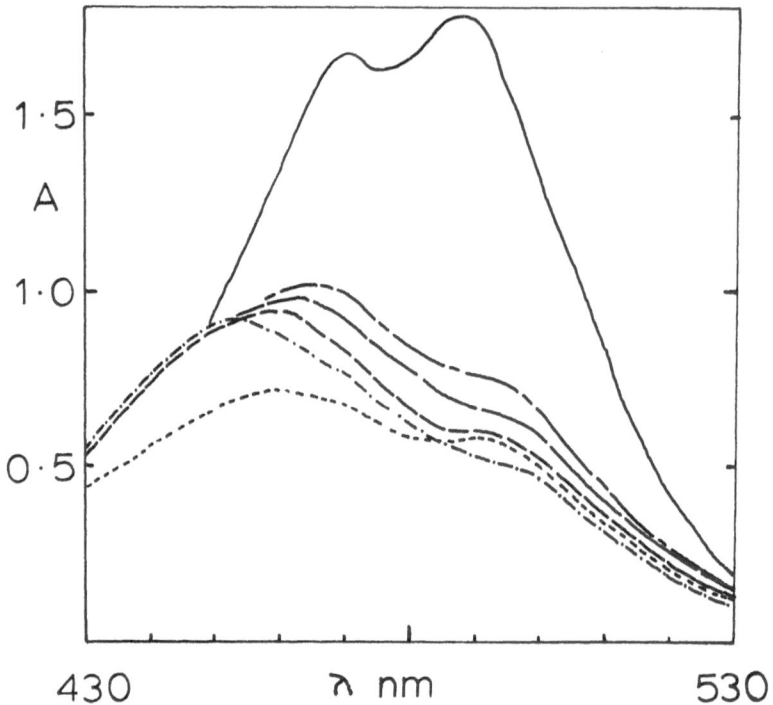

<u>Fig. 1</u> Varying S/D for CMC 4M, (AO)5 x 10^{-5}M, (NaCl)1 x 10^{-3}M
25° S/D $\cdots\cdots$ 2.5, $\cdot-\cdot-\cdot$ 10,$---$50,$-\!\!-\!\!-$75,$-\cdot\cdot-$150,$-\!\!-\!\!-$DYE ALONE

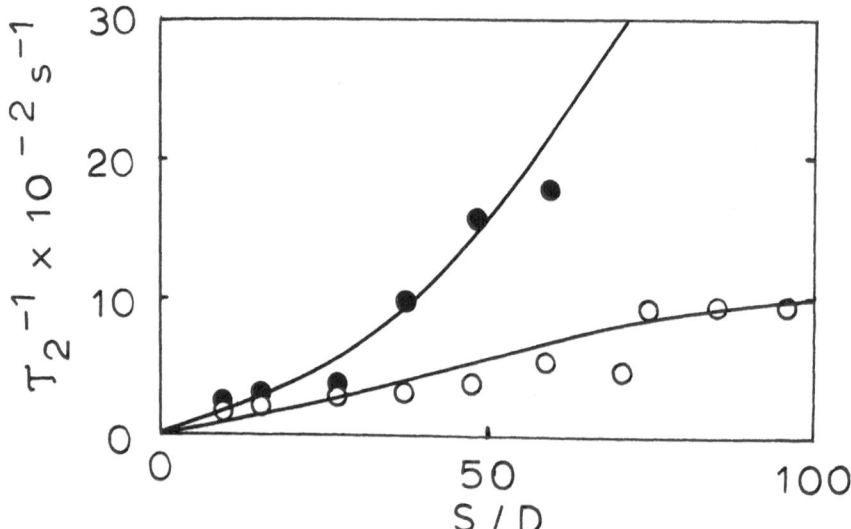

<u>Fig. 2</u> τ_2^{-1} versus S/D for CMC 4M-0 and 7M-●, (AO) = 1 x 10^{-4}M
(NaCl) = 1 x 10^{-3}M

metry observed at 492 nm, the monomer absorption λ_{max} for AO.
The degree of substitution of the CMC samples, as referred to in
the text, by the carboxymethyl group were; 4,7,9 and 12 per 10
sugar units. For the CMC samples of degree of substitution 7
(CMC 7), three molecular weights (weight average) were used;
7L (8×10^4), 7M (2.5×10^5) and 7H (7×10^5).

RESULTS AND DISCUSSION

Figure 1 shows typical absorption spectra of CMC 4M-AO complexes. The concentration of AO is kept constant and the S/D
ratio is increased. The characteristic monomer absorption peak
at 492 nm is replaced by an absorption at 456-460 nm, depending
on the S/D ratio of the complex. The short wavelength peak is
ascribed to formation of dye aggregates, on the CMC at low S/D
ratios. The reappearance of the 492 nm peak is evident at high
S/D ratios as the dye becomes diluted on the polyanion.

Relaxation spectra for CMC-AO complexes were not single
exponentials, and were analysed in terms of a fast and slow relaxation time τ_1 and τ_2. The relaxation times were determined from
a log signal amplitude vs time plot, by calculation of the longer
relaxation time from the linear portion of the curve followed by
extension of the linear portion to shorter times and subtracting
this from the experimental curve to yield a straight line from
which the shorter time can be determined. The possible phenomena
which can give rise to the observed relaxational effects are association of free dye molecules with polymer bound dye molecules
(intermolecular) or association of polymer bound monomer dye molecules to give polymer bound aggregates (intramolecular).

τ Dependence of Site/Dye Ratio of Complex

In Figures 2 and 3, the reciprocal relaxation times for τ_2
are plotted as a function of S/D, for a concentration of AO,
10^{-4}M. The overall profiles for CMC samples of degrees of substitution 4,7,9 and 12 are similar, in that a rise in the value
of τ_2^{-1} with increase in polyanion concentration or S/D ratio
(it may also be viewed as diluting the dye on the polymer) is observed. The increase levels off by S/D = 80 for CMC4M, but a
steady increase is still observed at this level of S/D for the 7M
9M and 12M samples. It could be that the value of τ_2^{-1} would also
level off at higher S/D ratios in the case of 7,9 and 12 d.s.
CMC's but at higher S/D ratios the amplitude of the relaxation
drops off making analysis difficult. Similar trends are observed
for plots of τ_1^{-1} vs S/D, Figure 4. The increase of τ^{-1} with increasing S/D may be either accommodated by intermolecular and/or
intramolecular mechanisms. If a dye molecule binds rapidly to a
site on the polymer followed by a slow intramolecular rearrange-

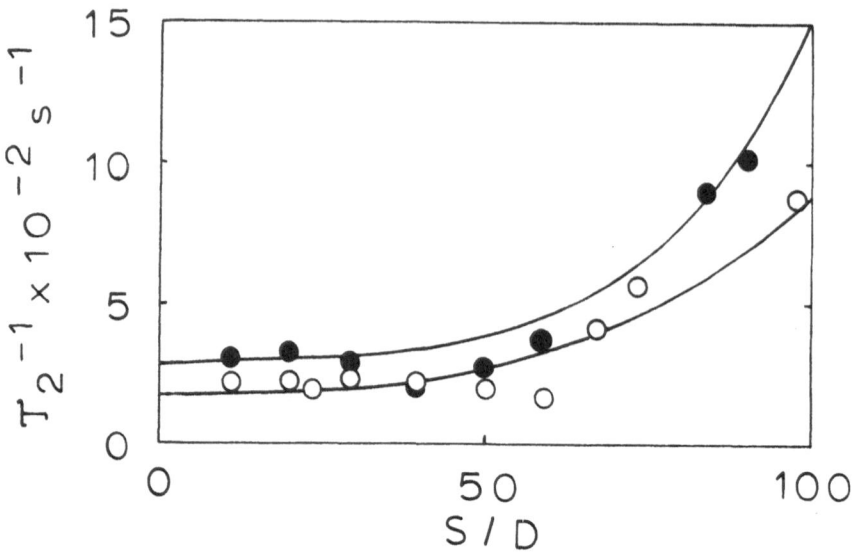

<u>Fig. 3</u> τ_2^{-1} versus S/D for CMC 9M-O, and 12M-●, (AO) = 1 x 10^{-4}M
(NaCl) = 1 x 10^{-3}M

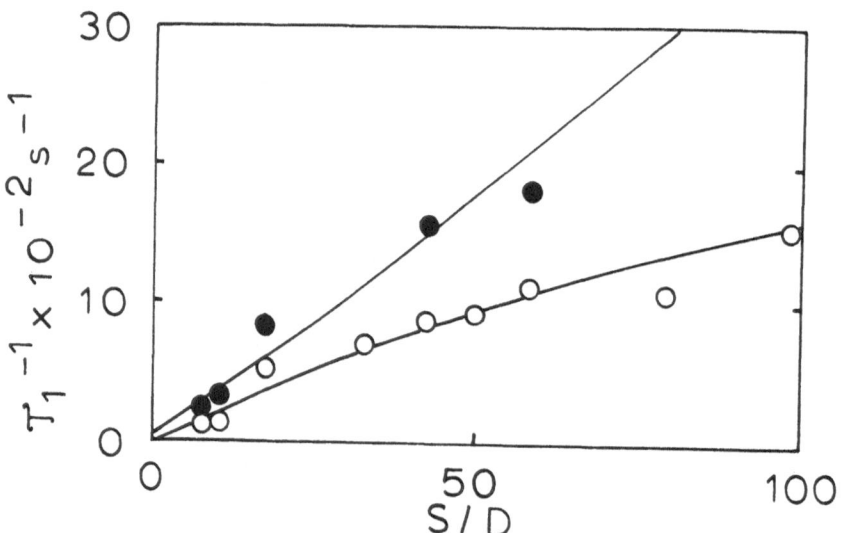

<u>Fig. 4</u> τ_1^{-1} versus S/D for CMC 4M-O, and 9M-●, (AO) = 1 x 10^{-4}M,
(NaCl) = 1 x 10^{-3}M.

ment to form aggregates, the number of original binding sites, and hence nucleation points for aggregation, will be a function of polymer concentration.

τ Dependence on Degree of Substitution

Figure 5 demonstrates that at a relatively low S/D ratio, 10/1, there is a direct proportionality between the degree of substitution and the value of τ_2^{-1}. This tends to indicate that even the relaxational effect described by τ_2 has some intermolecular character under these conditions. However, as the S/D ratio is increased to 40/1 this effect becomes less apparent, showing almost constancy of τ_2^{-1} with degree of substitution. Similar behaviour was found for the pH dependence of the relaxational processes associated with an AO/poly α-L-glutamic acid complex (1), in which an increase in the degree of dissociation, (d.d.) induced by increasing the pH of the buffer system, caused τ_1^{-1} and τ_2^{-1} to increase. In the poly α-L-glutamic acid system there is a well characterised (4) α-helix to random coil transition at high ionic strength (0.1M) on changing the pH from <5 to >7. Our findings indicate that caution must be observed in relating changes in relaxation times for such dye/polyion systems to specific conformational effects. Even if poly α-L-glutamic acid did not form an α-helix, τ_1^{-1} and τ_2^{-1} relationship to the d.s. of carboxymethyl-celluloses, where there is no conformational change predict a high d.d. dependence of τ_1^{-1} and τ_2^{-1}. Thus, at low S/D ratios the results appear consistent with a mechanism involving some intermolecular character even at the rate determining step, but as the S/D ratio is increased, the slow process appears to involve intramolecular processes, which may be desolvation of the hydrophilic part of the dye molecule or displacement of counterions.

τ Dependence on Dye Concentration

It can be seen in Figure 6 that under conditions where the S/D ratio is as high as 100/1, for CMC 9M, the value of τ_2^{-1} is proportional to dye concentration. Measurement at the lowest dye concentration, 1.25×10^{-5}M, for S/D = 100/1, could not be made because of loss of sensitivity under these conditions. Hammes et al (1) proposed that such behaviour could possibly fit with a model in which dimer stacks were formed,

$$P_n R' + mP + R \; \rightleftarrows \; P_{n+m} R_2 \; \rightleftarrows \; P_{n+m} R_2' \qquad (1)$$

where P represents the polyanion molecule, n the number of polyanion molecules associated with a dye molecule (most probably (1)) and R represents the dye molecule. Schwarz (5) derived an

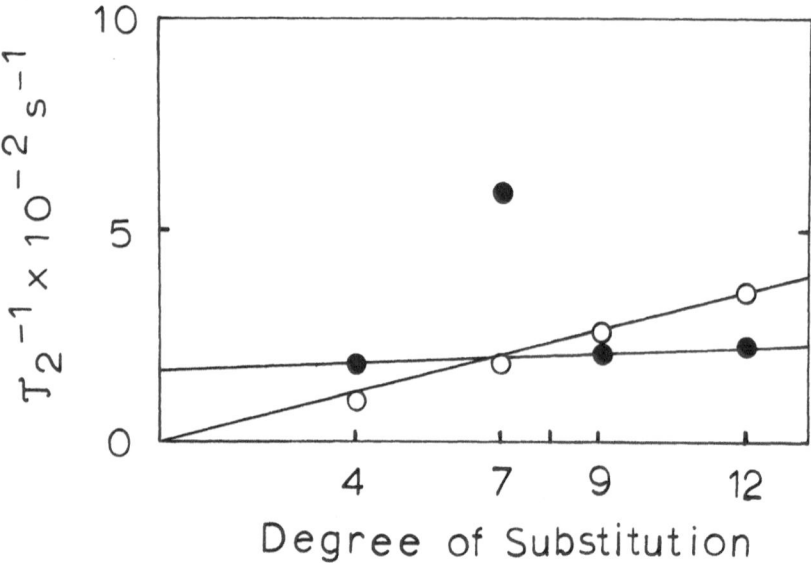

<u>Fig. 5</u> τ_2^{-1} versus degree of substitution of CMC, for S/D O = 10, ● = 40, (AO) = 1 x 10^{-4}M, (NaCl) = 1 x 10^{-3}M.

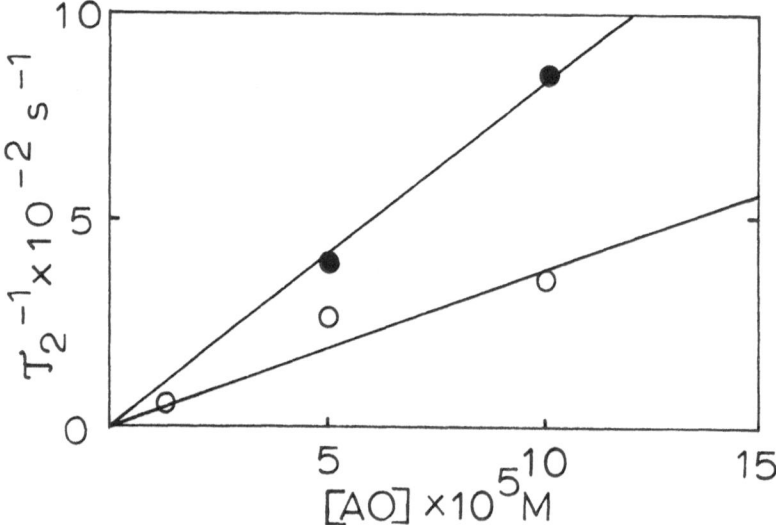

<u>Fig. 6</u> τ_2^{-1} versus concentration of AO for CMC 9M at S/D O = 50, ● = 100.

expression for the reciprocal mean relaxation time $(1/\tau^*)$, eq (2)

$$\frac{1}{\tau^*} = 2\varkappa_r \sqrt{\sigma\{(D/S)\ (1-(D/S))\}}(S/D)D \tag{2}$$

applicable in the case of very highly co-operative binding, where σ is a function of the stacking co-efficient, q^6, S and D have their usual meanings and $\varkappa_r$ is the associative rate constant for a free dye molecule associating with a bound dye molecule, to give a polymer bound dimer. This mechanism, intermolecular in nature, will give rise to a linear variation of $1/\tau^*$ with dye concentration, passing through the origin, for a constant S/D ratio. It is valid to compare τ^* to τ_2 due to the closeness of τ_1 and τ_2 values and their similar variation with S/D ratio, in order to make the point that the results indicate a mechanism with intermolecular character for association of an AO molecule with a polymer bound AO molecule. The alternative mechanism proposed by Schwarz (5) for the case of very strong binding and fairly weak co-operativity when a negligible amount of free dye (monomeric or aggregated) is in solution, is intramolecular in nature, and would give rise to τ^* independent of dye concentration, determined only by the S/D ratio. This should be true for τ_2 also. Obviously this is not occurring for the CMC9M–AO system at S/D ratio of 100/1 and is even less likely to occur for lower degrees of substitution under comparable conditions.

τ Dependence on Molecular Weight of CMC

In Figure 7, the molecular weight dependence of τ_1^{-1}, and τ_2^{-1} is shown for two values of the S/D ratio of the complex, each S/D ratio being examined kinetically at three different molecular weights of CMC, with a constant degree of substitution, 7. Under both sets of S/D ratios, both τ_1^{-1} and τ_2^{-1} are observed to increase with increase in molecular weight. The influence of viscosity effects can be discounted in the light of recent work on diffusion controlled reactions in CMC solutions (7). With reference to Scott's treatment (8) of the binding of ionic dyes to a polymeric ion, where affinity of the dye for the polyion is expressed in terms of critical electrolyte concentration which is the concentration of a simple electrolyte required to cause complete and critical dissociation of the dye/polyion complex, one expects the affinity of the dye for the polyion to increase with increase in molecular weight, up to a certain molecular weight, then to become independent of molecular weight. If the polyion P carries z negative charges, the dye/polyion complex is $p^z\ zR^+$ where R^+ is the dye cation and M^+ is the counterion to the polymer,

$$p^{z-}\ zM^+ + z\ R^+ \rightleftharpoons p^{z-}\ zR^+ + zM^+$$

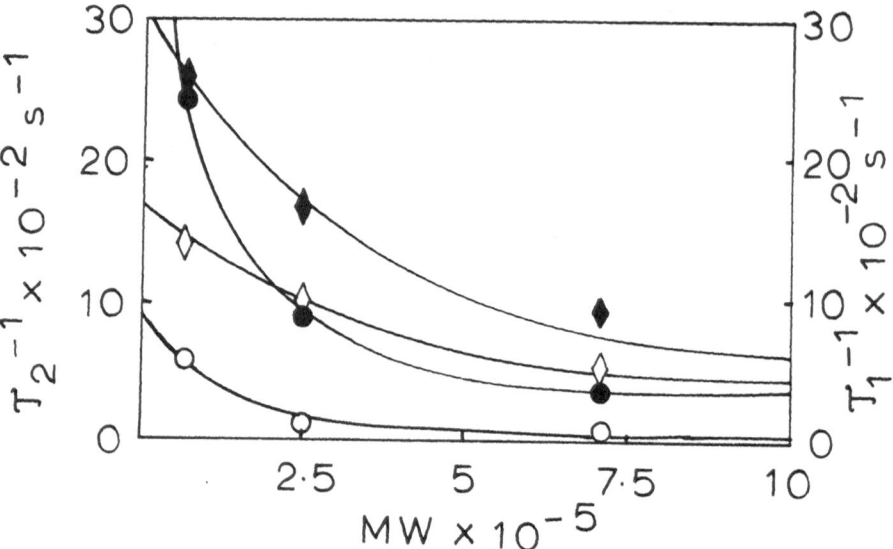

<u>Fig. 7</u> τ_1^{-1} and τ_2^{-1} versus molecular weight of CMC. τ_1^{-1} at S/D ◊ = 10, ♦ = 40, τ_2^{-1} at S/D 0 = 10, 0 = 40.   (AO) = 1 x 10⁻⁴M, (NaCl) = 1 x 10⁻³M.

and $$K = \frac{\{p^{z-} zR^+\}\ \{M^+\}^z}{\{p^{z-} zM^+\}\ \{R^+\}^z}$$

Since K, the constant describing the equilibrium distribution of dye between the polyanion phase and solution phase can be approximately expressed as consisting of an associative and dissociative rate constant, n_1 and n_{-1} respectively.

$$K = \frac{n_1}{n_{-1}}$$

if the equilibrium constant of the dye/polyanion complex increases with increase in molecular weight (to a certain maximum value), we should observe either an increase of τ_1^{-1} or a constancy of τ_1^{-1} with molecular weight, if τ_1 is the relaxation time characterising the initial association of the dye molecule with the polyanion. As we observe the opposite effect, in that τ_1^{-1} decreases with an increase of molecular weight of CMC (d.s. = 7) from 8 x 10⁻⁴ to 2.5 x 10⁵ showing a much smaller decrement on changing the molecular weight of CMC from 2.5 x 10⁵ to 7 x 10⁵, then τ_1 may not be associated with the initial binding of the dye to the polyanion but rather an intramolecular process of dye aggregation.

A possible mechanism which would accommodate these findings

is one in which a very rapid initial dye to polyanion interaction occurs, followed by two consecutive intramolecular processes occurring which represent the faster (τ_1) and slower (τ_2) relaxation processes.

$$P + R \rightleftharpoons PR \rightleftharpoons PR' \rightleftharpoons PR''$$

Thus the evidence from this preliminary study would tend to indicate that describing the relaxation process in the CMC-AO system by two simple equilibria is only an approximation as the mechanism appears to have quite a mixed character, possessing both intermolecular and intramolecular characteristics depending on the make up of the dye-polyanion complex.

REFERENCES

1. G. G. Hammes and C. D. Hubbard, J.Phys.Chem., 70, 1615 (1966).
2. G. G. Hammes and C. D. Hubbard, J.Phys.Chem., 70, 2889, (1966).
3. S. P. Moulik, S. Ghosh and A. R. Das, Ind.J.Chem., 14A, 306 (1976).
4. P. A. Doty, A. Wada, J. T. Yang and E. R. Blout, J.Poly.Sci., 23, 851 (1957).
5. G. Schwarz, Eur.J.Biochem., 12, 442 (1970).
6. V. Vitagliano, J.Phys.Chem., 77, 1922 (1973).
7. O. I. Micic, B. H. Milosavljevic, G. O. Phillips and D. J. Wedlock, J.C.S. Faraday II, in press.
8. J. E. Scott, Biochem.Soc.Trans., Second BDH Lecture 787 (1973).

FAST KINETICS BY NMR : BINDING AND RELEASE OF CATIONS BY BIOMOLECULES

J. Grandjean, A. Cornelis, and P. Laszlo

Institut de Chimie Organique B6 - Université de Liège
Sart-Tilman par 4000 Liège, Belgium

Because of its high sensitivity, sodium-23 nuclear magnetic resonance (NMR) is useful for the study of ion-substrate
interactions[1]. Complex formation induces a relaxation enhancement of the nuclear spins manifested by line broadening whereas
the position of the peak is little affected. Under some conditions, such relaxation measurements can be used to determine
the forward and the reverse rate constants for complex formation. In this paper we present examples of the determination of
kinetic parameters by NMR.

Relaxation of nuclear spins occurs under two modes :
longitudinal and transverse relaxation characterized respectively by time constants T_1 and T_2. The quadrupolar interaction i.e.
the interaction of the quadrupolar moment of the nucleus with
the electric field gradient present at the sodium site, provides
the predominant relaxation mechanism. Due to the rather symmetrical environment of the solvated sodium ion - for instance
$(Na^+)6H_2O$ - a weak electric field gradient is expected. Upon
complexation by a biomolecule, the alkali ion coordinates both
with solvent and substrate molecules, which both increase the
electric field gradient, and decreases by a considerable amount
the cation rotational motion. Both these effects generate a
relaxation rate enhancement.

In a two-sites system, where the ion exchanges between its
free (solvated) and its bound form, the observed relaxation rate
$1/T_2$ is given by [2] :

$$\frac{1}{T_2} = \frac{p_A}{T_2^A} + \frac{0,6\ p_B}{T_2'^B + \tau_B} + \frac{0,4\ p_B}{T_2''^B + \tau_B}$$

415

W. J. Gettins and E. Wyn-Jones (Eds.). Techniques and Applications of Fast Reactions in Solution. 415–418.
Copyright © 1979 by D. Reidel Publishing Company.

where p_A and p_B are the populations of the solvated and complexed ions, T_2^A, $T_2'B$, $T_2''B$ refer to the transverse relaxation times in the solvated state A and in the bound state B whereas τ_B is the life time of the complex. The chemical shift contribution to the relaxation is usually negligible.

Since the sodium relaxation rate is fast, kinetic informations are only available for large τ_B values ($> 10^{-4}$ s). This occurs with strong cation-substrate interactions. The rate constants for sodium complexation by crown compounds or by monensin have been determined (3,4), in this manner.

On the other hand, when the exchange rate is faster than the relaxation rates in the bound form, the life time of the complex can also contribute to the relaxation enhancement. Both the component relaxation times $T_2'B$ and $T_2''B$ depend on the life time τ_B (2); through a correlation time τ_c :

$$\frac{1}{T_2'B} = \frac{\pi^2}{5} \chi^2 \; [\tau_c + \frac{\tau_c}{1 + \omega^2 \tau_c^2} \;]$$

$$\frac{1}{T_2''B} = \frac{\pi^2}{5} \chi^2 \; [\frac{\tau_c}{1 + 4\omega^2\tau_c^2} + \frac{\tau_c}{1 + \omega^2\tau_c^2} \;]$$

where χ is the quadrupolar constant and ω the sodium resonance frequency.

In the absence of an internal motion, the correlation time τ_c is related to the rotation rate of the complex $1/\tau_R$ and the exchange rate $1/\tau_B$ of the sodium between the two sites :

$$\frac{1}{\tau_c} = \frac{1}{\tau_R} + \frac{1}{\tau_B}$$

Similar expressions have also been deduced for the longitudinal relaxation rates (2) :

$$\frac{1}{T_1} = \frac{p_A}{T_1^A} + \frac{0.8}{T_1'B + \tau_B} + \frac{0.2}{T_1''B + \tau_B}$$

$$\frac{1}{T_1'B} = \frac{2\pi^2}{5} \chi^2 \; [\frac{\tau_c}{1 + 4\omega^2\tau_c^2}]$$

$$\frac{1}{T''^{B}_1} = \frac{2\pi^2}{5} \chi^2 \left[\frac{\tau_c}{1 + \omega^2\tau_c^2}\right]$$

In the fast exchange limit ($\tau_B \ll T'^{B}_{1,2}$, $T''^{B}_{1,2}$) the correlation time τ_c can be obtained either from the T_1/T_2 ratio or from the ratio of the transverse relaxation rates determined at two different frequencies.

This is the method applied in our study of the sodium interaction with half-decalcified parvalbumin. The alkali ion binds to the calcium-binding site of this muscular protein vacated by the Ca^{++} ion. The rate constants for the equilibrium (5) :

$$Na^+ + \text{protein} \; \underset{k_-}{\overset{k_+}{\rightleftarrows}} \; \text{complex}$$

are $k_+ = 0.6 \; 10^9 \; M^{-1} \; s^{-1}$ and $k_- = 2.10^7 \; s^{-1}$. The forward rate constant falls in the expected range for diffusion-controlled binding of the cation. Likewise, the rate constant for the binding of calcium is estimated to be about 10^8-10^9 $(M.s.)^{-1}$. Therefore the 10^5 greater stability constant for the formation of the divalent ion complex results from the off rate constant k_-. Fuller dehydration of the calcium ion upon binding to the protein could account for such a large difference in the stability constants. Complementary information comes from the observation that the sodium ion remains on the protein a time long as compared to the rotational correlation time of the complex ($\tau_R \simeq 3.3 \; 10^{-9} \; s.$). This, together with a calculated quadrupolar constant of 1.4 MHz are uniquely consistent with site binding, but not with atmospheric binding of the alkali ion.

Another relevant example of kinetic parameters determined by this method comes from the antibiotic field. Assuming a two sites model, the sodium-gramicidin A complex is characterized by the following rate constants (6) :

$$k_+ = 2.2 \; 10^9 \; M^{-1} \; s^{-1} \quad \text{and} \quad k_- = 5.5 \; 10^8 \; s^{-1}.$$

As the ion binding site is unknown, an internal motion cannot be discarded; therefore these values must be taken as upper limits on the actual binding and release rates. Here again the life time of the complex and a calculated quadrupolar constant of 1.7 MHz are in agreement with site-binding of the sodium ion on the antibiotic.

REFERENCES

(1) P. Laszlo, Angew. Chem., 90, 271 (1978); Angew. Chem. Int.
    Ed. Engl., 17, 254 (1978).
(2) T.E. Bull, J. Magn. Res., 8, 341 (1972).
    T.E. Bull, J. Andrasko, E. Chiancone, S. Forsén, J. Mol.
    Biol., 73, 2581 (1973).
(3) E. Schori, J. Jagur-Grodzinski, Z. Luz, M. Shporer, J. Am.
    Chem. Soc., 93, 7133 (1971).
    E. Schori, J. Jagur-Grodzinski, M. Shporer, J. Am.Chem.
    Soc., 95, 3842 (1973).
(4) H. Degani, Biophys. Chem., 6, 345 (1977).
(5) J. Grandjean, P. Laszlo, Ch. Gerday, F.E.B.S. Lett., 81,
    376 (1977).
    J. Grandjean, P. Laszlo in "Protons and Ions involved in
    Fast Dynamics Phenomena", P. Laszlo, Ed., Elsevier,
    Amsterdam, p. 373 (1978).
(6) A. Cornélis, P. Laszlo, Biochem, in press.

THE REACTIVITY OF TRYPTOPHAN RESIDUES IN PROTEINS AS STUDIED BY THE FLUORESCENCE STOPPED-FLOW TECHNIQUE.

B.F. Peterman and K.J. Laidler

Department of Chemistry, University of Ottawa, Ottawa
Ontario K1N 9B4, Canada.

Protein residues of a given type do not show the same
reactivity, and in many cases higher reactivity can be related
to greater accessibility of the residue in the bulk aqueous
environment. The reactivity and distribution of tryptophan
residues are conveniently studied by measuring the quenching of
the fluorescence. In the present investigation the quenching
of the tryptophan fluorescence by N-bromosuccinamide (NBS) was
used to obtain information about the accessibility of tryptophan
residues in apocytochrome c and α-chymotrypsin. Details of the
work are in course of publication (Peterman and Laidler, 1979).

With excitation at 280 nm the fluorescence of the indole
ring in tryptophan is emitted with a maximum near 350 nm, and we
followed the quenching of this fluorescence on reaction with NBS.
The kinetic measurements were carried out using a Durrum stopped-
flow instrument equipped with a 75 W Xenon-mercury lamp, and
measurements were made of absorbance and of fluorescence.

REACTION BETWEEN NBS AND N-AcTrpNH$_2$

We first studied a model compound, N-acetyl-L-tryptophana-
mide (N-AcTrpNH$_2$). In sodium phosphate buffer, at pH 7.0, a
temperature of 23°C and an ionic strength of 0.05 M, 1.1 $\pm$ 0.1
mol of NBS reacts with 1 mol of N-AcTrpNH$_2$. With a large excess
of NBS the fluorescence of N-AcTrpNH$_2$ decreased exponentially
with time. Rate measurements at various NBS concentrations
showed that the first-order rate coefficients are linear in the
NBS concentration (Figure 1), and from the slope the second-order
rate constant for the quenching process was found to be

W. J. Gettins and E. Wyn-Jones (Eds.), Techniques and Applications of Fast Reactions in Solution. 419–422.
Copyright © 1979 by D. Reidel Publishing Company.

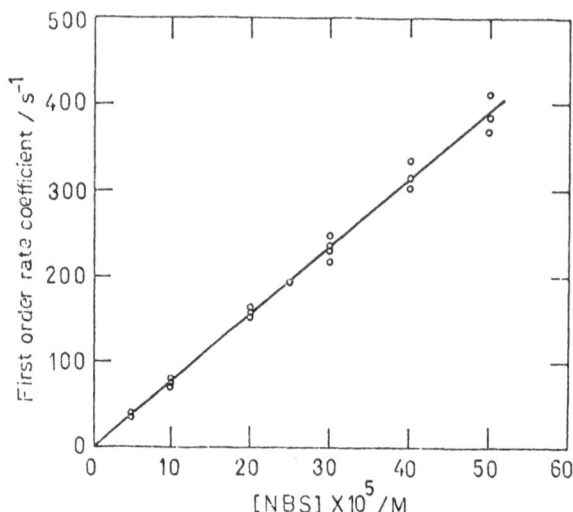

Figure 1. First-order rate coefficients plotted as a function
of NBS concentration. The NBS concentration was 5.0×10^{-6}M.
Ionic strength = 0.05 M, pH = 6.0, T = 23.0°C.

$(7.8 \pm 0.8) \times 10^5$ dm^3 mol^{-1} s^{-1}. There was no ionic strength
dependence in the range from 0.01 to 0.2 M. Near neutral pH the
rate is practically independent of pH, but there is a strong
increase when the pH is below 5.

The non-fluorescent intermediate formed in the quenching
reaction appears to be a bromohydrin compound.

REACTION BETWEEN NBS AND Gly-Trp-Gly

A study with the tripeptide Gly-Trp-Gly yielded similar
results, the second-order rate constant being $(8.8 \pm 0.8) \times 10^5$
dm^3 mol^{-1} s^{-1}.

REACTION BETWEEN NBS AND APOCYTOCHROME c

A molecule of horse-heart cytochrome c contains only one
tryptophan residue. When the heme group is removed, to form apocytochrome c, the polypeptide chain is unable to assume a globular
conformation, and exists as a random coil, with the tryptophan
largely exposed to the solvent (Stellwagen et al., 1972). The

fluorescence quenching by NBS followed a second-order law with a rate constant of $(3.7 \pm 0.3) \times 10^5$ dm^3 mol^{-1} s^{-1}. This small value, compared with that for N-AcTrpNH$_2$, is presumably due at least in part to shielding of the tryptophan residue by the protein matrix. To obtain support for this hypothesis we studied the rate of collisional quenching of apocytochrome c by acrylamide, and again found the rate to be about one half that for N-AcTrpNH$_2$.

REACTION BETWEEN NBS AND α-CHYMOTRYPSIN

The X-ray analysis of α-chymotrypsin (Birktoft and Blow, 1972) shows that the molecule has six tryptophan residues at the surface and two buried in the interior of the molecule. Three of the six surface residues have the indole rings protruding out of the molecule, and the remaining three are pointing inwards.

Our study of the fluorescence quenching by NBS showed that the decay involved three exponential terms (Figure 2). The three second-order rate constants and the corresponding relative amplitudes (normalized to a total of 8) are

(1) 1.74×10^5 dm^3 mol^{-1} s^{-1} 2.8 ± 0.3
(2) 0.56×10^5 dm^3 mol^{-1} s^{-1} 3.0 ± 0.3
(3) 0.11×10^5 dm^3 mol^{-1} s^{-1} 2.2 ± 0.2

The amplitudes are consistent with the 3:3:2 ratio found in the X-ray studies. Rate constant (1) presumably relates to the three surface tryptophans which have their indole rings protruding out of the molecule and which have the highest reactivity with NBS. Rate constant (2) is attributed to the three surface tryptophans whose indole rings are pointing inwards and which therefore have intermediate reactivity. Rate constant (3) is presumably associated with the two tryptophan molecules which are buried in the molecule and which are least reactive.

These conclusions are supported by the results of some very preliminary measurements we have made of the activation energies for the three processes. The activation energy for reaction (1) is slightly but significantly less than for (2), which is less than that for (3). The entropies of activation fall in the same order. These temperature studies are being extended and will be published in due course.

CONCLUSIONS

This work has shown that the fluorescence quenching of tryptophan residues is a useful tool for studying the availabi-

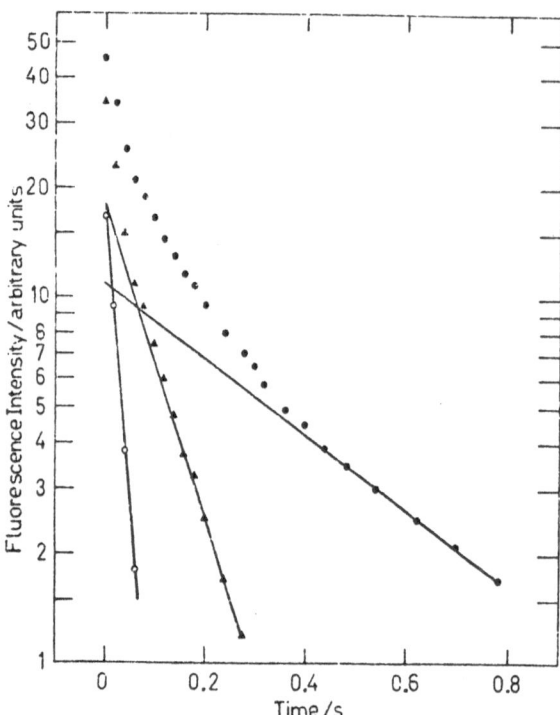

Figure 2. A plot for the quenching of α-chymotrypsin fluorescence, showing the three exponential terms, corresponding to
first-order rate coefficients of 31 s^{-1}, 9.8 s^{-1} and 2.4 s^{-1}.

lity of these residues and relating it to the position of the
residues in the protein matrix. An important advantage of the
technique is that very short times(< 100 ms) are involved, so
that conformational changes are unlikely.

We are extending this work to a number of other peptides
and proteins.

REFERENCES

Birktoft, J.J. and Blow, D.M. (1972), J. Mol. Biol. **68**, 187.

Peterman, B.F., and Laidler, K.J. (1979), Biochim. Biophys.
Acta, in press.

Stellwagen, E., Rysary, R. and Babul, G. (1972), J. Biol.
Chem., **247**, 8074.

ANAEROBIC STOPPED-FLOW AND QUENCHED-FLOW STUDIES ON NITROGENASE.
THE COUPLING OF MgATP HYDROLYSIS TO PROTEIN-PROTEIN ELECTRON
TRANSFER AND THE DETECTION OF A DINITROGEN-HYDRIDE INTERMEDIATE
IN N_2 REDUCTION.

Roger N.F. Thorneley, Robert R. Eady and David J. Lowe

Agricultural Research Council, Unit of Nitrogen Fixation,
University of Sussex, Brighton, BN1 9RQ, U.K.

Nitrogenase catalyses the six electron reduction of
dinitrogen (N_2) to ammonia and is a key enzyme in the nitrogen
cycle. The enzyme from Klebsiella pneumoniae comprises two
redox proteins. The Fe-protein (mol.wt. 68,000 daltons) contains
a single Fe_4S_4 cluster and acts as an electron donor to the Mo-Fe
protein (mol.wt. 218,000 daltons) which contains 2 Mo atoms and
about 32 Fe atoms, some of which are present as Fe_4S_4 clusters.
Substrate reduction requires both proteins and is obligatory
coupled to the hydrolysis of MgATP to MgADP and P_i. Stopped-flow
spectrophotometry at 420 nm conveniently monitors the electron
transfer from the Fe-protein (Kp2) to the Mo-Fe protein (Kp1).
The first order rate constant at 10°C ($k_{obs} = 24$ s^{-1}) for this
electron transfer reaction is independent of both protein
concentrations (5-50 µM) and is essentially identical to the
first order rate constant for a pre-steady-state 'burst' of P_i
formation measured with the quenched flow technique (Fig. 1). It
is concluded that MgATP not only induces the electron transfer
from Kp2 to Kp1 but that it is also hydrolysed to MgADP + P_i in
this partial reaction of the catalytic cycle.

A five fold molar excess of Kp2 over Kp1 protein was used
since this gave a maximum amplitude for the increase in
extinction at 420 nm due to oxidation of Kp2 by a fixed
concentration of Kp1. The amplitude of the 'burst' phase for P_i
formation indicates that 2.3 mole equivalent of MgATP are
hydrolysed per mole of Kp1 reduced under these conditions.
Uncertainties with regard to the percentage of the Kp1 and Kp2
proteins which are active as electron acceptors and donors
respectively, the stoichiometry of the active protein complex
between Kp1 and Kp2 and the number of electrons transferred by

W. J. Gettins and E. Wyn-Jones (Eds.). Techniques and Applications of Fast Reactions in Solution. 423–427.
Copyright © 1979 by D. Reidel Publishing Company.

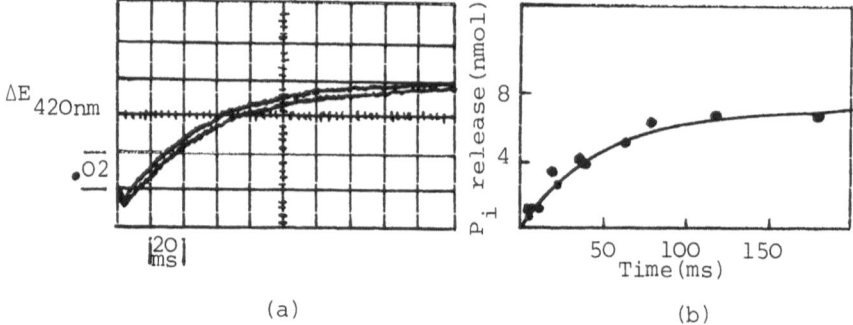

(a)   (b)

Figure 1. Correlation between electron transfer from
the Fe-protein (Kp2 to the Mo-Fe protein (Kp1) and
release of phosphate from ATP for nitrogenase of
Klebsiella pneumoniae
Kp1 (10 µM) at 10°C in pH 7.4 25 mM-Hepes [4-(2-
hydroxyethyl)-1-piperazine-ethanesulphonic acid]/NaOH
buffer in the presence of $Na_2S_2O_4$ (10 mM), ATP (5 mM)
and $MgCl_2$ (10 mM).
(a) Stopped-flow oscillograph of the absorbance change
at 420 nm, time constant 42±3 ms.
(b) time course of phosphate release. The curve is
the best fit exponential, time constant 44±4 ms (Eady,
Lowe and Thorneley 1978).

Kp2 makes it difficult to calculate the number of electrons
transferred per molecule of MgATP hydrolysed. This ratio, as
calculated from steady state kinetic data, is between 12 and 14
moles of MgATP hydrolysed per mole of N_2 reduced. If the
electron transfer from Kp2 to Kp1 is the only step where MgATP
is hydrolysed, then a catalytic cycle involving single electron
transfers from Kp2 to Kp1, each coupled to the hydrolysis of 2
mole equivalents of MgATP, would be consistent with these data.

A detailed discussion of the distribution of electrons over
the Fe-S and Mo centres of Kp1 under various conditions is
beyond the scope of this short article and the reader should
consult Lowe, Eady and Thorneley (1978) for the most recent data
and a literature review. However, at some stage in the catalytic
cycle, electrons and protons are accepted by N_2 to give NH_3. The
chemistry of certain Mo-complexes suggests an enzymic mechanism
involving the stepwise addition of electrons and protons to N_2
coordinated to Mo to yield NH_3 via partially-reduced dinitrogen-
hydride species (Thorneley et al. 1979).

Evidence for a partially reduced dinitrogen hydride intermediate which remains bound to nitrogenase has been obtained using
the quenching agent p-dimethylaminobenzaldehyde (PDMAB) Fig. 2.

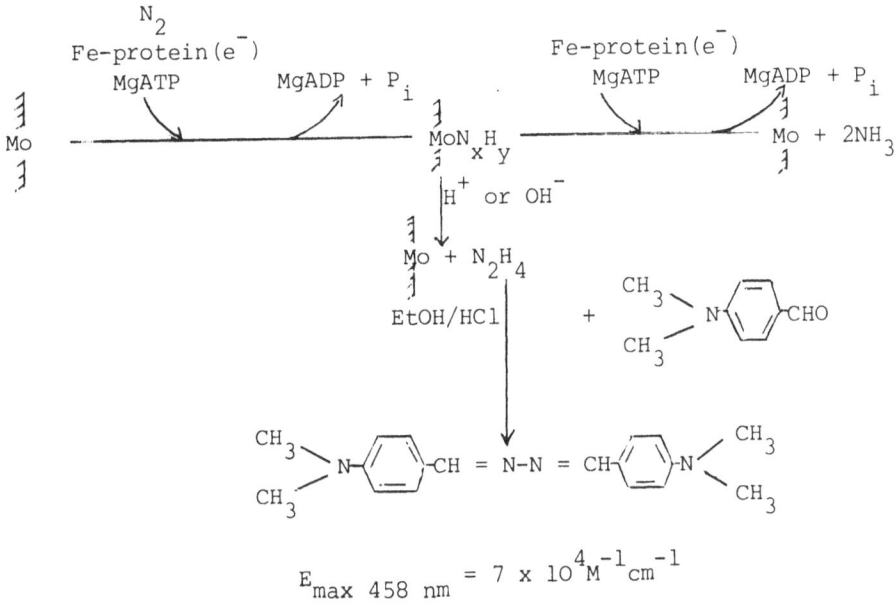

Figure 2. Trapping of a dinitrogen hydride intermediate by a
          quenched flow technique used by Thorneley, Eady and
          Lowe (1978)

The technique relies upon the high sensitivity and specificity
of PDMAB for the detection of hydrazine (N_2H_4). The correlation
between the concentration of intermediate and the rate of NH_3
production is shown in Figure 3. The maximum concentration of
intermediate is reached with a time constant of 3s, which is close
to the turnover time for the enzyme under these conditions. The
concentration of intermediate decreases at long times due to the
inhibition of nitrogenase activity by MgADP. The concentration
of the intermediate also depends on the degree of saturation of
the enzyme with N_2 and the concentration of inhibitors of N_2
reduction such as C_2H_2 and CO. Since the maximum concentration
of the intermediate is only 10% of the total Kp1 protein present,
the rate limiting step in N_2 reduction is unlikely to be the
further reduction of this intermediate.

The chemistry of W and Mo dinitrogen complexes (Chatt et al.
1977) suggests that enzyme-bound dinitrogen-hydrides other than
hydrazine (N_2H_4), could undergo non-enzymic hydrolysis on alkali
or acid quenching of the functioning enzyme to give the N_2H_4
which is detected as its PDMAB derivative. The hydrazido(2-)
species (=N-NH_2) is a particularly attractive alternative to
coordinate hydrazine (-NH_2-NH_2) since in a W complex containing
this ligand about equal yields of N_2H_4 are obtained after
treatment with acid or alkali. This is also observed with the

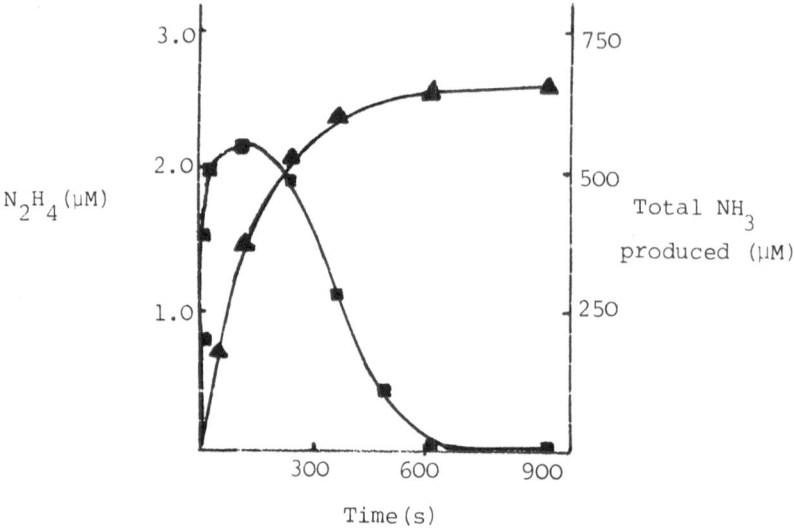

Figure 3. Time course for concentration of N_2H_4 (■) and NH_3 (▲) produced during catalytic reduction of N_2 by K.pneumoniae nitrogenase at pH 7.4 and 30°C. Assays (0.55 ml) contained Mo-Fe protein 20 µM, Fe protein 18 µM, $MgCl_2$ 20 mM, ATP 18 mM, creatine phosphate 18 mM, $Na_2S_2O_4$ 27 mM, creatine kinase 50 µg, and HEPES 25 mM, N_2H_4 was determined in assays quenched with 2 ml ethanol containing HCl 1 M and para-dimethylaminobenzaldehyde 0.07 M. After centrifuging to remove precipitated protein, the N_2H_4 concentration was determined spectrophotometrically at 458 nm in a 4-cm path length cell using an experimentally determined calibration curve. The points at times less than 1 min were obtained with a rapid-quench apparatus when 0.4 ml of mixed solutions were shot into 2 ml para-dimethylaminobenzaldehyde solution (Thorneley, Eady & Lowe 1978).

the functioning enzyme but not with the unprotonated dinitrogen complexes of Mo or W. Hydrazine (N_2H_4) is a poor substrate for nitrogenase, with low affinity ($K_m \sim 15$ mM), hence at the concentrations of intermediate detected in the enzyme system (1-2 µM), N_2H_4 might be expected to dissociate from the enzyme and accumulate in solution. This is not observed and for this reason N_2H_4 bound to the enzyme is not thought to be the source of the N_2H_4 trapped as its PDMAB derivative after quenching.

The reduction cycle in Fig. 4 is consistent with the data from quenching functioning nitrogenase and the chemistry of known complexes of Mo and W with dinitrogen and its hydrides. It involves the stepwise addition of electrons and protons to N_2 coordinated to Mo, causing a progressive degradation of the triple bond of dinitrogen. A concomitant increase in bond order between Mo and the α-nitrogen atom leads to a nitride intermediate which is subsequently hydrolysed to give the second molecule of NH_3.

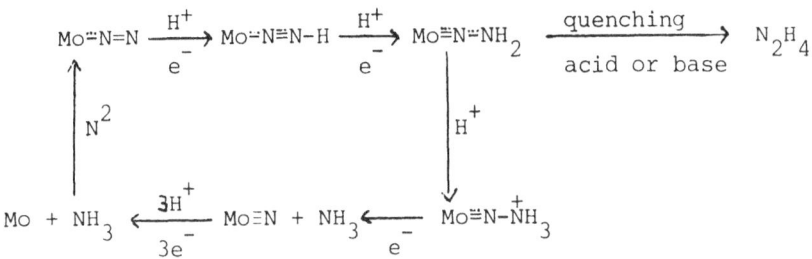

Figure 4. Possible reduction cycle of dinitrogen on molybdenum in nitrogenase (Thorneley et al. 1979)

The first detection of a dinitrogen hydride intermediate in N_2 reduction by nitrogenase and the demonstration of the coupling of electron transfer to MgATP hydrolysis in the same enzyme relied on 'rapid reaction' techniques. Both reactions are important for understanding the mechanism of biological nitrogen fixation. In addition, one of them provides a system of well defined components, amenable to study by a variety of techniques for the investigation of one of the most important problems in biology today, that of energy transduction.

References

Chatt, J., Pearman, A.J. & Richards, R.L. (1977) J.Chem.Soc. (Dalton) 1852-1860.
Eady, R.R., Lowe, D.J. & Thorneley, R.N.F. (1978) FEBS Lett. 95 211-213.
Lowe, D.J., Eady, R.R. & Thorneley, R.N.F. (1978) Biochem.J., 173 277-290.
Thorneley, R.N.F., Eady, R.R. & Lowe, D.J. (1978) Nature, 272 577-558.
Thorneley, R.N.F., Chatt, J.C., Eady, R.R., Lowe, D.J., O'Donnell, M.J., Postgate, J.R., Richards, R.L. & Smith, B.E. (1979) Proceedings of the Steenbock-Kettering International Symposium on Nitrogen Fixation. Edit: W.H. Orme-Johnson and W.E. Newton, University Park Press, Wisconsin.

FAST REACTIONS IN SOLUTION UNDER HIGH PRESSURE : STOPPED-FLOW STUDIES ON SOME BIOLOGICAL REACTIONS

K. Heremans, F. Ceuterick, J. Snauwaert and J. Wauters

Department of Chemistry, Katholieke Universiteit te Leuven, Belgium.

1. INTRODUCTION

A recent review and a book (1,2) on the state of the art in high pressure chemistry, indicate that the use of pressure as an experimental technique for the study of biological reactions is growing. Besides the use of pressure to simulate deep sea conditions (3), the effect of pressure on rates and equilibria is studied to obtain activation and reaction volumes.

The introduction of pressure as a variable in fast reaction studies goes back to Brower who designed a pressure-jump technique for work under pressure. The temperature-jump technique has been adapted by a few research groups. The field has been reviewed recently (4).

The stopped-flow technique proved to be more difficult to adapt. In this paper we describe an instrument which has been used for some time in our laboratory (5). We report results on the formation of intermediates in chymotrypsin (CT), on the formation of the salt bridge in CT and on the alkaline isomerization of cytochrome c (Cyt).

2. DESCRIPTION OF HIGH PRESSURE STOPPED-FLOW

The design of our instrument is based on the simplest approach. The stopped-flow unit, i.e. syringe driving mechanism, mixing chamber, observation chamber and stopping syringe are constructed as a single unit which can be placed into a high pressure vessel. The syringe driving mechanism is a single step motor whose central axis turns 90° when voltage is applied. This rotation is translated to a translation of the syringe plungers. A

W. J. Gettins and E. Wyn-Jones (Eds.), Techniques and Applications of Fast Reactions in Solution. 429–432.
Copyright © 1979 by D. Reidel Publishing Company.

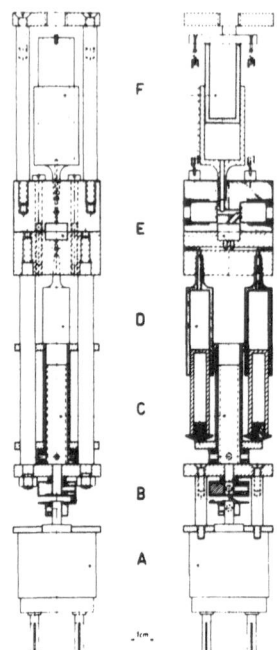

Fig.1 : Outline of stopped-flow mechanism for experiments under pressure. A,B,C : syringe driving mechanism; D : feed syringes; E : mixing chamber; F : waste syringe. The whole unit is placed into a high pressure chamber.

free wheel mechanism disconnects the rotation of the motor from the translation of the syringes when the motor is in the off position. An outline is given in Fig.1. A more detailed description is given elsewhere (5). The amplifier signal from the optical detection is recorded both on a Tektronix 5D13N storage scope for visual inspection and on a transient recorder (Biomation 802). The digital 1 kbyte memory from the transient recorder is stored on a Racal P 70 tape recorder for later processing on an interfaced HP 9815 A desk-top calculator.

3. THE ACYLATION OF CHYMOTRYSPIN

The following mechanism has been proposed for the CT catalyzed hydrolysis of specific substrates :

$$E + S \overset{k_{23}}{\rightleftharpoons} ES \underset{\searrow}{\overset{k_{34}}{\rightarrow}} EP_2 \rightarrow E + P_2$$
$$P_1$$

where ES represents the Michaelis complex, EP_2 the acylenzyme intermediate, P_1 and P_2 are products i.e. alcohol and acid respectively. In the present experiments N-acetyl leucine methylester was used as a substrate. Proflavin was used to detect the formation of the acetylenzyme. We have previously shown that the binding of proflavin to CT is pressure independent (6). Using the method of Hess et al. (7) and assuming that the binding of the ester to CT is pressure independent, we find that $\Delta V_{23}^{\#} = -7$ ml. Neumann et al. (8) recently obtained negative activation volumes for the formation of the tetrahedral intermediate in the base catalyzed hydrolysis of esters.

4. FORMATION OF THE SALT BRIDGE IN CHYMOTRYSPIN

Fersht (9) has shown that CT exist in solution as an equilibrium between active and inactive enzyme. The active molecule contains an intact salt bridge close to the substrate binding site. Alkaline pH destabilizes the salt bridge. Proflavin, an active site inhibitor of CT, binds only to the active molecule. The rate of reconstitution of the salt bridge is in the stopped-flow range. From the amplitude the equilibrium constant of the process can be obtained. The system can be described by two coupled equilibria

$$E_i \rightleftharpoons E_a \stackrel{P}{\rightleftharpoons} E_a P$$

This mechanism has two relaxation times. The binding of proflavin is too fast to be resolved by stopped-flow. The slow relaxation time is coupled to the fast equilibrium. The kinetics of the reconstitution is measured by pH-jump. From the pressure dependence of the rate constant we find $\Delta V^{\#} = 23$ ml for the formation of the salt bridge. From the change in amplitude with pressure we obtain $\Delta V = 32$ ml. It is difficult to isolate the contribution to ΔV from the salt bridge and the change in conformation. However if the environment is predominantly hydrophobic the ΔV is largely due to the salt bridge as a consequence of the lower dielectric constant of the environment.

The present results must be taken into account when measuring steady-state activity of CT under pressure. They also explain our previous findings (6) that the binding of proflavin to CT is pressure independent.

5. THE ALKALINE ISOMERIZATION OF CYTOCHROME C

In a recent review we have summarized our results on the effect of pressure on the redoxreactions of Cyt at neutral pH (4). At alkaline pH, Cyt is converted to a nonreducible form. The kinetics of this process have been studied by Davis et al. (10).

Detailed proposals have been made about the mechanism by Lambeth
et al. (11). Met-80 which is linked to the porphyrin is substituted at alkaline pH by Lys-79. The reducibility is correlated
with the possession of a 695 nm absorption band. At pH 8.3 we
find that pressure restores the 695 band with ΔV = 45 ml, in a-
greement with the finding of Ogunmola et al. (12) at acid pH.
The high spin structure (open crevice) is converted into a low
spin structure (closed crevice) under pressure. The activation
volume for this process is +20 ml. This is characteristic for
a dissociative mechanism (13). A possible model system for this
process has recently been studied (14).

ACKNOWLEDGEMENTS

     The authors thank Professor Leo De Maeyer for encouragement
and support. The Belgian National Science Foundation provided
an instrumentation grant to K.H.

REFERENCES

1.   T.Asano and W.J. le Noble, Chem.Rev., 78, 407 (1978).
2.   H. Kelm (Ed.), High Pressure Chemistry, Reidel, Dordrecht,
     (1978).
3.   A.G. Macdonald, Physiological aspects of deep sea biology,
     C.U.P., Cambridge, 1975.
4.   K. Heremans, Fast reactions in solution, in ref. 2.
5.   K. Heremans, J. Snauwaert and J. Rijckenberg, Proc. 6th Int.
     Conf. High Pressure, Boulder, in press, 1977.
6.   K. Heremans et al., Proc. 4th Int. Conf. High Pressure,
     Kyoto, 623, 1974.
7.   J. McConn et al., J. Biol.Chem., 246, 2918 (1971).
8.   Neuman et al., J.Am.Chem.Soc., 98, 6975 (1976).
9.   A.R. Fersht, J.Mol.Biol., 64, 497 (1972).
10. L.A. Davis et al., J.Biol.Chem., 249 2624 (1974).
11. D.O. Lambeth et al., J.Biol.Chem. 248, 8130 (1973).
12. G.B. Ogunmola et al., Proc.Nat.Acad.Sci.USA, 74, 1 (1977).
13. D.A. Palmer and H. Kelm, in ref. 2.
14. T.R. Sullivan et al., J.Chem.Soc.Dalton, 1460 (1977).

KINETIC AND ISOTROPIC STUDIES OF ACID-BASE CATALYSIS IN
PEROXIDASE REACTIONS

H. B. Dunford

Department of Chemistry, University of Alberta,
Edmonton, Alberta, Canada, T6G 2G2

Horseradish peroxidase, HRP, is a remarkably stable enzyme; it
is resistant to denaturation from heat, changes in pH and
photo-oxidation.[1] With the aid of pH jump experiments, its
reactions can be studied over the pH range of 3.0 to 11.5.
A comprehensive transient-state kinetic study of all of its
reactions in the cycle HRP + H_2O_2 → HRP-I, HRP-I + AH → HRP-II
+ A·, HRP-II + AH → HRP + A· has been undertaken to test for the
possibility of acid-base catalysis.[2] (HRP-I and HRP-II refer
to compounds I and II, which contain two and one oxidizing
equivalents in excess of that of the ferric ion; AH is a reducing substrate and A· is a free radical). In HRP-I an acid
group with pK of 5.1 is important, in HRP-II one of pK 8.6.
Both of these are distal groups, but their identity is unknown.
Depending upon the substrate, either acid or base catalysis
can occur in HRP-I reactions. For HRP-II reactions only acid
catalysis is important: for substrates which are oxidized
with difficulty, protonation of the proximal imidazole group
in the fifth coordination position of the $Fe^{(IV)}$ of HRP-II
is important. (The pK is 0!) The amount of this hyperreactive
form cannot be detected spectroscopically at physiological pH,
but it is the only kinetically active species for inorganic
substrates such as iodide, nitrite or bisulfite.[2]

The formation of HRP-I from the native enzyme and hydrogen peroxide exhibits a more subtle form of prototropy in that
the proton which is transfered to the enzyme may be transfered
back to the substrate in the course of the same reaction.[3]
Because of the complicated nature of the reaction for the formation of compound I, it would appear worthwhile to compare
this reaction for different peroxidases with "simple" ligand

W. J. Gettins and E. Wyn-Jones (Eds.). Techniques and Applications of Fast Reactions in Solution. 433–437.
Copyright © 1979 by D. Reidel Publishing Company.

Table I[a,b,c]

Association rate constants, $M^{-1}s^{-1}$

	H_2O_2	HF	HN_3	HCN
CcP	4.5×10^7	5.1×10^7	–	1.1×10^5
HRP	2.0×10^7	5×10^7	2×10^7	1.1×10^5
TuP$_1$	1.2×10^7	1.4×10^7	–	8.7×10^4
TuP$_7$	1.6×10^6	1.0×10^6	–	1.2×10^4
LP	1.0×10^7	9.7×10^2	–	1.3×10^6
MetMy	1.4×10^2	2.0×10	9×10^5	5×10^2

[a]Abbreviations: for peroxidases, CcP cytochrome c,
HRP horseradish, TuP$_1$ turnip isoenzyme 1, TuP$_7$ turnip
isoenzyme 7, LP lacto-; MetMy metmyoglobin.

[b]References for Tables I and II: CcP 4-7, HRP 8-14,
TuP 15-16, LP 17-19, MetMy 20-23. Original references
should be consulted for details and experimental errors.
Interpretations given here may differ from those in the
literature.

[c]– indicates constants which have not or cannot be determined.

binding reactions which may mimic a part of the overall process
for compound I formation. Such comparisons constitute the
remainder of this paper.

Ligand binding rate constants and rate constants for
compound I formation are listed in Table I along with enzyme
abbreviations. In the case of metmyoglobin, compound II is
formed directly with no compound I formation. Both peroxidase
compounds I[24,25] and II[26] contain one oxygen atom derived
from peroxide which is covalently bonded to iron. The compounds II derived from peroxidases and metmyoglobin appear to
be very similar.[27] It follows therefore that one product of
the reaction of H_2O_2 with metmyoglobin is an OH radical or its
oxidizing equivalent.

It can be seen that yeast and the plant peroxidases
which all contain ferriprotoporphyrin IX behave similarly with
a consistent trend in all results. Lactoperoxidase with a
heme group of unknown structure behaves much differently in
its ligand binding reactions; and metmyoglobin (which also
contains ferriprotoporphyrin IX) cannot be fit into any peroxidase pattern. In all cases for the peroxidases it is the
electrically neutral (protonated) form of the ligand which
reacts (see Appendix). For the yeast and plant peroxidases
the results for H_2O_2, HF and HN_3 indicate that a dissociative

mechanism is dominant, in which loss of water from the sixth coordination position of the ferric ion[28] is rate limiting. In the case of HRP, k_H/k_D data for compound I formation are consistent with a partially rate limiting proton transfer. Hydrocyanic acid does not fit the pattern of the other ligands and its behavior may be rationalized as follows. The H_2O_2 molecule is bent which could facilitate a concerted process involving proton transfer and binding of the OOH^- residue. This implies a large shift in the pK value of H_2O_2 upon bind- ing, which is the essence of acid-base catalysis.[29] The other molecules are linear with acid strengths $HF>HN_3>HCN$. The comb- ination of linearity and weak acidity of HCN may be sufficient to interfere with a dissociative mechanism. The ratio of k_H/k_D is unity within experimental error for HCN binding to HRP which is consistent with the binding process being nearly complete before proton transfer can occur.

The kinetic results, obtained over maximum pH ranges, have been analyzed for pK values of acid groups with a strong influence on the binding rates. Many of these pK values have been confirmed independently. The results are shown in Table II.

The self-consistent trend in pK values for yeast and plant peroxidases may indicate that the same acid groups are present in all enzymes, in which case electronic withdrawing effects which increase the acidities operate in the order $TuP_7>TuP_1$ $>HRP>CcP$. It is suggested that a major difference between the yeast and plant peroxidases and metmyoglobin is that the peroxi- dases have an acid group appropriately positioned to accept the peroxide proton from one oxygen atom and to transfer it back to the other.[8]

Table II[a]

Important Enzyme pK Values

| | Acid | | | Alkaline | | |
	H_2O_2	HF	HCN	H_2O_2	HF	HCN
CcP	5.5	5.5	5.4	–	–	–
HRP	4.0	4.1	4.1	11.0	–	10.8
TuP_1	3.7	3.9	3.9	10.0	–	10.0
TuP_7	2.5	–	–	9.0	8.5	8.5
LP	–	–	6.3	–	–	7.6

[a]See footnotes to Table I.

Appendix

A choice between two mechanisms which otherwise appear indistinguishable may be made if one of them has a rate constant exceeding that for diffusion control.[30] One mechanism may also be favored if it leads to a consistent trend of results for a family of reactions. The prime evidence for a dissociative mechanism is that a family of reactions has approximately the same rate constants. All of the above factors were used in the assignment of peroxide binding reactions to the electrically neutral ligand.

References

1. Y.-J. Kang and J.D. Spikes, Biochem. Biophys. Res. Comm. 74, 1160 (1977).
2. H.B. Dunford and J.S. Stillman, Coord. Chem. Rev. 19, 187 (1976).
3. P. Jones and H.B. Dunford, J. Theor. Biol. 69, 457 (1977).
4. S. Loo and J.E. Erman, Biochemistry 14, 3467 (1975).
5. J.E. Erman, Biochemistry 13, 34 (1974).
6. B. Lent, C.W. Conroy and J.E. Erman, Arch. Biochem. Biophys. 177, 56 (1976).
7. J.E. Erman, Biochemistry 13, 39 (1974).
8. H.B. Dunford, W.D. Hewson and H. Steiner, Can. J. Chem. 56, 2844 (1978).
9. H.B. Dunford and W.D. Hewson, Biochemistry 16, 2949 (1977).
10. D. Dolman, G.A. Newell, M.D. Thurlow and H.B. Dunford, Can. J. Biochem. 53, 495 (1975).
11. W.D. Hewson and H.B. Dunford, Can. J. Chem. 53, 1928 (1975).
12. H.B. Dunford and R.A. Alberty, Biochemistry 6, 447 (1967).
13. W.D. Ellis and H.B. Dunford, Biochemistry 7, 2054 (1968).
14. I. Morishima, S. Ogawa, T. Yonezawa, H. Nakatani, K. Hiromi, T. Sano and T. Yasunaga, Biochem. Biophys. Res. Comm. 83, 724 (1978).
15. D. Job, J. Ricard and H.B. Dunford, Can. J. Biochem. 56, 702 (1978).
16. D. Job and J. Ricard, Arch. Biochem. Biophys. 170, 427 (1975).
17. R.J. Maguire, H.B. Dunford and M. Morrison, Can. J. Biochem. 49, 1165 (1971).
18. R. Segal, H.B. Dunford and M. Morrison, Can. J. Biochem. 46, 1471 (1968).
19. D. Dolman, H.B. Dunford, D.M. Chowdhury and M. Morrison, Biochemistry 7, 3991 (1968).
20. T. Yonetani and H. Schleyer, J. Biol. Chem. 242, 1974 (1967).
21. E. Antonini and M. Brunori, "Hemoglobins and Myoglobins and their Reaction with Ligands," North-Holland, Amsterdam (1971) pp. 215-285.

22. D.E. Goldsack, W.S. Eberlein and R.A. Alberty, J. Biol. Chem. 240, 4312 (1965).
23. D.E. Goldsack, W.S. Eberlein and R.A. Alberty, J. Biol. Chem. 241, 2653 (1966).
24. G.R. Schonbaum and S. Lo, J. Biol. Chem. 247, 3353 (1972).
25. L.P. Hager, D.L. Doubek, R.M. Silverstein, J.H. Hargis and J.C. Martin, J. Am. Chem. Soc. 94, 4364 (1972).
26. M.L. Cotton, H.B. Dunford and J.M.T. Raycheba, Can. J. Biochem. 51, 627 (1973).
27. P. George, Advan. Catal. 4, 367 (1952).
28. T.L. Poulos, S.T. Freer, R.A. Alden, N.H. Xuong, S.L. Edwards, R.C. Hamlin and J. Kraut, J. Biol. Chem. 253, 3730 (1978).
29. W.P. Jencks, Accts. Chem. Res. 9, 425 (1976).
30. H.B. Dunford, J. Theor. Biol. 46, 467 (1974).

TEMPERATURE AND PRESSURE RELAXATION STUDIES OF PROTEIN ASSEMBLY : TUBULIN AND GLUTAMATE DEHYDROGENASE

Yves Engelborghs

Katholieke Universiteit Leuven, Laboratorium voor Chemische en Biologische Dynamica, Celestijnenlaan 200D B-3030 Heverlee, Belgium

Tubulin is a protein (Mw = 110.000) present in all eucariotic cells, where it forms microtubules. These are tubular structures of 25 nm diameter and variable length. Microtubules form together with other biopolymers e.g. actin, the cytoskeleton. They are also involved in cell motility and several transport mechanisms e.g. the formation of the spindle figure for transport of chromosomes, the transport of neuronal vesicles, secretory granules etc. Their presence in the cell is strictly controlled in time and space (1).

Microtubules can be formed in vitro, provided several conditions are met, such as the presence of GTP. High tubulin concentrations are necessary unless glycerol (4M), DMSO (10%) or protein factors are present. The following experiments are done in the presence of these factors which copurify with tubulin from pig brains. The polymerisation equilibrium is very temperature and pressure sensitive (2). At 4°C no microtubules are formed but only small oligomers with a ring structure.

The kinetics of tubulin assembly are studied with temperature jump experiments, measuring turbidity at 350 nm. It was previously shown that turbidity or light scattering is a good measure of the weight concentration of the microtubules (3). The experiments are performed in a slow-T-jump cell, specially designed for high light throughput in the Cary 118 spectrophotometer (Fig. 1A). Changing temperature is done with a single stopcock, constructed from two polished glass disks (Fig. 1B). The time constant of the cell is 0.4/s.

A T-jump from 4 to 25°C results in a sigmoidal increase in turbidity. A subsequent jump to 30 or 35°C results in a single first order relaxation process. The same relaxation process can also be observed by the addition of a small amount of unpolyme-

W. J. Gettins and E. Wyn-Jones (Eds.), Techniques and Applications of Fast Reactions in Solution. 439–442.
Copyright © 1979 *by D. Reidel Publishing Company.*

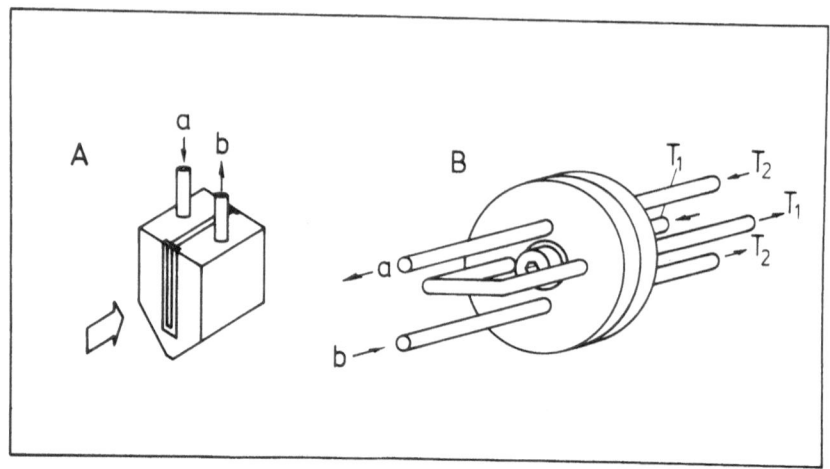

Fig.1.   A) Slow-T-jump cell used in Cary 118. The optical path
is 2 cm and the width is 2 mm. The heat exchanging wall is a
gold plate of 0.5 mm. B) 2 x 4 way stopcock. Turning one half
by 90° allows exchange of termostat fluids.

rised tubulin to preformed microtubules. A quantitative analysis
reveals that the reciprocal relaxation time depends linearly on
the concentration of microtubules, while the amplitude is determined by the change in free monomer concentration (Fig. 2A).The
following mechanism can be proposed :

$$T_i + T \rightleftharpoons T_{i+1}$$

$$- d[T]/dt = k_+ m [T] - k_- m \longrightarrow k_+ m = \tau^{-1}$$

With k_+, k_- the on and off rate constants of growth and m the
concentration of microtubules. The contribution of nucleation
is negligible. The rate constants are determined with m calculated from the average length, as obtained from electron microsco-
py (4).
      Here we want to compare the assembly kinetics of tubulin
with the results of Glutamate Dehydrogenase (GDH), as studied
by the P-jump technique. The instrument is described by Davis
and Gutfreund (5). It is optimized for light scattering measurements by the introduction of diaphragms to reduce reflections,
and corrected for lamp fluctuations (200 W XHg lamp, Hanovia)
by measuring the ratio of scattered to transmitted light, after
attenuation of the latter with two polarizers. The results with
GDH are in agreement with the mechanism proposed by Thusius et
al. (6). The concentration dependence of the square of the reciprocal relaxation time is shown in Fig. 2B.

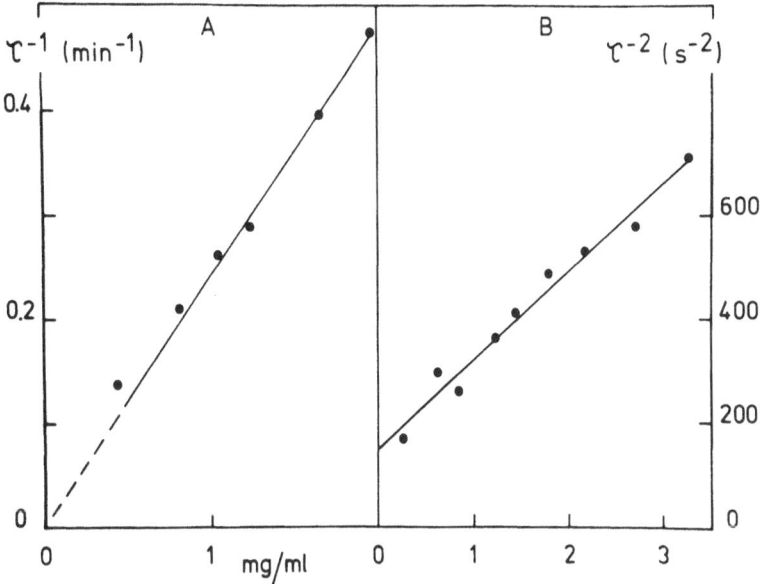

Fig. 2. A) Concentration dependence of the reciprocal relaxation time for growth of microtubules. B) Concentration dependence of the square of the reciprocal relaxation time for GDH.

The mechanism of open polymerisation can be proposed and the rate constants can be determined from the quadratic plot (6) :

$$\tau^{-2} = k_-^2 + 4k_+k_- C_{GDH}$$

It is interesting to note that for tubulin as for GDH, both k_+ and k_- are of comparable magnitude :

	k_+	k_-
Tubulin (25°C)	$3 \times 10^6 \ M^{-1} s^{-1}$	$5 \ s^{-1}$
GDH (15°C)	$1 \times 10^6 \ M^{-1} s^{-1}$	$12 \ s^{-1}$

However, the relaxation process occurs in the min range for tubulin and in the ms range for GDH. This is due to the presence of the nucleation process in the case of tubulin, which drastically reduces the concentration of growth sites.

GDH assembly was also studied at temperature lower than 5°C, where negative activation energy was found for k_+ (To be published elsewhere).

Physicochemical and electron microscopical studies (7) show that depolymerised microtubules at 4°C consist of tubulin monoers as well as ringlike oligomers (36S), formed by the association of tubulin and the protein factors. Several types of ring oligomers are found depending on the buffer conditions used. This oligomer formation is presently under study with the P-jump technique, as described above. Multiple relaxation processes are found revealing a complex mechanism with several intermediates.

Acknowledgement

Y. E. is Bevoegdverklaard Navorser of the Belgian National Science foundation (N.F.W.O.) and wishes to thank this organization for financial support, and Prof. L. De Maeyer for valuable discussions.

References

(1) Olmsted, J.B. and Borisy, G.G.
Ann. Rev. Biochem. 42 (1973) 507 - 540

(2) Engelborghs, Y., Heremans, K., De Maeyer, L., and Hoebeke, J.
Nature 259 (1976) 686 - 689

(3) Gaskin, F., Cantor, C.R., and Shelanski, M.L.
J. Mol. Biol. 89 (1974) 737 - 758

(4) Engelborghs, Y., Overbergh, N., and De Maeyer, L.
FEBS Letters 80 (1977) 81 - 85

(5) Davis, J. and Gutfreund, H.
FEBS Letters 72 (1976) 199 - 207

(6) Thusius, D., Dessen, P., and Jallon, J.M.
J. Mol. Biol. 92 (1975) 413 - 432

(7) Marcum, J.M. and Borisy, G.G.
J. Biol. Chem. 253 (1978) 2825 - 2857

ULTRASONIC ABSORPTION DETECTION OF STRUCTURAL FLUCTUATIONS IN
VIRAL CAPSIDS. A DISCUSSION OF THE PRESENTLY AVAILABLE EVIDENCE.

Roger Cerf

Laboratoire d'Acoustique Moléculaire (ERA CNRS),
Université Louis Pasteur, 4 rue Blaise Pascal,
67070 Strasbourg, France

In collaboration with the Virology Laboratory of the Institut
de Biologie Moléculaire et Cellulaire in Strasbourg, the Laboratory of Molecular Acoustics has studied the ultrasonic absorption
of two small icosahedral viruses, bromegrass mosaic virus (BMV)
and tomato bushy stunt virus (TBSV). Measurements carried out
between 0.6 and 40 MHz showed that ultrasonic waves in the MHz
range are absorbed much more by capsids than by dissociated protein. Examples of ultrasonic spectra of solutions of BMV protein
capsids and of subunit-dimers are shown in Fig. 1 (at 5 mg/ml in
10 mM Na cacodylate, 1.00 M NaCl, at 23°C ; capsids : •, pH = 4.80 ;
dimers : o, pH = 6.60). Spectra obtained for BMV (Δ, pH = 5.00,
at the same particle concentration as for the BMV capsids, otherwise similar conditions) and for the pure solvent (——) are also
shown.

We have previously suggested that the excess absorption is
due to movements that are characteristic of the assembled system
and consist of cooperative structural changes involving either a
few protein molecules or else the whole shell (1,2). The main
argument in favor of this interpretation is that the contribution
from protein-solvent interactions to the absorption should vary
according to the surface of protein exposed and should therefore
decrease when the protein dimers assemble into capsids or into
virions. Since an excess of absorption was observed, there must
have been a relaxation process that existed only in the assembled
particle.

Subsequent work aimed to test this proposal. Two results
(A and B below), which will be published in detail (3), confirm
the preceding interpretation and will be utilized in the following

443

W. J. Gettins and E. Wyn-Jones (Eds.). Techniques and Applications of Fast Reactions in Solution. 443–445.
Copyright © 1979 by D. Reidel Publishing Company.

discussion, in which questions raised at the Fast Reaction
Meeting are answered.

A. Swollen virions absorbed ultrasound less than compact
virions. BMV was swollen by dialysis against 50 mM sodium cacodylate buffer, pH 7.4. Swelling the virion reduced to less than
half the amount by which the absorption by the virion exceeded
the absorption by the protein-dimer.

B. Crosslinking the protein in TBSV with glutaraldehyde
diminished the absorption. (BMV, at the particle concentration
used for these experiments, could not be crosslinked).

DISCUSSION

These results show that the ultrasonic absorption increases
when subunits become more tightly packed, which supports the proposed assignment of the excess absorption to structural fluctuations that exist only in the assembled particles. We consider
this explanation more likely than two others.

First, the excess of absorption exhibited by assembled
systems in the frequency range investigated might result simply
from the increase in relaxation frequency due to assembly of the
protein. However, in this case, one would expect the relaxation
frequency to increase monotonically as a function of some characteristic of the assembly, such as its tightness. The absorption,
measured at each frequency, would then vary monotonically in the
following order : swollen virion < virion < crosslinked virion.
What we actually observed was : swollen virion < virion > crosslinked virion.

Second, one might argue that protein-solvent interactions
could be modified by the close approach of protein subunits to
one-another. In this case, excess absorption would still be produced by self-assembly, but would not necessarily reflect movements within the viral particle itself. If the ultrasonic absorption altered when the H_2O was replaced as solvent by D_2O (as
observed in quite different systems in Evan Wyn-Jones Laboratory),
a   change in protein-solvent interactions upon self-assembly
would indeed appear probable. However, measurements carried out
on BMV by Bernard Michels after the Aberystwyth meeting have
shown no change.

In conclusion, a mechanism which does not exist in dissociated protein and consists in structural fluctuations of the
assembled system, appears highly probable. The observed effects
are consistent with a process where relaxation amplitude decreases when the tightness of the protein-assembly increases.
Because the ultrasonic absorption of BMV virions is no greater
than the sum of the contributions of their component protein
shells and RNA, the excess of absorption seems to be ascribable

mainly to the protein shell. Possible biological implications
are discussed in reference (3).

Extending the measurements to lower frequencies should make
it possible to fully separate the contribution of the new mechanism from the absorption due to the dimer-solvent system.

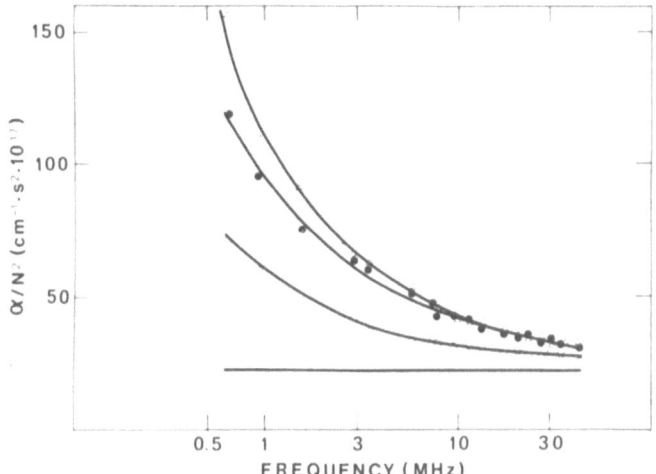

Fig. 1 : Ultrasonic absorption of solutions
of BMV, BMV protein capsids and subunit-dimers.
See text and reference (3).

REFERENCES

1. R. Cerf, (1976), Third Conference of the USSR on Problems of
   Ultrasonic Spectroscopy, Vilnius ; in (1977), *Ultragarsas*, 9,
   39-42.

2. R. Cerf, (1976), 16th Solvay Conference on Chemistry, Brussels ;
   in (1979), *Advances in Chemical Physics*, 39, 242-243.

3. R. Cerf, B. Michels, J.A. Schulz, J. Witz, P. Pfeiffer and
   L. Hirth, (1979), *Proc. Natl. Acad. Sci. U.S.A.*, in the press.

THE MECHANISM OF SOUND ABSORPTION IN SOLUTIONS OF NUCLEOTIDES AND NUCLEOSIDES

Paul Hemmes

Department of Chemistry, Rutgers University, Newark, New Jersey 07102

Abstract: The various mechanisms which can lead to concentration independent relaxation frequencies are discussed. Results are presented of base stacking studies of 6-methylpurine in water urea. It is found that 7M urea does not totally eliminate stacking. The major effect of urea is to decrease the rate of formation of stacks. The most probable mechanism for the concentration independent relaxation at low frequencies in nucleotides is shown to be an anion desolvation.

The biochemical significance of the nucleotides and nucleosides is unquestioned. In view of the increasing use of diagnostic ultrasonics in medicine some interest has developed in the study of the acoustic properties of such molecules. The pH dependence of the acoustic absorption has been studied by Zana and coworkers (1). Our interest has been confined to the study of sound absorption in approximately neutral solutions of these molecules. Sound velocity studies have been of interest also (2)(3)(4) but will not be discussed here.

We can divide the mechanisms for sound absorption by nucleotides and nucleosides into unimolecular processes (those with concentration independent relaxation times) and processes of higher order. Among the latter, the processes which are known to produce ultrasonic relaxations in the megahertz region are stacking, and ion pair formation in nucleotides, and protonation or deprotonation. This last process has been well discussed in reference 1 and references therein. Ion pair formation in nucleotides shows no peculiar features and will not be considered in this discussion. Base stacking has been studied by

W. J. Gettins and E. Wyn-Jones (Eds.). Techniques and Applications of Fast Reactions in Solution. 447–451.
Copyright © 1979 by D. Reidel Publishing Company.

ultrasonic absorption techniques previously (5)(6)(7)(8). Our major
concern was the influence of urea on the stacking process in 6 methyl-
purine. Urea is widely believed to cause destacking. However our
data (Table 1) show that the ultrasonic relaxation persists even in 7M
urea.

Table 1

Concentration of Urea (M)	$\mu max \times 10^4$
0.0	2.03
0.25	1.96
0.5	1.74
1.0	1.55
2.0	1.35
3.0	0.98
5.0	0.77
7.0	0.30

Concentration of 6-me purine is 0.075 M in all cases.

From these results we can conclude that water structure cannot
be the dominant factor in stabilizing stacks. In addition we are now
looking at the kinetics of stacking in 3M urea. Our preliminary
results show that the influence of urea is to decrease the rate of
stack formation and hardly change the rate of dissociation. Once
again this supports the idea of only a small effect of water structure
on stacking.

The unimolecular processes are far more difficult to study. If
we simply list the possible unimolecular processes possible in nucleosides
we have

1. rotations about the glycosidic bond (syn-anti isomerization)
2. ribose ring puckering
3. molecular desolvation processes
4. vibrational relaxations.

In addition in nucleotides we have

5. rotation about the P-O-C bonds
6. rotations about the P-O-P bonds
7. cation desolvation within ion pairs
8. anion desolvation within ion pairs

The syn-anti isomerization was proven for adenosine by Rhodes
and Schimmel (9). Since pyrimidine nucleosides show no relaxation

above 10 MHz and neither do simple sugars, we can rule out process 2 as a common mechanism for ultrasonic relaxations for this type of molecule. For nucleotides the data we have obtained (Table 2) show that no new effects are found in di and tri nucleotides which rule out process 6.

Table 2

Low Frequency Relaxation in the Series of Ribose Phosphate Compounds

Compound	Conc.	f_r(MHz)	$A \times 10^{17}$	$A/m \times 10^{17}$
ATP	0.2M	35	84	420
ATP-Mg	0.1M	35	62	620
CTP-Mg	0.1M	35	62	620
ADP	0.2M	35	35	175
ADP-Mg	0.2M	35	70	350
5'-AMP	0.2M	35	15	75
5'-AMP + Mg^{++}	0.2M	35	80=	400
5"-AMP + Ma^{++}	0.1M	35	--	---
RP + Mg^{++}	0.2M	35	35	175
$P_2O_7^{-2}$ + K^+	0.1M	25		

Furthermore, the independence of the 35 MHz effect on the base and the cation used as counter ion rule out syn-anti and cation desolvations as the mechanism. In order to further characterize the effect we measured ATP in 1 M NaOH solution. In this medium one of the ribose hydroxyls is dissociated producing a negatively charged group in a position where rotational motion would bring the oxygens or the phosphate into proximity. This should certainly increase the energy barrier to that rotation. The effect remains close to 35 MHz, however, indicating that the mechanism is not a rotation such as process 5. The possibility of the process being a vibrational relaxation was considered. Despite a number of interesting correlations between the magnitude of the effect and the number of P-O bonds it is not possible to fit the process to the relaxation of one or two vibrational modes. It may be possible to fit the data with a great many modes but the number of parameters is then so large that such agreement would not be convincing proof of mechanism. It remains to show if a desolvation mechanism of either the molecule or the anion would account for the data. Since CTP and ATP give very similar results, desolvation of the base seems unlikely. Desolvation of the sugar moiety and anion desolvation remain. Since pyrophosphate and triphosphate show similar relaxations in the low megahertz region while uncharged ribose species do not, we feel that phosphate desolvation is the most probable mechanism.

In order to obtain some more support for this hypothesis we have conducted experiments in mixed solvents. This procedure was dictated by the following considerations. For a two step mechanism of the usual Eigen type

$$M^{+2} + L^{-4} \underset{k_{-1}}{\overset{k_1}{\rightleftarrows}} M,L^{-2} \underset{k_{-2}}{\overset{k_2}{\rightleftarrows}} ML + W$$

where

M is the free cation
L is the free anion
M,L is the solvent separated ion pair
ML is the contact pair
W is a solvent molecule

the slower relaxation time is given by the expression (10) $\tau_2^{-1} = k_{-2}(W)^n$ where n is the order of the reaction with respect to solvent and (W) is the activity of the solvent. In pure solvent this term is approximately unity but not in mixed solvents. Therefore by variation of the solvent activity we may be able to prove the mechanism for this process. Of course this procedure presents many dangers. At the moment we have found that the relaxation frequency in ATP solutions decreases sharply with addition of ethylene glycol or ethanol. Further work is being done to confirm the dependence of τ_2^{-1} on the water activity.

Acknowledgement: The author wishes to thank the General Medical Science Institute of the National Institute of Health for financial support of this work

References

(1)   S. Yiv, J. Lang and R. Zana, "Protons and Ions Involved in Fast Dynamic Phenomena", Elsevier Publ. Co. Amsterdam, 1978 p. 33.

(2)   A. P. Sarvazyan, V. A. Buckin and P. Hemmes, submitted for publication.

(3)   V. A. Buckin, A. P. Sarvazyan, I. I. Dudchenko and P. Hemmes, submitted for publication.

(4)   P. Hemmes, V. A. Buckin, A. A. Mayeuski and A. P. Sarvazyan, submitted for publication.

(5)   D. Porschke and F. Eggers, Eur. J. Biochem., 26, 490 (1972).

(6)   F. Garland and R. Patel, J. Phys. Chem., 78, p. 848 (1974).

(7)   F. Garland and S. Christian, J. Phys. Chem., 79, p. 1247 (1975).

(8)   M. P. Heyn, C. Nicola and G. Schwarz, J. Phys. Chem., 81, p. 1611 (1977).

(9)   L. M. Rhodes and P. R. Schimmel, Biochemistry, 10, p. 4426 (1971).

(10) P. Hemmes, J. N. Costanzo and F. Jordan, J. Phys. Chem., 82, p. 387 (1978).

ELEMENTARY STEPS IN THE NUCLEOPHILIC ADDITION OF AMINES
TO OLEFINS

Claude F. Bernasconi, David J. Carré,
and John P. Fox

Thimann Laboratories, University of California,
Santa Cruz, California 95064

Abstract. Nucleophilic addition of amines to olefins which
are activated by electron withdrawing substituents occurs
readily in aqueous dimethylsulfoxide. The reaction comprises
two steps: (1) nucleophilic addition to form a zwitterionic
complex; (2) removal of the ammonio proton of the zwitterion
by a base. In most cases the first step is rate limiting but
in some cases proton transfer is rate limiting. The latter
situation prevails either when the reverse of the nucleophilic
attack step is very rapid, as in the reaction of morpholine with
benzylidenemalononitrile, or when the rate of proton transfer
is depressed by a steric effect, as in the reaction of morpholine with 1,1-dinitro-2,2-diphenylethylene. The steric effects
in this latter system are among the most dramatic ones reported
to date. Our data also show that the kinetic barrier to
nucleophilic attack is substantially higher for nitro than
for cyano activated olefins. This effect seems to be related
to the well known fact that proton transfers involving nitro
activated carbon acids are much slower than those of cyano
activated carbon acids.

INTRODUCTION

We have recently started a research program in which we
intend to explore questions of mechanism and reactivity in
reactions of nucleophiles with activated olefins. Scheme I
shows the typical reaction steps for an amine nucleophile in
an aqueous medium; X and Y are electron withdrawing groups.

W. J. Gettins and E. Wyn-Jones (Eds.), Techniques and Applications of Fast Reactions in Solution. 453–462.
Copyright © 1979 by D. Reidel Publishing Company.

Scheme I

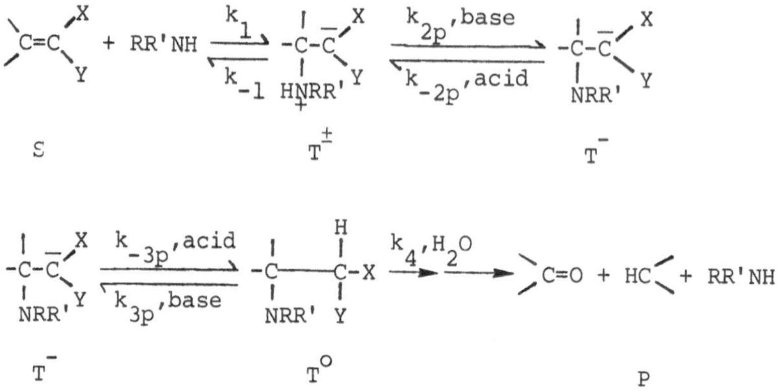

$$\ce{>C=C<^X_Y} + \text{RR'NH} \underset{k_{-1}}{\overset{k_1}{\rightleftharpoons}} \ce{-\underset{H\overset{+}{N}RR'}{\overset{}{\underset{|}{C}}}-\underset{|}{\overset{X}{C}}<} \underset{k_{-2p}, \text{acid}}{\overset{k_{2p}, \text{base}}{\rightleftharpoons}} \ce{-\underset{NRR'}{\overset{}{\underset{|}{C}}}-\underset{|}{\overset{X}{C}}<}$$

S $T^{\pm}$ T^-

$$\ce{-\underset{NRR'}{\overset{}{\underset{|}{C}}}-\underset{|}{\overset{X}{C}}<_Y} \underset{k_{3p}, \text{base}}{\overset{k_{-3p}, \text{acid}}{\rightleftharpoons}} \ce{-\underset{NRR'}{\overset{}{\underset{|}{C}}}-\underset{|}{\overset{H}{C}}-X} \overset{k_4, H_2O}{\longrightarrow} \ce{>C=O} + \ce{HC<} + \text{RR'NH}$$

T^- T^O P

Our aim is to find suitable systems where the individual reaction steps of Scheme I can be studied separately. In as much as some of these steps are also relevant to nucleophilic vinylic substitutions (1), ElcB elimination reactions (2) and proton transfer reactions involving both normal acids (3) and carbon acids (4) we hope to contribute to a better understanding of these reactions as well.

In this paper we summarize some of our most striking findings. A more detailed account will be given elsewhere (5-7).

RESULTS

Some of the olefins which we studied are shown as 1-3. As

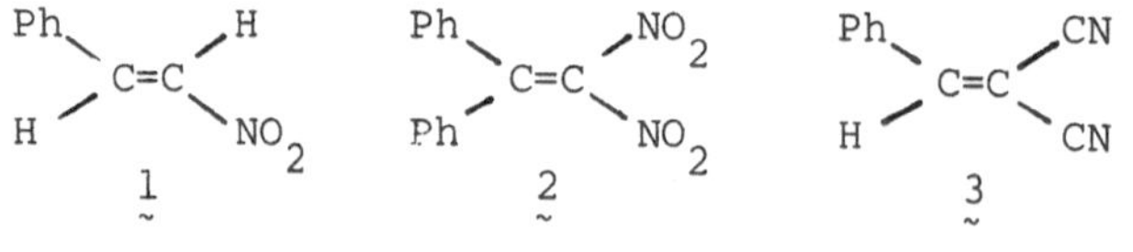

$$\ce{^{Ph}_{H}>C=C<^H_{NO_2}} \qquad \ce{^{Ph}_{Ph}>C=C<^{NO_2}_{NO_2}} \qquad \ce{^{Ph}_{H}>C=C<^{CN}_{CN}}$$
$$\underset{1}{} \qquad\qquad\qquad \underset{2}{} \qquad\qquad\qquad \underset{3}{}$$

nucleophiles we used piperidine, morpholine, n-butylamine and aniline. All reactions were conducted in 50% Me_2SO-50% water (v/v) at 20°C and constant ionic strength of 0.5 M (KCl).

In a typical experiment a solution of the olefin was mixed with an amine solution in the stopped-flow apparatus. By monitoring the reaction at a wavelength where T^- absorbs strongly we observed two relaxation effects with relaxation times τ_1 and τ_2 which are well separated. In the reactions of benzylidenemalononitrile (3) τ_1 turned out to be too short for the stopped-flow method and the temperature-jump method had to be used.

The short relaxation time, τ_1, can be attributed to the first two steps in Scheme 1, eq 1, where by $T^{\pm}$ is a steady

$$S + RR'NH \underset{k_{-1}}{\overset{k_1}{\rightleftharpoons}} T^{\pm} \underset{k_{-2p}}{\overset{k_{2p}}{\rightleftharpoons}} T^- \qquad (1)$$

state intermediate; τ_2 is due to the last two steps in Scheme I with T^O being a steady state intermediate in most cases. In this paper we shall only deal with reaction 1.

All kinetic runs were conducted under pseudo first order conditions ([RR'NH] >> [S]), thus τ_1^{-1} is given by

$$\tau_1^{-1} = \frac{k_1 k_{2p}}{k_{-1} + k_{2p}}[RR'NH] + \frac{k_{-1} k_{-2p}}{k_{-1} + k_{2p}} \qquad (2)$$

where k_{2p} and k_{-2p} refer to proton transfer and are defined as

$$k_{2p} = k_{2p}^W + k_{2p}^{OH}[OH^-] + k_{2p}^A[RR'NH] + k_{2p}^B[B] \qquad (3)$$

$$k_{-2p} = k_{-2p}^H[H^+] + k_{-2p}^W + k_{-2p}^{AH}[RR'NH_2^+] + k_{-2p}^{BH}[BH^+] \qquad (4)$$

with k_{2p}^W, k_{2p}^{OH}, k_{2p}^A and k_{2p}^B being the rate coefficients for the deprotonation of $T^{\pm}$ by the solvent, the hydroxide ion, the amine and another base (buffer) which sometimes was added to the reaction solution, respectively, and k_{-2p}^H, k_{-2p}^W, k_{-2p}^{AH} and k_{-2p}^{BH} being the rate coefficients for the protonation of T^- on nitrogen by the hydronium ion, the solvent, and the conjugate acids of the amine and of B, respectively.

Plots of τ_1^{-1} vs free amine concentration were found to either consist of a series of parallel straight lines as shown for a representative example in Fig. 1, or to be curved as shown for a representative example in Fig. 2. Behavior according to Fig. 1 indicates rapid proton transfer ($k_{2p} >> k_{-1}$) with rate limiting nucleophilic attack, with eq 2 simplifying to

$$\tau_1^{-1} = k_1[RR'NH] + k_{-1}\frac{[H^+]}{K_a^{\pm}} \qquad (5)$$

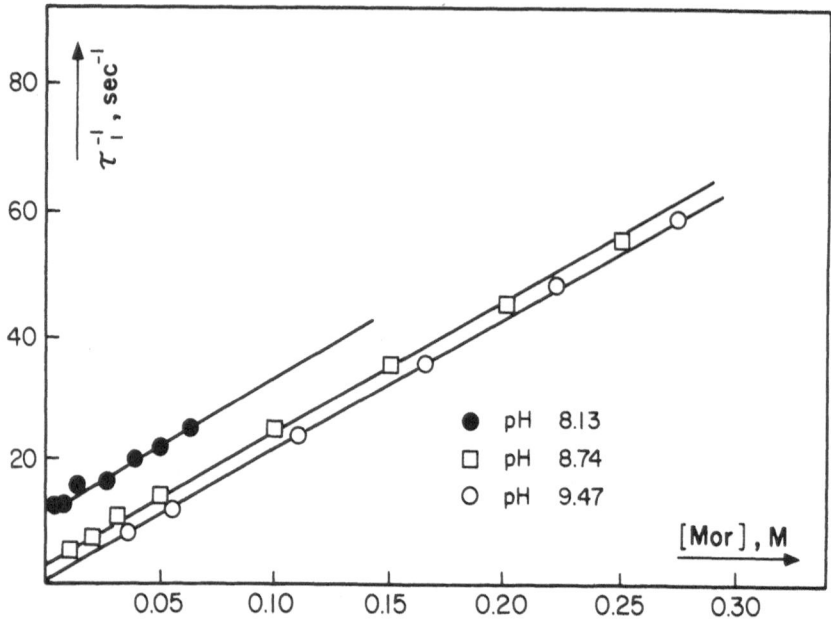

<u>Figure 1</u> τ_1^{-1} for the reaction of morpholine with β-nitro-styrene (1).

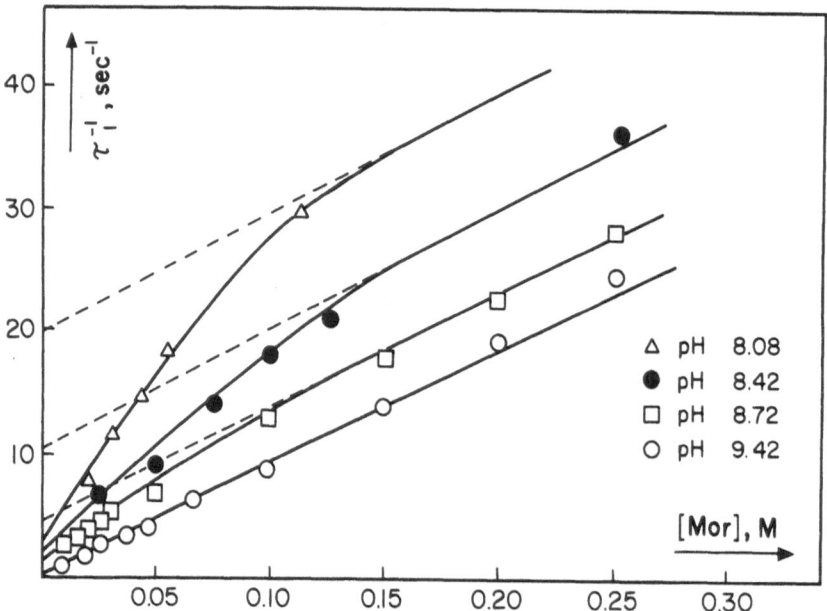

<u>Figure 2</u> τ_1^{-1} for the reaction of morpholine with 1,1-dinitro-2,2-diphenylethylene (2).

where $K_a^{\pm}$ is the acid dissociation constant of $T^{\pm}$. Examples for which this behavior was found are the reactions of 1 with piperidine, morpholine, n-butylamine and aniline (6), the reactions of 2 with piperidine and n-butylamine (5), and the reaction of 3 with piperidine (7).

Behavior according to Fig. 2 shows a transition from rate limiting proton transfer at low amine concentration to rate limiting nucleophilic attack at high amine concentration. This comes about because the $k_{2p}^A[RR'NH]$ term in eq 3 is dominant and thus k_{2p} changes from $k_{2p} < k_{-1}$ at low to $k_{2p} \gg k_{-1}$ at high amine concentration. This behavior is exemplified by the reactions of 2 with morpholine and aniline (5), and by the reaction of 3 with morpholine (7).

Further evidence for our interpretation comes from experiments in the presence of added buffers of low nucleophilic reactivity which catalyze proton transfer ($k_{2p}^B[B]$ and $k_{-2p}^{BH}[BH^+]$ terms in eqs 3 and 4, respectively). For example, in the reaction of 2 with morpholine, addition of p-cyanophenoxide ion, DABCO (1,4-diaza-2,2,2-bicyclooctane) or N-methylmorpholine accelerates the reaction at low morpholine concentration where proton transfer is rate limiting, but has no effect at high morpholine concentration where nucleophilic attack is rate limiting.

Using procedures and performing additional kinetic experiments described elsewhere (5-7) we were able to evaluate numerous rate coefficients referring to reaction 1. Some of them are summarized in Tables I and II.

DISCUSSION

1. Rate Limiting Proton Transfer

Even though nucleophilic attack is rate limiting in the majority of the reactions, it is interesting to discuss the reasons why proton transfer is rate limiting in some cases. For proton transfer to be rate limiting we need $k_{2p} < (\ll) k_{-1}$. This can come about either as a consequence of a very high k_{-1} value, or because of a rather low k_{2p} value. Inspection of Tables I and II reveals that in the reaction of morpholine with 3 ($k_{-1} \approx 4 \times 10^5$ sec^{-1}) and the reactions of aniline with 2 ($k_{-1} \approx 5 \times 10^6$ sec^{-1}) the k_{-1} values are indeed very high, high enough to make $k_{2p} < k_{-1}$ at low base concentrations even if deprotonation of $T^{\pm}$ were diffusion controlled or nearly so.

Table I. Rate and Equilibrium Constants for the Reactions of
          Amines with $\underset{\sim}{1}$, $\underset{\sim}{2}$ and $\underset{\sim}{3}$ in 50% Me$_2$SO-50% water (v/v)
          at 20°.

pK_a^{AH} [a]	Piperidine 11.0	Morpholine 8.72	n-Butylamine 10.65	Aniline 4.25
β-Nitrostyrene ($\underset{\sim}{1}$)				
$k_1 (M^{-1}s^{-1})$	1.14×10^3	2.17×10^2	31	5
$k_{-1} (s^{-1})$	36	10^3	1.25	3.8×10^6
$K_1=k_1/k_{-1} (M^{-1})$	31.8	0.22	24.8	1.3×10^{-6}
$pK_a^{\pm}$	8.30	6.15	8.62	2.2
1,1-Dinitro-2,2-diphenylethylene ($\underset{\sim}{2}$)				
$k_1 (M^{-1}s^{-1})$	6.8	0.95	40	≈ 1
$k_{-1} (s^{-1})$	10^2	2.4×10^3	0.36	5×10^6
$K_1=k_1/k_{-1} (M^{-1})$	6.8×10^{-2}	4×10^{-4}	1.1×10^2	$\approx 2\times10^{-7}$
$pK_a^{\pm}$	6.22	3.94	5.91	≈ -0.5
Benzylidenemalononitrile ($\underset{\sim}{3}$)				
$k_1 (M^{-1}s^{-1})$	1.7×10^5	5×10^4		
$k_{-1} (s^{-1})$	$\approx 10^4$	$\approx 4\times10^5$		
$K_1=k_1/k_{-1} (M^{-1})$	≈ 17	≈ 0.12		
$pK_a^{\pm}$	≈ 10.9	≈ 8.6		

[a] pK_a of RR'NH$_2^+$.

    On the other hand, in the reaction of morpholine with $\underset{\sim}{2}$
k_{-1} = 2.4 × 10^3 sec^{-1} is relatively low; here it is the
unusually low proton transfer rate constants (entries 6-9 in
Table II) which make $k_{2p} < k_{-1}$. The reasons for the slow proton
transfer are discussed below.

Table II. Rate Coefficients for Proton Transfer Between $T^{\pm}$
and T^- in 50% Me_2SO-50% water (v/v) at 20°.

#	Process	Substrate/Amine	Rate Constant	ΔpK^a
1	T^-+H^+	2/morpholine	4.2×10^6	5.38
2	T^-+H^+	2/aniline	$\approx3\times10^6$	0.94
3	$T^{\pm}+RR'NH$	2/aniline	$\approx6\times10^6$	4.79
4	$T^{\pm}+AcO^-$	2/aniline	$\approx7\times10^7$	6.24
5	$T^{\pm}+RR'NH$	3/morpholine	$>>10^6$	≈0.1
6	$T^{\pm}+RR'NH$	2/morpholine	3.8×10^4	4.78
7	$T^{\pm}+p-CN-C_6H_4O^-$	2/morpholine	5×10^5	4.76
8	$T^{\pm}+DABCO$	2/morpholine	1.1×10^4	4.86
9	$T^{\pm}+N$-methylmor.	2/morpholine	2×10^3	3.62

$^a\Delta pK = pK_a$ (Acceptor) $- pK_a$ (Donor)

2. Nucleophilic Reactivity

In the reactions of β-nitrostyrene, 1, the k_1-values follow
the normal order of amine nucleophilic reactivity (8), in particular piperidine >> n-butylamine. This contrasts with the
unusual order found in the reactions of 1,1-dinitro-2,2-diphenyl-
ethylene, 2, in particular piperidine < n-butylamine; this is
undoubtedly due to steric hindrance by the two bulky phenyl
groups in 2.

An even more striking manifestation of the steric effect
is the fact that, despite the much stronger electronic activation of 2, the reactivity of the less activated 1 towards the
two bulky amines piperidine and morpholine is considerably
higher than that of 2. For example, with piperidine we have
$k_1(1)/k_1(2)$ = 168, $k_{-1}(1)/k_{-1}(2)$ = 0.36 and $K_1(1)/K_1(2)$ = 468.
With the less bulky n-butylamine the reactivities of 1 and 2 are
comparable, indicating an approximate compensation of the steric
by the electronic effect: $k_1(1)/k_1(2)$ = 0.79; $k_{-1}(1)/k_{-1}(2)$ =
3.5, and $K_1(1)/K_1(2)$ = 0.225.

Another interesting comparison is that between the reactions of 1 and 3. The equilibrium constant for piperidine
addition to 1, K_1 = 31.8, is almost the same as that for
piperidine addition to 3, $K_1 \approx 17$, but the rate constants k_1
and k_{-1} differ by more than two orders of magnitude (k_1 =

1.14×10^3 M^{-1} s^{-1} for $\underset{\sim}{1}$, 1.75×10^5 M^{-1} s^{-1} for $\underset{\sim}{3}$; $k_{-1} = 36$ s^{-1} for $\underset{\sim}{1}$, 10^4 s^{-1} for $\underset{\sim}{3}$), indicating a higher kinetic barrier in the nitro compared to the cyano compound. These findings are reminiscent of proton transfer involving carbon acids activated by nitro and cyano groups, respectively (4). Nitro activated alkanes are deprotonated very slowly because the negative charge in the anion is being delocalized into the nitro group(s) which necessitates an extensive reorganization of the solvent structure, whereas cyano activated alkanes are deprotonated very rapidly because the negative charge remains essentially localized (4). Since the structural and solvational changes which occur upon nucleophilic addition are very similar to those in the deprotonation of these carbon acids, it appears that the same rationale can be invoked in explaining why there is a higher kinetic barrier towards nucleophilic attack in nitro activated compared to cyano activated olefins.

3. Acidity of $T^{\pm}$ ($pK_a^{\pm}$)

Despite their negative charge the $\overline{C}HNO_2$ and $\overline{C}(NO_2)_2$ moieties are quite strongly electron withdrawing as can be seen from the $pK_a^{\pm}$ values which are 2-2.5 units lower than the pK_a of the parent $RR'NH_2^+$ in the β-nitrostyrene series, and almost 5 units lower in the 1,1-dinitro-2,2-diphenylethylene series (Table I).

In contrast, the $pK_a^{\pm}$ values are about the same as those for the parent $RR'NH_2^+$ in the benzylidenemalanonitrile series. This contrasting behavior between the nitro and cyano derivaties may again be related to the fact that the negative charge is strongly delocalized in the nitro derivatives and hence far removed from the amino group in $T^{\pm}$, while in the cyano derivative the charge resides essentially on carbon and hence can influence the acidity of $T^{\pm}$ by its close proximity to the amino group.

4. Proton Transfer Between $T^{\pm}$ and T^{-}

A selection of proton transfer rate constants is summarized in Table II. Our results for the $\underset{\sim}{2}$/morpholine system are particularly striking. All rate constants refer to thermodynamically strongly favored (ΔpK >> 0) proton transfers between normal acids and bases and thus would be expected to have values in the order of 10^8 to 10^9 M^{-1} s^{-1} ($\sim 10^{10}$ for $T^- + H^+$)(3). However the measured rate constants are enormously depressed, by as much as 10^5 to 10^6 fold in some cases (entries 8 and 9 in Table II). This is another manifestation of the great steric bulkyness of $T^{\pm}$ and T^- which makes the (de)protonation site very inaccessible. The decreasing trend of the rate constants with increasing bulkyness of the base (p-CN-C$_6$H$_4$O$^-$ < morpholine < DABCO < N-methylmorpholine) is consistent with this interpretation (9).

Steric effects on proton transfers between normal acids and bases are not uncommon (3, 4a, 10) although they are usually less dramatic than the ones reported here (11). On the other hand, steric hindrance to protonation of a nitrogen base by the <u>hydronium ion</u> is virtually unheard of except for recent reports of relatively modest rate depressions with 2,6-di-t-butylpyridine (12) and dipicrylamide (13). In this light our low value of 3.6×10^6 M^{-1} s^{-1} for $T^- + H^+ \to T^{\pm}$ in the 2/morpholine system is quite remarkable.

In the reaction of aniline with $\underset{\sim}{2}$ the rates are somewhat depressed but much less so than in the morpholine reaction (entries 3 and 4 in Table II). This is undoubtedly because here T^- is a secondary rather than a tertiary amine which reduces the steric bulk, and the deprotonating bases are also relatively small. The rate constant for the reaction $T^- + H^+ \to T^{\pm}$ seems unexpectedly low, though, but this must be partly due to the small ΔpK in this reaction.

The proton transfer rates in the reactions involving benzylidenemalononitrile could not be measured but lower limits can be given, for example $k_{2p}^A \gg 10^6$ for the reaction $T^{\pm} + RR'NH \to T^- + RR'NH_2^+$. This indicates that proton transfer rates in these systems are normal (3) or nearly so.

REFERENCES AND NOTES

1. Recent reviews: (a) Rappoport, Z.: 1969, Adv. Phys. Org. Chem. 7, p. 1; (b) Modena, G.: 1971, Acc. Chem. Res. 4, p. 73.

2. Recent reviews: (a) McLennan, D. J.: 1967, Quart. Rev. 21, p. 490; (b) More O'Ferrall, R. A.: 1973, in "The Chemistry of the Carbon-Halogen Bond," part 2, Patai, S., Ed., Wiley Interscience, New York, p. 609.

3. Eigen, M.: 1964, Angew. Chem. Int. Ed. Engl. 3, p. 1.

4. (a) Crooks, J. E.: 1975, in "Proton Transfer Reactions," Caldin, E. and Gold, V., Eds., Wiley, New York, p. 153; (b) Hibbert, F.: 1977, Comprehensive Kinetics 8, p. 97.

5. Bernasconi, C. F. and Carré, D. J.: 1979, J. Am. Chem. Soc. 101, in press.

6. Bernasconi, C. F. and Carré, D. J.: in preparation.

7. Bernasconi, C. F. and Fox, J. P.: in preparation.

8.   (a) Hall, H. K.: 1964, J. Org. Chem. 29, p. 3539; (b)
     McDowell, S. T. and Stirling, C. J. M.: 1967, J. Chem.
     Soc. B, p. 343; (c) Jencks, W. P. and Gilchrist, M.:
     1968, J. Am. Chem. Soc. 90, p. 2622; (d) Bernasconi, C. F.:
     1973, MTP Int. Rev. Sci.: Org. Chem., Ser. One 3, p. 33.

9.   Other possible interpretations for the low proton transfer
     rate constants such as intramolecular hydrogen bonding or
     the involvement of the aci-form have been discussed and
     rejected (5).

10. Bernasconi, C. F.: 1978, Acc. Chem. Res. 11, p. 147.

11. An exception is the extremely slow proton transfer involving
     certain proton cryptates, Cheney, J. and Lehn, J. M.: 1972,
     J. Chem. Soc., Chem. Comm., p. 487.

12. Bernasconi, C. F. and Carré, D. J.: 1979, J. Am. Chem. Soc.,
     in press.

13. Strohbusch, F., Marshall, D. B., and Eyring, E. M.: 1979,
     this volume.

THE TRANSITION STATE OF THE ACID CATALYZED ALKOXIDE ION DEPARTURE FROM MEISENHEIMER COMPLEXES (1)

Claude F. Bernasconi and Joseph R. Gandler

Thimann Laboratories, University of California,
Santa Cruz, California 95064

Abstract. Alkoxide ion departure from Meisenheimer complexes of the 1,1-dialkoxy-2,6-dinitro-4-X-cyclohexadienate type is catalyzed by pyridinium ions and by the hydronium ion. Bronsted α values which vary between 0.35 and 0.65 decrease when the alkoxy group becomes more electron withdrawing and increase when the X-substituent is made more electron withdrawing. This is inconsistent with a mechanism where proton transfer is rate limiting but in complete agreement with a concerted mechanism of acid catalysis as shown with the help of More O'Ferrall-Jencks energy diagrams.

The rates of alkoxide ion departure from Meisenheimer complexes such as 1 and 2 were measured spectrophotometrically in aqueous acid, either in the stopped-flow apparatus or in a conventional spectrophotometer. In the presence of pyridine, γ-picoline and nicotineamide buffers the pseudo-first-order rate

$$1 \quad R = CH_3CH_2,\ CH_3,\ CH_3OCH_2CH_2,\ ClCH_2CH_2,$$
$$HC{\equiv}CCH_2$$

463

W. J. Gettins and E. Wyn-Jones (Eds.). Techniques and Applications of Fast Reactions in Solution. 463–468.
Copyright © 1979 by D. Reidel Publishing Company.

$$2 \quad X = CF_3, \; CN, \; SO_2CH_3, \; NO_2, \; SO_2CF_3$$

constant obeys eq 3 where k_o refers to the "water reaction,"

$$k_{obs} = k_o + k_{BH^+}[BH^+] + k_{H^+}[H_3O^+] \qquad (3)$$

while k_{BH^+} and k_{H^+} are the rate constants for the pyridinium (or substituted pyridinium) ion and the hydronium ion catalysis, respectively. The rate constants for general acid catalysis obey the Bronsted relation very well, with k_{H^+} falling on the same line as the k_{BH^+} values. The Bronsted α values are summarized as a function of the R-substituent (eq 1) and the X-substituent (eq 2) in Table I.

Table I. Bronsted α Values for Reactions 1 and 2

R	α	X	α
CH_3CH_2	0.65±0.02	CF_3	0.55±0.01
CH_3	0.62±0.02	CN	0.58±0.01
$CH_3OCH_2CH_2$	0.53±0.01	SO_2CH_3	0.59±0.01
$ClCH_2CH_2$	0.47±0.02	NO_2	0.61±0.02
$HC\equiv CCH_2$	0.35±0.01	SO_2CF_3	0.63±0.02

The two most obvious mechanisms which would account for the observed general acid catalysis are (a) rate limiting protonation of the complex 1 or 2 on one of the alkoxy groups followed by rapid alcohol depature, and (b) a concerted reaction with transition state 3. Note that in 3, depending on whether C-O bond breaking is ahead or lags behind proton transfer the leaving oxygen carries some negative or some positive charge, as indicated by the symbol ±δ.

$$\underset{\underset{\sim}{3}}{}$$

Rate limiting protonation of 1 or 2 is inconsistent with both the magnitude of α and the increasing trend of α with more electronwithdrawing R-groups: since protonation of the ketal type oxygen in 1 and 2 is thermodynamically disfavored (2), the α values should all be close to one and should be highest for the most electronwithdrawing R group (least basic ketal oxygen)(3).

On the other hand the values of α which lie in the range of 0.35 to 0.65 are typical for a concerted reaction (4) and the trend in α with changes in R as well as in X are consistent with the concerted mechanism (5, 6). This latter point is most easily appreciated if the reaction is discussed in the framework of the energy diagrams made popular by More O'Ferrall (7) and Jencks (5, 6). Fig. 1 shows such a diagram where corners D and A refer to reactants and products, respectively, corners B and C to intermediates of the (hypothetical) step-wise processes, the horizontal axes refer to proton transfer and the vertical axes to C-O bond cleavage/formation. The reaction coordinate of the concerted reaction is located inside the square; for an idealized concerted process the reaction coordinate would be represented by a diagonal as shown in the figure.

Substituent effects on the transition state of the concerted pathway can be understood as perturbations of the energy surface at the location of the transition state (5-7). These perturbations arise from changes in the relative energy of the four corners of the diagram. Changes in the relative energies of corners A and D are transmitted along the reaction coordinate and induce the transition state to shift toward the corner whose relative energy has been raised. This is the well known "Hammond effect" (8). Changes in the relative energies of corners B and C are called perpendicular effects and induce a shift of the transition state toward the corner whose energy

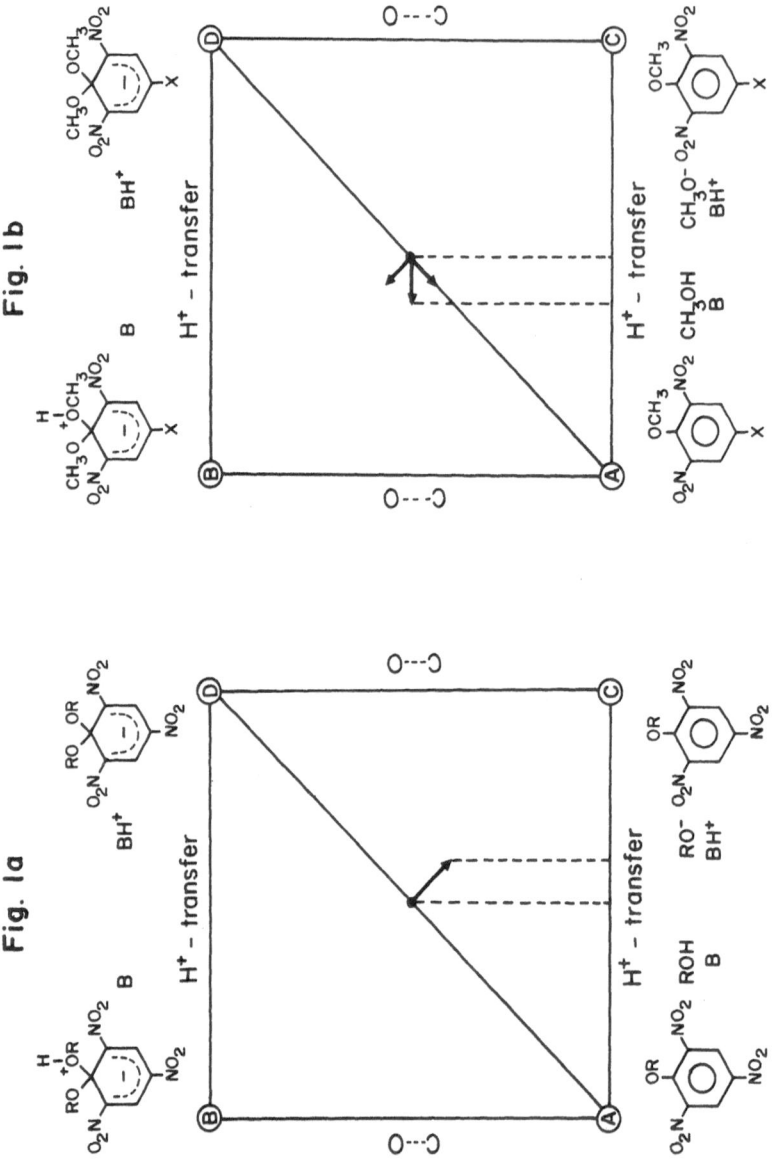

Figure 1   More O'Ferrall-Jencks Energy Diagram: (a)   Effect of Making R More Electron Withdrawing; (b)   Effect of Making X More Electron Withdrawing.

has been <u>lowered</u>. This is sometimes referred to as the "Thornton effect" (9) and is easily understood if one realizes that with respect to the B-C diagonal the transition state lies in a potential well.

Let us now discuss how changes in R or X affect the transition state of the concerted reaction. The effect of making R more electronwithdrawing is to stabilize RO^- (corner C) and to destabilize the zwitterionic complex in corner B. This raises the energy of corner B over that of corner C and causes a perpendicular shift of the transition state toward corner C, as shown in Fig. 1a. On the other hand, since there is no charge on the oxygen in ROH (corner A) or the complex (corner D), the relative energies of these two corners do not change significantly and hence there is no significant effect on the transition state along the reaction coordinate. Thus the overall effect is essentially completely determined by the perpendicular effect whose horizontal component indicates that there is less proton transfer in the transition state (Fig. 1a). This should manifest itself by a decrease in α which is in fact what we observe (Table I). Our results seem typical for reactions in which an alkoxide or aryloxide ion is expelled from an addition compound by a concerted acid catalyzed pathway (10-13).

The effect of making X more electron withdrawing is to stabilize the complexes in the corners B and D relative to A and C. As a result the transition state is shifted towards corner A ("Hammond effect") and toward corner B ("Thornton effect") as shown in Fig. 1b. The net result of the two effects is to shift the transition state in the direction of more proton transfer and α should increase. This is again borne out by the results; a similar trend in α was found for the phenoxide ion departure from substituted benzaldehyde methyl phenyl acetals (13).

The above discussion shows that the More O'Ferrall-Jencks diagrams provide an excellent framework for a qualitative understanding of the reactions reported in this study as well as several others. There are however some quantitative problems which are less well understood and which are discussed elsewhere (1).

REFERENCES

1. For a more detailed account of this work see Bernasconi, C. F. and Gandler, J. R.: 1978, J. Am. Chem. Soc. 100, 8117.

2. Crampton, M. R. and Willison, M. J.: 1974, Chem. Soc.,
    Perkin Trans. 2, p. 1686.

3. Eigen, M.: 1964, Angew. Chem. Int. Ed. Engl. 3, p. 1.

4. Jencks, W. P.: 1976, Acc. Chem. Res. 9, p. 425.

5. Jencks, W. P.: 1972, Chem. Rev. 72, p. 705.

6. Jencks, D. A. and Jencks, W. P.: 1977, J. Am. Chem. Soc.
    99, p. 7948.

7. More O'Ferrall, R. A.: 1970, J. Chem. Soc. B, p. 274.

8. Hammond, G. S.: 1955, J. Am. Chem. Soc. 77, p. 334.

9. Thornton, E. R.: 1967, J. Am. Chem. Soc. 89, p. 2915.

10. Gravitz, N. and Jencks, W. P.: 1974, J. Am. Chem. Soc. 96,
    p. 507.

11. Sayer, J. M. and Jencks, W. P.: 1977, J. Am. Chem. Soc.
    99, p. 464.

12. Funderbunk, L. H., Aldwin, L., and Jencks, W. P.: 1978,
    J. Am. Chem. Soc. 100, p. 5444.

13. Capon, B. and Nimmo, K.: 1975, J. Chem. Soc., Perkin Trans.
    2, p. 1113.

THE ROLE OF MEISENHEIMER COMPLEXES IN AROMATIC NUCLEOPHILIC SUBSTITUTION REACTIONS

Roger Bird

Royal Military College of Science, Shrivenham, Swindon, Wilts.

The intermediates in the reactions of picryl halides with base have been studied by stopped-flow spectrophotometry. A mechanism for the base assisted hydrolysis which can explain the order of reactivity is proposed.

The order of reactivity of picryl halides is the reverse of that observed in the reactions of the corresponding alkyl halides.[1] Alkyl fluorides react only slowly with nucleophiles whereas picryl fluoride reacts extremely rapidly with hydroxide ions to produce picric acid. Addition of sodium hydroxide to a picryl halide in aqueous dioxan produces a transient red intermediate. The present study has been concerned with the measurement of the rates of formation and decay of the intermediates derived from picryl fluoride, chloride and bromide using stopped-flow spectrophotometry. The spectra of the intermediates are similar to those attributed to Meisenheimer complexes. Picryl fluoride reacts to give this intermediate in a few milliseconds. The rate of appearance of the intermediate was therefore only measurable at very low base concentrations (less than 6×10^{-3} M hydroxide). Under these conditions the rate was found to be first order in both hydroxide and picryl fluoride. Apart from a

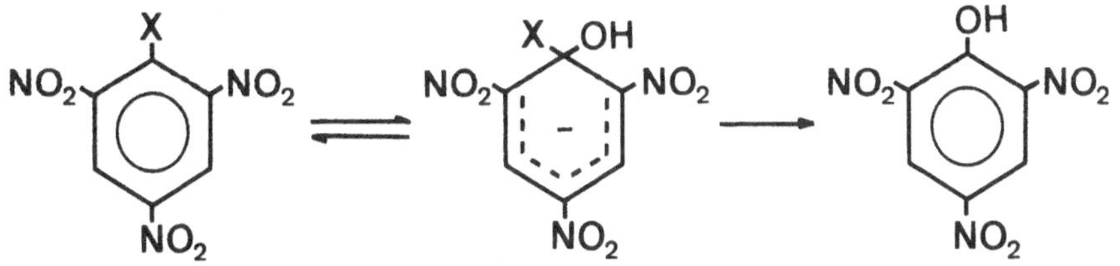

Figure 1

W. J. Gettins and E. Wyn-Jones (Eds.). Techniques and Applications of Fast Reactions in Solution. 469–472.
Copyright © 1979 by D. Reidel Publishing Company.

small initial deviation the first order plot for the disappearance of the intermediate showed good linearity. These results
can be explained by the simple mechanism shown in figure 1 in
which the rapidly formed Meisenheimer complex slowly loses
fluoride ion to give the product.

Unfortunately neither picryl chloride nor bromide gave first
order plots for the disappearance of the intermediate. Analogue
computer analysis showed that this was not simply due to
competition between rates of formation and disappearance. Good
agreement between observed and analogue computer simulated rates
was obtained if a second intermediate was introduced into the
model. The most likely species corresponding to this intermediate
is the Meisenheimer complex produced by addition of hydroxide at
the 3-position. This model produces the scheme shown below:

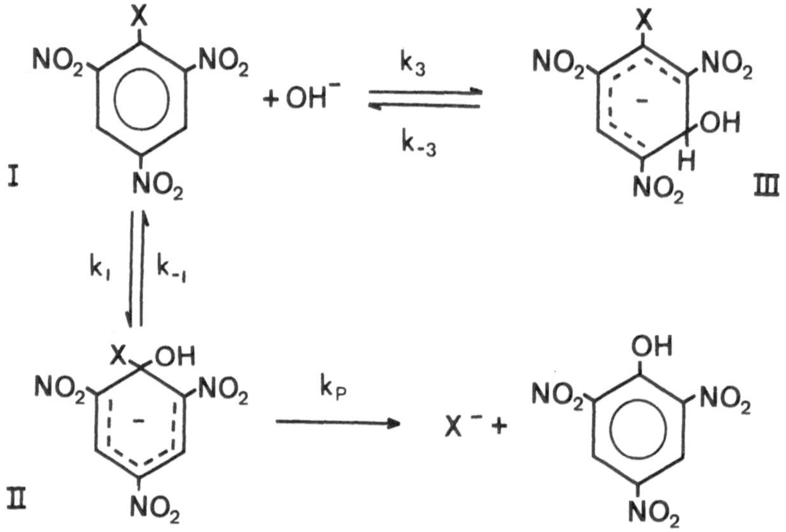

The presence of two intermediates would be difficult to detect
as the spectra of II and III are similar[2]. The equilibrium
reaction between I and III effectively reduces the standing
concentration of I. Corrected values for the rate of disappearance of I as a function of hydroxide concentration show that
this reaction is first order in both picryl halide and hydroxide
(figure 2). This base dependence shows that the formation of the
1-adduct (II) is slow compared with chloride or bromide loss
from II. It is therefore not surprising that the reactivities
of the picryl halides have little in common with their alkyl
analogues. The rate constants for all three halides are shown
in table 1.

Figure 2

Graph of corrected rate against hydroxide concentration

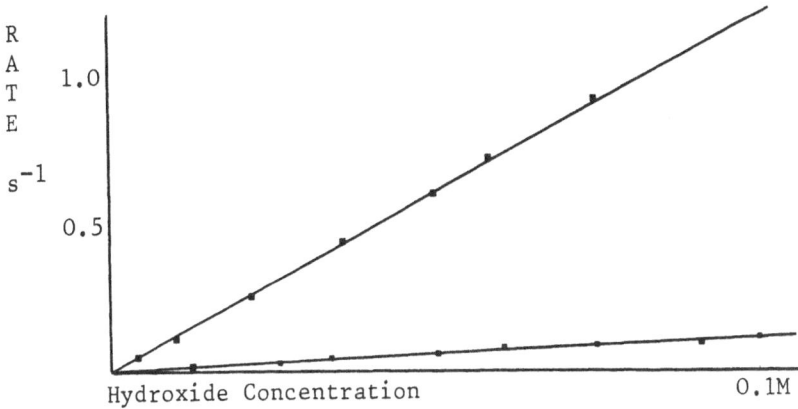

Table 1

Rates of formation and disappearance of intermediates derived
from picryl halides

Halide	$k_1/1 \ mol^{-1} \ s^{-1}$	$k_3/1 \ mol^{-1} \ s^{-1}$	k_{-3}/s^{-1}	kp/s^{-1}
Fluoride		5300		0.112
Chloride	11.8	61.0	0.153	fast
Bromide	0.89	28.6	1.25	fast

Examination of the equilibrium constants for the formation of III
shows that picryl chloride and bromide are quickly converted to
the inactive adduct III and that this is the coloured species
observed (very little II is present). As the rates of formation
of I from III (k_{-3}) and of II from I ($k_1 \cdot [OH^-]$) are comparable
in these cases the 'rate determining step' is a combination of
these two reactions. Apart from making meaningless any
discussion of the relative values of the overall rates of
hydrolysis it underlines the dangers involved in complex
equilibrium reactions. The values of the rates and equilibrium
constants for the formation of III might be taken to indicate a
'fast pre-equilibrium' reaction followed by a slow reaction to
give products. This is clearly not the case as shown by the
marked curvature found in a pseudo first order plot.

The simple model originally derived for picryl fluoride must
also be reconsidered in the light of the more complicated model
for the bromide and chloride. Inspection of the ratio of the
rate constants for the formation of the 1- and 3- Meisenheimer
complexes (k_1/k_3) for the bromide (0.03) and chloride (0.19)
leads to the conclusion that this ratio may be close to unity for
the fluoride. The colour observed in the reactions of picryl

fluoride would therefore be due to the presence of both species (II and III). The rate of disappearance would be complex and this could explain the initial fast reaction reported earlier if the rate of loss of fluoride ion from II (kp) is comparable with the formation of picryl fluoride from III (k_{-3}). Certainly the observed rate constant for the disappearance is similar to the value of k_{-3} for the chloride and kp for the fluoride is likely to be much slower than for the chloride or the bromide.

Once again the simple analysis, which neglected the slight curvatuve at the beginning of the plot, appears to have given misleading results and it seems likely that all three halides share a common mechanism. The reactivity of the halide depends only upon its ability to promote the formation of the 1-adduct (II) and not on the 'leaving group' ability of the halide ion.

1.   R. E. Parker and T. O. Read, J.C.S., 1969,9.

2.   M. R. Crampton, Adv. Phys. Org. Chem., 1969, 7, 211.

THE KINETICS AND MECHANISM OF THE REACTIONS OF CYCLO-1,5-OCTA-
DIENECHLOROAMINEIRIDIUM(I) WITH 2,2'-BIPYRIDYL AND 1,10-PHENAN-
TROLENE

Wynand J. Louw and Carol C. Hepner

Council for Scientific and Industrial Research,
Pretoria, South Africa

The kinetics of the reactions

$$(cod)IrCl(am) + am' \xrightarrow[\text{N}_2]{\text{methanol}} (cod)Ir(am') + Cl^- + am \qquad (1)$$

have been studied at $20^\circ C$ (am = 2-picolene (pic) or pyridine
(py), am' = 2,2'-bipyridyl (bipy) or 1,10-phenantrolene (phen)).
The reactions were fast and had to be followed by a stopped
flow technique. It was found that the equilibrium

$$(cod)IrCl/_2 \xrightleftharpoons{\text{methanol}} 2(cod)IrCl(S) \qquad (2)$$

lies far over to the right (S = methanol). Methanolic "dimer"
solutions were therefore used to study the reactions

$$(cod)IrCl(S) + am' \longrightarrow codIr(am')^+ + Cl^- + S \qquad (3)$$

In order to establish the effect of the outgoing ligand
concentration on the observed rate of reaction (1), am was
added to the reaction solutions. Convincing evidence was
found for the existance of $(cod)IrCl(am)_2$ in solution.

For reaction (1) the kinetic data suggest that the solvent
intermediate (cod)IrCl(S) is in a steady state during the
reaction and that there is competition for reaction with this
specie between am and am'.

The following reaction scheme was thus postulated:

W. J. Gettins and E. Wyn-Jones (Eds.), Techniques and Applications of Fast Reactions in Solution. 473–475.
Copyright © 1979 by D. Reidel Publishing Company.

$$(cod)Ir\underset{am}{\overset{Cl}{\underset{am}{\big\langle}}}\underset{pic}{\overset{am}{\rightleftharpoons}} K_e (cod)Ir\underset{Cl}{\overset{am}{\underset{k_{-s}}{\big\langle}}}\overset{k_s}{\rightleftharpoons} (cod)Ir\underset{Cl}{\overset{MeOH}{\big\langle}} + am$$

$$am' \overset{}{\underset{-am}{\Big\backslash}} k_1 \quad k_2 \overset{}{\underset{-MeOH}{\Big/}} am' \qquad (4)$$

$$(cod)Ir(am')^+ + Cl^-$$

The rate law derived from this scheme;

$$k_{obsd} = \frac{k_1 [am']}{1 + K_e [am]} +$$

$$\frac{k_s k_2 [am']}{k_{-s}[am] + k_2[am'] + k_2 K_e[am'][am] + k_{-s}K_e[am]^2} \qquad (5)$$

conforms to all of the experimental data, viz∫

(a) REACTION OF (cod)IrCl(pic) + bipy.

A plot of k_{obsd} vs. [bipy] when no pic was added showed two distinct regions. At high bipy a straight line plot was obtained while at low [bipy] the plot curved towards zero.

As the pic concentration was increased the distinction between the two parts of the plot became less well defined, approaching a straight line through the origin.

This behavior can be explained by rate law 5 as follows: at low [pic] and high [bipy] the k_{-s} [pic], $k_{-s}K_e$ [pic]^2 and $k_2 K_e$ [bipy][pic] terms can be neglected relative to the k_2 [bipy] term and rate law (5) simplifies to the classical two-term rate law ($k_{obsd} = k_s + k_1$ [bipy]) which is represented by the straight line portion of the plot from which k_s (2.7 s^{-1}) and k_1 (2.0 s^{-1} M^{-1}) were calculated. At higher [pic] the competition between the forward reaction (k_2 [bipy]) and the back reaction (k_{-s} [pic]) results in a curved plot. At very high [pic] the equilibrium K_s lies far over to the left, making the contribution of the (cod)IrCl(MeOH) species to the overall reaction very small and the plot approaches a straight line described by $k_{obsd} = k_1$ [bipy]/K_e [pic].

Rate law (5) can be rewritten as:

$$\frac{1}{k_{obsd}'} = \frac{k_{-s}[pic]}{k_2 k_s[bipy]} + \frac{1}{k_s} + \frac{K_e[pic]}{k_s} + \frac{k_{-s}K_e[pic]^2}{k_2 k_s[bipy]} \quad (6)$$

were $k_{obsd}' = k_{obsd} - k_1[bipy]/(1 + K_e[pic])$.

Plots of $1/k_{obsd}'$ vs. $1/[bipy]$ for each [pic] should yield straight lines with slopes and intercepts both dependent on [pic]. Plotting these intercepts ($1/k_s + K_e[pic]/k_s$ against [pic], linear graphs should be obtained with an intercept ($1/k_s$) and a slope (K_e/k_s) and K_e could therefore be recalculated.

Since K_e is not originally known, an estimated K_e value was used and changed in a computer program untill this value corresponded to the recalculated K_e values (40 M^{-1}). A plot of the slopes ($k_{-s}[pic]/k_2 k_s + k_{-s}K_e[pic]^2/k_2 k_s$) vs. [pic] gave a parabolic curve. For a plot of slope/[pic] vs. [pic] a straight line was obtained.

From the latter slope, $k_{-s}K_e/k_2 k_s$ the ratio k_{-s}/k_2 could be calculated. The k_2 value (3.33×10^4 s^{-1} M^{-1}) was obtained from reaction (3), and the k_{-s} value (13.5×10^4 s^{-1} M^{-1}) could be calculated.

(b) REACTION OF (cod)IrCl(pic) + phen.

A key point in the proposed rate law (5) is that k_s, k_{-s} and K_e should be independent of the nature and reactivity of the incoming nucleophile. Thus the study of the reaction of (cod)IrCl(pic) with another amine, phen, was undertaken. The reaction with phen was slower than the reaction with bipy and was found to proceed totally via the solvent path. All the rate and equilibrium constants for this system were determined as for the bipy system using rate law (6) ($k_{obsd.} = k_{obsd}'$) and k_s, k_{-s} and K_e were found to agree, within experimental error, with those obtained for the reaction with bipy ($k_s = 2.8$ s^{-1}, $k_{-s} = 7 \times 10^4$ s^{-1} M^{-1}, $K_e = 67$ M^{-1}.

(c) REACTION OF (cod)IrCl(py) WITH bipy.

The reaction proceeded entirely through the k_1-path (rate law (5)) and no k_s path contribution was seen ($k_1 = 555$ $s^{-1}M^{-1}$, $K_e = 69$ M^{-1}).

THE KINETICS OF DONOR-ACCEPTOR COMPLEX FORMATION, STUDIED BY A MICROWAVE TEMPERATURE-JUMP METHOD.

J.P. Field and E.F. Caldin.

University of Kent, Canterbury, Kent, U.K.

THE MICROWAVE TEMPERATURE-JUMP APPARATUS

Recent modifications to our apparatus (1) have greatly increased its range of application. (a) The sensitivity has been improved about 50-fold by signal-averaging. While the use of a microwave pulse to effect a temperature-jump has the advantage that it is applicable to any polar solvent, its disadvantage has been that the temperature rise is only a few tenths of a degree, much smaller than with Joule or laser heating, and the resulting signal change may be as small as 0.01%. The defect can be remedied, however, by repetitive pulsing and signal-averaging (2), to which the technique is well suited. This is now easily arranged, since commercial signal-averagers are available. With the magnetron set to deliver 20 pulses per second, less than two minutes is required to produce 2000 traces; signal-averaging should then improve the signal-to-noise ratio by a factor of $(2000)^{\frac{1}{2}}$ or about 45. In practice we have found that even when a single trace shows no detectable relaxation a good exponential curve may be obtained by signal-averaging; the effect is the same as if the temperature-rise were several degrees. A flowing solution must be used, to maintain temperature control. (b) The use of low temperatures is desirable when one is studying a very fast reaction with an equilibrium constant which is small at room temperature but increases as T decreases. We have been able to make rate measurements down to $-100^{\circ}C$ by making the sample flow (by means of a peristaltic pump) in a closed circuit, through a coil immersed in a thermostat bath and then through the reaction cell; the temperature in the cell is constant to $\pm$ 0.1 degrees.

W. J. Gettins and E. Wyn-Jones (Eds.). Techniques and Applications of Fast Reactions in Solution. 477–478.
Copyright © 1979 by D. Reidel Publishing Company.

DONOR-ACCEPTOR COMPLEXES

The donor-acceptor complexes formed between electron acceptors (such as iodine or tetracyanoethylene) and electron donors (such as aliphatic amines or aromatic hydrocarbons) have been extensively studied from various points of view - spectroscopic, structural and thermodynamic (3). Kinetic investigation has lagged behind, because the reactions are extremely fast and the equilibrium constants small; the rate constant of one such reaction had been measured by means of our microwave apparatus (without the recent improvements), at -83°C (4). We have now determined rate constants, in 1-chlorobutane, as solvent for the reactions of zinc tetraphenyl-porphyrin with pyridine and 2-methyl pyridine, which have been well characterised by spectrophotometric methods (5, 6). They are less than the diffusion-controlled value by an order of magnitude. Further investigations are in progress.

REFERENCES

1.   E.F. Caldin and J.E. Crooks, J. Sci. Instr., 1967, 44, 449.
2.   H.T. Witt, Methods in Enzymology, Volume XVI, "Fast Reactions", edited by Kenneth Kustin, Chapter IX, Academic Press, 1969, p. 337; J. Aubard, this volume.
3.   R. Foster, "Charge Transfer Complexes", Academic Press, London 1969.
4.   E.F. Caldin, J.E. Crooks, D. O'Donnell, D. Smith and S. Toner, J.C.S. Faraday Trans. I., 1972, 68, 849.
5.   J.R. Miller and G.D. Dorough, J.A.C.S., 1952, 74, 3977.
6.   C.H. Kirtzsey, P. Hambright and C.S. Storm, J. Inorg. Chem., 1969, 8, 10, 2142.

TUNNELLING IN H-TRANSFER REACTIONS; APPLICATION OF
FAST-REACTION METHODS

E.F. Caldin

University of Kent at Canterbury, U.K.

INTRODUCTION

The rate of a reaction in solution such as (1), between a
base B and a carbon acid written as R_3 CH, whose rate- determining
step is the transfer of a proton from acid to base, is usually
reduced when deuterium is substituted for hydrogen. The

$$R_3CH + B \underset{\longleftarrow}{\overset{k^H}{\longrightarrow}} R_3C^- + HB^+ \qquad (1)$$

explanation that is often adequate is that the energy-barrier is
lower for proton-transfer than for deuteron-transfer, because the
zero-point energy for the C-H bond in the acid is higher than
for the C-D bond. The highest value of the ratio of the rate
constants k^H/k^D at 25°C that could be explained in this way,
however, is about 7, whereas there are a number of reactions for
which the value is much larger. The differences in the parameters
of the Arrhenius equation ($k = A \exp(-E_a/RT)$) are similarly
anomalous for some of these reactions; most notably, the value
of A^D/A^H may be much larger than 1. Hydrogen-atom transfers
and hydride-ion transfers show similar behaviour.

These interesting anomalies can be understood in terms of
tunnelling. (For reviews see refs. 1-5.) The proton is much
the lightest species concerned in the reactions, and its wave
properties are important. The wavelength corresponding to its
small mass ($\lambda = h/mv$) at ordinary temperatures is comparable
with the expected width of the energy barrier for the reaction.
In consequence, it is able to "tunnel through" the energy barrier.

479

W. J. Gettins and E. Wyn-Jones (Eds.). Techniques and Applications of Fast Reactions in Solution. 479–489.
Copyright © 1979 by D. Reidel Publishing Company.

The usual kinetic treatment based on the model of a particle
surmounting the barrier will then be incorrect; in particular
the reaction will be faster than it predicts, especially at low
temperatures. Deuteron transfer, however, is much less affected
by tunnelling, because of the much larger mass; the isotope
effects may therefore be larger than classical theory can explain.
Detailed calculations based on the model of a wave incident on
an energy-barrier confirm these conclusions, and show that the
effects of tunnelling will depend on the height and width (or
curvature) of the barrier. Conversely, light may be thrown on
these properties of the barrier by experimental determination of
the isotope effects. Until relatively recently, only a few
reactions were known that clearly showed effects of tunnelling,
but a review by R.P. Bell in 1974 (3) listed over 20 such reactions,
and several more have since been investigated.

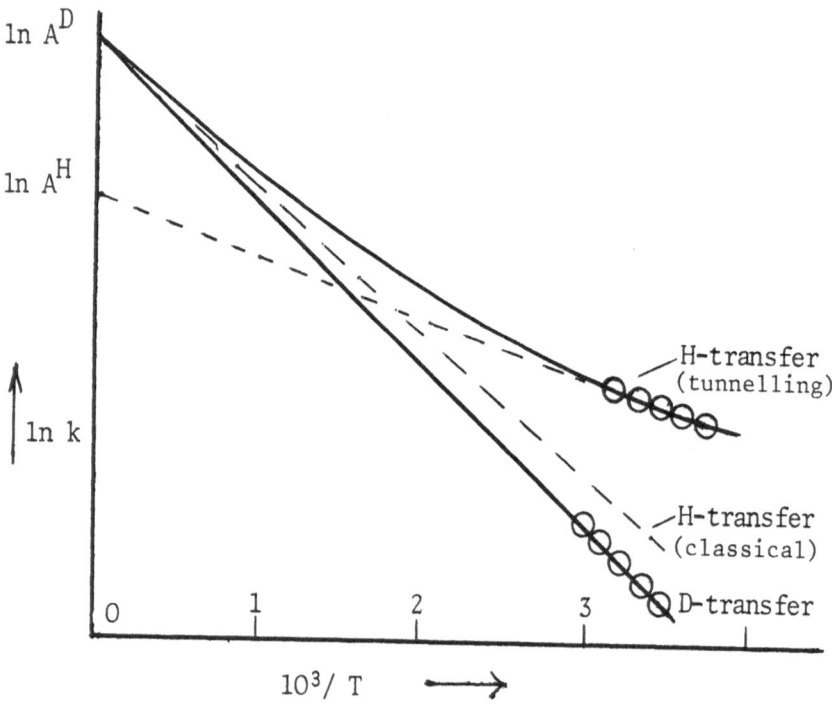

Fig. 1. Arrhenius plots (schematic) for H and D
transfer reactions, illustrating effects
of tunnelling.

KINETIC EFFECTS OF TUNNELLING

Four consequences of tunnelling may be distinguished. They may be illustrated by reference to fig. 1, which shows schematically the kind of Arrhenius plots of ln k^H and ln k^D against 1/T that may be expected if tunnelling is important. The circles indicate experimental points, which necessarily cover a reciprocal-temperature range much shorter than the complete range down to 1/T = 0.

(1) The Arrhenius plot for H-transfer deviates upwards at low temperatures from the straight line expected for the classical particle model; the plot for D-transfer deviates much less.

(2) Since the plots for H and D transfer diverge more than the classical model predicts, the value of k^H/k^D at a given temperature (corresponding to the vertical distance between the plots) will be anomalously large.

(3) The Arrhenius activation energy for H-transfer (E_a^H), derived from the slope of the plot in the experimental temperature range, will be less than the classical value, which corresponds to the height of the energy-barrier (E^H), i.e. $E_a^H < E^H$; while for D-transfer the two will be nearly equal, $E_a^D \simeq E^D$. Hence the difference of the experimental activation energies ($E_a^D - E_a^H$) will be greater than the difference in the heights of the energy barriers ($E^D - E^H$), which is attributed to the difference in zero-point energies of C - H and C - D bonds. The maximum difference, corresponding to complete loss of translational zero-point-energy in the transition state, is about 1.1 kcal mol^{-1}; the corresponding value of k^H/k^D, given by exp $(E^D - E^H)$/RT, is about 7 at 25°C. Experimental values greater than these suggest the possibility of tunnelling.

(4) The experimental rates for H-transfer at ordinary temperatures will give an intercept below the classical value. This intercept represents the A-factor in the Arrhenius equation (ln k = ln A - (E_a/R)(1/T)). The apparent value of A^H will therefore be less than the classical value, whereas that of A^D will approximate to it; thus A^D will be greater than A^H. This is the most striking effect predicted by the tunnelling model.

EXPERIMENTAL WORK

Fast-reaction methods applied to tunnelling. Some very fast reactions, such as the recombination of the ions of weak acids, probably involve tunnelling, but for a systematic study more interest attaches to reactions with an appreciable energy barrier, since tunnelling effects increase with the barrier height. Many of the reactions which show clearly the importance of tunnelling proceed at conventional rates. An example is the base-catalysed bromination of a ring ketone (2-carbethoxycyclopentanone) catalysed by fluoride ion (6), for which Bell and his co-workers   found $A^D/A^H = 24 \pm 4$; this was the first published instance where A^D is much greater than A^H, and is still one of the best-established. When single-step reactions such as (1) were investigated, they were found to be too fast for conventional methods, but could be followed by stopped-flow methods at ordinary temperatures, or by conventional means at low temperatures, or sometimes only by a combination of stopped-flow and low-temperature methods. These are the main fast-reaction techniques that have been used in the systematic investigation of tunnelling in reactions of carbon acids.

Some reactions involving exchange of H atoms or protons in a ring system have recently been studied by n.m.r. or e.s.r. methods, and large isotopic effects observed. The ring may be a porphin ring (7), a porphyrin ring (8), or a H-bonded chelate ring (9). Here the H is 'bound', moving in a double-minimum potential well; the theory is therefore somewhat modified, but leads to broadly similar conclusions. Such a case is described elsewhere in this volume (7), and is representative of the special contribution of nuclear magnetic resonance measurements to this subject (cf. below). Electron-spin resonance methods also give special information, on radical reactions (see below, and ref. 9).

Reactions of carbon acids with ethoxide in ethanol. Carbon acids containing electron-attracting groups react with ethoxide ion in ethanol (or other OR^- - ROH systems) producing carbanions whose deep colours enable the reaction to be followed spectrophotometrically.

$$R_3CH + EtO^- \rightleftharpoons R_3C^- + EtOH \qquad (2)$$

The reverse reaction can also be effected by a range of weak acids such as acetic acid:-

$$R_3C^- + HA \rightarrow R_3CH + A^- \qquad (3)$$

In a series of acids the rate constant is related roughly linearly to the pK of the acid, as expected for a proton transfer; so the rate can be adjusted to some extent by appropriate choice of the acid. All these reactions are 'fast' at room temperature. Isotopic effects can be measured for reaction 2, by using R_3CD, but not conveniently for reaction 3, because the deuterated acid DA would rapidly exchange with a solvent such as EtOH.

Acids which lend themselves to this kind of investigation (10-15) are exemplified by 4-nitrobenzylnitromethane,

$NO_2C_6H_4CH_2NO_2$ (4-NPNM). Others are 4-nitrobenzylcyanide,
$NO_2C_6H_4CH_2CN$ (4-NBC); 2, 4, 6-trinitrotoluene, $C_6H_2(NO_2)_3CH_3$ (T.N.T.);
di-(4-nitrophenyl) methane, $(NO_2C_6H_4)_2CH_2$; and tri-(4-nitrophenyl)
methane, $(NO_2C_6H_4)_3CH$ (TNTPM). The first of the reactions of
such acids to yield evidence of tunnelling was the reaction of the
anion of TNT with acetic acid (10), which gave a curved Arrhenius
plot below -100°C. The reaction with HF also showed deviations
below about -50°C (11). The reaction of 4-nitrobenzyl cyanide
with ethoxide ion, studied down to -124°C, gave a curved plot
(12); moreover the isotopic effect gave $A^D/A^H \approx 5$, considerably
greater than 1, and the two sets of data were in satisfactory
agreement as to the barrier dimensions (13). The reaction of
TNTPM with alkoxide ions (14) also gave isotopic effects large
enough to suggest tunnelling, although the Arrhenius plot was
linear from 25°C down to -60°C.

When these results were reviewed (4) along with those of
Bell (6), E.S. Lewis (16), J.R. Jones (17) Shiner and Martin (18),
and others, it was noticed that the tunnelling effects were
smaller (i.e. the calculated barrier width was larger) for the
reactions in alcohols than for the reactions in water. It might
also have been noticed that there was a tendency for the larger
tunnelling effects to be associated with higher activation
energies (see Table VII of ref. 4). The only other clue to the
factors controlling tunnelling was the observation by Lewis that
the introduction of sterically-hindering methyl groups greatly
increased the isotopic rate ratio k^H/k^D for the reaction of
2-nitropropane with pyridine $(Me_2NO_2CH + py \rightarrow Me_2NO_2C^- + Hpy^+)$,
from 10 to 24; he suggested that the effect of steric repulsion
was to raise the height of the barrier, i.e. to increase its
curvature.

Reactions of carbon acids with nitrogen bases in aprotic
solvents. In order to broaden the range of solvents and bases
that could be studied, it was decided to use nitrogen bases in
place of alkoxide ions. Since the reactants are then both neutral,
they can be dissolved in solvents of low as well as high polarity.
The reaction then produces an ion pair:-

$$AH + B \rightleftharpoons A^-HB^+.. \qquad\qquad (4)$$

The stopped-flow method has generally been used; it has the
advantages that it can be used with any solvent, that the product
need not be stable for more than a few seconds, that low
temperatures can be used, and measurements can be done very
quickly. Aliphatic amines, amidines $(HN=CR^1(NR_2))$,
tetramethylguanidine $(HN=C(NMe_2)_2)$ and its substituted derivatives,
and cyclic amidines have all been used as bases (19-26). With
4-NPNM or TNT as acid, they nearly all give values of k^H/k^D
greater than 10, suggesting that tunnelling is of importance.

The largest effects have been obtained with tetramethylguanidine (TMG) as base and 4-NPNM ($NO_2C_6H_4CH_2NO_2$) as acid, in solvents of low polarity; for this reaction in toluene, Caldin and Mateo (20) reported $k^H/k^D = 45 \pm 2$ at 25°C and $A^D/A^H = 32 \pm 13$. Calculations from the theoretical equations for tunnelling through a parabolic barrier, assuming a mass $m^H = 1$ atomic unit for the particle transferred, gave reasonable barrier widths of about 0.8 Å. In highly polar solvents, much smaller effects were observed; for acetronitrile (dipole moment 3.37 D), the values were $k^H/k^D = 11.8 \pm 0.3$ and $A^D/A^H = 1.0 \pm 0.2$. Part of the difference could be attributed to a lower activation energy, but it was also noticeable that the calculated barrier width was larger, about 1.0 Å. One would expect, however, that the barrier width would be about the same in all solvents, and indeed the results can be interpreted equally well on the alternative assumption that this is so and that the effective mass m^H is greater than 1 atomic unit. This will be the case if the motion of the proton is coupled with motions of heavier particles, such as would be involved in the reorganisation of solvent molecules. Since the reaction produces ions, an increase of solvation may be expected, which will require reorientation of solvent molecules. (In accordance, the reaction has a negative activation entropy and the rate constant increases with the polarity of the solvent). It was therefore suggested that in acetonitrile, for instance, the polar solvent molecules rotate in the field of the highly polar activated complex, thus increasing the effective mass and decreasing the effects of tunnelling; while in low-polar solvents, such as toluene, the solvent molecules do not rotate until the proton transfer is effectively complete. The activation volumes in low-polarity solvents are all about equal, in accordance with this view (25).

It has since been pointed out by Rogne (27-29) that spuriously large isotopic effects might result from the isotopic exchange which is possible in this type of reaction, in which a base (BH) containing the HN=C< grouping reacts with the deuterated acid AD_2, producing ADH and eventually AH_2:-

$$AD_2 + BH \rightleftharpoons AD^-DBH^+ \rightleftharpoons AD^-HBD^+ \rightleftharpoons ADH + BD;$$

$$ADH + BH \rightleftharpoons AH^-DBH^+ \rightleftharpoons AH^-HBD^+ \rightleftharpoons AH_2 + BD. \qquad (5)$$

Assuming that the rearrangement of the ion-pair ($AD^-DBH^+ \rightleftharpoons AD^-HBD^+$) is fast, Rogne's detailed examination gives for the rate ratio k^H/k^D for TMG in toluene at 30°C a value of only about 11. On re-investigating the deuteron-transfer reaction, however, with special attention to the exclusion of extraneous water, Mateo (30) found nearly the same high value of k^H/k^D as before, and also found that deuterated base (BD) gave the same value, even though it cannot take part in reactions such as equation 5. This

suggests that the rearrangement of the ion-pair may be slow
enough to reduce the rate of solvent-exchange and its effect
on the kinetics. The question must at present be regarded as not
finally settled.

Proton transfer exchange in a porphin ring. The intramolecular
transfer of protons within a porphin ring reported by Limbach (7)
is of interest, both because it shows the special types of
information accessible by n.m.r. methods (31) and because the
protons migrate between nitrogen atoms at a fixed distance
(equation 6), so that this is an instance of a "bound" proton

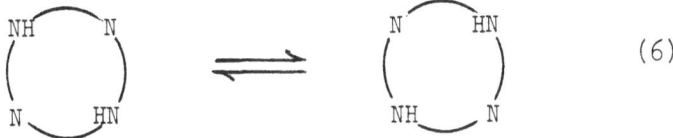

(6)

moving in a symmetrical double-minimum potential. The characteristic
feature of n.m.r. line-broadening or relaxation-time measurements
is that they can show which atoms are concerned in the reaction,
as well as giving a value of the rate constant. In the present
instance, it is possible to use ^{15}N and ^{13}C as well as ^{1}H, and to
show that the reaction observed is intramolecular. The rate
appears to be independent of solvent, presumably because solvent
molecules are excluded by the ring system from the reaction site.
The isotopic rate ratio is 11 ± 3, and the difference of
activation energies $(E_a^D - E_a^H)$ is 14 ± 2.6 kJ mol^{-1}; both values
suggest that tunnelling is important. Moreover the Arrhenius plot
for the proton-transfer reaction deviates from linearity at low
temperatures, and indeed below about 200 K it becomes horizontal;
this indicates that tunnelling occurs from one energy-level only.
The appropriate model for such an intramolecular exchange within
a rigid ring system, with solvent excluded, is a symmetrical
double-minimum potential with discrete energy levels. This
differs significantly from the "free-proton" model we have used
for intermolecular reactions. The kinetic effects, however, are
broadly similar, i.e. large isotopic effects and a curved
Arrhenius plot for H-transfer. The non-involvement of the solvent
is in sharp contrast to its role in the reactions considered in
the preceding paragraph.

H-atom transfer in the reactions of phenols with radicals.
If there is anything in the idea that tunnelling effects may be
masked by the coupling of proton transfer with the resulting
reorganisation of solvent molecules, we should look for evidence
of tunnelling in H-atom transfer reactions, which involve no
transfer of charge and hence much less change of solvation. Some
high isotopic rate ratios have in fact been observed. The
abstraction of H or D from dihydrophenanthrene by molecular oxygen
in hydrocarbon solvent, studied at low temperatures, gave rates

corresponding to $k^H/k^D \approx 30$ at 25°C, and the Arrhenius plot for
H-transfer is curved (32). A systematic study of the reactions
of numerous substituted phenols with polyvinylacetate radicals
in various solvents has been made by Simonyi (33); several of
the values of k^H/k^D at 25°C are above 20, and the highest is
57 ± 23 for 2,4,6-trimethyl phenol in ethyltrifluoroacetate
solution at 50°C. There have been several investigations of
H-transfer reactions of phenols with deeply-coloured stable free
radicals such as DPPH and galvinoxyl (34-36); with some phenols
the isotopic rate ratio is in the region of 10-15 at 25°C.
Reactions of thiophenols have been studied by Lewis (37); the
isotopic rate ratios are increased by substitution near the
reaction site, suggesting (since the solvent takes little part)
that steric repulsion does increase the barrier height, rather
than merely exclude solvent (cf. above). An intramolecular
reaction of this type is the exchange of H atoms in the
phenoxyl-type radical formed from 1,2-dihydroxybenxene (equation 7),
which has been studied by e.s.r. line-broadening methods (9); the

$$; \quad OH..O \rightleftharpoons O..HO \qquad (7)$$

e.s.r. spectra provide evidence both on the mechanism and on the
rate of the reaction, which is fast ($k^H \approx 10^5$ s^{-1} at room
temperature). The reported value of k^H/k^D at 25°C is about 100,
with a lower limit of 57; but the authors attribute this high
value partly to other contributions besides that of the OH
stretching.

Intramolecular H-atom transfer in phenyl radical. A reaction
recently studied by K.U. Ingold and co-workers (38,39) which
shows strikingly all the effects of tunnelling – anomalous
isotopic effects, and curvature of the Arrhenius plot, even for
D-transfer – is the intramolecular isomerisation of the 2,4,6-tri-
t-butyl-phenyl radical. This involves the transfer of H (or D)
from a methyl group to a ring:-

The radicals and the mechanism are identified by e.s.r. methods,
and the reaction is monitored by the e.s.r. signal down to low
temperatures in hydrocarbon solution. The Arrhenius plots are
both curved, that for H-transfer more sharply than that for D-transfer
of course. The Arrhenius parameters give high values for A^D/A^H

(10 to 10^2, depending on temperature) and $E_a^D - E_a^H$ (3 to 4 kcal mol^{-1}). The isotopic rate ratio becomes enormous at low temperatures (ca. 10^4 at 123 K); at 25°C it is about 20. The data can be successfully fitted to the theoretical equations for tunnelling of a "free" proton. The rates are the same in a solid matrix as in solution (39); this confirms that the reaction is intramolecular, and also that solvent reorganisation is negligible. This example shows clearly the advantages of determining isotopic effects down to low temperatures, and of studying well-characterised H-atom transfers.

CONCLUSION

Factors favourable to tunnelling are (a) a high narrow energy-barrier and (b) a minimum of solvent reorganisation during formation of the activated complex. As regards (a), not much is known about the relative widths of energy-barriers, but it is sometimes possible to vary the height, by choice of substituents. As regards (b), transfer of an H atom will normally produce less reorganisation of solvent molecules than transfer of a proton, which alters the charge distribution; and proton transfers of the type $AH + B^- \rightarrow A^- + HB$ (equation 3) will produce less reorganisation than those of the type $AH + B \rightarrow A^-HB^+$ (equation 4). Non-polar solvents, and solvents with small mobile molecules such as water, may be expected to reduce tunnelling effects in proton-transfer reactions less than polar solvents. Some pointers towards a systematic study of tunnelling can therefore be seen, but experimental exigences will often frustrate a promising programme.

It is clear that the study of tunnelling is likely to advance our knowledge of energy-barriers in chemical reactions, particularly the ways in which their width and height depend upon the electron-distribution at the reacting centres and upon the steric effects of bulky groups near the reaction site. New light may be expected on the role of the solvent and on the question of how far the reorganisation of solvent molecules is coupled to changes of bonding, and which of these processes occurs first. Such information is of great interest for studies of reaction mechanism, as well as for the physical aspects of chemical kinetics.

REFERENCES

1. R.P. Bell, "The Proton in Chemistry", 2nd edn., Chapman and Hall, London (1973), chap. 12.
2. E.S. Lewis in "Proton-transfer Reactions", ed. E.F. Caldin and V. Gold, Chapman and Hall, London (1975), Chap. 10.
3. R.P. Bell, Chem. Soc. Rev., 1974, 3, 513.
4. E.F. Caldin, Chem. Rev., 1969, 69, 135.
5. E.F. Caldin in "Reaction Transition States", ed. J.E. Dubois Gordon and Breach, Paris, 1972, pp. 247-255.

6. R.P. Bell, J.A. Fendley and J.R. Hulett, Proc. Roy. Soc., A, (1956) 235, 453.

7. J. Hennig and H.H. Limbach, this volume.

8. S.S. Eaton and G.R. Eaton, J. Amer. Chem. Soc., 1977, 99, 1604.

9. K. Loth, F. Graf and H.H. Gunthard, Chemical Physics 1976, 13, 95.

10. E.F. Caldin and E. Harbron, J. Chem. Soc., 1962, 3454; cf. J.B. Ainscough and E.F. Caldin, J. Chem. Soc., 1960, 2407; E.F. Caldin and G. Long, Proc. Roy. Soc. A., 1955, 228, 263.

11. E.F. Caldin and M. Kasparian, Discuss. Faraday Soc., 1965, 39, 25; cf. E.F. Caldin and E. Harbron, J. Chem. Soc., 1962, 2314.

12. E.F. Caldin, M. Kasparian and G. Tomalin, Trans. Faraday Soc., 1968, 64, 2802.

13. E.F. Caldin and G. Tomalin, Trans. Faraday Soc., 1969, 64, 2814.

14. E.F. Caldin, E. Dawson, R.M. Hyde and A. Queen, J.C.S. Faraday Trans. I, 1974, 70, 528; cf. E.F. Caldin and J.C. Trickett, Trans. Faraday Soc., 1953, 49, 772.

15. A. Jarczewski, P. Pruszynski and K.T. Leffek, Can. J. Chem., 1975, 53, 1176.

16. E.S. Lewis and L.H. Funderburk, J. Amer. Chem. Soc., 1967, 89, 2322.

17. J.R. Jones, Trans. Faraday Soc., 1965, 61, 95, 2456; 1967, 63, 993, 111.

18. V.J. Shiner and B. Martin, Pure Appl. Chem., 1964, 8, 371.

19. E.F. Caldin, A. Jarczewski, and K.T. Leffek, Trans. Faraday Soc., 1971, 67, p. 110.

20. E.F. Caldin and S. Mateo, Chem. Comm., 1973, 854; 1975, J. Chem. Soc., Faraday Trans. I, 1975, 71, p. 1876; 1976, 72, 112.

21. E.F. Caldin, S. Mateo, and C.J. Wilson, Faraday Symp. Chem. Soc., 1975, 10, p. 121.

22. E.F. Caldin, O. Rogne and C.J. Wilson, J.C.S. Faraday Trans. I, 1978, 74, 1796.

23. E.F. Caldin and O. Rogne, J.C.S. Faraday Trans. I, 1978, 74, 2065.

24. I. Heggen, J. Lindstrøm and O. Rogne, J.C.S. Faraday Trans. I, 1978, 74, 1263.

25. C.D. Hubbard, C.J. Wilson, and E.F. Caldin, J. Amer. Chem. Soc., 1976, 98, p. 1870.

26. P. Pruszynski and A. Jarczewski, Roc. Chim., 1977, 51, 2171.

27. O. Rogne, Chem. Comm., 1977, 695.

28. J.H. Blanch and O. Rogne, J.C.S. Faraday Trans. I, 1978, 74, 1254.

29. O. Rogne, Acta Chem. Scand., 1978, A32, 559.

30. S. Mateo, unpublished work at the University of Kent.

31. For a brief account see e.g. D.N. Hague, "Fast Reactions", Wiley, London, 1971, chap. 2; E.F. Caldin, "Fast Reactions in Solution", Blackwell, Oxford, 1964, chap. 11.

32. A. Bromberg et al., Chem. Comm., 1968, 1352; J Amer. Chem. Soc., 1969, 91, 2860; J.C.S. Perkin II, 1972, 588.

33. M. Simonyi et al., Adv. Phys. Org. Chem., 1970, 9, 127;
     J.C.S. Perkin II, 1978, 405.
34. P.B. Ayscough and K.R. Russell, Can. J. Chem. 1965, 43, 3039;
     J.A. Howard and K.U. Ingold, Can. J. Chem., 1962, 40, 1851;
     1964, 42, 2324.
35. D.A. Palmer and H. Kelm, Aust. J. Chem., 1977, 30, 1229.
36. S.P. Dagnall and E.F. Caldin, this volume.
37. E.S. Lewis et al., J. Amer. Chem. Soc., 1976, 98, 2254,
     2260, 2264, 2268; cf. also ref 2, p. 334, and ref. 16.
38. G.D. Brunton, D. Griller, L.R.C. Barclay, and K.U. Ingold,
     J. Amer. Chem. Soc., 1976, 98, 6803.
39. G.D. Brunton. J.A. Gray, D. Griller, L.R.C. Barclay and
     K.U. Ingold, J. Amer. Chem. Soc., 1978, 100, 4197.

KINETIC STUDY OF HYDROGEN TUNNELLING IN MESO-TETRAPHENYL
PORPHINE BY NMR-LINESHAPE ANALYSIS AND SELECTIVE $T_{1\rho}$ -
RELAXATION TIME MEASUREMENTS*

Jürgen Hennig and Hans-Heinrich Limbach

Institut fur Physikalische Chemie der Universitat
D-7800 Freiburg i. Br., W. Germany

The kinetics of the hydrogen migration between the degenerate tautomers of meso-tetraphenylporphine (TPP-H_2) and of the
deuterated species (TPP-D_2)

both dissolved in various media has been studied over a wide
range of temperatures by NMR-lineshape analysis and by measurements of the longitudinal relaxation times, $T_{1\rho}$, in the rotating
frame of selective lines.

As proven by the ^{1}H-NMR-spectra of TPP-$^{15}N_4H_2$, the rate constants obtained are entirely due to a random walk of the inner
hydrogen atoms between the four nitrogen atoms and not due to
intermolecular proton transfer. As shown in Fig. 1, in the slow
exchange range a doublet splitting of the inner proton signal
shows that each proton is coupled with one ^{15}N-spin with a coupling constant of $J_{^{15}N-H}$=101 Hz. However, in the fast exchange

*To be published in J.C.S. Faraday II, 75, 1979. The
figures are reproduced with permission of the editor.

491

W. J. Gettins and E. Wyn-Jones (Eds.). Techniques and Applications of Fast Reactions in Solution. 491–494.
Copyright © 1979 by D. Reidel Publishing Company.

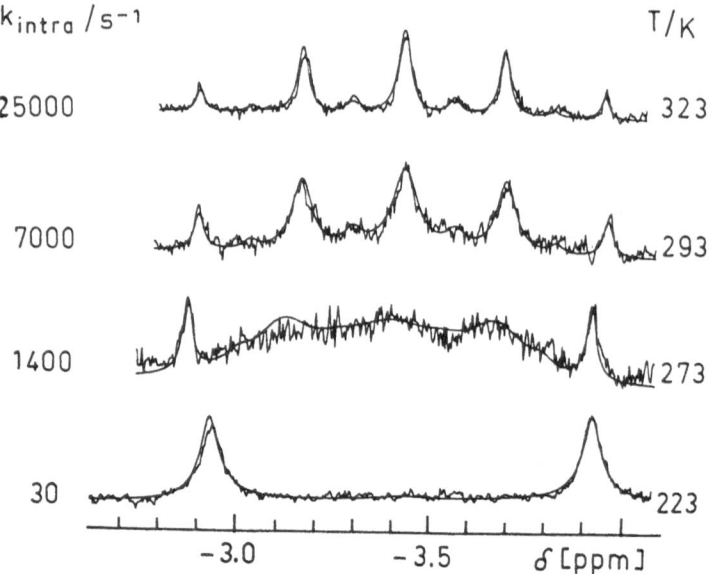

Figure 1 ^{1}H-NMR-signal of the inner hydrogens in TPP-^{15}N$_4$H$_2$ as a function of the temperature. The quartet between the pentet lines are due to TPP-^{15}N$_3$^{14}NH$_2$.

range a pentet splitting is observed with a line distance of $J_{15N-H}/4$. Consequently, in the fast exchange range, within the NMR-timescale, the inner protons are localised with equal probability at each of the four nitrogen atoms due to rapid random walk. The rate constants do not depend on the type of the solvent used. Their dependence on the temperature is given by

$$k^H = k^H_{to} + k^H_{t1} \exp(-E^H_1/RT), \quad 160 \leqslant T \leqslant 323 \text{ K},$$

$$k^D = k^D_{t2} \exp(-E^D_2/RT), \quad 213 \leqslant T \leqslant 305 \text{ K, where}$$

$$k^H_{to} = 5.0 \pm 0.5 \text{ s}^{-1}, \quad \ln k^H_{t1} = 26.5 \pm 1, \quad E^H_1 = 43.4 \pm 1.3 \text{ kJmol}^{-1}$$

and

$$\ln k^D_{t2} = 29.7 \pm 0.7, \quad E^D_2 = 57.3 \pm 1.3 \text{ kJmol}^{-1}.$$

The Arrhenius diagram is shown in Figure 2.
The results can not be explained in terms of the transition state theory but only in terms of a vibrational model of tunnelling in a symmetric double minimum potential with quantized vibrational levels along the reaction co-ordinate. The rate constants are related to the tunnel splittings, k_{ti}, of barrier-separated degenerate states. The hydrogen motion proceeds by tunnelling between the ground states with the tunnel frequency, k^H_{to}, and between the first excited vibrational NH-stretching states with the frequency, k^H_{t1}. It must be the normal mode of the stretching vibration that becomes the reaction co-ordinate because the energy of activation, E^H_1, is very close to the

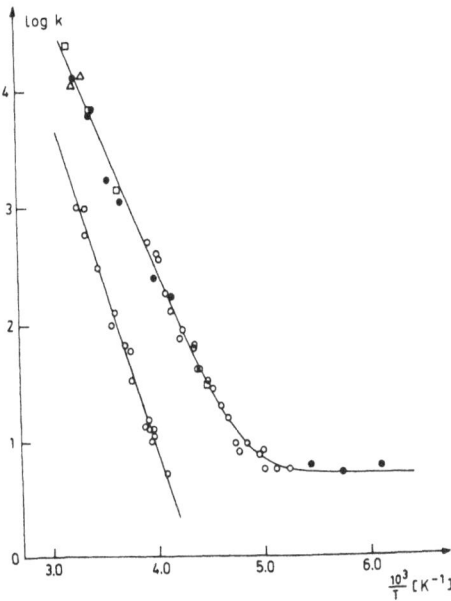

Figure 2   Arrhenius diagram of the deuterium migration (left)
             and of the hydrogen migration (right) in TPP. Open
             symbols:lineshape analysis, filled symbols: $T_{1\rho}$-
             relaxation experiments.

energy of the NH-stretching vibration ($\tilde{\nu}_{NH}$ = 3315 cm^{-1} = 40
kJmol^{-1}). The Arrhenius parameters of the deuterium motion must
arise from the reaction between the second excited ND-stretching
levels because E_2^D is about twice the energy of the ND-stretching
vibration ($\tilde{\nu}_{ND}$ = 2478 cm^{-1} = 30 kJmol^{-1}). The tunnel frequencies
k_{to}^D and k_{t1}^D do not contribute in any significant way to the
reaction rate in the temperature range where rate constants have
been observed. Fig. 3 gives a picture of the vibrational model
of tunnelling in TPP.

     Semiempirical calculations of the tunnel frequencies
confirm this vibrational model and show that hydrogen tunnelling
in TPP is a synchronous process like hydrogen tunnelling during
the inversion of ammonia.

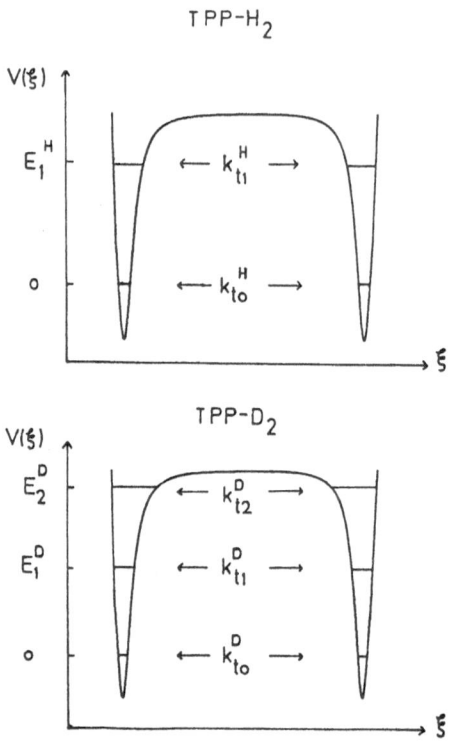

<u>Figure 3</u>   Vibrational model of tunnelling in TPP. ξ is the
           reaction co-ordinate and $V(\xi)$ the potential (schemat-
           ically).   Each level i is split by $\Delta E_i = h\, k_{ti}$, not
           shown here on account of its smallness.

HYDROGEN AND DEUTERIUM ATOM TRANSFER REACTIONS FROM SUBSTITUTED PHENOLS TO GALVINOXYL.

S. P. Dagnall and E. F. Caldin

University of Kent at Canterbury, U.K.

The rate constants and Arrhenius parameters for the hydrogen and deuterium atom transfer reactions (equation 1) from some substituted phenols (ArOH), such as I, to the stable free radical glavinoxyl (1, 2) in aprotic solvents have been determined, in the hope of gaining a better understanding of the factors that control tunnelling.

$$ArOH + G\cdot \rightarrow ArO\cdot + GH \qquad (1)$$

I = 2,4,6-tri-t-butyl phenol =

G· = galvinoxyl =

That the overall reaction is an H-atom transfer is established by the observed change of the e.s.r. spectrum from that of galvinoxyl to that characteristic of the tri-t-butylphenoxy radical. There is a considerable kinetic deuterium isotope effect, indicating that the rate-limiting step involves the transfer of H.

The reactions were studied under first-order conditions, the phenol being greatly in excess of the galvinoxyl, in a stopped-flow apparatus or (for slower reactions) a uv-vis spectrophotometer. Rate constants were determined by a curve-

W. J. Gettins and E. Wyn-Jones (Eds.). Techniques and Applications of Fast Reactions in Solution. 495–496.
Copyright © 1979 by D. Reidel Publishing Company.

matching (analog) technique (3). The experimental traces
were accurately exponential and reproducible to better than
± 0.5%. Linear plots of observed rate constant against
the concentration of phenol were obtained, from which second-
order constants could be determined. The rate-limiting step
is the transfer of the H atom. Complications from side-reactions
(oxidation and dimerisation) can be minimised by careful choice
of substituents (2). The stability of the phenoxy radical
depends upon the substituents; bulky groups prevent coupling
reactions.

The steric and electronic effects of the phenol
substituents influence the rate of the reaction. The rates
of H-atom transfer in various solvents differ slightly, showing
an order which follows the ability of the solvent to form hydrogen
bonds with phenol. Arrhenius plots were determined over the
range 50°C to -60°C, for two of the phenols(2,4,6-tri-t-butyl
and 2,6-di-t-butyl), and show positive deviations from
linearity which may be attributable to quantum-mechanical
tunnelling of the hydrogen atom.

REFERENCES

1.   Coppinger, G.M.J.: 1957, J.Amer.Chem.Soc., 79, 501.
2.   Forrester, A.R., Hay, J.M. and Thomson, R.H.: 1968,
     "Organic Chemistry of Stable Free Radicals," Academic
     Press, London.
3.   Caldin, E.F. Rogne, O. and Wilson, C.J.: 1978, J.C.S.
     Faraday I, 74, 1796.

THERMODYNAMICS AND KINETICS OF INTRAMOLECULAR PROTON-TRANSFERS. TEMPERATURE-JUMP STUDIES.

O. Bensaude, G. Dodin and M. Dreyfus

Institut de Topologie et de Dynamique des Systèmes de l'Université Paris VII, associé au C.N.R.S., 1, rue Guy de la Brosse, 75005 PARIS - France.

Since the publication, in 1953, by Watson and Watson and Crick[1] of the double-helix model for DNA, it has been suspected that the tautomerism of the nucleic acid bases might interface with the efficiency of the replication processes in leading to incorrect base-pairing.

The purpose of this paper is to illustrate how T-jump can be used in the study of tautomerism.

Typical relaxation signals are exponentials and contain two informations: the amplitude and the relaxation time. The amplitude reflects the thermodynamics of the systems, whereas the relaxation time is used for mechanistic purposes.

The amplitude for a simple tautomeric equilibrium HM $\rightleftharpoons$ MH is given by:

$$A = . \frac{\ell . \Delta T}{RT^2} . C. \frac{K}{(1+K)^2} . \Delta H. (\varepsilon_{HM} - \varepsilon_{MH})$$

where ℓ the optical length, ΔT the temperature jump, C = (HM)-(MH) the total substrate concentration T the final temperature, R the thermodynamic constant are characteristics of the experiments or physical constants, and where ΔH_{AB}, the enthalpy of reaction, K the tautomeric equilibrium constant, $(\varepsilon_{MH} - \varepsilon_{HM})$ the difference in the extinction coefficients of the tautomers at the wavelength of the study, are three unknowns.

W. J. Gettins and E. Wyn-Jones (Eds.), Techniques and Applications of Fast Reactions in Solution. 497–500.
Copyright © 1979 by D. Reidel Publishing Company.

The difference in the extinction coefficients is used for attribution of the relaxation signals to tautomerism, by comparing the variations of the amplitude with wavelength to the differential spectrum of model compounds (usually the methylated derivatives)[2,3] Moreover, from this spectrum, an estimate for the difference of the extinction coefficients is obtained. Turning back to the relaxation amplitude, when the tautomeric equilibrium is highly shifted as is often the case), the expression AT^2 is proportional to K and, therefore, according to Van't Hoff's law, the plot of ln AT^2 versus l/T give us the enthalpy of reaction ΔH.[3a]

It is then possible to deduce an estimation for the tautomeric equilibrium constant. This way, we have measured tautomeric equilibrium constants for several nucleic acid-base derivatives:[2-5] we were able to determine a tautomeric equilibrium constant as small as 1/400 for the N-1H/N-3H equilibrium of cytosine,[3a] and to demonstrate that 1-methyl adenine[4] (a starfish hormone) existed under the _imine_ form to the extent of 1% in aqueous solution !

The second information we can obtain is _mechanistic_. From the variations of the _relaxation time inverse_ with pH, in _aqueous_ solutions, it was shown that tautomeric interconversion is cata- lyzed by acids and bases;[6] the substrates exchange a proton with the catalysts thus forming a protonated or deprotonated species common to both tautomers and this species reprotonates or depro- tonates to give either tautomer (mechanism D - scheme I). As "normal" acids and bases are usually involved, the proton-transfer steps are rate-encounter controlled[7] when thermodynamically favo- rable. Therefore, with the knowledge of the acid-base equilibrium constants and that of the tautomeric equilibrium constant, the interconversion kinetics is entirely predictible! Conversely, from the kinetics, the tautomeric equilibrium may be estimated. This second estimation is totally independent of that obtained from the amplitude. However, when the proton is transferred between very close groups, like the lactim/lactam equilibrium of 2-pyridones, the previous mechanistic scheme is insufficient.[6] It is then necessary to postulate the existence of a so-called non- dissociative mechanism. As we were interested in this mechanism, we started working in _aprotic solvents_. In these media, the relaxation time inverse is proportional to the substrate concen- tration. This result was interpreted in terms of a mechanism in which the interconversion occurs within a cyclic dimer[8,9] (scheme II-upper part). The dimerization step is rate-encounter control- led.[10] Upon addition of water to the media,[8,11] the dimerization pathway is inhibited by the hydration of the substrate, but it can be shown that water also catalyzes the interconversion (scheme II- lower part). It was therefore, suggested that the non-dissocia- tive mechanism responsible for part of the tautomeric intercon- vertion of 2-pyridones is _bifunctionally catalyzed by the water_.

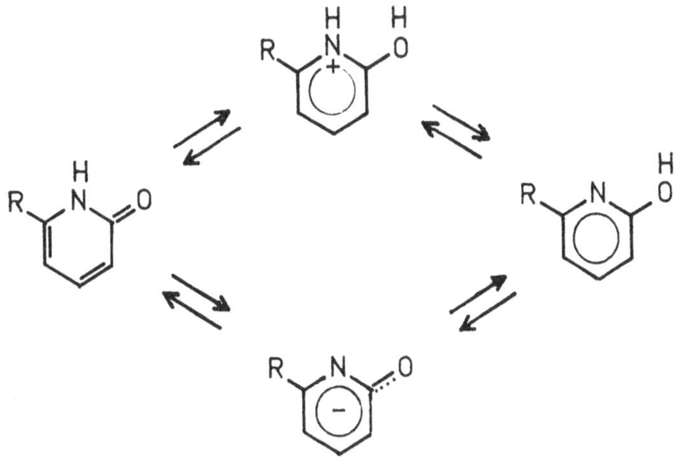

Scheme I - Dissociative Tautomeric Interconversion Mechanisms.

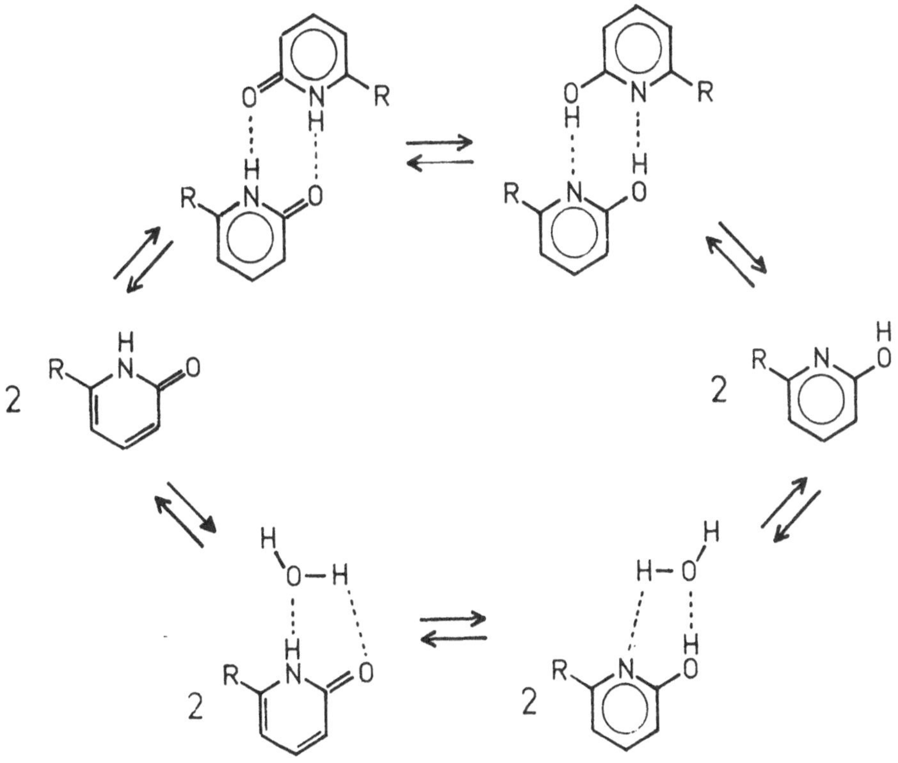

Scheme II - Non-Dissociative Tautomeric Interconversion Mechanisms.

In conclusion to our work, it is now possible to predict the lifetimes of the nucleic acid-base tautomers. Since the abnormal lactim and imine forms are expected to be present in a ratio proportion of 10^{-4} to 10^{-5}, we expect, on the basis of the pKs, their lifetimes to be shorter than 10 μs. Researches investigating nucleic acid-base recognition might find this result interesting.

REFERENCES

1. J.D. Watson and F. Crick, Cold Spring Harbor Symp. Quant. Biol., 18, 23, (1953).
2. M. Dreyfus, G. Dodin, O. Bensaude and J.E. Dubois, J.Am.Chem. Soc., 97, 2369, (1975).
3. a) M. Dreyfus, O. Bensaude; G. Dodin and J.E. Dubois, J.Am.Chem.Soc., 98, 6338, (1976) ; b) O. Bensaude, J. Aubard, M. Dreyfus, G. Dodin and J.E. Dubois, J.Am.Chem.Soc., 100, 2823, (1978).
4. M. Dreyfus, G. Dodin, O. Bensaude and J.E. Dubois, J.Am.Chem. Soc., 99, 7027, (1977).
5. a) G. Dodin, M. Dreyfus, O. Bensaude and J.E. Dubois, J.Am.Chem.Soc., 99, 7257, (1977); b) O. Bensaude, M. Chevrier and J.E. Dubois, Tetrahedron, 2259, (1978).
6. O. Bensaude, M. Dreyfus, G. Dodin and J.E. Dubois, J.Am.Chem. Soc., 98, 4438, (1976).
7. M. Eigen, Angew.Chem. (Int.Ed.) 3, 1, (1964).
8. O. Bensaude, in "Protons and Ions Involved in Fast Dynamic Phenomena" P. Laszlo ed. Elsevier, Amsterdam (1978).
9. O. Bensaude, M. Chevrier and J.E. Dubois, J.Am.Chem.Soc., 100, 7055 (1978).
10. a) G.C. Hammes and H.O. Spivey, J.Am.Chem.Soc., 88, 1621, (1966) ; b) G.C. Hammes and Lillford, J.Am.Chem.Soc., 92, 7578, (1970).
11. O. Bensaude, M. Chevrier and J.E. Dubois, J.Am.Chem.Soc., 101, in press (1979).

HIGH FIELD PERTURBATION AND CHEMICAL RELAXATION OF 2,4,6-TRICHLOROPHENOL AND TRIETHYLAMINE MIXTURES IN CYCLOHEXANE

M. De Maeyer, P. Wolschann and L. Hellemans

Department of Chemistry, University of Leuven, Celestijnenlaan 200 D, 3030 Heverlee, Belgium

INTRODUCTION

The admittance composed of conductivity σ and permittivity $\varepsilon_o\varepsilon$ regulates the steady-state current flowing through a dielectric sample subjected to an alternating voltage. The quantity is given in complex notation by

$$y = \sigma + j\omega\varepsilon_o\varepsilon \qquad (1)$$

with ω the circular frequency of the voltage. The permittivity itself is a complex quantity ($\varepsilon = \varepsilon' - j\varepsilon''$) to allow for relaxation effects. The admittance varies at high field density by several mechanisms. The orientation of polar substances is no longer random, so that the permittivity decreases due to saturation[1]. The dissociation of weak electrolytes is enhanced thus increasing the conductivity[2]. The equilibrium between reaction partners of different polarity is shifted in favor of the most polar species according to[3]

$$d\ln K/dE = \Delta M/RT \qquad (2)$$

where ΔM represents the molar change of electric moment. The effect constitutes a chemical mode of polarization contributing to the overall permittivity. Dielectric dispersion and absorption of the increment reveal chemical relaxation.

For rapidly rotating molecules with permanent moment μ_i the following proportionality holds

$$\Delta M \sim \Sigma \, v_i \mu_i^2 E/3kT \qquad (3)$$

501

W. J. Gettins and E. Wyn-Jones (Eds.), Techniques and Applications of Fast Reactions in Solution. 501–507.
Copyright © 1979 by D. Reidel Publishing Company.

wherein v_i is the stoichiometric coefficient. It is clear now from eqn. 2 that in the limit of zero field no chemical perturbation can occur. From the definition of permittivity as the derivative of the displacement D with respect to the field, one finds the increment for a field-dependent equilibrium to be[3]

$$\varepsilon_o \Delta\varepsilon(\omega) = (\delta D/\delta\xi) \ (\delta\xi/\delta E) \ \phi(\omega) \qquad (4)$$

with ξ the extent of reaction and ϕ a chemical relaxation function. Finally, the amplitude of the effect is

$$\Delta\varepsilon = \Gamma(\Delta M)^2/\varepsilon_o RT \qquad (5)$$

in which the reaction capacity Γ is recognized. It can be noted that terms of the fourth power of μ_i appear; moreover, the effect is proportional to the square of E.

The motive for the present investigation is the question whether high fields promote proton jumping in acid-base adducts and whether the reaction partners can be regarded as tautomers. According to conventional dielectric measurements by Polish authors such an equilibrium would be well-poised for the title compounds[4]. Malecki analyzed the nonlinear dielectric effects in this system without concern for relaxation[5].

EXPERIMENTAL

Small changes of the permittivity are measured with a modulated resonance method[6]. The sample liquid is enclosed in a symmetrical capacitor to which inductors are connected in parallel. Depending on the inductor chosen, the circuit becomes resonant in the frequency range of 0.1 to 100 MHz. It is excited by a tuned gene-

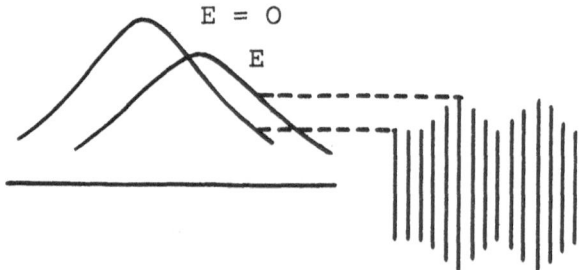

Fig. 1. Resonant voltage as function of frequency at zero and at peak high field, and origin of amplitude modulation illustrated at one particular frequency of the swept band.

rator which can be made to sweep slowly through the resonance band, while a periodic high field of 85 Hz is
applied to the sample. The essence of the experiment
consists in recording the difference of the resonance
characteristics at zero and at peak high field, appearing as amplitude modulation of the resonant voltage.
Fig. 1 shows that the modulation depth is related to both
frequency shift and voltage change. These quantities are
connected to field-induced capacitive and conductive
changes by means of the circuit parameters. Further
details on the experimental technique and its application
can be found elsewhere[6-9]. The observables are the chan-
ges $\Delta\varepsilon'/\varepsilon'$ and $\Delta\tan\delta$, a compound quantity equal to
$\Delta\varepsilon''/\varepsilon' + \Delta\sigma/\omega\varepsilon_0\varepsilon'$.

RESULTS AND DISCUSSION

     Typical results are shown in Fig. 2. The most
striking feature is chemical relaxation of the nonlinear
dielectric effect. The loss data include a contribution

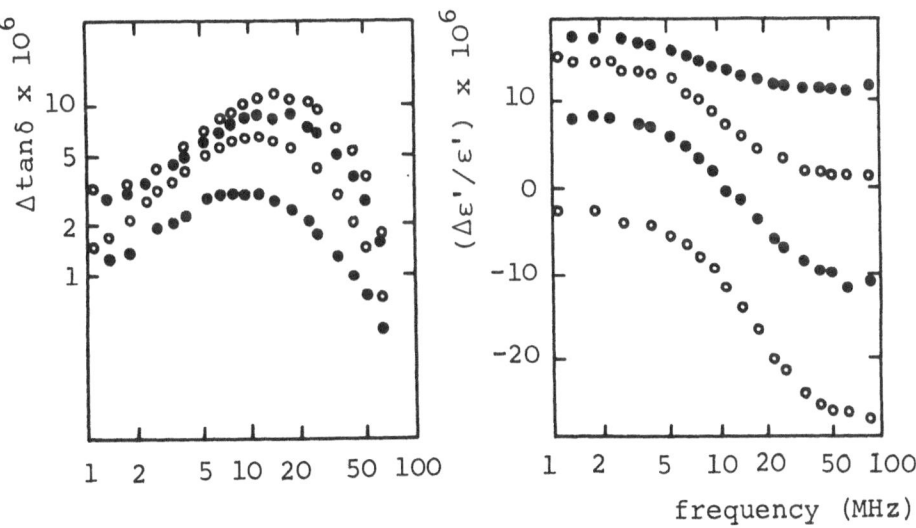

Fig. 2. Nonlinear change of loss angle and permittivity
of equimolar solutions of 2,4,6-trichlorophenol and tri-
ethylamine in cyclohexane as function of frequency at
25 °C and E = 176 kV/cm. The concentration is with increasing amplitude of the effect: 4.98, 8.40, 13.3 and
16.9 mM. For the neat solvent $\Delta\varepsilon'/\varepsilon'$ = 14.6 ppm. The
orientation relaxation time of the polar adduct is
210 ± 15 ps at 25 °C.

from enhanced dissociation of the ion-pairs, visible at
low frequency. The normal saturation term L on account
of the polar adduct approaches the rotation relaxation
region at about 1 GHz and causes $\Delta\tan\delta$ to become negative at high frequency[8,9]. The dispersion of the increment is superposed on the negative contribution L. From
the amplitude of both nonlinear effects the overall
equilibrium constant defined as

$$K = k_f (1 + K_2)/k_b \qquad (6)$$

and average dipole moment can be estimated according to
the reaction scheme for the acid-base reaction

$$AH + B \underset{k_b}{\overset{k_f}{\rightleftharpoons}} AH\text{--}B \overset{K_2}{\rightleftharpoons} A^-\text{--}HB^+ \qquad (7)$$

The results are collected in Table I. UV-spectroscopic
measurements confirm the value of the equilibrium constant.

Table I. Overall equilibrium constant $K_1(1 + K_2)$ and
         average dipole moment of the acid-base adduct.

$T(°C)$	$K(M^{-1})$	$\mu(D)$
7	649 $\pm$ 75	6.3 $\pm$ 0.1
16	392 $\pm$ 50	6.3 $\pm$ 0.2
25	170 $\pm$ 2?	6.0 $\pm$ 0.2
36	60 $\pm$ 13	5.7 $\pm$ 0.2

Appropriate functions including a single Debye relaxation term for the chemical increment are fitted to
the data. The relaxation time changes with concentration
as given by

$$1/\tau = k_f (\bar{c}_{AH} + \bar{c}_B) + k_b/(1 + K_2) \qquad (8)$$

where $\bar{c}_i$ is the equilibrium concentration. The graph in
Fig. 3 shows the relation to hold at different temperatures for equimolar solutions. The slope gives $k_b/(1 + K_2)$.
We conclude the proton transfer to be much faster than
the rotation of the adduct, so that it cannot be affected by the high field[10]. The reaction partners of the
field-sensitive, one-step equilibrium are the hardly
polar acid and base on the one hand, and the polar
rapidly tautomerizing adduct on the other hand. With
unequal amounts of acid and base the mechanism becomes
more involved.

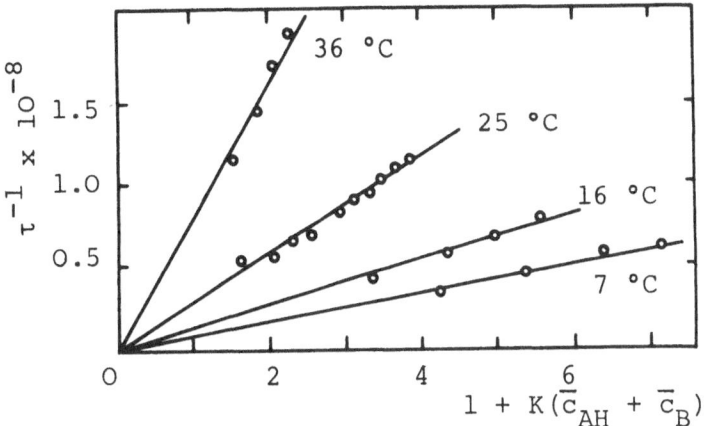

Fig. 3. Inverse relaxation time changing with concentration according to eqn. 8 for equimolar solutions of acid and base in cyclohexane at different temperatures. Values of K are taken from Table I.

The average dipole moment $\bar{\mu}$ of the adduct varies but slightly with temperature. The value is used to determine K_2 according to standard μ values for adducts where complete (μ_1) or no transfer (μ_2) took place[4]. In view of the rapid jumping process we write, correcting the procedure of ref. 4a

$$\bar{\mu} = X_1\mu_1 + (1 - X_1)\mu_2 \qquad (9)$$

with X_1 the time fraction the proton spends with the acceptor. In cyclohexane K_2 is 0.58 at 25 °C. The forward rate constant is $(5.3 \pm 0.2) \times 10^9$ $M^{-1}s^{-1}$ at 25 °C, which is half the limit of the diffusion-controlled process. It is faster than observed for similar acid-base systems studied in chlorobenzene[11], perhaps on account of the inert solvent.

From Fig. 4 it is seen that the apparent Arrhenius activation energy is negative ($E_A = -5.5 \pm 4.2$ kJ/mol) and significantly smaller than the activation energy for viscous flow ($E_A = 12$ kJ/mol). This result cannot be explained by the disruption of the internal hydrogen bond of the phenol. Small and negative activation energies have been obtained in other laboratories for analogous systems as well[12,13]. Caldin assumes the intermediate formation of an exothermic loose complex (AH,B) to account for the observations. However, it is not obvious that in spite of the rapid rotation of the partners at encounter the opportunity for hydrogen bonding would be missed.

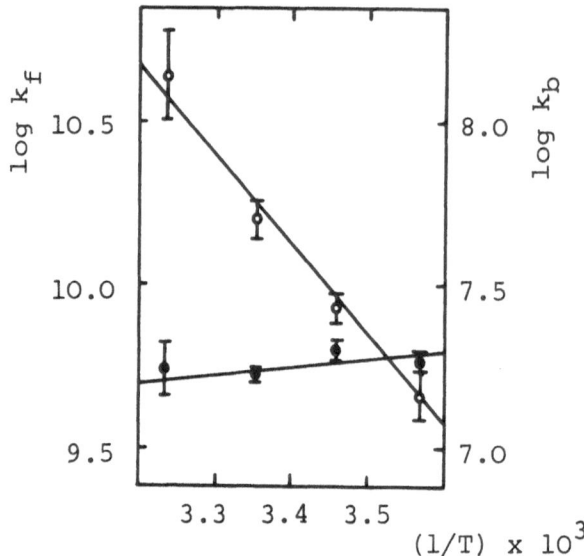

Fig. 4. Arrhenius plot of rate constants k_f (●) and k_b (o).

        At any rate, it is challenging to find that on this
time scale there is need to attribute additional fine-
structure to a so-called elementary step.

ACKNOWLEDGEMENT

This work is sponsored by F.K.F.O. grant 2.0051.77 and
a Belgian government grant Convention 76/81-II.4.
P.W. is on leave from the Institut für Theoretische
Chemie und Strahlenchemie der Universität Wien, Austria.
L.H. is research associate of the Belgian Research Coun-
cil (Nationaal Fonds voor Wetenschappelijk Onderzoek).

REFERENCES

1 C. Böttcher,"Theory of Electric Polarization", Vol. I,
   Elsevier, Amsterdam (1973) p. 289.
2 L. Onsager, J. Chem. Phys., 2, 599 (1934).
3 K. Bergmann, M. Eigen and L. De Maeyer, Ber. Bunsenges. Phys. Chem., 67, 819 (1963).
4a J. Jadzyn and J. Malecki, Acta Phys. Polonica , A41,
   599 (1972).
 b H. Ratajczak and L. Sobczyk, J. Chem. Phys., 50, 556
   (1969).
5 J. Malecki, J. C. S. Faraday II, 72, 104 (1976).

6   L. Hellemans and L. De Maeyer, J. Chem. Phys., 63, 3490 (1975).

7   A. Persoons and L. Hellemans, Biophys. J., 24, 119 (1978).

8   R. Nackaerts, M. De Maeyer and L. Hellemans, J. Electrostat. (in press).

9   M. De Maeyer, R. Nackaerts, R. Ooms and L. Hellemans, "The Nonlinear Dielectric Effect and Chemical Relaxation: a Happy Marriage", in Nonlinear Behaviour of Molecules, Atoms and Ions in Electric, Magnetic or Electromagnetic Fields, L. Néel ed., Elsevier, Amsterdam (in press).

10 G. Schwarz, J. Phys. Chem., 71, 4021 (1967).

11 E. F. Caldin, J. E. Crooks and D. O'Donnell, J. C. S. Faraday I, 69, 993 (1973).

12 K. J. Ivin, J. J. McGarvey, E. L. Simmons and R. Small, Trans. Faraday Soc., 67, 104 (1971).

13 E. F. Caldin and K. Tortschanoff, J. C. S. Faraday I, 74, 1804 (1978).

ELECTRON TRANSFER REACTIONS

Joseph F. Holzwarth

Fritz-Haber-Institut der Max-Planck-Gesellschaft,
Faradayweg 4-6, D-1000 Berlin 33, W. Germany

The most important theoretical ideas concerning adiabatic
outer sphere electron transfer reactions in solution are summarized. The kinetics of the reduction of a series of different
tris-1,10,-phenanthroline complexes of Fe(III) by $Fe(CN)_6^{4-}$ were
measured in order to test the influence of the redox-potential on
these reactions. The electron exchange rate of the complexes
$Fe(dipy)_3^{3+}$, $Ru(dipy)_3^{3+}$ and $Os(dipy)_3^{3+}$ was derived from the study
of their reduction by $Fe(CN)_6^{4-}$. Using the redox reaction between
the anionic complexes of $Fe(CN)_6^{4-}$ and $IrCl_6^{2-}$ the effect of added
cations was investigated. It was observed that the cations not
only shield the coulombic interaction but also exert an additional
accelerating effect, which increases in the order Li^+, Na^+, K^+,
NH_4^+, Rb^+ and Cs^+. Monovalent cations with an ionic radius of
0.23 nm were found to cause the maximum rate enhancement.

1. INTRODUCTION

Electron transfer (ET) reactions, which are also called oxi-
dation-reduction (redox) reactions, form the simplest class of
chemical reactions. Experimental and theoretical work in this
field has distinguished two types of mechanism, called inner-
sphere and outer-sphere. In inner-sphere reactions there is
strong interaction in the transition state and a bridging group
normally connects the two reactant centres; the bridging ion
or molecule may or may not be transferred during the reaction.
If the interaction between the ractants in the activated complex
is weak enough that it can be neglected in the calculation of
the activation energy, but strong enough to ensure a transition

W. J. Gettins and E. Wyn-Jones (Eds.), Techniques and Applications of Fast Reactions in Solution. 509–521.
Copyright © 1979 by D. Reidel Publishing Company.

probability of unity, and the reactants are only changing their
oxidation state, we speak of adiabatic outer-sphere ET reactions.
The development and testing of quantitative theories has been
most successful in recent years in the field of outer-sphere ET
reactions of stable transition metal complexes: we will concentrate
on this area in the following sections.

2. OUTER-SPHERE ELECTRON TRANSFER REACTIONS

The progress of theories in outer-sphere ET reactions between transition metal complexes has been due mainly to work by
R. A. Marcus (1), N. Sutin (2), V. G. Levich and R. R. Dogonadze
(3). If we take substitution-inert transition metal complexes in
solution as the reactants in an outer-sphere ET reaction it is
helpful to consider three areas of the reaction complex: the
metal ion in the centre, the strongly coordinated substitution
inert ligands in the inner-sphere, and the solvent molecules and
other ions in the outer-sphere. The time scales of the processes
which have to occur are: 10^{-15} s for electron transfer between
the centres, 10^{-13} s for vibrations of the coordinated ligands
to change bond lengths, and 10^{-11} s for the concerted rotations
of solvent molecules. The slowest process governs the rate constant.

Because the very fast movement of electrons allows for no
movement of the nuclei during transfer all other rearrange-
ments have to be completed before ET can occur (Franck-Condon restriction). To calculate the energy of the transition state it is
necessary to know the electric field which is adequate to the
situation where ET occurs. Marcus calculated the energy of the
transition state by considering an artificial static field where
the electron is only partly (m) transferred, so that he could calculate the positions for the different charged nuclei, which gave
the same potential energy for the system, whether the electron was
remaining on the reactants or was transferred to the products.

A detailed mechanism for the steps leading from separated reactants to separated products is the following:

$$A + B \rightleftharpoons A..B \qquad \text{(1) encounter complex of reactants}$$

$$A..B \rightleftharpoons (A..B)^*_r \qquad \text{(2) activated complexes of reactants or products}$$

$$(A^-..B^+)^*_p \rightleftharpoons A^-..B^+ \qquad \text{(3)}$$

$$A^-..B^+ \rightleftharpoons A^- + B^+ \qquad \text{(4) encounter complex of products}$$

Step 1 and step 4 are the formation or dissociation of encounter complexes from the separated reactants or products, respectively. In the encounter complex the distance between the reaction centre is approximately the same as in the activated complex. Step 2 and step 3 are the formation or dissociation of the activated complexes of the reactants or products, respectively.

If we apply the formalism of the transition state theory to such a mechanism we can write for the rate constant,

$$k = \kappa \cdot Z \exp(-\Delta G^*/RT) \tag{I}$$

In an adiabatic electron transfer reaction in aqueous solution the transfer coefficient κ is 1 and the collision frequency between two uncharged reactants is 10^{11} M^{-1} s^{-1} (1). If other solvents are used, Z may increase in value because it depends on the time of the concerted rotations of solvent molecules. If κ is smaller than 1 we speak of nonadiabatic reactions: their theoretical treatment is difficult and will not be attempted here (4). The free energy of activation (ΔG^*) can be divided into four important parts:

$$\Delta G^* = \Delta G_i^* + \Delta G_o^* + \Delta G_c^* + \Delta G_e^* \tag{II}$$

ΔG_i^* stands for the free energy required to change the inner co-ordination sphere (change in bond lengths)

ΔG_o^* gives the change in free energy of the outer-sphere (rotation of solvent dipoles or ion movement)

ΔG_c^* is the free energy required to bring the reactants from infinite distance to their separation in the activated complex (depending on the ionic strength)

ΔG_e^* expresses the free energy change caused by electronic changes (spin flipping); this term is not important in most reactions.

Equations to calculate ΔG_i^* from the force constants of bonds and ΔG_o^* from the static and the optical dielectric constants of the solvent and the size of the reactants are given in the literature (1,2). These parameters are, however, not well known in the transition state, and calculations of rate constants, using them, are therefore not very satisfactory.

Two relations which can be tested by experiment are given in the following part. Marcus (1) has derived an expression for the dependence of ΔG^* on the difference in redox potential ΔG^o (here the free energy change of the reaction corrected by the difference in Coulomb intraction between reactants and products), for the reaction of reactant A(ox) and B(red):

$$\Delta G^{*}_{AB} = \lambda_{AB}/4 + \Delta G^{o}_{AB}/2 + (\Delta G^{o}_{AB})^{2}/4\lambda_{AB} + \Delta G^{*}_{cAB} \qquad (III)$$

The Coulombic interaction term ΔG^{*}_{c} can be calculated using Debye-Hückel theory or can be neglected if the ionic strength and the dielectric constant of a solution are high and large reactant ions are used. λ has the dimensions of energy and is a measure of the reorganization of the inner- and outer coordination sphere, necessary to reach the transition state; its value is four times the activation energy for thermal electron transfer ΔG^{*} if ΔG^{o} is equal to zero and ΔG^{*} neglected. Details can be found elsewhere (1,2,4). Assuming that λ_{AB} is combined in a simple additive manner from the λ values of the reactants the following equation holds:

$$\lambda_{AB} = \lambda_{AA}/2 + \lambda_{BB}/2$$

or   (IV)

$$\lambda_{AB} = 2\Delta G^{*}_{AA} + 2\Delta G^{*}_{BB}$$

ΔG^{*}_{AA} is the activation energy required to transfer an electron from the oxidized form of reactant A to its reduced form; the same is valid for ΔG^{*}_{BB} and reactant B. These reactions are called homonuclear electron transfer or electron exchange reactions; here ΔG^{o} and the differences in the Coulomb interactions between reactants and products are zero. The physical interpretation of equ. IV arises from the fact that ΔG^{*}_{AA} is the reorganisation energy required to reach the transition state $(A(ox):A(red))^{*}$, ΔG^{*}_{BB} is the same energy required to reach $(B(ox):B(red))^{*}$, and ΔG^{*}_{AB} is the energy required to reach $(A(ox):B(red))^{*}$. (In the latter complex only one A and one B are reorganized in their inner and outer coordination sphere in contrast to the former where two A and two B are involved.)

The relations given in this section are suitable for predicting trends in ET reactions if we change the free energy ΔG^{o} of reactions or the size of reactants or if we try to divide the activation energy ΔG^{*}_{AB} into the parts caused by the reactants A and B.

3. EXPERIMENTAL

All rate constants reported in this article were measured using the continuous flow method with integrating observation (5). The complexes used were prepared according to literature cited elsewhere (6). $HClO_{4}$, $H_{2}SO_{4}$, HCl and all alkali chlorides were p. a. or suprapure and purchased from Merck AG. Only triply distilled water from a quartz apparatus was used to prepare the solutions. Only reactant concentrations smaller than 10^{-5} M were used.

4. RESULTS

A series of tris-phenanthroline complexes of Fe(III) was used
to test the relationship between the free energy ΔG^{o}, and the activation energy ΔG^{*}, in redox reactions. Different substituents
in the 5 and 6 positions of the 1,10-phenanthroline cause different
redox potentials of its Fe(III)-complexes, without a marked change
in their size and charge. It is therefore reasonable to assume that
the configuration in the transition state of these Fe(III)complexes,
with the same electron donor complex, is similar, so that only the
change in the free energy of these ET reactions should influence
the activation barrier. In Table 1 the measured ET rate constants
k_{exp} and their values, corrected for diffusion (k_{12}), of the series
$Fe(CN)_6^{4-}$ + $Fe(x-phen)_3^{3+}$ are given. The influence of diffusion on the
rate constant was calculated, using the steady state equation (6):

$$k_{exp}^{-1} = k_{ET}^{-1} + k_{dif.}^{-1} , \text{ with } k_{dif.} = 3.2 \times 10^9 \text{ M}^{-1}\text{s}^{-1} \text{ (see Fig. 2) (V)}$$

The activation energy ΔG_{12}^{*} was obtained from equ. I , λ_{12} was
calculated using equ.III (ΔG_{12}^{o} was calculated from the measured
differences in redox-potential ΔE_{12}^{o}), and $m_{12} = \Delta G_{12}^{*} \cdot \lambda_{12}^{-1}$ (1).
The values of the reorganization energy λ_{12} are similar within the
accuracy of measurement (5 %), showing that the assumptions
which were made about the similarity of the Fe(III) complexes
are justified.

k_{exp} $M^{-1}s^{-1}$	ΔE^{o}_{12} eV	Red. 1	Ox. 2	k_{12} (1/Mol·sec)	ΔG_{12}^{*} (kcal/Mol)	λ_{12} (kcal/Mol)	m_{12}
			$Fe(DMP)_3^{3+}$	$5,0 \cdot 10^8$	3,10	23,6	-0,36
$4,3 \cdot 10^8$	0,28		$Fe(MP)_3^{3+}$	$9,8 \cdot 10^8$	2,70	23,6	-0,34
$7,5 \cdot 10^8$	0,33						
$1,3 \cdot 10^9$	0,37	$Fe(CN)_6^{4-}$	$Fe P_3^{3+}$	$2,2 \cdot 10^9$	2,25	22,9	-0,32
$1,7 \cdot 10^9$	0,42		$Fe(CP)_3^{3+}$	$3,6 \cdot 10^9$	1,95	23,1	-0,29
$2,3 \cdot 10^9$	0,56		$Fe(NP)_3^{3+}$	$8,2 \cdot 10^9$	1,50	24,8	-0,23

Table 1: Influence of the difference in redox potential
(ΔE) on the free energy of activation ΔG_{12}^{*} after Marcus.
Temp. 296 K, in 0.83 M $HClO_4$; P = 1,10-phenanthroline,
DMP = 5.6-dimethyl-P, MP = 5-methyl-P, CP = 5-chloro-P,
NP = 5-nitro-P.

The plot of the free energy of the reaction ΔG_{12}^{o} against the activation energy ΔG_{12}^{*} modified by the quadratic term $(\Delta G_{12}^{o})^2/4\lambda_{12}$ shows a linear dependence with a slope of 0.5, as expected from equ. III (Fig. 1). Coulomb intractions could be neglected in this experiment because the ionic strength was 0.83 M in $HCLO_4$. This is demonstrated in Fig. 2 for several completely diffusion controlled ET reactions, which show the same rate constant at an ionic strength around 1 M. Another result obtained from these measurements is the rate constant of purely diffusion controlled reactions of complex ions comparable to those used in Fig. 1.

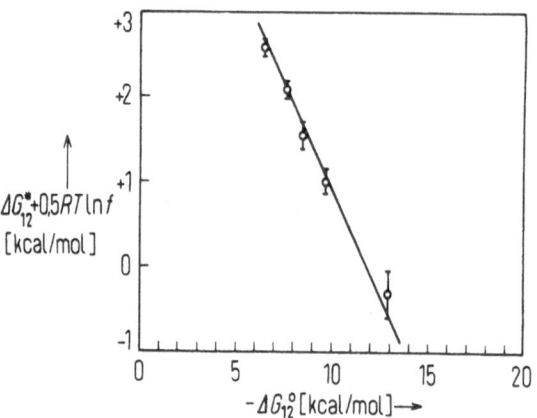

Fig. 1: Free energy of activation ΔG_{12}^{*} modified with the quadratic term in equ. III plotted against the standard free energy change ΔG_{12}^{o} for the reduction of the various $Fe(x-P)_3^{3+}$ in Table 1 by $Fe(CN)_6^{4-}$; $\ln f = -(\Delta G_{12}^{o})^2/(4RT\lambda_{12})$.

In Table 2 the results are summarized which were obtained from ET experiments between the same electron donor $(Fe(CN)_6^{4-})$ and three different tris-dipyridyl-complexes, of Fe(III), Ru(III) and Os(III), as electron acceptors. In the latter the size of the central metal atom is increased without changing the structure and the charge of these complexes. One can expect that the increase in size causes a decrease of the inner sphere reorganization energy ΔG_i^{*}, on account of weaker ligand central-metal interactions, as well as a decrease in the outer sphere reorganization energy ΔG_o^{*}, because of the solvent dipoles, which are further away from the reaction centre. Thus the change in the electro-

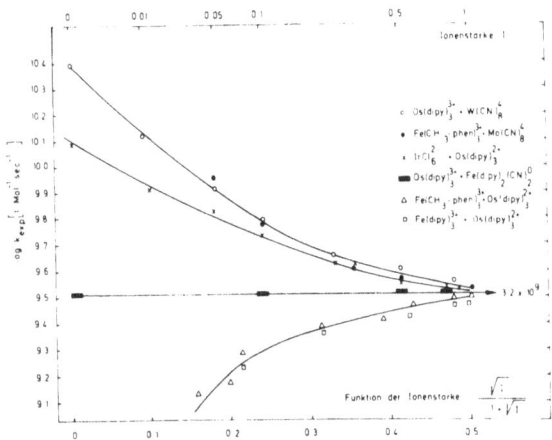

Fig. 2: Dependence of the experimentally observed rate
constants, k_{exp} of purely diffusion controlled ET reactions, on the ionic strength I, using complexes
with different charge products, in aqueous $NaClO_4$
(o ▬ x) or $HClO_4$ (o ● x △ □); Temp. 297 K.

Red.(1)	Ox.(2)	heteronuklear				homonuklear Ox	
		k_{ET} (1/ M ·s)	ΔG^*_{12} (kcal/M)	ΔG^o_{12} (kcal/M)	λ_{12} (kcal/M)	k_{ET} (1/Mol·s)	λ_{22} (kcal/M)
$Fe(CN)_6^{4-}$	$Fe(dipy)_3^{3+}$	$2.2 \cdot 10^9$	2.25	-8.76	23.2	$6.8 \; 10^9$	3.15
	$Ru(dipy)_3^{3+}$	$2.2 \cdot 10^{10}$	0.89	-12.68	21.4	$3.1 \; 10^{10}$	1.35
	$Os(dipy)_3^{3+}$	$1.6 \cdot 10^8$	3.8	-2.54	20	$\geqslant 10^{11}$	0

Table 2: Calculation of homonuclear electron exchange
rate constants from heteronuclear ET reactions. Temp.
297 K, in 1 M H_2SO_4.

static field felt by the solvent dipoles during ET is weaker,
and the energy required to reorient these dipoles is less. The
heteronuclear rate constants k_{ET} are corrected for the effect
of diffusion and were used to calculate λ_{12} (equ. III) from ΔG^*_{12}
(equ. I) and the measured difference in redox-potential ΔG^o_{12}.

The addition of 1 M of H_2SO_4 justfies the neglect of the coulombic term ΔG_C^* (Fig. 2). From the heteronuclear λ_{12}, together with the λ_{11} value of the $Fe(CN)_6^{-3/-4}$ complex (40.1 kcal.mol^{-1} (7)), λ_{22} of the homonuclear electron exchange could be calculated using equ. IV. The k_{ET} values of the homonuclear ET reactions of $Fe(dipy)_3^{+3/+2}$, $Ru(dipy)_3^{+3/+2}$ and $Os(dipy)_3^{+3/+2}$ were obtained from equ. I using the relationship $\Delta G_{22}^* = \lambda_{22}/4$. An increase in the homonuclear electron exchange rate constants of these tris-di-pyridyl-complexes is observed with increasing size, as expected from the Marcus theory. The absolute numbers of k_{ET} might be uncertain by a factor of three.

In many ET reactions reported in the literature high concentrations of "inert electrolyte" are used to shield electrostatic interactions. The measurements described now were aimed at separating the primary salt effect from further possible effects, the "inert electrolyte" might have on the ET process. The reactants $Fe(CN)_6^{4-}$ and $IrCl_6^{-}$ were chosen because they are known to be inert to ligand substitution. Their ET reaction can therefore be classified as "outer sphere" type, and is primarily determined by the reorganization of the outer coordination sphere. Since cations of the supporting electrolyte are likely to be present in this region we can expect any influence to be felt here. Fig. 3 is a scale diagram of the reacting species. It shows how the presence of cations can modify the surrounding ionic and solvent region. Remarkable dif-

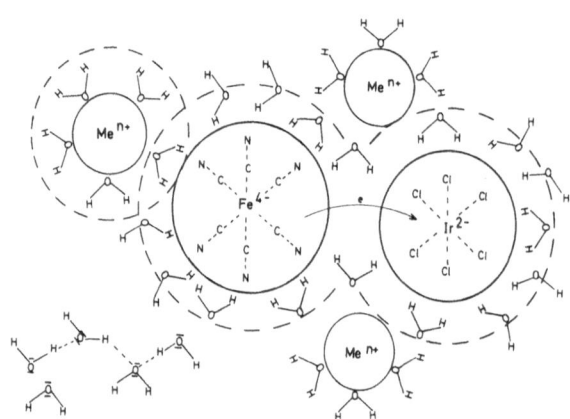

Fig. 3: Scale diagram of the situation in the ET reaction shown, if cations as supporting electrolyte are added; (change in water structure).

ferences between the cations can be seen in the experimental results, given in Fig. 4. An ET rate constant of 10^6 M^{-1} s^{-1} is measured at a reactant concentration of 5 x 10^{-6} M. At low concentrations of the cations added, k_{ET} is increasing, with increasing
charge, if we compare rates measured at the same ionic strength I.
Using higher concentrations of Me^{2+} and Me^{3+} the rates of ET level
off rapidly towards a constant value which agrees with the maximum
rate enhancement of 3 x 10^2, calculated according to Debye theory.
The concentrations of Me^{2+} or Me^{3+}, at which the values of k_{ET}
start levelling off, correspond to the formation of ion associates with the reactants. The behaviour of acidic solutions is
determined by the protonation of Fe(CN)$_6^{4-}$, and the strong change
in the redox-potential caused by this association. Details of
this ET reaction will be given in a forthcoming publication.
The rate constants in the presence of Li$^+$ can be explained according to the Debye-Hückel theory (primary salt effect). All
other k_{ET} values, measured in the presence of Me$^+$, are higher
than can be expected from pure coulombic interactions.

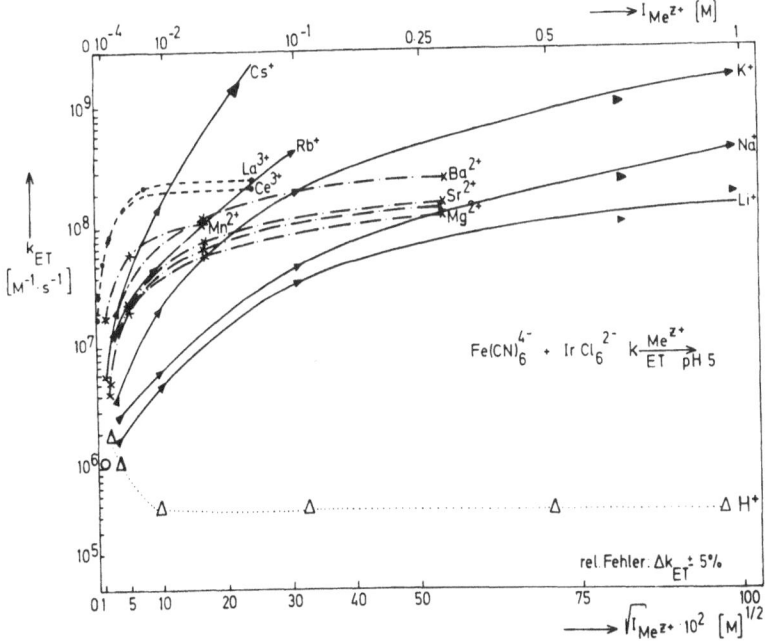

Fig. 4: Relation between the rate constant of ET and
the ionic strength, I, adjusted with the chlorides of
different cations. Medium H$_2$O plus additives; conc. of
Fe(CN)$_6^{4-}$ and IrCl$_6^{2-}$, 10^{-5} to 10^{-6} M; Temp. 297 K.

Tab. 3 summarizes these k_{ET} values, corrected for the effect of
diffusion. In Tab. 4 the same results for NR_4^+ ions are given. The
change in the free energy of the reactions, ΔG_{12}^o, were measured,
and λ_{12}, as well as k_{ET} ($\Delta G_{12}^o = o$) values, were calculated according to Marcus theory (equ. I and equ. III). An ionic strength
of 0.1 M was chosen to avoid precipitation of the reactants.

Ionic strength [mol dm^{-3}]	Me$^+$	experimental results		calculated(after Marcus)	
		$k_{ET}(\Delta G_{12}^o \neq 0)$ [M^{-1} s^{-1}]	ΔG_{12}^o [kcal mol^{-1}]	λ_{12} [kcal mol^{-1}]	$k_{ET}(\Delta G_{12}^o =0)$ [M^{-1} s^{-1}]
I = 10^{-1}	Li$^+$	$3{,}5 \cdot 10^7$	12,7701	40,34	$4{,}0 \cdot 10^3$
	Na$^+$	$4{,}7 \cdot 10^7$	12,7701	39,56	$5{,}6 \cdot 10^3$
	K$^+$	$2{,}0 \cdot 10^8$	12,5627	35,38	$3{,}2 \cdot 10^4$
	NH$_4^+$	$1{,}9 \cdot 10^8$	12,5627	35,52	$3{,}1 \cdot 10^4$
	Rb$^+$	$4{,}1 \cdot 10^9$	12,3322	26,49	$1{,}4 \cdot 10^7$
	Cs$^+$	$6{,}1 \cdot 10^9$	12,0786	24,93	$2{,}7 \cdot 10^7$
I = 1	Li$^+$	$1{,}8 \cdot 10^8$	11,7559	34,47	$4{,}7 \cdot 10^4$
	Na$^+$	$3{,}9 \cdot 10^8$	11,7559	32,38	$1{,}1 \cdot 10^5$
	K$^+$	$1{,}7 \cdot 10^9$	11,5254	27,95	$7{,}5 \cdot 10^5$
	NH$_4^+$	$2{,}5 \cdot 10^9$	11,5254	26,84	$1{,}2 \cdot 10^6$

Table 3: Influence of different alkali-cations on the
rate constant of ET at the same ionic strength, observed in the reaction $Fe(CN)_6^{4-} + IrCl_6^{2-}$. Temp. 297 K
in H_2O, anion Cl$^-$, reactant conc. 10^{-5} to 10^{-6} M.

Corrected Electron Transfer Rate Constants for
Quarternary Ammonium Ions

$$IrCl_6^{2-} + Fe(CN)_6^{4-} \xrightarrow[NR_4^+]{k_{ET}}$$

Ion	experimental values		calculated (after Marcus)	
	$k_{ET}(\Delta G_{12}^o \neq 0)$ [M^{-1} s^{-1}]	ΔG_{12}^o [kcal mol^{-1}]	λ_{12} [kcal mol^{-1}]	$k_{ET}(\Delta G_{12}^o =0)$ [M^{-1} s^{-1}]
NH$_4^+$	$1{,}9 \cdot 10^8$	12,4474	35,35	$3{,}3 \cdot 10^4$
(CH$_3$)$_4$ N$^+$	$1{,}9 \cdot 10^8$	13,2542	36,54	$2{,}0 \cdot 10^4$
(C$_2$H$_5$)$_4$ N$^+$	$5{,}1 \cdot 10^7$	13,8305	40,94	$3{,}1 \cdot 10^3$
(C$_4$H$_9$)$_4$ N$^+$	$1{,}9 \cdot 10^7$	14,1762	44,08	$8{,}2 \cdot 10^2$

Table 4: Influence of quartary ammonium chlorides on
the ET rate constant between $Fe(CN)_6^{4-}$ and $IrCl_6^{2-}$. Temp.
297 K in H_2O, reactant conc.: 10^{-5} to 10^{-6} M.

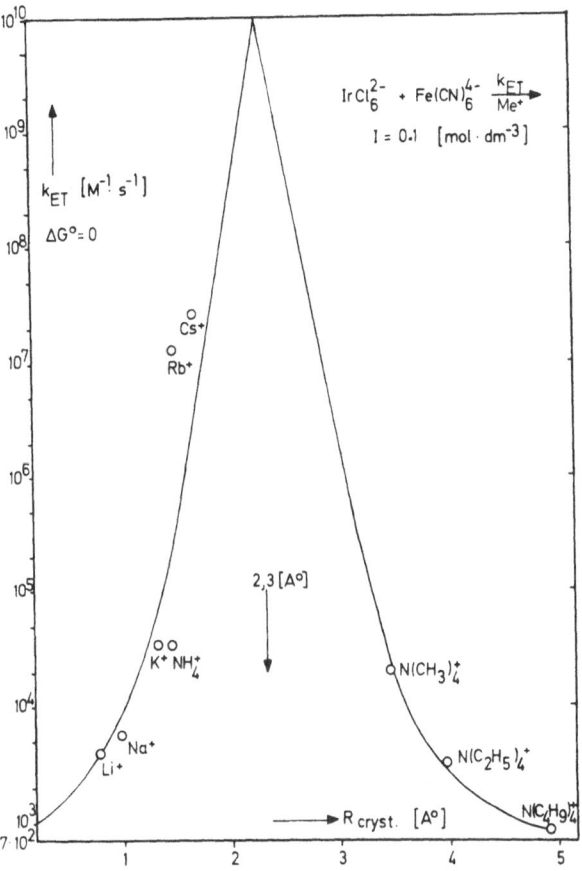

Fig. 5: Relationship between the crystallographic
radii of monovalent cations and the rate constant of
ET, calculated for the situation of no change in the
free energy of the reaction $Fe(CN)_6^{4-}$ + $IrCl_6^{2-}$ (Data
from Table 3 and 4).

Fig.5 shows the dependence of the electron transfer rate constants,
k_{ET}, on the crystallographic radii of Me^+. A situation was simulated,
using Marcus theory, where the change in the free energy of reaction,
ΔG°, is zero. This is necessary to obtain rate constants, which can
be compared, because they are only determined by the reorganization
energy λ. If we extrapolate from the measured values to larger
or smaller radii we find the optimum radius for monovalent cations
to 0.23 nm. The corresponding k_{ET} of 10^{10} $M^{-1} s^{-1}$, is in excellent
agreement with the Z value of equ. I. Z represents the maximum in
the rate of ET possible in aqueous solution for uncharged reactants.

In Fig. 5 an ionic strength of 0.1 M was used so that coulombic repulsion of the reactants is not completely shielded. This effect accounts for a factor of 10 as shown by the measurements in Table 3 (I = 1 M). These results give clear evidence that cations are exerting a strong catalytic effect on ET reactions between anionic reactants, depending on their size, charge and concentration. This cannot be explained by shielding of the Coulomb interaction. Strong ion associates are not accelerating ET apart from a pure primary salt effect. Measurements of the temperature dependence of the above mentioned ET reactions in the presence of 0.1 M Me^+ (Fig. 6) show no linear relationship. This is an indication for a complex reaction mechanism. The same temperature dependence measured by Shporer et al. show artificial results, caused by the technique applied (7).

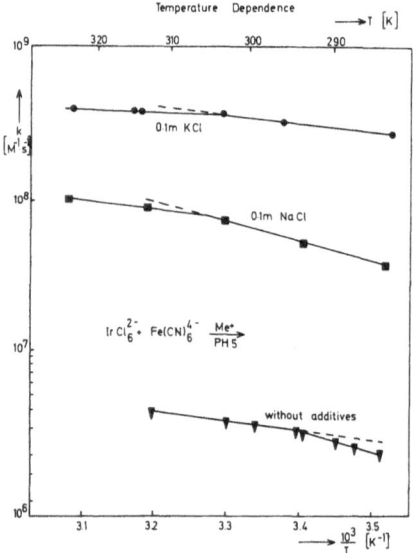

Fig. 6: Temperature dependence of the observed rate constant of the ET reaction shown in the presence of 0.1 M NaCl, and KCl, as well as without supporting electrolyte. Conc. Na_2IrCl_6 and $Na_4Fe(CN)_6$ 10^{-5}-10^{-6} M in H_2O.

5. CONCLUSION

Modern electron transfer theories, especially the Marcus theory, are very useful to account for the effect of the difference in redox potential in outer sphere electron transfer reactions. Outer sphere and inner sphere reorganization effects can be distinguished. In heteronuclear ET reactions the contribution of each reactant might be calculated if one homonuclear ET rate constant is known. There is no way of calculating correct ET rate constants only from the size and charge of the reactants and the dielectric constants as well as the electrolytic content of the medium used. This is mainly due to a lack of information about the reactant configuration in the transition state and the structure of water in the surrounding solution. A problem such as the catalytic effect of cations on ET between anionic reactants is far too complicated to be solved by existing ET theories. One can only speculate that the replacement of the water dipoles connected through hydrogen bridges by suitable cations, in the vicinity of the transition complex, promotes the polarization state, necessary for ET to occur (8). Strong ion associates show no catalytic effect. A type of interaction in which the hydration sphere of the cations are not or only weakly influenced can therefore be postulated.

ACKNOWLEDGEMENT

I am indebted to my coworkers, H. Jürgensen, H. Bruhn and S. Nigam who performed most of the experiments reported in this article.

REFERENCES

(1) Marcus, R. A.: 1964, Ann. Rev. Phys. Chem., 15, pp. 155.
(2) Sutin, N.: 1968, Accounts Chem. Res. 1, pp. 225.
(3) Levich, V. G., and Dogonadze, R. R.: 1959, Dokl. Akad. Nauk.
     SSR, 124, pp. 123; Proc. Acad. Sci. USSR, Phys. Chem. Sect.
     English Transl., 124, pp. 9.
(4) Lumry, R. W., Reynolds, W. L.: 1966, "Mechanisms of Electron
     Transfer", The Ronald Press Comp., N. Y.
(5) Holzwarth, J. F.: 1979, Fast Continuous Flow, this volume.
(6) Holzwarth, J. F., Strohmaier, L., and Gerischer, H.: 1972,
     Ber. Bunsenges. Phys. Chem., 76, pp. 1048;
     Holzwarth, J. F., and Jürgensen, H.: 1974, Ber. Bunsenges.
     Phys. Chem., 78, pp. 526.
(7) Shporer, M., Ron, G., Loewenstein, A., and Navon, G.: 1965,
     Inorg. Chem., 4, pp. 361.
(8) Holzwarth, J., and Strohmaier, L.: 1973, Ber. Bunsenges.
     Phys. Chem., 77, pp. 1145.

COMBINED STOPPED-FLOW/CONTINUOUS-FLOW ARRANGEMENT FOR KINETIC
MEASUREMENTS IN THE SECOND TO MICROSECOND RANGE: INVESTIGATION
OF ELECTRON-TRANSFER REACTIONS WITH METAL-PORPHYRIN COMPLEXES

H. Bruhn, J. Westerhausen, J. F. Holzwarth and
J. H. Fuhrhop*

Fritz-Haber-Institut der Max-Planck-Gesellschaft
Faradayweg 4-6, D-1000 Berlin 33, W. Germany.

*Institut Org. Chem., Freie Universität, Takustr. 3,
D-1000 Berlin 33, W. Germany.

A combined stopped-flow, continuous-flow method is described
which allows the measurement of half lives of first and second
order reaction between 10^{-1} s and 10^{-5} s if the continuous flow
mode is applied and 10^{-2} s to 10 s if the stopped flow mode is used.
This special flow arrangement was employed to investigate the oxidation of substitution-inert transition metal complexes such as
$Fe(CN)_2(phen)_2^0$, $Ag(II)$-protoporphyrin, $Ag(II)$-tetracetylporphyrin
and cytochrome(II)c, with $Os(dipy)_3^{3+}$ and $IrCl_6^{2-}$. In aqueous
solutions we found second order electron transfer rate constants
between 10^6 and 10^9 $M^{-1}s^{-1}$, depending on the ionic strength of the
solutions. Measurements in the presence of the surface active substances sodium-dodecylsulfate (SDS) or dodecyltrimethylammonium-
chloride (DTAC) have shown: The rates of oxdiation of $Fe(CN)_2(phen)_2^0$
by $Os(dipy)_3^{3+}$ or $IrCl_6^{2-}$ were strongly decreased if the iron complex
is incorporated into micelles of SDS. Incorporation of the porphyrin
complexes into micelles of SDS or DTAC causes an additional barrier
for oxidation of $Ag(II)$-protoporphyrin, which is not observed in
the same electron transfer reactions of $Ag(II)$-tetracetylporphyrin.
This might be explained by a reaction path through the edges of the
porphyrin ring, which is consistent with the oxidation of cyto-
chrome(II)c by $Os(dipy)_3^{3+}$ or $IrCl_6^{2-}$.

W. J. Gettins and E. Wyn-Jones (Eds.). Techniques and Applications of Fast Reactions in Solution. 523-534.
Copyright © 1979 by D. Reidel Publishing Company.

1. INTRODUCTION

The rapid mixing of two reactants affords the most general and flexible approach to the kinetic study of fast reactions. Until recently only very special flow arrangements incorporate mixing devices which allow measurements of reactions with half lives around 1 ms (1). We have developed a continuous flow apparatus with a time resolution of 10 µs for first and second order reactions (2). To extend the application of this "Continuous Flow Method with Integrating Observation" (CFMIO) to longer half lives we have inserted another detection channel to use the same measuring cell in a stopped-flow mode up to half lives of 10 s and higher; (for details see EXPERIMENTAL).

We have used this combined flow method to investigate the electron transfer behaviour of the complex ions of silver(II)-porphyrin and cytochrome(II)c with the very fast oxidizing agents $Os(dipy)_3^{3+}$ and $IrCl_6^{2-}$. These reactions are of interest because complexes containing a porphyrin ring system represent a class of compounds which participate in important life processes, for example, in photosynthesis, in mitochondrial respiration and as oxygen carriers in the transport system of blood.

In nature these complexes are connected with proteins and often they are situated inside membranes. By using the silver-porphyrin complexes as a simple model for the more complex systems occurring in nature we can compare their electron transfer rate constants in aqueous solutions with those in solutions containing micelles which incorporate our model complexes. In this way we are trying to get more information about the effect of hydrophobic barriers on electron transfer. Well understood oxidation reduction reactions of transition metal complexes which are reacting with an outer sphere mechanism are measured under equal conditions to separate pure electrostatic effects from other influences.

Cytochrome c is investigated as an example of an iron-porphyrin complex incorporated into a big protein molecule.

2. EXPERIMENTAL

The principle of the combined stopped-flow, continuous-flow measuring cell is shown in Fig. 1. The "Continuous Flow Method with Integrating Observation", CFMIO (2), uses a light beam in the direction of flow to observe the progress of reaction. In this way it is possible to integrate the mixing chamber into the observation tube as well as to increase the thickness of the observed layer up to some cm. Rectangular to the flow tube we have inserted light guides to use the same measuring cell in a stopped flow (SF) mode. This cell allows the measurement of half lives of first and second order reactions between 10^{-1} s and 10^{-5} s if the CFMIO mode is applied and 10^{-2} s to 10 s if the SF mode is used. The measuring cell and the syringes transporting the re-actants are thermostated and only glass and teflon is used for the parts containing the reactant solutions. A photograph of the cell is included in Fig. 1, details of the complete arrangement are given in reference (2).

PRINCIPLE OF COMBINED
(SF) STOPPED FLOW - CONTINUOUS FLOW (CF)

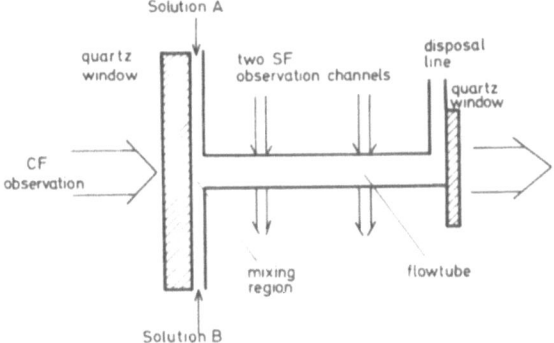

Fig. 1: Principle of the combined CF, SF arrangement including a photograph of the measuring cell.

The relaxation trace of the complexation reaction of Ni^{2+} with pyridine-2-azo-p-dimethylaniline (PADA), shown in Fig. 2, was achieved by applying the SF mode. The measured time constant is in good agreement with the literature (3).

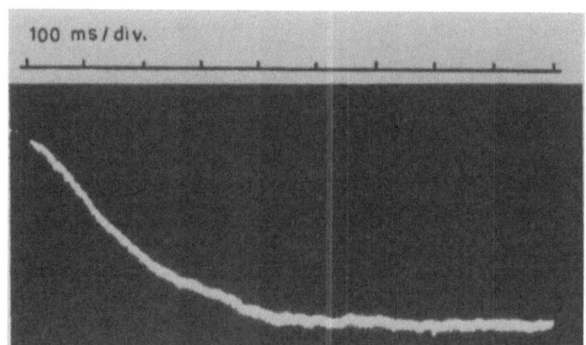

Fig. 2: Relaxation trace of the reaction Ni^{2+} + PADA using the SF mode; conc.: $Ni^{2+} = 10^{-2}$ M, PADA = 10^{-5} M, temp. 292 K, $k_{12} = 1.1 \times 10^{3}$ $M^{-1}s^{-1}$ in H_2O.

Most of the materials used are mentioned in reference (3) and (4). Cytochrome c (purissimum) from horse heart and SDS (purissimum) was supplied by Fluka, and DTAC was obtained from Eastman Kodak. All solutions were prepared immediately before the experiments, using triply distilled water from a quartz apparatus. For achieving the reduced form cytochrome(II)c see reference (5). All reactions including porphyrin complexes were measured under pseudo-first-order conditions so that the exact concentrations of the porphyrin compounds were not necessary to calculate the rate constants. This avoids uncertainties caused by aggregation of the porphyrin complexes.

3. RESULTS

The outer sphere second order electron transfer reactions between the substitution inert complexes $Os(dipy)_3^{3+}$ or $IrCl_6^{2-}$ as oxidizing agents and $Fe(CN)_2(phen)_2$ as the electron donor compound, which have been investigated in aqueous solutions before (6), were chosen to examine the effect of micelles on the rate of reaction. In Fig. 3 the observed rate constants of ET in the presence of different concentrations of SDS without additional salt and with 0.5 M NaCl to shield coulombic forces are shown. In the measurement without additional NaCl we observed:
Below the cmc of SDS (8×10^{-3} M), a small decrease of the ET rate of $IrCl_6^{2-}$ with increasing amount of SDS, which is caused by weak association of SDS-monomers with $Fe(CN)_2(phen)_2$; a greater decrease in the rate for the reaction with $Os(dipy)_3^{3+}$ at low SDS concentrations, which changes to an increase just below the cmc. The special

behaviour of the latter is caused by the strong association of SDS and $Os(dipy)_3^{3+}$ (4). Above the cmc both reactions are strongly decreasing, their differences in rate can be explained by opposite electrostatic interaction with the negative surface charge of the micelles.

If 0.5 M NaCl is added to the same solutions electrostatic interactions disappear almost completely and the cmc is shifted to 5×10^{-4} M: No change in rate is observed below the cmc but starting just before the cmc, k_{ET} of both reactions is almost linearly decreasing with micelle concentration and reaches a value two orders of magnitude below the rate in aqueous solution. (All observed rates above 10^8 have to be corrected for diffusion to calculate the electron transfer rate constants.) A detailed mechanism and exact calculations as well as a proof for the incorporation of $Fe(CN)_2(phen)_2^o$ into micelles of SDS are reported in reference (4).

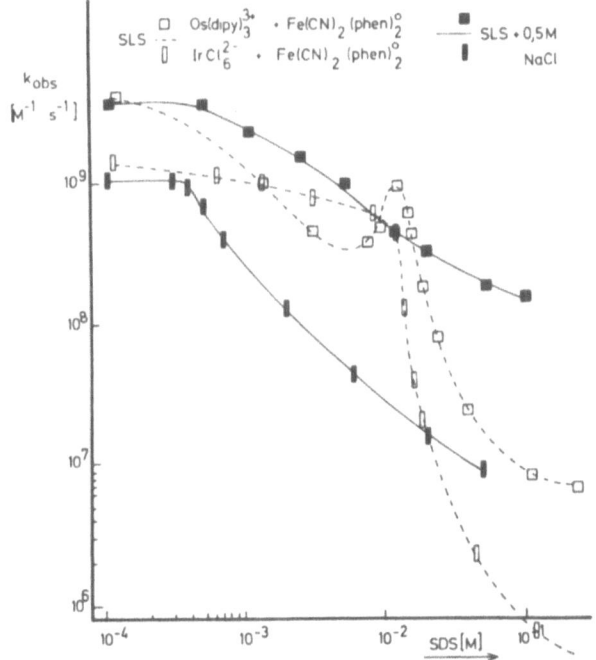

ET in Sodiumdodecylsulfat (SDS) Solutions

Fig. 3: Dependence of the rate of the ET-reactions shown on the concentration of SDS in the presence and absence of further electrolyte. Temp. 298 K, reactant concentrations between 2×10^{-5} and 10^{-6} M.

528

We can conclude from the results in Fig. 3 that the incorporation of one reactant $(Fe(CN)_2(phen)_2^0$ into the micelles of SDS causes a decrease of more than 10^2 in the rate of ET either with $IrCl_6^{2-}$ or with $Os(dipy)_3^{3+}$, if electrostatic interactions are shielded by the addition of O.5 M NaCl. In solutions without NaCl special association reactions complicate the interpretation of the results.

Fig. 4 shows the absorption spectra of Ag(II)PP and Ag(II)TCP in aqueous and micellar solutions of SDS. The differences in the groups attached to the porphyrin ring can be seen from the structure included in Fig. 4. A marked shift in the so called Soret-band around 400 nm in SDS solutions in comparison to H_2O indicates that the porphyrin complexes are incorporated into micelles. This is in agreement with NMR data and observations of Simplicio (7). The increase in the intensity of the Soret-band by a factor of two caused by micelles is an indicator that mainly monomers of the porphyrin complexes are incorporated and that dimers or higher aggregates are existing in aqueous solution. The true aggregation number in both systems is yet not fully known. Further investigations are necessary to clarify this question. We have therefore performed all ET experiments shown in Table 1-4 with an excess of oxidizing agent, so that a knowledge of the concentrations of the porphyrin complexes is not necessary in order to calculate the rate constants.

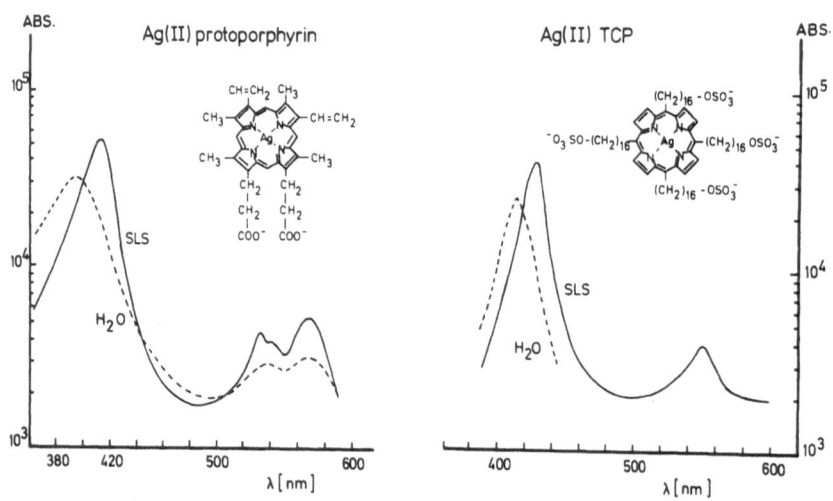

Fig. 4: Absorption spectrum of Ag(II)PP and Ag(II)TCP in aqueous and 2×10^{-2} M SDS containing solutions at 298 K.

In Table 1 the results of the oxidation of $Ag(II)PP^{2-}$ by $IrCl_6^{2-}$ are summarized. The rate of ET in aqueous solution is two orders of magnitude below the diffusion controlled limit, even the addition of cations does not enhance the rate. A possible explanation can be given by assuming that the shielding of electrostatic repulsion by the cations is overcompensated by an increase in the aggregation number, so that the sides of the vinyl groups are blocked by the propionic acid groups and ET is becoming more difficult. The addition of SLS concentrations higher than the cmc causes a decrease of the reaction rate by a factor of 10. This barrier can also be explained by blocking of the vinyl groups inside the micelle, with the propionic acid groups stacking in the Stern layer. The addition of 0.1 M NaCl can not change this behaviour ($k_{ET} = 2.4 \times 10^6$ $M^{-1} s^{-1}$ in 10^{-2} SLS + 0.1 M NaCl).

Reaction of Ag(II)-protoporphyrin

$$[Ag(II)\,protoporphyrin]^{2-} + IrCl_6^{2-} \xrightarrow{k_{12}}$$

Medium H_2O + Additives [M]	k_{12} [M^{-1} s^{-1}]	
0	$2,1 \times 10^7$	electrostatic
0,5 NaCl	$1,1 \times 10^7$	
10^{-3} SLS	$1,9 \times 10^7$	surfactant effect
5×10^{-3} SLS	$5,2 \times 10^6$	
10^{-2} SLS	$2,9 \times 10^6$	
2×10^{-2} SLS	$1,7 \times 10^6$	
10^{-2} LiCl	$5,5 \times 10^6$	cation
10^{-2} NaCl	$8,2 \times 10^6$	
10^{-2} CsCl	$8,1 \times 10^6$	
10^{-2} CaCl$_2$	$2,2 \times 10^6$	

Table 1: ET rate constants of $Ag(II)PP + IrCl_6^{2-}$ at 297 K in aqueous solutions with different additives; conc. $Ag(II)PP \leq 10^{-6}$ M, conc. $IrCl_6^{2-} = 10^{-5}$ M.

The ET rate constants of the reactions of Ag(II)PP^{2-} with Os(dipy)$_3^{3+}$ are given in Table 2. At low ionic strength the ET rate is still two orders below complete diffusion control if one takes the electrostatic attraction into account; this is demonstrated by the decrease of the rate to 1.4×10^7 M^{-1}s^{-1} in solutions containing 1 M NaCl; 0.1 M HCl is more effective because the propionic acid groups are protonated. Addition of SLS concentrations below the cmc causes aggregation or precipitation of Os(dipy)$_3^{3+}$ with SLS, these results are therefore not representative. Above the cmc a small barrier for ET is observed. If micelles of the cationic surfactant DTAC are present an ET barrier equal to the results in Table 1 is measured. Again blocking of the vinyl-groups inside the micelles might be the reason.

Reaction of Ag(II) protoporphyrin

$$Ag(II)protoporphyrin^{2-} + Os(bipy)_3^{3+} \xrightarrow{k_{12}} Ag(III)pp^- + Os(bipy)_3^{2+}$$

Medium H$_2$O + Additives [M]	k_{12} [M^{-1} s^{-1}]		
0	$2,1 \times 10^8$	electrostatic	
0,1 NaCl	$1,4 \times 10^8$		
1 NaCl	$1,4 \times 10^7$		
0,1 HCl	$8,5 \times 10^6$		
10^{-2} SLS	$6,8 \times 10^5$	anionic	surfactant effect
0,1 NaCl + 10^{-2} SLS	$7,4 \times 10^6$		
0,5 NaCl + 10^{-2} SLS	$6,7 \times 10^6$		
2×10^{-2} DTAC	$1,4 \times 10^6$	cationic	
0,5 NaCl + 2×10^{-2} DTAC	$1,5 \times 10^6$		

Table 2: ET rate constants of Ag(II)PP + Os(bipy)$_3^{3+}$ at 297 K in aqueous solutions with different additives; conc. Ag(II)PP $\leq 10^{-6}$ M, conc. Os(bipy)$_3^{3+}$ = 10^{-5} M.

Using $Ag(II)TCP^{4-}$, with 4 long carbon chains attached to the porphyrin ring instead of $Ag(II)PP^{2-}$ with two vinyl and two propionic acid groups, in the ET reaction with $Os(dipy)_3^{3+}$ (Table 3) a behaviour similar to Table 2 is observed in solutions without surfactant. The precipitation or aggregation of $Os(dipy)_3^{3+}$ in the presence of SLS without further electrolyte falsifies these results. Solutions containing anionic micelles and electrolyte (2 x $10 M^{-2}$ SLS and 0.5 M NaCl) show no precipitation of $Os(dipy)_3^{3+}$ but also no marked barrier for ET.

Reactions of Ag(II) TCP

$$[Ag(II)TCP]^{4-} + Os(bipy)_3^{3+} \xrightarrow{k_{12}} [Ag(III)TCP]^{3-} + Os(bipy)_3^{2+}$$

Medium H_2O + Additives [M]	k_{12} [M^{-1} s^{-1}]		
0	$2,6 \times 10^8$		electrostatic
0,5 NaCl	7×10^7		effect
5×10^{-3} SLS	$4,3 \times 10^7$	anionic surfactant	
10^{-2} SLS	$1,5 \times 10^6$		
2×10^{-2} SLS	$5,8 \times 10^5$		
5×10^{-2} SLS	$2,4 \times 10^5$		
0,5 NaCl + 2×10^{-2} SLS	$5,9 \times 10^7$		

Table 3: ET rate constants of $Ag(II)TCP + Os(bipy)_3^{3+}$ at 298 K in aqueous solutions with different additives; conc. $Ag(II)TCP \leq 10^{-6}$ M, conc. $Os(bipy)_3^{3+} = 10^{-5}$ M.

If we investigate the reaction $Ag(II)TCP^{4-} + IrCl_6^{2-}$ (Table 4)
we measure an ET rate which is a factor of 3-5 lower than with
$Os(dipy)_3^{3+}$ in aqueous solution. The electrostatic repulsion and
the slower electron-exchange rate of $IrCl_6^{2-/3-}$ might be the reason.
Similar to Table 3 the addition of SLS above the cmc shows no
marked barrier for ET. If a SLS concentration higher than the
cmc containing 0.5 M NaCl is used again almost the same rate constant as in an equal solution without SLS is measured. There are
two explanations for this result, either Ag(II)TCP is not incorporated into micelles of SLS or more reasonable the long C_{16}-
chains of Ag(II)TCP are forming only loose aggregates with the
C_{12}-chains of SLS. These aggregates are not blocking the edges
of the porphyrin ring.

$$[Ag(II)\,TCP]^{4-} + IrCl_6^{2-} \xrightarrow{k_{12}} [Ag(III)\,TCP]^{3-} + IrCl_6^{3-}$$

Medium H_2O + Additives [M]	k_{12} [M^{-1} s^{-1}]	
0	9×10^6	electrostatic
0,5 NaCl	$1,3 \times 10^7$	effect
10^{-3} SLS	$8,4 \times 10^6$	
5×10^{-3} SLS	$8,3 \times 10^6$	anionic surfactant
8×10^{-3} SLS	$5,9 \times 10^6$	
10^{-2} SLS	$5,2 \times 10^6$	
$0,5$ NaCl $+ 10^{-2}$ SLS	$5,6 \times 10^6$	

Table 4: ET rate constants of $Ag(II)TCP + IrCl_6^{2-}$ at
296 K in aqueous solutions with different additives;
conc. $Ag(II)TCP \leq 10^{-6}$ M, conc. $IrCl_6^{2-} = 10^{-5}$ M.

4. DISCUSSION

Comparing the results of Table 1-4 we notice that Ag(II)PP
and Ag(II)TCP are showing similar results in aqueous solutions
without surfactants if electrostatic effects are compensated.
The rate of ET with the very fast oxidizing agents $IrCl_6^{2-}$ and
$Os(bipy)_3^{3+}$ are two orders below diffusion control. The compensation
of the difference in the redox potential between $IrCl_6^{2-}$ (0.96 eV;
$k_{EX} = 10^7$ $M^{-1}s^{-1}$) and $Os(dipy)_3^{3+}$ (0.85 eV; $k_{EX} = 10^{11}M^{-1}s^{-1}$) by

the faster electron-exchange rates (6) explains the similarity in
the reactions of both oxidizing species. An ET through the edges of
the porphyrin system and not along the axial positions of Ag(II)
might be the reason why these reactions are not diffusion controlled. The ET reactions of cytochrome(II)c with the same electron
acceptors (Table 5) at high ionic strength showing almost the same
rate constants are consistent with these assumptions. In cyt.c

reaction		k $[M^{-1}s^{-1}]$	Tris-Puffer pH 7
Cyt c(II)$^{7+}$	+ IrCl$_6^{-2}$	1,8 x 10^9	—
		6 x 10^7	0,5 NaCl
	+ Os(dipy)$_3^{3+}$	1,3x 10^6	—
		8 x 10^6	0,5 NaCl
	+ Fe(CN)$_6^{3-}$	3 x 10^8	—
		3,5 x 10^7	0,5 NaCl

Table 5: ET rate constants of cytochrome(II)c with
IrCl$_6^{2-}$, Os(dipy)$_3^{3+}$ and Fe(CN)$_6^{3-}$ in aqueous solutions
with different additives at 296 K, conc. cyt(II)c
= 10^{-6} M, conc. oxidizing agent = 10^{-5} M.

the axial positions of the porphyrin system are burried in the
interior of the surrounding protein (5). (The measurements with
cytochrome(II)c at low ionic strength are superposed by electrostatic interactions.) In the presence of micelles the similarity
between Ag(II)PP and Ag(II)TCP disappears. A possible explanation
is the following: The interaction of Ag(II)PP with micelles is
strong enough to keep the edges of the porphyrin ring inside the
micelles which creates a barrier for ET similar to cytochrome c.
The aggregates formed by Ag(II)TCP with SDS are not normal micelles.
In such aggregates the edges of the porphyrin ring can be attacked
by the oxidizing agent in a similar way like in H$_2$O.

5. CONCLUSION

To sum up these results we find that the ET between the
electron donors Ag(II)PP^{2-} or Ag(II)TCP^{4-} and the electron acceptors IrCl$_6^{2-}$ or Os(dipy)$_3^{3+}$ is not diffusion controlled in aqueous
solutions; this observation is made at high and low ionic strength.
The incorporation of the porphyrin complexes into cationic (DTAC)
and anionic (SLS) micelles creates a barrier for ET in the case of
Ag(II)PP. This is explained by blocking the side of the porphyrin

ring with the vinyl groups of Ag(II)PP inside the micelle and
stacking of the propionic acid groups in the Stern layer. In such
an arrangement the edges of the porphyrin ring which can transfer
electrons are situated inside the micelle and ET is more difficult.
Ag(II)TCP forms different aggregates with surfactants. Its four
symmetrical C_{16}-chains do not allow the formation of normal micelles, the electron transferring edges can still be attacked in
the same way as in solutions without surfactant. The comparison
of the ET-reactions of the porphyrin complexes with the oxidation
of the uncharged $Fe(CN)_2(phen)_2$ in the presence of micelles with
the latter showing the higher barrier gives also support to the
idea of ET through the edges of the porphyrin ring. These edges
are nearer to the surface of the micelles than the axial positions
so that the distance for ET is lower and the rate constants are
higher. These assumptions are confirmed by similar experiments
with cytochrome(II)c.
There are still some uncertainties in the explanation of the reactions of Ag(II)PP and Ag(II)TCP. Especially the state of aggregation in water is not known. Further experiments are in progress
to solve this problem.

APPENDIX: The abbreviations used are:

dipy or bipy	-	2,2'-dipyridine or 2,2'-bipyridine
phen	-	1,10-phenanthroline
cyt.(II)c, cyt.(III)c	-	ferrocytochrome c, ferricytochrome c
Ag(II)TCP	-	silver(II)tetracetylprophyrin
Ag(II)PP	-	silver(II)protoporphyrin
SDS or SLS	-	sodium dodecylsulfate or sodium laurylsulfate
DTAC	-	dodecyltrimethylammoniumchloride
cmc	-	critical micelle concentration
ET, EX	-	electron transfer, electron exchange
CF, SF	-	continuous flow, stopped flow

REFERENCES

(1) Chance, B.: 1974, in "Investigation of Rates and Mechanisms
    of Reactions" Part II, ed. G. G. Hammes, John Wiley, New
    York, pp. 5
(2) Holzwarth, J. F.: 1979, "Fast Continuous Flow" this volume
(3) Holzwarth, J. F., Knoche, W., and Robinson, B. H.: 1978,
    Ber. Bunsenges. Phys. Chem., 82, pp. 1001
(4) Holzwarth, J. F., and Bruhn, H.: 1978, Ber. Bunsenges. Phys.
    Chem., 82, pp. 1006
(5) Sutin, N., and Christman, D. R.: 1961, J. Am. Chem. Soc.,
    83, pp. 1773
(6) Holzwarth, J. F., and Jürgens, H.: 1974, Ber. Bunsenges.
    Phys. Chem., 78, pp. 526
(7) Simplicio, J.: 1972, Biochemistry, 11, pp. 2525

ELECTRON-TRANSFER REACTIONS OF NITROAROMATIC RADICAL-ANIONS OBSERVED BY PULSE RADIOLYSIS

Peter Wardman and Eric D. Clarke

Cancer Research Campaign, Gray Laboratory, Mount Vernon Hospital, Northwood, Middx. HA6 2RN, England

ABSTRACT

Nitroaromatic radical-anions react with oxygen to produce superoxide ion in water at pH$\sim$8. The rate constants of this one-electron transfer reaction, measured by pulse radiolysis, were 1 to 4 orders of magnitude below the diffusion-controlled limit. Measurements of the accompanying free-energy changes (from the equilibrium constants of reactions with redox indicators) revealed a linear free-energy relationship between the logarithm of the rate constant and ΔG^o. The Eyring activation energy $\Delta G^{\neq}$ was also dependent upon ΔG^o.

INTRODUCTION

Nitroaromatic compounds (e.g. nitroimidazoles and nitro-furans) are useful against anaerobic microbial infections (1). They are also differentially toxic towards anaerobic or hypoxic mammalian cells (2) and selectively radiosensitize hypoxic cells, offering a potential new use for nitro compounds in cancer radio-therapy (3,4). The selective toxicity, i.e. the protection by O_2, may arise via the one-electron transfer reaction (I), since the nitro radicals, RNO_2^- have been observed in anaerobic micro-somal incubations (5) and the anaerobic reduction products such as hydroxylamines probably lead to cytotoxicity by binding to DNA and RNA (6):

$$RNO_2^- + O_2 \longrightarrow RNO_2 + O_2^- \qquad (I)$$

Pulse radiolysis of solutions containing formate (to con-

W. J. Gettins and E. Wyn-Jones (Eds.). Techniques and Applications of Fast Reactions in Solution. 535–538.
Copyright © 1979 by D. Reidel Publishing Company.

vert OH and H to the reducing agent CO_2^-) is a facile method of
generating RNO_2^- in water and monitoring reactions such as (I)
on the microsecond timescale (7). A preliminary report has been
published (8).

RESULTS

The decay of the absorption of RNO_2^- at 400-550 nm was exponential and the rate constant was first-order in $[O_2]$. Values
of k_1 were obtained for about 30 compounds, some of which have
been described (8). Typically, for 5-nitrofurans (e.g. nitro-
furantoin), $k_1 \simeq 2 \times 10^5$; for 2-nitroimidazoles (e.g. misonida-
zole), 4×10^6; for 5-nitroimidazoles (e.g. metronidazole),
7×10^6; and for some 4-nitroimidazoles, $k_1 \simeq 2 \times 10^7$ dm^3 mol^{-1}
s^{-1} in water at pH$\sim$8 and $\sim$295 K. A preliminary experiment (8)
designed to obtain spectral proof that O_2^- was a product of
reaction (I), accompanied by the regeneration of RNO_2, was repeated with improved precautions against scattered light. Excellent agreement was obtained between the final product spectrum
and that for O_2^-.

Values of ΔG^o_1 were estimated from measurements of the
equilibrium constants of the one-electron transfer reactions
between RNO_2^- and redox indicators such as quinones or viologens
(9,10), and the known reduction potential of the O_2/O_2^- couple
(11,12). If we convert to the (non-) standard state of 1 mol
dm^{-3} reactants and products, values of ΔG^o_1 ranged from ca. -5
kJ mol^{-1} for 5-nitrofurans to -40 kJ mol^{-1} for 4-nitroimidazoles,
i.e. $K_1 \simeq 10^1$ to 10^7. Data for additional compounds have confirmed the reported (8) linear free energy relationship between
$\log k_1$ and ΔG^o_1, with $d(\log k_1)/d(\Delta G^o_1) \simeq -0.06$ mol kJ^{-1} and
$k_1 \simeq 2 \times 10^5$ dm^3 mol^{-1} s^{-1} when $\Delta G^o_1 = 0$ $(K_1 = 1)$.

The temperature-dependence of reaction (I) between $\sim$275 and
350 K followed Arrhenius behaviour, with E_a in the range 26 to 40
kJ mol^{-1} for 6 compounds $(\log_{10} A/dm^3$ mol^{-1} $s^{-1} = 11.8$ to 13.0).
Although there was no apparent relationship between E_a and ΔG^o_1,
calculation of the Eyring activation energies $\Delta G^{\ddagger}_1$ (32 to 40 kJ
mol^{-1}) revealed a linear dependence of $\Delta G^{\ddagger}_1$ upon ΔG^o_1, with
$d(\Delta G^{\ddagger}_1)/d(\Delta G^o_1) \simeq 0.3$.

It is interesting to compare the temperature-dependence of
reaction(I) with a temperature-independence found for the rate
constants for oxidation of 3 different nitro radicals by Fe
$(CN)_6^{3-}$ at pH$\sim$8 (data for one compound at $\sim$295 K have been published) (13). For all 3 compounds, $\Delta G^o \simeq -75$ kJ mol^{-1}, $k \simeq 1$ to
3×10^8 dm^3 mol^{-1} s^{-1} and $E_a \leqslant 1$ kJ mol^{-1}.

DISCUSSION

The most striking feature of electron-transfer reactions involving RNO_2^- is their relative slowness. Thus values (9,10) of rate constants for the oxidation of RNO_2^- by quinones are 10^2 - 10^3 higher than those for reaction (I) involving similar ΔG^o. The major contributing factor to this anomaly is probably the extent of solvent reorganisation involved, reflected in the low value (ca. 10^5 dm^3 mol^{-1} s^{-1}) of k_2 extrapolated from our measurements and

$$RNO_2^- + RNO_2 \rightleftharpoons RNO_2 + RNO_2^- \qquad (2)$$

derived from e.s.r. line broadening data (14). Whillans et al. (15) measured $k < 10^6$ dm^3 mol^{-1} s^{-1} for electron transfer from one nitro compound to another involving $\Delta G^o = -3.3$ kJ mol^{-1}.

Meisel has demonstrated (16) good agreement between the predictions of the Marcus theory (17) and the rate data for electron-transfer reactions involving quinones. The approximate dependence of k_1 upon ΔG^o_1 may be predicted from the Marcus expressions:

$$k = Z \exp(- \Delta G^*/RT) \qquad (3)$$

$$\Delta G^* = (\lambda/4)(1 + \Delta G^o/\lambda)^2 \qquad (4)$$

where $Z \approx 10^{11}$ dm^3 mol^{-1} s^{-1} and λ is a reorganisation parameter, approximately constant at 75 kJ mol^{-1} for quinones (16). If λ is constant then from (3) and (4) we have: $d(\log k_1)/d(\Delta G^o_1) \approx$ $-0.088 (1 + \Delta G^o_1/\lambda) = -0.06$ mol kJ^{-1} as observed if for RNO_2 with $\Delta G^o_1 \approx -25$ kJ mol^{-1} we also have $\lambda = 75$ kJ mol^{-1}. However, from (3) and (4) and these values of ΔG^o_1 and λ we calculate a value of k_1 about 3 orders of magnitude higher than observed.

It is readily shown that the Marcus activation energy ΔG^* is 7.5 kJ mol^{-1} higher than the Eyring energy ΔG^+. Using (3) we expect $\Delta G^+ \approx 35$ kJ mol^{-1} from the measured k_1 for $\Delta G^o_1 = -25$ kJ mol^{-1}, as observed. Further, from (4) if λ is constant then we have: $d(\Delta G^+)/d(\Delta G^o_1) \approx 0.5 + \Delta G^o_1/2\lambda \approx 0.3$ as observed for ΔG^o_1 = -25 kJ mol^{-1} if $\lambda = 75$ kJ mol^{-1}. Thus some aspects of our rate data for reaction(I) may be accounted for satisfactorily by the Marcus theory, but the absolute rate constants are much slower than expected.

This work is supported by the Cancer Research Campaign.

REFERENCES

1. E. Grunberg and E. H. Titsworth, Ann.Rev.Microbiol., 1973, 27, 317.
2. R. M. Sutherland, Cancer Res., 1974, 34, 3501.
3. G. E. Adams, J. F. Fowler and P. Wardman (eds.), Hypoxic Cell Sensitizers in Radiobiology and Radiotherapy. Br.J. Cancer, 1978, 37, Suppl. III.
4. P. Wardman, Curr. Topics Radiat.Res. Q., 1977, 11, 347.
5. R. P. Mason and J. L. Holtzman, Biochemistry, 1975, 14, 1626.
6. J. H. Weisburger and E. K. Weisburger, Pharmacol.Rev., 1974, 25, 1.
7. P. Wardman, Rep.Prog.Phys., 1978, 41, 259.
8. P. Wardman and E. D. Clarke, Biochem.Biophys.Res.Commun., 1976, 69, 942.
9. D. Meisel and P. Neta, J.Amer.Chem.Soc., 1975, 97, 5198.
10. P. Wardman and E. D. Clarke, J.C.S. Faraday Trans. I, 1976, 72, 1377.
11. P. M. Wood, F.E.B.S. Lett., 1974, 44, 22.
12. D. Meisel and G. Czapski, J.Phys.Chem., 1975, 79, 1503.
13. P. Wardman, Int.J.Radiat.Biol., 1975, 28, 267.
14. T. Layloff, T. Miller, R. N. Adams, H. Fah, A. Horsfield and W. Proctor, Nature, 1965, 205, 382.
15. D. W. Whillans, G. E. Adams and P. Neta, Radiat. Res., 1975, 62, 407.
16. D. Meisel, Chem.Phys.Lett., 1975, 34, 263.
17. R. A. Marcus, Ann.Rev.Phys.Chem., 1964, 15, 155.

FREE RADICAL REACTIONS WITH PEPSIN STUDIED BY PULSE RADIOLYSIS

R H Bisby, R B Cundall, G E Adams* and M L Posener*

Department of Chemistry and Applied Chemistry,
University of Salford, Salford M5 4WT, U.K.
*Gray Laboratory, Mount Vernon Hospital, Northwood,
Middlesex, HA6 2RN, U.K.

The technique of pulse radiolysis (1) has found extensive application in the study of fast free radical reactions with compounds of biological interest (2). The high energy, electron pulse of short duration (typically 0.2 μs) produced by a microwave linear accelerator forms products, including free radical species, from the radiolysis of water:

$$H_2O \rightsquigarrow \cdot OH, \ e^-_{aq}, \ \cdot H, \ H_2O_2, \ H_2, \ H_3O^+ \qquad - (1)$$

The reactions of these free radicals with solutes in an aqueous solution can then be followed with fast spectrophotometric detection. The primary radicals ($H \cdot$, e^-_{aq} and $\cdot OH$) react at, or very near, the diffusion controlled rate with many different sites within a protein molecule. More selective and less reactive secondary radicals (X_2^-) can be formed by scavenging of OH by halides or pseudo-halides (X^-) in N_2O-saturated solutions:

$$X^- + \cdot OH \rightarrow X \cdot + OH^- \qquad - (2)$$

$$X \cdot + X^- \rightarrow X_2^- \qquad - (3)$$

Such inorganic radical anions (e.g. Cl_2^-, Br_2^-, $(SCN)_2^-$ and I_2^-) have been used previously to identify essential residues in the active sites of enzymes (3-5). The present report concerns reactions of radical-anions with pepsin, necessarily in acidic solutions because of the stability of the enzyme, where such reactions have previously been less well characterised.

The reactions of Br_2^- and $(SCN)_2^-$ with pepsin can be followed

W. J. Gettins and E. Wyn-Jones (Eds.). Techniques and Applications of Fast Reactions in Solution. 539–543.
Copyright © 1979 by D. Reidel Publishing Company.

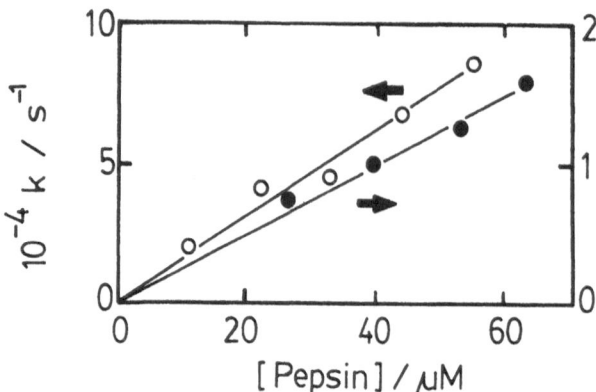

<u>Figure 1</u>   First order dependence on pepsin concentration on the
decay rate of radical-anion absorption: O 0.2MKBr, pH4.3;
● 5mMKSCN, pH5.9.

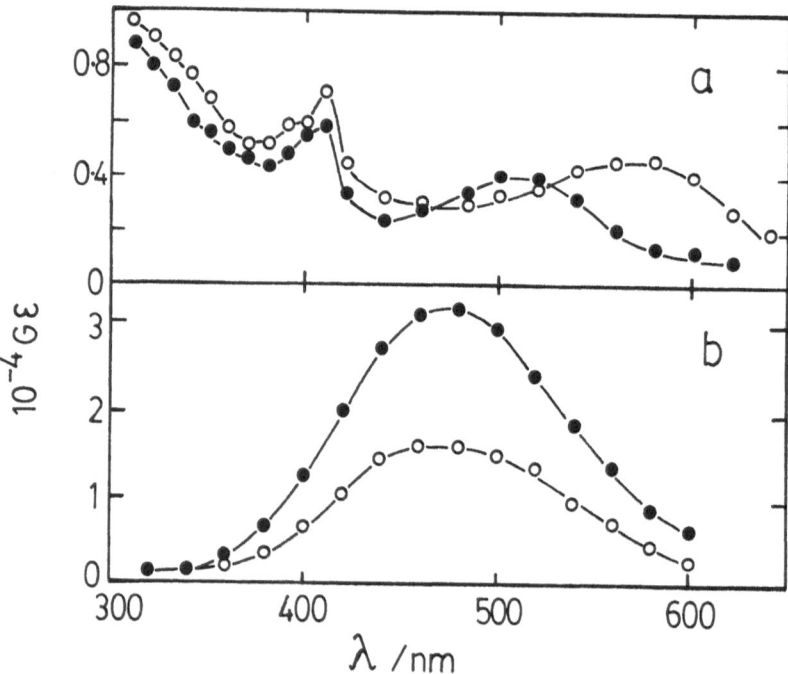

<u>Figure 2</u>   Transient spectra from pulse radiolysis of N_2O
saturated solutions of pepsin containing KSCN.
a) 2mg/ml pepsin, 2mM SCN , measured 200 μs after the pulse.
   O pH 2.1; ● pH 4.3.
b) 1mg/ml pepsin, 1M SCN , pH 4.5.   Measured at: ● 80 μs;
   and O 200 μs after the pulse.

by the decay of the radical anion absorptions at 360 nm and 480 nm
respectively. Figure 1 shows the pseudo-first order rate for
both Br_2^- and $(SCN)_2^-$ decay as a function of pepsin concentration.
The second order rate constants obtained from these data are
$(1.6 \pm 0.2) \times 10^9$ dm^3 mol^{-1} s^{-1} for reaction of Br_2^- with pepsin
at pH 4.3, and $(2.5 \pm 0.2) \times 10^8$ dm^3 mol^{-1} s^{-1} for reaction of
$(SCN)_2^-$ with pepsin at pH 5.9. These rates are similar to those
at neutral pH for other enzymes containing both tyrosine and
tryptophan residues (4,5).

Following the decay of the radical-anion absorption, other
secondary transient product spectra are observed (Figures 2a and 3).
The principal features of these product spectra can be
identified as the one-electron oxidation of tyrosine and
tryptophan by comparison with spectra obtained on pulse radiolysis
of solutions of the free amino-acids (6). The phenoxyl radical
of tyrosine

$$R-\langle\bigcirc\rangle-OH + X_2^- \longrightarrow R-\langle\bigcirc\rangle-\overset{\bullet}{O} + 2X^- + H^+ \quad - (5)$$

is characterised by an absorption maximum at 410 nm. The
protonated tryptophan radical

$$\text{(indole)}-R + X_2^- \longrightarrow \text{(indole)}-R + 2X^- \quad - (6)$$

has absorption maxima at 320 nm and 575 nm, whereas for the
neutral tryptophan radical

$$\text{(indole)}-R \rightleftharpoons \text{(indole)}-R + H^+ \quad - (7)$$

the long wavelength absorption maximum shifts to 520 nm. For
the tryptophan radical pKa (7) = 4.3 (7,8) whereas the pKa of
the tryptophan residue radical of pepsin is 2.9, as measured
from the variation in transient absorption at 575 nm with pH,
on oxidation of pepsin with Br_2^- (Fig. 3).

At a high concentration of SCN^- (1 mol dm^3) pulse radiolysis
shows (Figure 2b) that $(SCN)_2^-$ does not react to produce a product
spectrum, but instead decays by a second-order radical-radical
reaction. However, Figure 2a shows that at a low SCN^-
concentration (2 mmol dm^3), $(SCN)_2^-$ does react with pepsin.

Similar behaviour of $(SCN)_2^-$ has been observed with free
tryptophan and explained by a mechanism (7) involving an
equilibrium step:

$$SCN^- + TrpH^{+\cdot} \rightleftharpoons SCN^\cdot + TrpH \qquad - (8)$$

within the overall scheme.   Since this step only occurs with
the protonated radical, its effect is only observed in acidic
solutions when an appreciable fraction of the radicals exist in
the protonated state.   Although pepsin is inactivated by the
reaction of $(SCN)_2^-$ at low SCN^- concentrations, higher
concentrations of SCN^- are protective, an effect which must be
explained by the above mechanism (9).   This result implies
an essential role for one or more tryptophan residues in the
active site of pepsin.

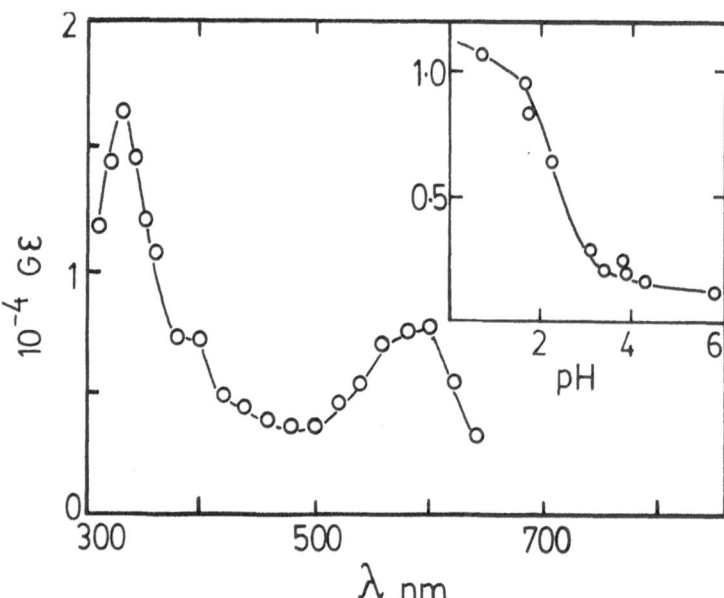

Figure 3   Transient spectrum from pulse radiolysis of aerated
solutions of pepsin (1mg/ml) containing KBr (50 mM) at pH 1.6.
Measured 100 μs after the pulse.   Inset: effect of pH on the
transient absorption at 575 nm.

References
1. Matheson, M.S. and Dorfman, L.M;
        'Pulse Radiolysis,' M.I.T. Press, Cambridge, Mass., (1969).
2. Wardman, P.; Rep.Prog.Phys., 41, 259 (1978).
3. Adams, G.E., Redpath, J.L., Bisby, R.H. and Cundall, R.B.;
        Isr.J.Chem., 10, 1079 (1972).
4. Adams, G.E., Redpath, J.L., Bisby, R.H., and Cundall,R.B.;
        J.C.S. Faraday Trans., 69, 1608 (1973).

5. Bisby, R.H., Cundall, R.B., Adams, G.E., and Redpath, J.L.:
   J.C.S. Faraday Trans., 70, 2210 (1974).
6. Adams, G.E., Aldrich, J.E., Bisby, R.H., Cundall, R.B.,
   Redpath, J.L., and Willson, R.E.; Radiat.Res., 49, 278 (1972).
7. Posener, M.L., Adams, G.E., Wardman, P., and Cundall, R.B.;
   J.C.S. Faraday Trans., 72, 2231 (1976).
8. Evans, R.F., Ghiron, C.A., Volkert, W.A. and Kuntz, R.,
   Chem.Phys.Lett., 42, 43 (1976).
9. Adams, G.E., Posener, M.L., Bisby, R.H., and Cundall, R.B.,
   and Key, J.R.; Int.J.Radiat.Biol., in press (1978).

INTERACTION OF STABLE NITROXYL RADICALS WITH RADIATION-INDUCED
SPECIES: A PULSE RADIOLYTIC STUDY.

Peter O'Neill and Terence C. Jenkins

Physics Department, Institute of Cancer Research,
Sutton, Surrey SM2 5PX (U.K.)

ABSTRACT

The persistent nitroxyl free radicals, TAN and TMPN, have
been shown to react with one-electron reduced viologens.   The
bimolecular rate constants decrease by about one order of magnitude for each 100mV increase in the one-electron reduction
potentials at pH7 ($E_7^{\frac{1}{2}}$) of the viologens.   In the reaction of
NPPN with one-electron reduced viologen and quinone species,
the rate constants were found to be independent of the $E_7^{\frac{1}{2}}$ values
of the parent compounds when $E_7^{\frac{1}{2}} <$ -200mV.   The reactions are
discussed in terms of an electron-transfer process.

From previous studies it has been demonstrated that nitroxyl
free radicals are reduced to the corresponding hydroxylamines by
α-hydroxyalkyl radicals (ref.1) and semi-quinone radical anions
(ref.2).   Contrastingly, simple alkyl and β-hydroxyalkyl radicals
have been shown to result in formation of diamagnetic $>$N-O-R
adducts (ref.1).   This paper is concerned with the mechanism of
reaction of nitroxyl free radicals with one-electron reduced
viologen and quinone species.

On electron pulse-irradiation at pH7 of deaerated aqueous
solutions containing 10^{-4} - 10^{-3}M viologen and 10^{-1}M formate,
quantitative reduction (refs.3,4) of the viologen (V^{2+}) occurs
as shown in reactions (1)-(4).   The lifetimes of the one-electron reduced viologens ($V^{.+}$) are $>$100ms as monitored at the
wavelength maxima of the optical absorption due to $V^{.+}$.   In the
presence of 10^{-4} - 5 x 10^{-4}M TMPN   (4-hydroxy-2,2,6,6-tetra-
methylpiperidine-1-oxyl), 10^{-4} - 5 x 10^{-4}M   TAN   (2,2,6,6-tetra-

W. J. Gettins and E. Wyn-Jones (Eds.). Techniques and Applications of Fast Reactions in Solution. 545–548.
Copyright © 1979 by D. Reidel Publishing Company.

$$H_2O \xrightarrow{\quad\text{/\/\/}\quad} e_{aq}^-,\ ^{\bullet}OH,\ H,\ (H_2O_2,\ H^+) \tag{1}$$

$$e_{aq}^- + V^{2+} \longrightarrow V^{\bullet +} \tag{2}$$

$$^{\bullet}OH(H) + HCO_2^- \longrightarrow H_2O(H_2) + CO_2^{\bullet -} \tag{3}$$

$$CO_2^{\bullet -} + V^{2+} \longrightarrow CO_2 + V^{\bullet +} \tag{4}$$

methyl-4-piperidone-1-oxyl), or $10^{-5} - 10^{-4}$M NPPN (norpseudo-pelletierine-N-oxyl),$V^{\bullet +}$ decays by first-order kinetics. From the dependence of the first-order rate constants on nitroxyl concentration the bimolecular rate constants for reaction of the nitroxyls with $V^{\bullet +}$ were determined and are presented in Table 1, together with the one-electron reduction potentials at pH7 $(E_7^{\frac{1}{2}})$ of the viologens (refs.3,4).

In the case of TMPN and TAN, the rate constants for reaction with $V^{\bullet +}$ decrease as the $E_7^{\frac{1}{2}}$ values increase. In order to extend the range of $E_7^{\frac{1}{2}}$ values of the substrates in the reaction with NPPN, a series of quinones ($\sim 3 \times 10^{-4}$M) were used, thereby extending the range of $E_7^{\frac{1}{2}}$ to 99mV (refs.5,6). At pH7 the quinones are reduced to semi-quinone radical anions ($Q^{\bullet -}$), by analogy with reactions (2) and (4). The bimolecular rate constants for reaction of NPPN ($10^{-4} - 10^{-3}$M) with $Q^{\bullet -}$ were determined from the enhanced rates of decay of $Q^{\bullet -}$, and are presented in Table 1 together with $E_7^{\frac{1}{2}}$ values of the quinones. The rate constants decrease on increasing $E_7^{\frac{1}{2}}$ of the quinones only when $E_7^{\frac{1}{2}}$ is > -200mV.

From the variation of the rate constants for reaction of the nitroxyls with $V^{\bullet +}$ and $Q^{\bullet -}$, it is suggested that the reaction mechanism involves electron transfer as shown:-

$$\gtrdot N \overset{\bullet}{-}\!\!O + V^{\bullet +}(Q^{\bullet -}) \underset{\xleftarrow{\quad\quad}}{\overset{k_1}{\rightleftharpoons}} \gtrdot N\text{-}O^- + V^{2+}(Q)$$
$$+ H^+ \updownarrow\, -H^+$$
$$\gtrdot N\text{-}OH$$

The protonation step has been included since the pK_a's of hydroxylamines derived from nitroxyls have been shown to be >12 (ref.1). Under the conditions used, no evidence was found for reversibility of the reaction; it is, therefore, concluded that the rate constants determined correspond to k_1.

Radical-radical combination with formation of a diamagnetic product(s) seems unlikely since it is expected that the rate constants for such a process would be $\sim$ independent of the $E_7^{\frac{1}{2}}$ values. Alternatively, oxidation of the nitroxyls to the corresponding oxoammonium ions could occur and the rates of reaction would be expected to be influenced by the second one-electron reduction potentials (E_7^{2}) of the substrates; however, no

Table 1 : Rate constants for reaction of TMPN, TAN and NPPN with one-electron reduced viologen and quinone species

Substrate S[a)	$E_7^{\frac{1}{2}}(S/S\cdot-)/mV$	10^{-5} k/M^{-1}s^{-1}		
		TAN	TMPN	NPPN
$4V^{2+}$	-635	25	13	8500
$3V^{2+}$	-548	9.0	2.8	7100
MV^{2+}	-447	2.0	0.36	9200
$2V^{2+}$	-356	0.83	0.11	7700
BV^{2+}	-354	1.0	0.18	7500
AQS$^-$	-375			6100
duroquinone	-235			3200
menadione	-203			4600
2,6-DMBQ	- 80			130
2,5-DMBQ	- 65			110
NQS$^-$	- 60			260
2-MBQ	+ 23			21
BQ	+ 99			6.9

a) $2V^{2+}$, $3V^{2+}$ and $4V^{2+}$ = 1,1'-ethano-, 1,1'-propano- and 1,1'-butano-2,2'-bipyridinium dibromide; MV^{2+} and BV^{2+} = methyl and benzyl viologen; AQS$^-$ = 9,10-anthraquinone-2-sulphonate; NQS$^-$ = 1,4-naphthoquinone-2-sulphonate; DMBQ, MBQ and BQ = dimethyl-, methyl- and 1,4-benzoquinone.

dependence is evident for those substrates whose E_7^2 values are known (refs.5,7).

The reaction Scheme presented is consistent with a linear dependence of log k_{obs} upon $E_7^{\frac{1}{2}}$ as predicted from the Marcus theory for electron-transfer processes (ref.8).

The invariance of k_{obs} with changing $E_7^{\frac{1}{2}}$(donor) in the quinone (or viologen)/NPPN system when $E_7^{\frac{1}{2}}$ <-200mV may be understood if these rate constants reflect the diffusion-controlled rate (k_{diff}) since

$$(k_{obs})^{-1} = (k_{act})^{-1} + (k_{diff})^{-1}$$

where k_{act} represents the activation-controlled rate constant.

ACKNOWLEDGEMENTS

We would like to thank MRC/CRC joint committee for financial support.

REFERENCES

1   S. Nigam, K-D. Asmus and R.L. Willson, J.C.S. Faraday I, 72, 2324 (1976).

2   R.C. Sealy, H.M. Swartz and P.L. Olive, Biochem. Biophys. Res. Comm., 82, 680 (1978).

3   R.F. Anderson, Ber. Bunsenges, Phys. Chem., 80, 969 (1976).

4   J.A. Farrington, M. Ebert and E.J. Land, J.C.S. Faraday I, 74, 665 (1978).

5   Y.A. Ilan, G. Czapski and D. Meisel, Biochem. Biophys. Acta, 430, 209 (1976).

6   D. Meisel and R.W. Fessenden, J. Amer. Chem. Soc., 98, 7505 (1976).

7   E. Steckhan and T. Kuwana, Ber. Bunsenges. Phys. Chem., 78, 253 (1974).

8   R.A. Marcus, Ann. Rev. Phys. Chem., 15, 155 (1964).

KINETICS OF OXIDATION OF BENZOPHENONE KETYL RADICALS: ABSOLUTE REACTION RATES AND OSCILLATORY BEHAVIOUR

Terence J. Kemp and Luis J.A. Martins

Department of Chemistry and Molecular Sciences, University of Warwick, Coventry CV4 7AL

Despite the enormous number of studies made of the oxidation of organic molecules by one-electron metal-ion oxidants (1), there are relatively few data for the fast intermediate step in these reactions, i.e. the oxidation of the intermediate organic radical by a second oxidant species to yield a presumed carbonium ion (2, 3). The sequence may be summarised for the case of a secondary alcohol:

$$R_2CHOH + M^{z+} \xrightarrow{k_1} R_2\overset{\bullet}{C}OH + M^{(z-1)+} + H^+ \tag{1}$$

$$R_2\overset{\bullet}{C}OH + M^{z+} \xrightarrow{k_2} R_2\overset{+}{C}OH + M^{(z-1)+} \tag{2}$$

$$R_2\overset{+}{C}OH + \text{nucleophile} \xrightarrow{k_3} \text{product} \tag{3}$$

Step (2) may involve further transient intermediates such as organo-metal ions, but in the present study we concentrate on direct measurement of the removal of $R_2\overset{\bullet}{C}OH$ (in our case, the visible-light absorbing $Ph_2\overset{\bullet}{C}OH$) (4).

A convenient procedure for the fast generation of $Ph_2\overset{\bullet}{C}OH$, which is not adversely affected by excited-state quenching species such as paramagnetic ions or dioxygen, is the photo-oxidation of benzilic acid $(0.214\,mol\,dm^{-3})$ by $[UO_2]^{2+}$ $(0.066\,mol\,dm^{-3})$ in acetone solution (5) following delivery of a 20 ns pulse of 347 nm laser radiation;

$$\begin{bmatrix} Ph_2C(OH)CO_2H \\ \downarrow \\ UO_2^{2+} \end{bmatrix} \xrightarrow[Me_2CO]{h\nu} \begin{bmatrix} Ph_2C(OH)CO_2H \\ \downarrow \\ UO_2^{2+} \end{bmatrix}^* \longrightarrow \begin{matrix} Ph_2\overset{\bullet}{C}OH + CO_2 + H^+ \\ + UO_2^+ \end{matrix}$$

$Ph_2\overset{\bullet}{C}OH$, monitored optically at $\lambda(\max)$ (542 nm), is formed completely during the pulse and decays slowly (reacting with $[UO_2]^{2+}$) with pseudo-

W. J. Gettins and E. Wyn-Jones (Eds.), Techniques and Applications of Fast Reactions in Solution. 549–553.
Copyright © 1979 by D. Reidel Publishing Company.

first-order kinetics, the exact rate depending on sensitiser
concentration. The decay is strongly accelerated by the presence of a
number of oxidising species, both inorganic and organic, to an extent
directly proportional to their concentration (up to certain limiting
values for the metal oxidants, above which non-linear behaviour is
found). This enables derivation of second-order rate constants for
step (2) in the scheme with R = Ph, as follows:

Table 1. k_2 for the reaction $R_2\overset{\bullet}{C}OH + M^{z+} \rightarrow R_2\overset{+}{C}OH + M^{(z-1)+}$

Oxidant	$k_2/dm^3\ mol^{-1}\ s^{-1}$		
	R = Ph	R = H	R = Me
Cu(II)	$(1.83\pm0.05)\times 10^6$	1.9×10^8 a 1.1×10^8 b	5.0×10^7 a 4.5×10^7 b
Fe(III)	$(4.91\pm0.26)\times 10^7$	$(8\pm2)\times 10^7$ a	$(5.8\pm0.3)\times 10^8$ a
Ag(I)	$(3.71\pm0.37)\times 10^6$	–	–
HgCl$_2$	$(6.50\pm0.50)\times 10^4$	–	–
$[UO_2]^{2+}$	$(1.34\pm0.04)\times 10^5$	–	–

a from Ref. 6 (for aqueous solutions)
b from Ref. 7 (for aqueous solutions)

Clearly a pronounced dependence of k_2 exists upon both R and the nature
of M^{z+}: the rate with Fe(III) is surprisingly low for R = Ph, considering
that simple electron-transfer is probably involved (8), although
solvent effects may be important. The rather lower rate of interaction
of $Ph_2\overset{\bullet}{C}OH$ with Cu(II) (compared with Fe(III)) conforms with Walling's
suggestion (8) of a Cu(III)-R_2COH intermediate. Interestingly, Ag(I)
is more reactive towards $Ph_2\overset{\bullet}{C}OH$ than Cu(II), although it is claimed
(9) to be weakly reactive towards hydroxycyclohexadienyl radicals.
The general pattern of rates for R = Ph suggests that Ph_2COH occupies
a position intermediate between the highly reactive $R_2\overset{\bullet}{C}OH$ (R = Me, H)
and the stable radicals (2) galvinoxyl and DPPH for which $k_2 \simeq 10 - 10^3$
$dm^3\ mol^{-1}\ s^{-1}$.

The reactivity of $Ph_2\overset{\bullet}{C}OH$ towards a few powerful organic
π-acceptors was measured (Table 2).

Table 2. k_2 for the reaction $R_2\overset{\bullet}{C}OH + \pi \rightarrow R_2\overset{+}{C}OH + \pi^{\overset{\bullet}{-}}$

Acceptor	$k_2/dm^3\ mol^{-1}\ s^{-1}$	
	R = Ph	R = Me a
1,3-dinitrobenzene	$(1.28\pm0.12)\times 10^6$	3.6×10^9
2-nitrobenzoic acid	$(1.00\pm0.20)\times 10^6$	5.4×10^8

a from Ref. 10.

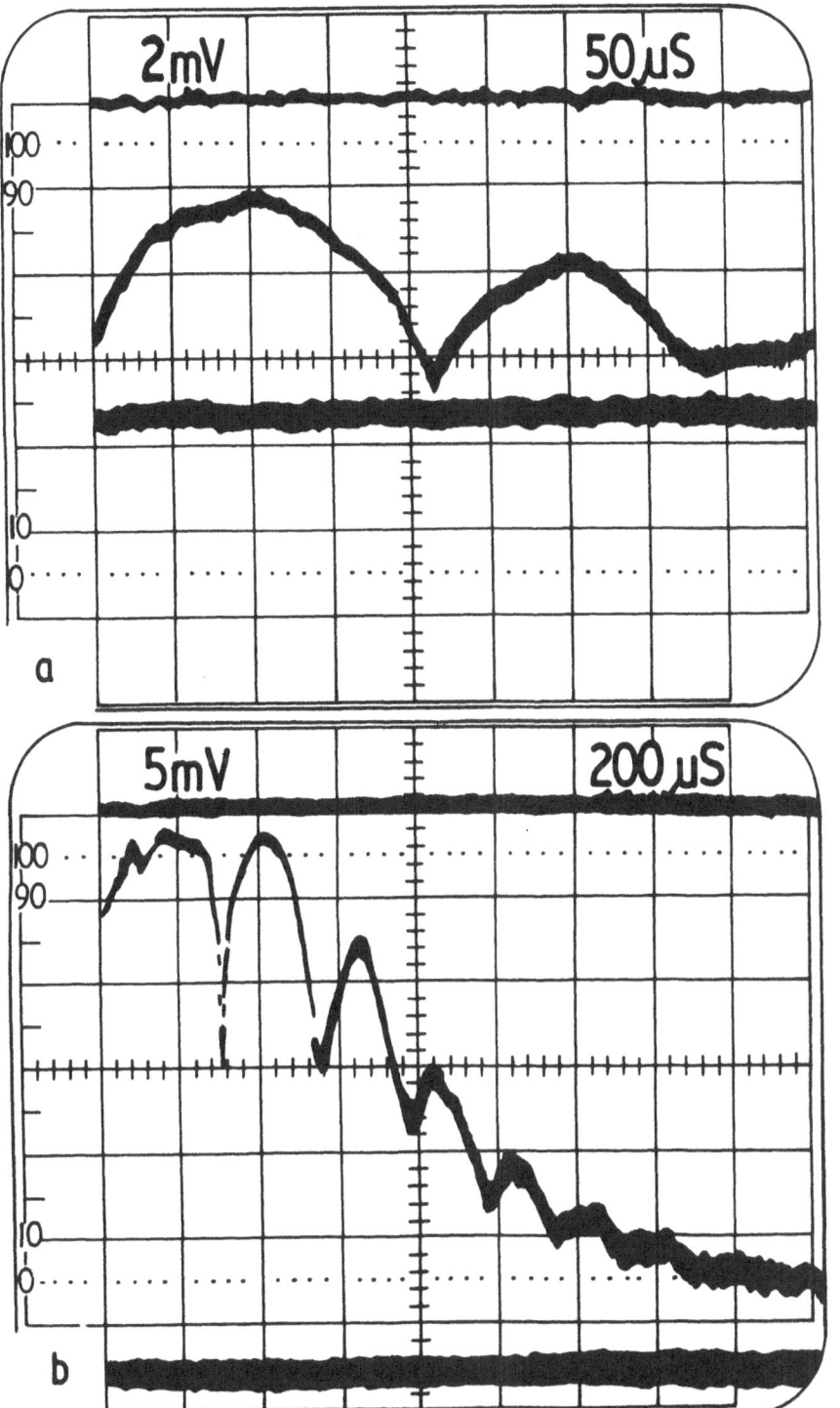

Figure 1. Oscillograms illustrating oscillations of the transient (λ = 545 nm) during the ns flash photolysis of an acetone solution of benzilic acid (0.214 mol dm^{-3}) containing AgClO$_4$ (0.016 mol dm^{-3}).

The values for k_2 are surprisingly low considering the stability of the Ph_2COH product, but reflect the rather low reactivity of Ph_2COH towards metal oxidants.

The most novel feature of this study emerged when metal ion oxidants were introduced into the benzilic acid-$[UO_2]^{2+}$-acetone system at high concentrations (when these were sufficiently high to give an absorbance > 0.1 at 347 nm then the $[UO_2]^{2+}$ sensitiser could be dispensed with). As the M^{z+} concentration is gradually increased, the decay profile of Ph_2COH gradually develops ripples, indicating damped oscillations, until at the highest concentrations (the values of which depend on M^{z+}), the decay is virtually replaced by large, damped oscillations which give way after a few hundred μs to rather more stable but less intense oscillations (Figure 1). A complete point-to-point spectrum taken at the highest metal ion concentrations as in Figure 2, indicates that the species responsible for the oscillation, while absorbing strongly at 542 nm, is not the Ph_2COH radical. Spectra taken at lower metal ion concentrations (below the critical value required for oscillations) or in the uranyl-benzilic acid system (which does not exhibit oscillations) are simply those of the Ph_2COH radical (4).

Currently we cannot offer a detailed mechanism, although it does seem that the oscillating species is derived from Ph_2COH, possibly as an adduct with a reactant or product species.

We thank INVOTAN (Lisbon) and the British Council for support of L.J.A.M.

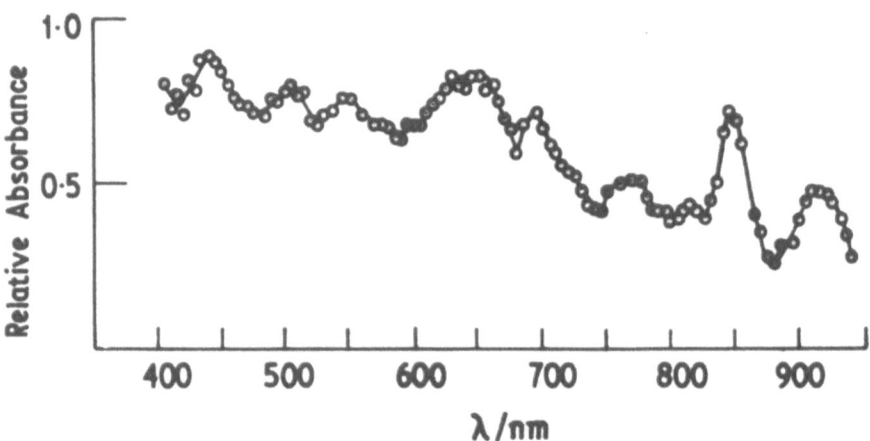

Figure 2. Spectrum of the transient produced under the conditions of Figure 1, recorded at the peak of the first oscillation (i.e. ca. 100 μs after the laser pulse).

References

1. T.J. Kemp in 'Comprehensive Chemical Kinetics', ed. C.H. Bamford and C.F.H. Tipper, Elsevier, Amsterdam, 1972, chap. 4.
2. I.V. Khudyakov and V.A. Kuz'min, Russ. Chem. Revs., 1978, 47, 22.
3. C. Walling, Accounts Chem. Research, 1975, 8, 125.
4. G. Porter and F. Wilkinson, Trans. Faraday Soc., 1961, 57, 1686.
5. H.D. Burrows, D. Greatorex and T.J. Kemp, J. Amer. Chem. Soc., 1971, 93, 2539.
6. G.V. Buxton and J.C. Green, J.C.S. Faraday I, 1978, 74, 697.
7. H. Cohen and D. Meyerstein, J. Amer. Chem. Soc., 1972, 94, 6944.
8. C. Walling, G.M. El-Taliawi and R.A. Johnson, J. Amer. Chem. Soc., 1974, 96, 134.
9. K. Bhatia and R.H. Schuler, J. Phys. Chem., 1974, 78, 2335.
10. P. Neta, M.G. Simic and M.Z. Hoffman, J. Phys. Chem., 1976, 80, 2018.

TIME RESOLVED POLARIZATION STUDIES USING SYNCHROTRON RADIATION

Cundall, R.B., Johnson, I., Jones, M.W., Munro, I.,*
and Thomas, E.W.

Department of Chemistry and Applied Chemistry,
University of Salford, Salford M5 4WT

*Daresbury Laboratory, Warrington WA4 4WT.

Nanosecond fluorescence polarization spectroscopy is a technique well suited for the analysis of the conformation and dynamic properties of macromolecules (1,2). Fluorescence polarization measurements have until recently been mainly carried out under steady state conditions, but in recent years the time dependence of fluorescence polarization has been successfully applied to the study of size, shape and segmental flexibility of macromolecules (3,4) with the added advantage that complex rotational motions can often be explicitly resolved on the time axis (5).

Although these measurements can be carried out fairly routinely using conventional single photon counting apparatus (6), the technique is restricted by the availability of light sources with good spectral profiles and are usually of the spark variety, fired by a gated thyratron. Such lamps are of low intensity, have low flash repetition rates with a full width at half maximum of around 2 ns. The exponential decay of the lamp profile is also followed by a long intensity tail which makes extraction of the lifetime of the sample somewhat difficult. The lamp profile one obtains is also very much a function of lamp design and expertise of the experimentalist in this field of research. A possible alternative light source would be a tunable picosecond laser system but these are restricted to the visible and near ultraviolet wavelength region and care is needed in interpretation of results where the high density of excitation energy might degrade the sample and where biphotonic processes are likely to occur.

It is therefore highly desirable to obtain an alternative

W. J. Gettins and E. Wyn-Jones (Eds.). Techniques and Applications of Fast Reactions in Solution. 555–560.
Copyright © 1979 by D. Reidel Publishing Company.

light source with subnanosecond characteristics and tunability
over a wide range of wavelengths in order to fully exploit the
technique of time resolved fluorescence polarization.   Such a
source is synchrotron radiation (7,8,9) which arises when
electrons are accelerated in a circular orbit confined by a
magnetic field.   The resulting centripetal acceleration results
in the emission of synchrotron radiation.   The features of
synchrotron radiation which make it ideally suited to lifetime
and time resolved polarization measurements are:

1. Subnanosecond lifetime ∼ 150 ps.
2. 100% polarized in plane of orbit.
3. Continuum from X-rays through to the infra red.
4. High repetition rate ∼ 1.28 MHz.

These features allow both lifetime and time resolved polarization
measurements to be extended into the subnanosecond region and
the high repetition frequency greatly facilitates rapid
accumulation of data.   Time resolved polarization measurements
are also made easier by the intrinsic polarization of the light
source.

A schematic layout of the Stanford positron electron
storage ring (SPEAR) optical path and detection system is shown
in Fig. 1.   The detection system uses conventional single photon
counting apparatus, with the 'start' pulse being obtained from
the signal photomultiplier while the 'stop' pulse is obtained
from a 'pick up electrode' situated inside the accelerator ring
and synchronised to the $(\pm)n^{th}$ passage of the electron bunch.
This also allows the detection system to count at the maximum
rate without overloading the detection system.

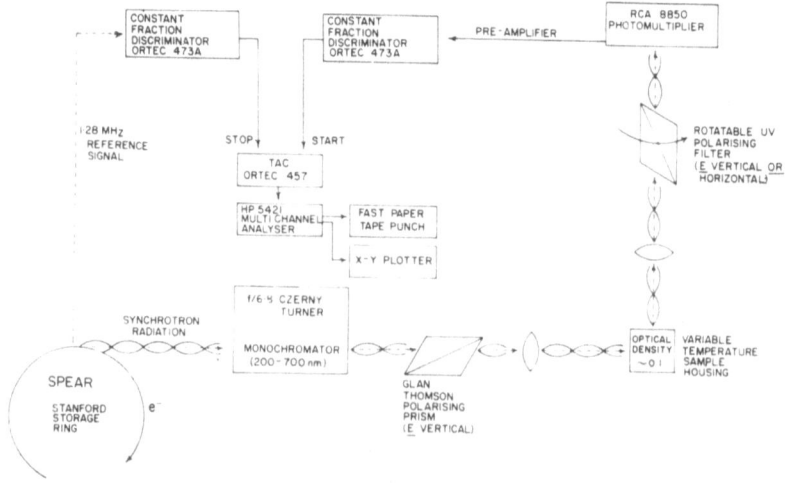

Figure 1.   Schematic layout of apparatus.

Principles of nanosecond polarization spectroscopy

A detailed account of the theory of nanosecond polarization spectroscopy has been given by Ta0 (10). The total emission as a function of time is given by:-

$$F(t) = Fy(t) + 2Fx(t) \qquad - (1)$$

where Fy(t) and Fx(t) are the vertical and horizontal components of fluorescence and emission anisotropy A(t) as

$$A(t) = \frac{Fy(t) - Fx(t)}{Fy(t) + 2Fx(t)} \qquad - (2)$$

For a chromophore with a single excited state lifetime rotating in common with a rigid sphere equations (1) and (2) are then given by:-

$$F(t) = Foe^{-t/\tau} \qquad - (3)$$

$$A(t) = Aoe^{-t/\phi} \qquad - (4)$$

where Fo is the initial fluorescence intensity and τ is the excited state lifetime. Ao is the initial anisotropy and can have a value ranging between -0.2 to 0.4 depending on the angle between the direction of absorption and emission given by:-

$$Ao = 0.4 \ (3\cos^2 \delta - 1) \ / \ 2$$

where ϕ is the rotational correlation time.

Analysis of the results on DPH and DAPH are based on a model which considers DPH to be a prolate ellipsoid. Equation (4) then becomes

$$A(t) = Aoe^{6D_\perp t} \qquad - (5)$$

where $D_\perp$ is the rotational diffusion constant for the long molecular axis. According to Perrin (1) assuming the two axes of the ellipsoid are 'a' and 'b' the equation of a prolate ellipsoid becomes:-

$$D = \frac{3kT}{16\pi\eta a^3} \left[2 \ln \frac{2a}{b} - 1 \right] \qquad - (6)$$

For our results an axial ratio of 6.0 was used based on the molecular orbital calculations of Lo and Whitehead (11) for this ground state of 1,6-diphenylhexatriene.

Results

    To test the technique, preliminary measurements were made
on 1,6 diphenyl hexa-1,3,5-triene (DPH) and its diacetamido
analog. (DAPH), the former being extensively used as a
fluorescence probe in studies of biological membrane structure
with the latter compound also showing promise in this respect.
The anisotropy as a function of time for DPH and DAPH in
propylene glycol are shown in Fig. 2.

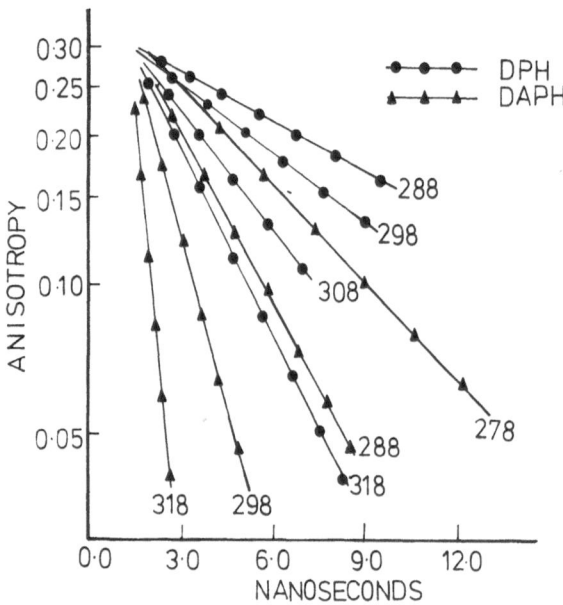

Figure 2.   Emission anisotropy kinetics of DPH and DAPH in
             propylene glycol excited at 350 nm.

    The correlation times are consistent with increasing
temperature or decreasing viscosity, Table 1.   The measure of
agreement in calculated parameters 'a' and 'b' as a function of
temperature is encouraging and demonstrates the consistency
available using this technique.   The correlation times for
DAPH are smaller however than those of DPH in contrast to what
one expects in view of the molecular dimensions of DAPH based on
the additional end groups.   It thus appears that the ellipsoid
model is not applicable and a refinement of the model is needed
for DAPH in the excited state.

T°K	DPH				DAPH			
	EXP (NS)	RADIUS A°	a A°	b A°	EXP (NS)	RADIUS A°	a A°	b A°
278	18.7	5.4	9.8	1.6	6.1	3.7	6.8	1.6
288	13.6	5.4	10.0	1.6	3.7	3.5	6.1	1.6
298	9.0	5.8	10.6	1.7	1.9	3.4	5.5	1.7
308	5.7	6	11.0	1.8	∼	∼	∼	1.8
318	3.6	6.1	11.2	1.8	0.68	3.5	5.7	1.8

Table 1.   Evaluation of 'a' and 'b' and radius 'r' for DPH and DAPH.

References

1. Perrin, F., 1934, J.Phys.Rad., 5, 497.
2. Weber, G., (1952), Biochem.J., 51, 145.
3. Yguerabide, J., 1972, Methods Enzymol., 26C, 498.
4. Yguerabide, J., Epstein, H.F., and Stryer, L., 1970.
5. Veatch, W.R., and Stryer, L., 1977, J.Mol.Biol., 117, 1109.
6. Ware, W.R., in Creation and Detection of the Excited State, 1971, 1A, 213.
7. Koch, E.E., Interaction of Radiation with condensed matter, 1977, 2, 225.
8. Watson, R.E., and Perlman, M.L., Science, 1978, 199, 1295.
9. Munro, I.H., 1978, Chemistry in Britain, in press.
10. Tao, T., Biopolymers, 1969, 8, 609.
11. Lo, D.H., and Whitehead, M.A., Canad.J.Chem., 1968, 46, 2041.

Acknowledgements

The authors wish to extend their gratitude to the Stanford Linear Accelerator Center (SLAC) for the opportunity to carry out this work.

SINGLE-PHOTON INFRARED PHOTOCHEMISTRY: WAVELENGTH AND TEMPERATURE DEPENDENCE OF THE QUANTUM YIELD FOR THE LASER-INDUCED IONIZATION OF WATER

David M. Goodall, Rodney C. Greenhow, Barry Knight, Joseph F. Holzwarth* and Wolfgang Frisch*

The Departments of Chemistry and Physics, University of York, York YO1 5DD, U.K.

*Fritz-Haber-Institut der Max-Planck-Gesellschaft, Faradayweg 4-6, 1000 Berlin 33, West Germany

Quantum yields Φ for the ionization of water to hydrogen and hydroxyl ions, present as transients in excess of their equilibrium concentration levels following irradiation by a pulse derived from a Q-switched laser, have been measured as a function of temperature and excitation wavelength. Irradiation wavelengths used were the ruby, neodymium, and iodine laser values 0.694, 1.06, and 1.31 µm, and stimulated Raman scattering shifts of these to 0.975 and 1.41 µm. Φ (298 K) increased from 9×10^{-9} to 9×10^{-6} with decrease of λ from 1.41 µm to 0.694 µm. The reaction occurred after absorption of a single photon into the excited O-H stretching vibrational levels: the energy of the laser photon E was in all cases in excess of the activation energy for reaction, which is shown to equal E_a for the self dissociation of water. The expression $\Phi = \Phi_o (1 - E_a/E)^s$ is based upon competition between reaction and relaxation, and correlates all the data with $s = 8.4 \pm 0.2$ and $\Phi_o = (1.7 \pm 0.3) \times 10^{-4}$. Values of s are interpreted with a hydrogen-bonded species $(H_2O)_4$ as the reactive entity. The ionization of water is the first reported example of an IR photochemical reaction in the liquid phase.

561

W. J. Gettins and E. Wyn-Jones (Eds.). Techniques and Applications of Fast Reactions in Solution. 561–568.
Copyright © 1979 by D. Reidel Publishing Company.

1. INTRODUCTION

In single-photon infrared photochemistry, the absorption of
one photon into the vibrational manifold of the electronic ground
state induces decomposition. Efficient IR photodecomposition has
been demonstrated for several matrix-isolated species for which
the activation energy was lower than the energy of the quantum
exciting one of the vibrational fundamental modes (1,2). Only
one photodecomposition has so far been reported in the liquid
state (3), where quantum yields are expected to be far lower than
in an inert gas matrix because of rapid thermalization of the
excitation energy. We refer to our initial communication of the
ionization of water at 296 K, induced by vibrational excitation
in the IR overtone region using a Nd:glass laser. In the present
paper we report an extension of this study, in which quantum
yields have been measured at a series of wavelengths and temperatures.

2. EXPERIMENTAL

The water used as reactant was continuously purified by re-
circulation through a closed loop containing the reference and
irradiation conductance cells and a thermostated ion exchange
column.

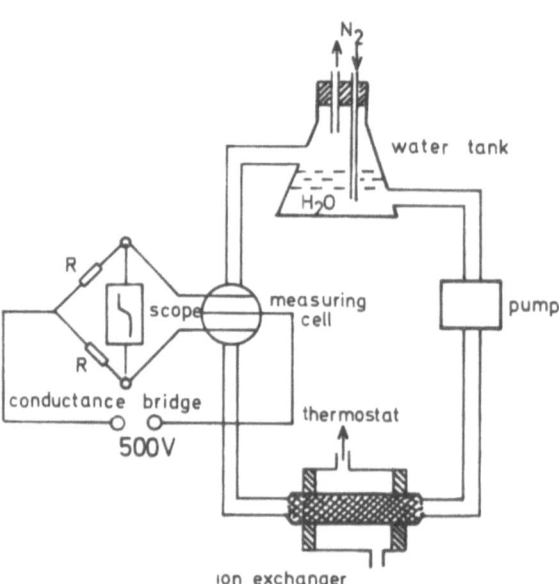

Fig. 1: Schematic diagram of the experimental
arrangement.

Quantum yields for photoionization were derived from the
transient conductance of excess hydrogen and hydroxyl ions (3),
with the limit of sensitivity of our DC conductance bridge allowing detection of concentration increments as low as 20 ppm
($\Delta[H^+]$ = 2 x 10^{-12} mol dm^{-3} at 298 K) Fig. 1. A Pockels cell Q-
switched solid state laser provided photolysis pulses at 0.694
μm or 1.06 μm. Stimulated Raman scattering shifting to 0.975 μm
was achieved by passing the condensed beam from the ruby laser
through pressurized H$_2$ gas, and 1.41 μm radiation was generated
similarly using the Nd:glass beam and liquid N$_2$. Full experimental details are given in (4). Experiments at 1.31 μm were conducted using an iodine laser (5). Irradiation wavelengths are shown
as vertical lines superimposed on the plot of decadic absorption
coefficients in Fig. 2.

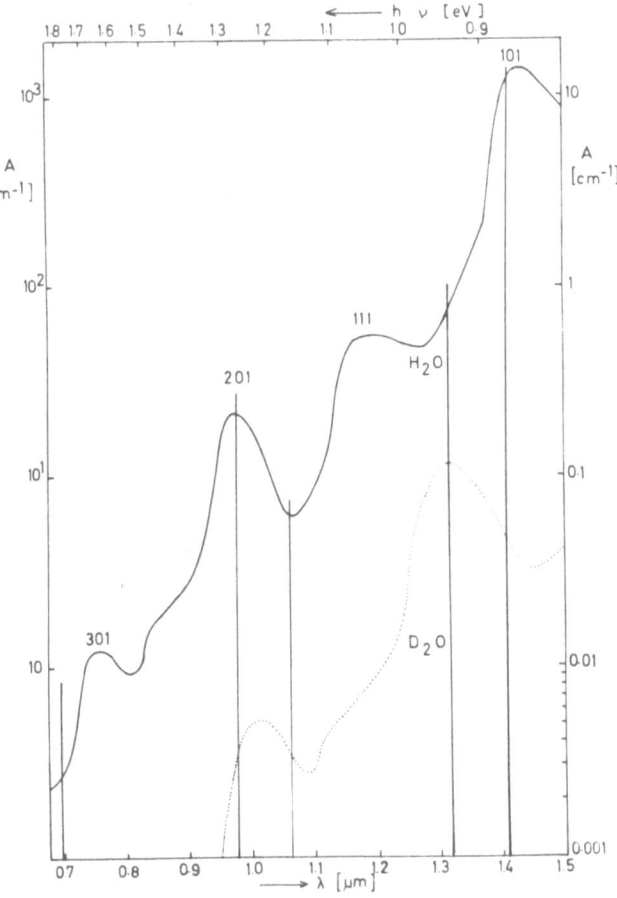

Fig. 2: Near IR spectrum of H$_2$O and D$_2$O at room
temperature.

3. EVIDENCE FOR SINGLE-PHOTON PHOTOCHEMISTRY

We shall initially summarize the evidence proving that absorption of one laser photon by water induced ionization to hydrogen and hydroxyl ions.

3.1 H^+ and OH^- were proved to be the transient intermediates by the agreement between observed and literature values of the recombination rate constants and activation energy (8). Our results are summarized in Table 1 and Fig. 3 for the measurements which were obtained with the iodine laser.

Table 1: Rate constants of the recombination of H^+ and OH^- in pure water at different temperatures.

$T \; \left[{}^{\circ}C \right]$	$1/T \times 10^3 \; \left[K^{-1} \right]$	$k_{12} \; \left[\dfrac{dm^3}{mol \; s} \right]$
11.9	3.508	0.98×10^{11}
20.1	3.410	1.13×10^{11}
27.7	3.325	1.27×10^{11}
32.0	3.278	1.34×10^{11}
38.6	3.208	1.50×10^{11}
47.8	3.116	1.65×10^{11}
58.8	3.012	1.90×10^{11}
65.6	2.925	1.99×10^{11}

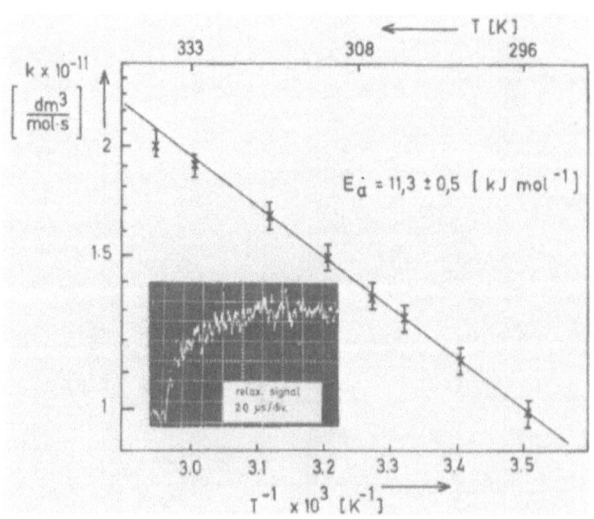

Fig. 3: Activation energy of the recombination $H^+ + OH^-$. The relaxation trace measured at 305.2 K is also shown.

3.2 The absorptive nature of the process was shown by the proportionality of the temperature jumps at 0.694 and 1.06 μm, calculated from the conductance increase after relaxation of the excess H^+ and OH^- ions, to the energy inputs calculated from the small signal absorption coefficients of H_2O. Scattering phenomena were not expected at the power densities used in these experiments, and would have exhibited identical non-linear behaviour in H_2O and D_2O. In fact the transient conductance change in D_2O at 1.06 μm was an order of magnitude less than in H_2O at 1.06 μm, as expected from the tenfold decrease in absorption coefficient at this wavelength.

3.3 A single photon process was confirmed by the invariance of the quantum yield Φ with input power density I_o. Φ was determined at 0.694 μm and 300 K over the range $3 < I_o < 250$ MW cm^{-2}, and showed no change with variation of either I_o, or filtration of the water. A threshold of 100 MW cm^{-2} has been reported for dielectric breakdown in carefully-purified liquids (6), so plasma formation cannot explain our observed photoconductivity.

4. VARIATION OF Φ WITH WAVELENGTH AND TEMPERATURE

Table 2 shows that Φ, though small, shows a dramatic increase with increasing photon energy, E.

Table 2: Quantum yield as a function of wavelength and photon energy at 298 K.

λ [μm]	E [kJ einstein^{-1}]	$10^7 (\phi \pm s_\phi)$
1.41	85	0.09 ± 0.08
1.31	91	0.02 ± 0.02
1.06	113	6.0 ± 0.4
0.975	123	12 ± 2
0.694	173	92 ± 10

E is always greater than the activation energy E_a for ionization of water, and the trend in Table 2 may be understood if vibrationally excited molecules may either enter the decomposition channel or relax to Boltzmann equilibrium with other degrees of freedom. By combining RRK theory with an exponential relaxation model Zimmerman (7) showed that (I) should account for the variation of Φ with λ in such a situation.

$$\phi = \phi_o (1 - E_a/E)^s \qquad (I)$$

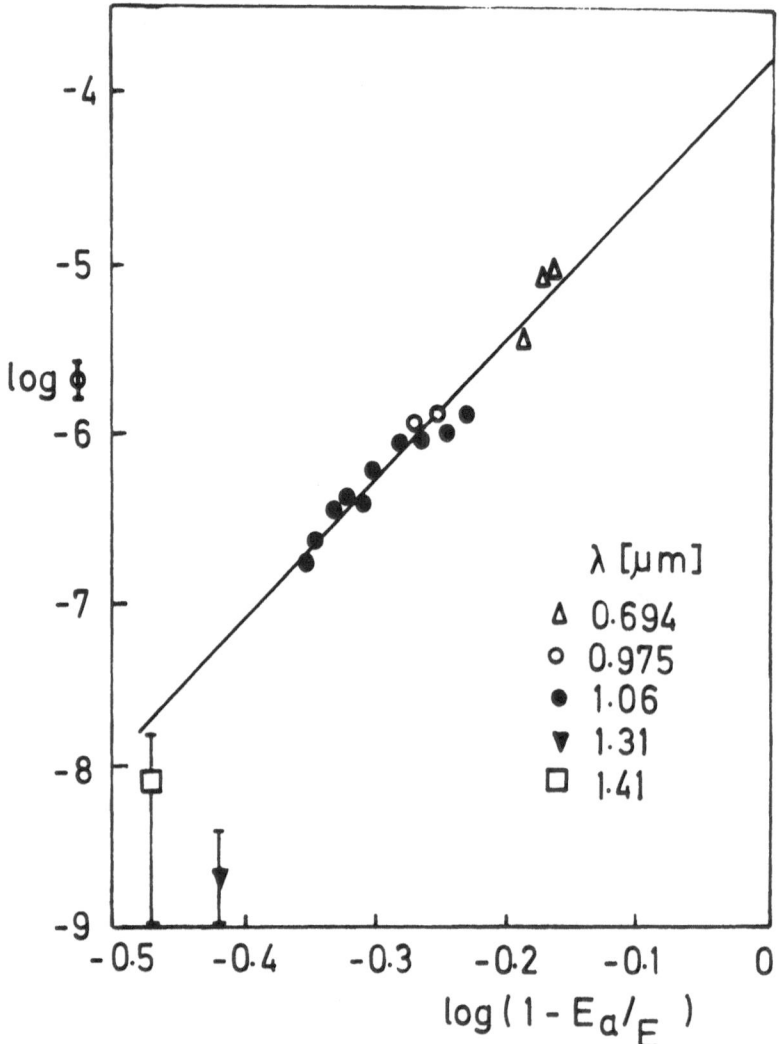

Fig. 4: Correlation of quantum yields for photoionization.

Fig. 4 shows that all data obtained with $\lambda \leqslant$ 1.06 μm fit to
(I), with best fit values for the number of degrees of vibrational
freedom pooling the excitation energy being s = 8.4 ± 0.2. E_a
was equated to $\Delta H^{\ominus}$ for the ionization of water, which decreases
from 63 to 46 kJ mol^{-1} in the temperature range investigated
(275-349 K).

The value of s is almost identical to the number of O-H bonds in a four H_2O complex, demonstrated by Eigen (8) to be the critical configuration for proton transfer in the direction of association of Fig. 5. We therefore speculate that for reaction to occur the vibrational excitation energy must be pooled in O-H stretching modes of a hydrogen-bonded $(H_2O)_4$ unit.

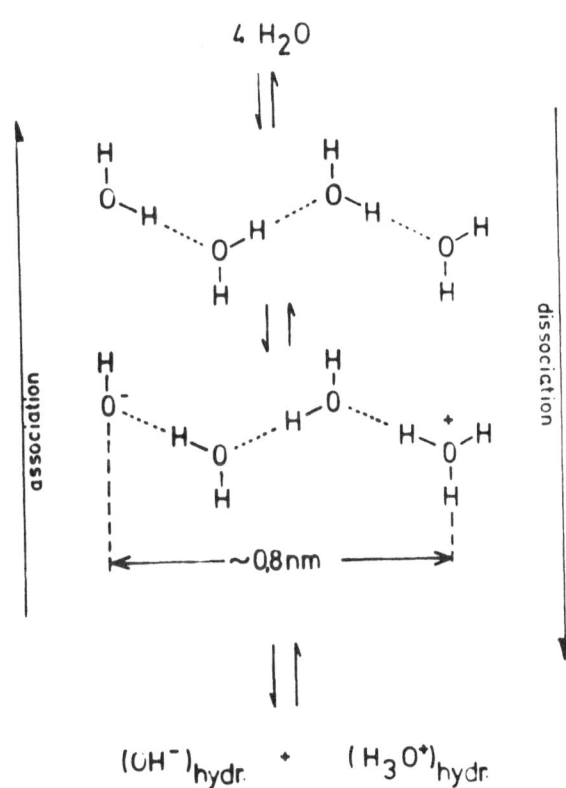

Fig. 5: Critical configuration for proton transfer.

Quantum yields at 1.31 and 1.41 μm, pumping in the envelope of the 101 combination band (9), are lower than expected from (I), which was parameterized with data from 201 and 301 combinations. Further experiments, ideally using a tunable near IR laser, are required to confirm whether this indicates that multiple excitation in the symmetric stretching mode is a prerequisite for this IR photochemical reaction.

5. ACKNOWLEDGEMENTS

We wish to thank the S.R.C. for a grant to purchase the solid-state laser, and N.A.T.O. for a travel grant.

REFERENCES

(1) Hall, R.T., and Pimentel, J.C.: 1963, J. Chem. Phys. 38, p. 1889.
(2) Davies, B., McNeish, A., Poliakoff, M., and Turner, J.J.: 1977, J. Am. Chem. Soc. 99, p. 7573.
(3) Goodall, D.M., and Greenhow, R.C.: 1971, Chem. Phys. Lett. 9, p. 583.
(4) Knight, B., Goodall, D., and Greenhow, R.C., manuscript submitted to J.C.S. Faraday II.
(5) Holzwarth, J.F., Schmidt, A., Wolff, H., and Volk, R.: 1977, J. Phys. Chem. 81, p. 2300.
(6) Dowley, M.W., Eisenthal, K.B., and Peticolas, W.B.: 1967, Phys. Rev. Lett. 18, p. 531.
(7) Zimmerman, G.: 1955, J. Chem. Phys. 23, p. 825.
(8) Eigen, M., Kruse, W., Maass, G., and De Maeyer, L.: 1964, Progr. Reaction Kinetics 2, p. 286;
     Eigen, M., and De Maeyer, L.: 1958, Proc. Roy. Soc. A247, p. 505;
     Barker, G.M., Fowles, P., Sammon, D.C., and Stringer, B.: 1970, Trans. Faraday Soc. 66, p. 1498.
(9) Buijs, K., and Choppin, G.R.: 1963, J. Chem. Phys. 39, p. 2035.

MOLECULAR MOTION OF POLYMERS IN SOLUTION

R.A. Pethrick

Department of Pure and Applied Chemistry
University of Strathclyde
295 Cathedral Street
Glasgow G1 1XL

The motion of a polymer molecule in solution can be subdivided into two classes: (1.1) whole molecule rotation and translation and (1.2) segmental and normal mode motion. The total dynamic spectrum of a polymer will be a composite of these processes.

1.1 Whole molecule rotation and translation

The simplest motions which a polymer may execute in solution are those of overall rotation and translation of the whole molecule. These processes are directly analogous to those observed in small rigid molecules and are influenced by changes in the moments of inertia and shape with molecular weight. Whole molecule rotation and translation are usually defined assuming the internal molecular geometry of the polymer remains constant during the period of observation. In certain rigid or highly regular polymers this approximation is valid. However, in the large majority of polymers the internal motion of elements of the chain may occur during the time of whole molecule rotation or translation.

1.2 Segmental and normal mode motion of a polymer chain

At equilibrium a polymer chain will experience a continuous bombardment from the surrounding solvent molecules. The collisions may be either elastic or inelastic. An elastic collision occurs when a solvent

W. J. Gettins and E. Wyn-Jones (Eds.), Techniques and Applications of Fast Reactions in Solution. 569–577.
Copyright © 1979 by D. Reidel Publishing Company.

molecule collides with the polymer and is scattered
with a possible change of momentum. The result of
such collisions will be similar to those between
molecules of the solvent and ultimately leads either
to whole molecule rotation or translation of the
polymer. In an elastic collision, momentum must be
conserved during the period of the collision.
Alternatively an inelastic collision is one in which
momentum is not conserved and provides a mechanism
for conversion of translational energy into internal
vibrational energy of the solvent or polymer. The
reverse process will also be possible and the
combination of the forward and reverse inelastic
processes is often referred to as vibrational to
translational relaxation. In an inelastic collision
conservation of energy is required. A vibrational
analysis of the polymer reveals that small angle
rotation or displacements of elements of the backbone,
such as torsional oscillation, are associated with the
lowest vibrational energy transitions. Inelastic
collisions will tend to activate preferentially the
lowest energy modes. Successive activation of
torsional oscillations may ultimately lead to changes
occurring in the conformations of the chain elements.(1)

In general the more
sterically crowded the
conformation the higher
its energy. Subsequent
inelastic collisions will
activate an energetically
unstable conformation
and enable the polymer
to return to its orig-
inal lowest energy
state. The rate of
internal rotation of
elements of the backbone
will be influenced by
the magnitude of the
intramolecular inter-
actions and the
efficiency of activation.
The intramolecular
interactions are usually
of a non-bonded form
and thus a function of
the chemical structure
of the polymer. A
detailed study of the activation of a particular
element of the backbone must also allow for

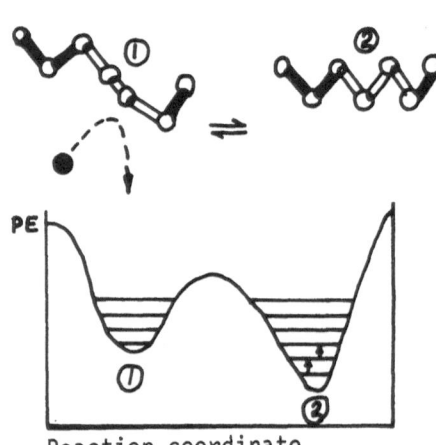

Reaction coordinate

Inelastic collisions

accumulation of energy in sub units of the chain by
vibrational energy transfer from neighbouring sites
as well as by direct collisional activation.   In
general the higher the temperature the more energetic
and hence efficient the collisions and subsequently
the greater the probability that internal rotation
will occur.   The Brownian nature of the motion of the
solvent molecules provides a random probability for
the activation of any given element of the polymer
backbone. (2)

     The random occurrence of inelastic collisions
may be expected to lead to a complex pattern of displacements of the polymer chain.   The overall motion
of the chain may be described by a superposition of
two forms of displacement: Segmental and normal mode
motion.   The former is usually associated with the
movement of a block of the chain relative to the
main backbone change from structure (2) to (1).   The
only restrictions placed on such a motion is that
sufficient energy should be available from inelastic
collisions for the intramolecular interactions res-
tricting free rotation of the polymer backbone to be
overcome.   It is implicit in this definition of
segmental motion that the relaxation process involves
direct activation from the lowest energy state via a
cascade mechanism which may lead to excitation of the
other thermal vibrational states.

     Normal mode behaviour may be considered as
that part of the molecular displacement which may be
described in terms of a phase coherent motion of the
whole polymer chain (Fig. 2).   The normal mode motion
is in fact an eigenvalue solution of the stochastic
perturbation of the flexible random chain and
requires a phase coherence of the overall displacement
in space and time of the macromolecule.   Using this
framework, segmental motion may be considered as the
phase incoherent movements of the microscopic elements
of the polymer backbone and usually is considered to
involve a random jump of one stable conformation to
another.   In high molecular weight polymers the
distinction between segmental motion and the higher
normal modes, whilst still being precise, may appear
to become blurred since a suitable summation of a
series of segmental motions may on average approximate
closely to a normal mode displacement.   This apparent
confusion is implicit in the stochastic formulation
but, as will be shown later, does notlead to any real
contradiction in terminology. (2)

The normal mode description of the polymer
motion is most frequently encountered in the discussion
of the displacement of a flexible polymer molecule
under the influence of a shear gradient. The effects
of establishing a shear gradient (Fig. 2) in a liquid
will lead to the production of a regular distribution
of momentum between the shearing boundaries. If this
shear field is further required to oscillate then the
direction of average momentum transfer will similarly
oscillate. In a shear field the distribution will be
biased in one direction and a net transfer of the
momentum will occur. The effect of a resultant in
terms of the momentum transfer will lead to a greater
probability of activation of a given conformation in
one region than in another. This probability will be
influenced by the magnitude and rate of change with
time of the shear gradient. If the period of
oscillation is low, the rates of activation and
deactivation are close to their equilibrium values,
little momentum is lost and the net effect is the
overall translation of the polymer. This motion is
often referred to as the zeroth normal mode. At

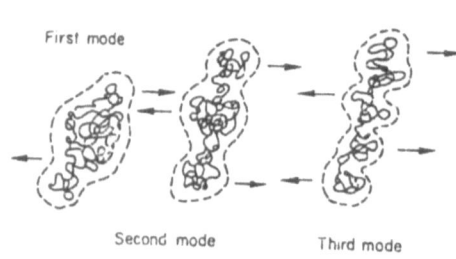

First mode

Second mode   Third mode

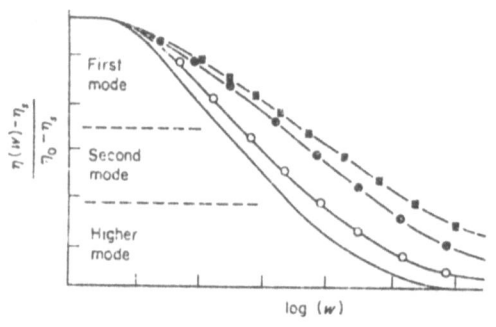

First
mode

Second
mode

Higher
mode

log (w)

higher rates of
oscillation, activated
conformations are unable
to return momentum to the
solvent during the period
of oscillation with the net
result that a distortion
of the polymer occurs and
momentum is lost.
Analysis of the complex
pattern of conformations
adopted by the polymer is
possible in terms of the
normal mode motion which
by definition requires a
certain phase coherence
of the overall distortion
of the polymer. Since
the degree of momentum
transfer is reflected in
the viscous drag which
the polymer exercises on
the solvent, the first
normal mode plays a major
role in determining the
polymer viscosity incre-
ment. At very high
frequencies of oscillation

Viscoelastic Relaxation
Fig. 2.

the change of direction of the momentum is so fast
that it is unable to influence the polymer conformations and the observed viscosity of the solution
approaches that of the solvent. (3- 5)

Some Aspects of Molecular Motion in Linear Polymer Chains

A survey of the data available on the motion
of polymers has enabled certain characteristic types
of behaviour to be identified. A polymer being a
long chain-like structure may be expected in the
molecular weight. The effects of increase in the
chain length of a polymer have been illustrated by
dielectric relaxation (6). At low molecular weight
the relaxation which reflects the rate of reorientation
of a dipole vector within the polymer will be
approximately proportional to the chain length (7).
This observation suggests that the relaxation is
associated with what is essentially whole molecule
rotation. For higher molecular weights the
relaxation becomes independent of the chain length
and is constant at a particular temperature for a
specific polymer sample. It is assumed that for the
high molecular weight
material the reorientat-
ion of the dipole is
achieved by a process
which does not involve
overall rotation of the
whole polymer chain-
segmental motion. The
value of the molecular
weight at which the
relaxation behaviour
changes from being
dominated by overall
rotational to becoming
determined by the ease of
segmental motion is a function of the magnitude of the
intramolecular potential governing rotation. It will
be appreciated that a molecular weight will exist
where overall rotation and segmental motion are
equally proable and a complex relaxation mechanism must
be used to describe the observed dynamic behaviour.
In general the dielectric relaxation behaviour of a
polymer molecule in solution will be dominated by the
energetically easiest process. In the case of a stiff
polymer, whole molecule rotation will dominate to high
values of the molecule weight; in a very flexible

polymer, solvent viscosity will inhibit whole molecule
rotation and as a consequence segmental rotation
dominates at low values of the molecular weight.(5-8)

    Molecular weight affects have been recognized
in a number of experiments (9-14).

    Similar effects to those described above have
been identified from ultrasonic studies of the normal
alkanes (14) polystyrene (11), and poly (α-methyl-
styrene-alkane) copolymers (15). Calculation of the
contribution to the
observed dispersion
behaviour would suggest
that normal mode processes
contribute typically 10-
20 per cent to the total
relaxation behaviour. At
high molecular weights
the normal mode processes
will have moved to low
frequency and will make a
negligible contribution to
the observed dispersion.
The molecular weight
dependence must therefore
be explained in terms of either changes in the
mobility of the polymer with the chain length or from
effects associated with intermolecular contacts.

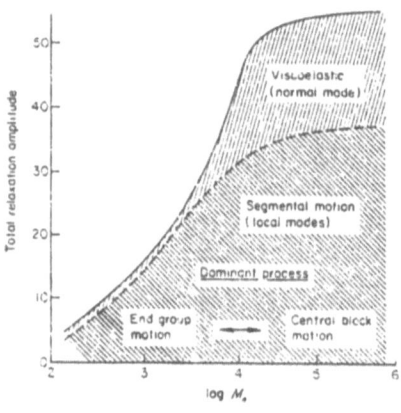

    Normal mode relaxation is usually studied by
the frequency dependence of the shear viscosity. An
extensive literature exists on this topic (16) and a
number of features have been successfully identified
which can be used to characterize such motion. For a
flexible polymer the shape of the relaxation curve and
its position will primarily be determined by the
strength of the interaction of the polymer with the
solvent, the length of the chain and its flexibility
as defined by the intramolecular interaction constants.
Theoretical analysis has been developed for the flexible
coil with virtually no interaction with the solvent,
with interaction of the solvent and the chain and also
for the partially rigid polymer. Numerous examples
of polymer-solvent combinations which possess the
desired characteristics can be found in the literature
(17-19). In general the theory predicts the initial
shape of the low frequency viscoelastic behaviour of
the polymer quite well but fails to describe the
limiting high frequency properties. This apparent

departure from theory is yet another manifestation of
the way in which the interplay of segmental and normal
mode relaxation influences the observed behaviour.

In an attempt to gain further insight into the
way in which normal mode and segmental relaxation
interact ultrasonic studies have been performed on
poly(2,4 dimethyl phenylene oxide) (PPO) (20) and
polydimethyl siloxane (PDS) (21). In solution in
toluene (PPO) exhibits a relaxation in the megahertz
frequency range, this is rather surprising since it
has been suggested that the end-to-end distance of
this polymer is independent of temperature in a theta
solvent which infers $\Delta H \simeq 0$. Analysis of the dependence
of the acoustic absorption amplitude with temperature
indicates that there is an energy difference of
approximately 1.6 kT/mole associated with this
relaxation. This observation is not inconsistent with
the light scattering data, since such a small value of
ΔH° would imply a very small change in the dimensions
with temperature. In fact the relaxation frequency
appears insensitive to temperature also suggesting an
activation energy for rotational isomerisation of less
than 12 kJ/mole. In PPO the viscoelastic relaxation
is almost completed at a frequency of 100 kHz, clearly
indicating that the observed relaxation is principally
segmental in origin. A further example of a distinct
separation of the normal mode and segmental relaxation
in a polymer is the behaviour of polydimethyl siloxane
(21). This is a very flexible polymer and the segmental relaxation at room temperature occurs at a
frequency of approximately 500 MHz. Viscoelastic
relaxation even in the highest molecular weight
materials is almost complete in the low megahertz
frequency range leading to the observation of a
distinct plateau in the acoustic absorption between
50-200 MHz.

In this brief review, an attempt has been made
to indicate how molecular weight and chemical structure
can influence the relaxation properties of polymer
molecules in solution. Other important effects which
can be considered are those of tacticity in polymers
such as a polyαMe styrene (22) and poly methyl
methacrylate (22) and solvent effects in segmental
relaxation.

The effects of tacticity can be clearly
identified from changes of the acoustic relaxation
amplitude with temperature tacticity will lead to

changes in the energy difference between stable
isomeric states and this leads to changes in the
value of the relaxing specific heat. In polymethyl
methacrylate the effects of tacticity are very
evident (22).

Solvent can in certain cases influence the
activation energy for segmental relaxation. If we
assume that the effective potential energy profile
is the sum of intra and inter molecular contributions,
then the variation of the apparent activation energy
should parallel the change in solvent viscosity.
Studies of this type have been a little limited,
however there appears to be sufficient evidence to
suggest such a hypothesis is valid in good solvents
(23).

References
1) Orville-Thomas W.J. (Ed) (1973) Internal Rotation in Molecules
Wiley ,New York. This volume contains a series of articles which
cover both the experimental and theoretical aspects of the study
of internal rotation in small molecules.
2) Stockmayer W.H.Gobush W. and Norvich R. (1971) Pure Appl Chem.
26 537
3) Rouse P.A. (1953) J.Chem Phys 21,1272; Zimm B.H. (1956)
J.Chem Phys 24, 269 Numerous modifications of this simple
theory have appeared but the final results are essentially
the same.
4) Kirkwood J.G. and Riseman J. (1948) J.Chem Phys 16,565
5) Paul E. and Mazo R.M.(1968) J.Chem Phys 48, 2378
6) Matsuo, K. (1972) Ph.D. thesis Dartmouth College.
7) North A.M. (1972) Chem Soc Rev. 1, 49
8) Williams G. (1972) Chem. Rev. 1
9) Schaefer J. (1969) Macromolecules 2, 210
10) Bullock A.T., Butterworth J.H. and Cameron G.G. (1971)
Europ. Polymer J. 7, 445
11) Cochran M.A. Dunbar J.H. North A.M. and R.A.Pethrick,
J.Chem Soc. Faraday Trans. II 70, 215 and references quoted
therein.
12) Connor, T.M. and Nartland A. (1968) Polymer 9, 591 and
references quoted therein.
13) Stockmayer W.H. (1967) Pure Appl Chem 15, 539; Davis J.E.
(1960) Ph.D. Thesis MIT; North A.M. and Phillips P.J. (1968)
Trans Faraday Soc. 64, 3235.
14) Cochran M.A. Jones P.B. North A.M. and R.A.Pethrick (1972)
J.Chem Soc. Faraday Trans II 68, 1719
15) North A.M. Rhoney I. and Pethrick R.A. (1974) J.Chem Soc.
Faraday Trans II 70, 223
16) Ferry J.D. (1973) Accounts Chem Res. 6,60 and references
therein

17) Lamb . J. and Matheson, A.J. (1964) Proc. Roy. Soc. A281, 207

18) Shen M.C., Hall W.F. and Wames R.E. (1968) J. Macromolecules Sci. C2,182

19) Moore, R.S. McSkimin H.J. Gieniewski C. and Andreatch P. (1969) J.Chem Phys 50,466 5088

20) B.Wandalt E.Evaman A.M.North and R.A.Pethrick J.Chem. Soc. Trans Faraday II 72, 1957 (1976) and references therein

21) B.T.Poh A.M.North and R.A.Pethrick Unpublished data

22) J.H.Dunbar A.M.North R.A.Pethrick and D.B.Steinhauer, J.Chem Soc. Faraday Trans 71, 1478 (1975)

23) B.T.Poh, W.Bell, A.M.North and R.A.Pethrick Unpublished data

KINETIC STUDIES ON GAS-LIQUID INTERFACE BY MEANS OF CAPILLARY
WAVE TECHNIQUE

T. Yasunaga and M. Sasaki

Department of Chemistry, Faculty of Science,
Hiroshima University, Hiroshima, 730, Japan

ABSTRACT.   A general theory for the propagation characteristics
of the capillary wave on gas-liquid interface, where the physco-
chemical equilibrium exists, was proposed on the basis of the
two-dimensional relaxation theory and the surface thermodynamics.
    The frequency dependences of the propagation characteristics
of the wave were measured in sodium dodecyl sulfate in order to
confirm the validity of the derived equation. A relaxation was
observed both in the propagation velocity and the damping
coefficient of the wave. The results were well interpreted by
the theoretical equations. From the concentration dependence of
the relaxation parameters obtained, the relaxation was attributed
to the adsorption-desorption reaction.

1. INTRODUCTION

    The measurements of the propagation characteristics of the
capillary wave, e.g., the propagation velocity and the damping
coefficient, are effective for the study of the dynamic properties of materials existing on the gas-liquid interface. The
theoretical studies for the insoluble monolayers have been
performed by Dorrestein[1], Mayer and Eliassen[2], and Mann and Du[3],
while those for the soluble monolayer have been performed by van
den Tempel and van de Riet[4], Hansen and Mann[5], and Lucassen and
Hansen[6]. The former has developed their theories taking account
of the surface rheologies, and the latter with the assumption
that the rate-determining step of surfactant transfer between
the surface and the bulk phase is the diffusion process.
    Provided that the physicochemical equilibrium on the

W. J. Gettins and E. Wyn-Jones (Eds.). Techniques and Applications of Fast Reactions in Solution. 579–586.
Copyright © 1979 by D. Reidel Publishing Company.

interface, e.g., the conformational change and the monomer-dimer reaction, is perturbed by the propagation of the wave, the relaxation effect concerned with the equilibrium may be expected. Though studies of such phenomena are valuable for the clarification of the dynamic properties of the equilibrium, theoretical studies have never been performed, while experimentally Davies and Vose[7] have observed the relaxation effect on the surfactant solution.

In the present paper the equation[8] derived on the basis of the relaxation theory and the surface thermodynamics will be checked by the experimental results.

2. EXPERIMENTAL

The apparatus used was a modification of that described by Davies and Vose[7]; the schematic diagram is shown in Fig. 1. The capillary wave was generated by a vibrator attached to a drive-unit of the trumpet speaker. The vibrator was made of Teflon, which has a weak affinity to all solutions. The flash of the stroboscope was synchronized with the signal of the oscillator, and the stationary image of focus was observed with a microscope. The propagation velocity and the damping coefficient were obtained as has been described in Brown's paper[9]. The frequency range of the apparatus was from 25 Hz to 4 kHz.

Sodium dodecyl sulfate (SDS) was prepared from Tokyo Kasei reagent-grade dodecyl alcohol (purity; 99.5%) according to the procedure of Dreger et al.[10] Octylamine hydrochloride (OAC) and dodecylamine hydrochoride (DAC) were prepared as follows. Octylamine and dodecylamine (Tokyo Kasei reagent-grade; Purities; 97.7 and 99.7% respectively) were neutralized by HCl in benzene solutions and were then recrystallized three times from benzene solutions and finally washed with petroleum ether. The values of the CMC were determined to be 8.3, 15.8, and 175 mM for the

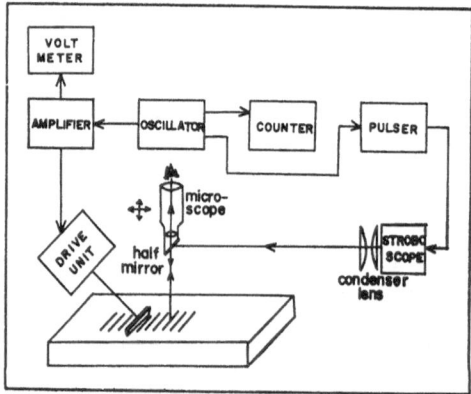

Fig. 1. Schematic diagram of the apparatus for measurements of the propagation characteristics of the capillary wave.

SDS, DAC, and OAC solutions respectively by the electric conductivity method at 25°C.

3. RESULTS AND DISCUSSION

The measurements were carried out on the surfaces of the SDS, DAC, and OAC solutions; the frequency dependences of α/f are shown in Fig. 2. As seen from this figure, the experimental values of α/f are greater than the theoretical ones of α_d/f, where α_d refers to the damping coefficient based on only the diffusion process between the surface and the bulk phase. According to the theory presented in the previous paper[8], the equations for the frequency dependences of c and α on the surface of relatively concentrated surfactant solutions are given by

$$c^2 = (\frac{g}{k} + \frac{\gamma k}{\rho})(1 + 2\varepsilon_{ela,re} + \frac{\gamma k}{\rho c^2} \frac{\omega'^2 \tau^2 \delta'}{1+\omega'^2\tau^2}) \quad , \tag{1}$$

$$\frac{\alpha}{f} = \frac{\alpha_d}{f} + \frac{2\pi\gamma k}{3\rho c^3} \frac{\omega'\tau\delta'}{1+\omega'^2\tau^2} \quad . \tag{2}$$

Equation 2 is, then, rearranged as follows:

$$\frac{3\rho c^3}{2\pi\gamma k}(\frac{\alpha}{f} - \frac{\alpha_d}{f}) = \frac{\omega'\tau\delta'}{1+\omega'^2\tau^2} \quad . \tag{3}$$

The frequency dependences of the l. h. s. of this equation are shown in Fig. 3. This figure shows that the excess damping can be expressed by a single relaxation equation. The relaxation phenomena were also observed in all the other solutions except the 60 mM OAC solution, where the apparent relaxation strength was negligibly small.

The relaxation parameters can also be obtained from the frequency dispersion of c, which was expressed by the following equation with the assumption that $\gamma k/\rho \gg g/k$:

$$\frac{c^3}{f} - \frac{g\lambda^2}{2\pi} = \frac{2\pi\gamma}{\rho} (1 + 2\varepsilon_{ela,re} + \frac{\gamma k}{\rho c^2} \frac{\omega'^2 \tau^2 \delta'}{1+\omega'^2\tau^2}) \quad . \tag{4}$$

The theoretical curves in Fig. 4 were calculated by means of this equation, using the values of the relaxation parameters. Unfortunately, the apparent relaxation strength was so small that the experimental results of the propagation velocity were not precise enough to give the relaxation parameters, only to refine the validity of those obtained from the damping coefficient.

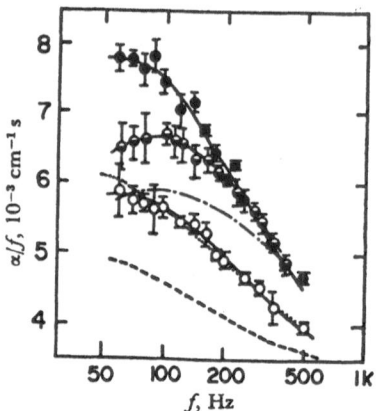

Fig. 2. The plots of α/f vs. f in SDS, DAC, and OAC solutions. The theoretical curves of α_d/f are shown; ----: 8 mM SDS, ····: 15 mM DAC, — · —: 121 mM OAC. o: 8mM SDS. ●: 15 mM DAC. ◉: 121 mM OAC.

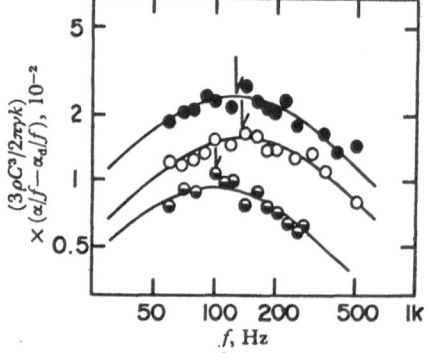

Fig. 3. The plots of $(3\rho c^3/2\pi\gamma k)\cdot(\alpha/f-\alpha_d/f)$ vs. f in SDS, DAC, and OAC solutions. The solid lines show the theoretical curves calculated by r. h. s. of Eq. 3. The arrows show the relaxation frequency $f_r = (4\pi\tau)^{-1}$. o: 8 mM SDS, ●: 15 mM DAC, ◉: 121 mM OAC.

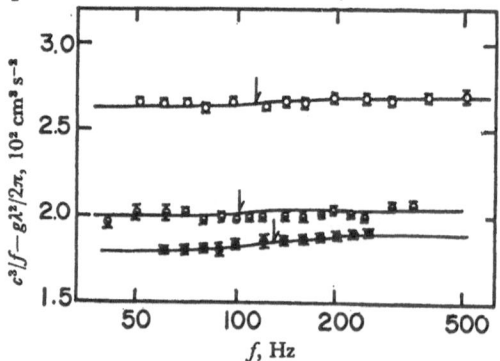

Fig. 4. The frequency dispersion of $(c^3/f-g\lambda^2/2\pi)$ in SDS, DAC, and OAC solutions. The solid lines show the theoretical curves calculated by Eq. 4. The arrows show the relaxation frequency obtained from the experimental results of the damping coefficient.

If the observed relaxation phenomena are based on the adsorption-desorption process of the surfactants, the concentration dependence of the relaxation time is expressed by[8]

$$\tau^{-1} = \frac{d_s k_a c_m \exp(-\frac{Ze\psi_{sub}}{k_B T})}{10^3 \Gamma_{max}} + k_d \, , \tag{5}$$

where c_m is the monomer concentration of the surfactants. In the concentrated surface concentration of ionic surfactants, most of the adsorbed surfactants are neutralized by the counter ions and the Stern layer is formed at the surface. Then, ψ_{sub} may be much smaller than $k_B T/e$. Under this condition, Eq. 5 is simplified to

$$\tau^{-1} = \frac{d_s k_a}{10^3 \Gamma_{max}} c_m + k_d = k_{a,\theta} c_m + k_d \, . \tag{6}$$

The plots of τ^{-1} vs. c_m are on the straight lines, as is shown in Fig. 5. The linearity of these plots suggests that the relaxation phenomena are based on the adsorption-desorption process of the surfactants. The values of $k_{a,\theta}$ and k_d were calculated from the straight lines in Fig. 5; they are listed in Table 1. The values of k_a were calculated from $k_{a,\theta}$ with the literature values of $d_s{}^{[11]}$ and $\Gamma_{max}{}^{[12]}$, they are also listed in Table 1.

Among the values in Table 1, $k_{a,\theta}$ is fairly dependent on the CMC, but k_d is appreciably independent of the CMC, as is seen from Fig. 6. As a result, the following relation was obtained:

$$\log k_{a,\theta} = (3.4\pm0.3) - (0.8\pm0.3)\log CMC \, . \tag{7}$$

On the other hand, the values of k_d were evaluated as follows. Since the state of surfactants on an adsorbed layer is similar

TABLE 1   Kinetic Parameters in SDS, DAC, and OAC solutions at 25°C

	$k_{a,\theta}$ $(10^5 M^{-1}s^{-1})$	k_a $(10^5 s^{-1})$	k_d $(10^2 s^{-1})$	ΔG $(-RT)$	$\Delta\gamma$ $(10^{10} dyn\ cm\ mol^{-1})^{c)}$	
SDS	1.1±0.5	4±2	9±1	6.1[a)] 6.8	10[b)]	6
DAC	0.6±0.3	3±1	8±2	5.9[a)] 6.3	8[b)]	7
OAC	0.1±0.03	0.6±0.2	—	—	3[b)]	7

a) The values were calculated by means of Szyszkowski's equation.
b) The values were calculated by means of $\Delta\gamma \approx (\gamma_{water}-\gamma_{CMC})/\Gamma_{max}$.
c) 1 dyn = 10^{-5} N.

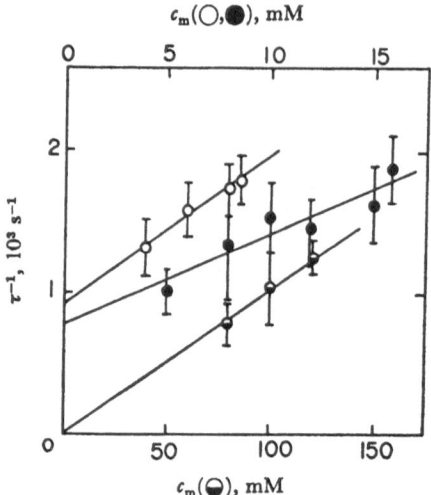

Fig. 5. The plots of τ^{-1} vs. c_m in SDS(o), DAC($\bullet$), and OAC ($\ominus$) solutions.

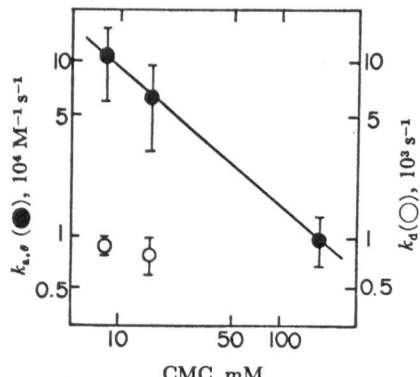

Fig. 6. The plots of $k_{a,\theta}$ (closed circles) and k_d (open circles) vs. CMC.

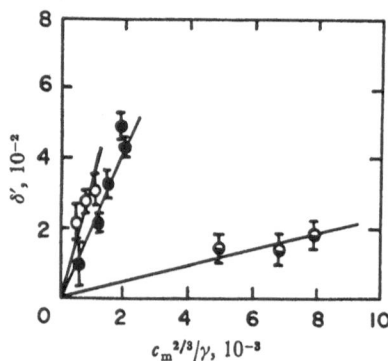

Fig. 7. The plots of δ' vs. $c_m^{2/3}/\gamma$ in SDS(o), DAC($\bullet$), and OAC($\ominus$) solutions.

to that in a micelle in a surfactant solution, the desorption
rate constant can be reasonably compared with the dissociation
rate constant of the monomer from the micelle. In an SDS solution, the value of the former falls in the same order of magnitude as that of the latter, 7.5×10^2 s^{-1}, obtained by means of
the pressure-jump method[13]. These facts support the idea that
the adsorption-desorption mechanism proposed is reasonable.

The adsorption-desorption energies, ΔG, were calculated by
means of k_a and k_d; $\Delta G = -RT\ln(k_a/k_d)$. They are listed in Table
1. The obtained values, however, cannot be referred to the
literature ones by the static methods, since the surface excess
near the CMC increases appreciably with c_m. Then, the effective
adsorption-desorption energy near the CMC was computed by means
of Szyszkowski's equation; it is listed in Table 1. The values
of ΔG obtained are in good agreement with the calculated ones.
This also suggests that the proposed mechanism is reasonable.

The concentration dependences of the apparent relaxation
strength obtained must also be interpreted by means of the
following equation:

$$\delta' = \frac{2(\Delta\gamma)^2 N^{-1/3} c_m^{2/3}}{3\times 10^2 RT\gamma} \tag{8}$$

The plots of δ' vs. $c_m^{2/3}/\gamma$ are on the straight lines, as is
shown in Fig. 7. The values of $\Delta\gamma$ calculated from the slopes of
these straight lines are in good agreement with those of
$\delta\gamma/\delta\Gamma \approx (\gamma_{water}-\gamma_{CMC})/\Gamma_{max}$ listed in Table 1, where γ_{water} and
γ_{CMC} are the surface tensions in water and the surfactant solu-
tion at the CMC respectively.

A preliminary kinetic investigation of the adsorption-
desorption on the surfaces of other surfactant solutions by means
of the capillary-wave method has shown that similar relaxation
phenomena exist in these systems. Further studies of these
systems will lead to a quantitative clarification of the adsorption-desorption phenomena.

REFERENCES

1. R. Dorrestein, Proc. Acad. Sci., B54, 260 (1950).
2. E. Mayer and J. D. Eliassen, J. Colloid Interface Sci., 37,
   228 (1971).
3. J. A. Mann and G. Du, J. Colloid Interface Sci., 37, 2 (1971).
4. M. van den Tempel and R. P. van de Riet, J. Chem. Phys., 42,
   2769 (1965).
5. R. S. Hansen and J. A. Mann, J. Appl. Phys., 35, 152 (1964).
6. J. Lucassen and R. S. Hansen, J. Colloid Interface Sci.,
   22, 32 (1966).

7. J. T. Davies and R. W. Vose, Proc. R. Soc. London, Ser. A, 286, 218 (1965).

8. M. Sasaki, T. Yasunaga, and N. Tatsumoto, Bull. Chem. Soc. Jpn., 50, 852 (1977).

9. R. C. Brown, Proc. Phys. Soc. Lond., 48, 312 (1936).

10. E. E. Dreger, G. I. Keim, and G. D. Miles, Ind. Eng. Chem., 36, 610 (1944).

11. W. D. Harkins and R. W. Motton, J. Colloid Sci., 1, 106 (1946).

12. M. Muramatsu, K. Tajima, and T. Sasaki, Bull. Chem. Soc. Jpn., 41, 1279 (1968).

13. K. Takeda, J. Sci. Hiroshima Univ. Ser. A, 40, 69 (1976).

KINETICS ON ADSORPTION-DESORPTION OF IONS AT INORGANIC
OXIDES-H$_2$O INTERFACE BY MEANS OF RELAXATION TECHNIQUES

T. Yasunaga, M. Sasaki, M. Ashida and K. Hachiya

Department of Chemistry, Faculty of Science,
Hiroshima University, Hiroshima 730, Japan

ABSTRACT. Double relaxations of the order of μsec and msec were
found in aqueous γ-Al$_2$O$_3$ suspension containing Pb^{2+} by means of
the electric field pulse and pressure-jump methods, respectively,
with electric conductivity detection. The reciprocal fast
relaxation time decreases with increasing the Pb^{2+} concentration,
but the slow one increases. The fast relaxation measured is
attributed to the adsorption-desorption process of Pb^{2+} on the
hydrous oxide surface group ≡Al-OH, and the slow one to the
deprotonation-protonation process induced by Pb^{2+} adsorbed. The
values of the forward and backward rate constants of the former
process were determined to be $k_1 = (1.4\pm0.4) \times 10^8$ mol^{-1} dm^3 sec^{-1}
and $k_{-1} = (1.0\pm0.4) \times 10^4$ sec^{-1}, and those of the latter process
$k_2 = (1.3\pm0.6) \times 10$ sec^{-1} and $k_{-2} = (1.5\pm0.1) \times 10^6$ mol^{-1} dm^3 sec^{-1},
respectively at 20°C.

1. INTRODUCTION

The catalytic reactions at solid-liquid interfaces of metal
oxides have been of great interest to colloid chemists. The
adsorption-desorption phenomena of various metal ions on oxides
such as γ-Al$_2$O$_3$ have been extensively investigated since
adsorption-desorption is a fundamental step in heterogeneous
catalytic reactions. Several mechanisms have been proposed by
many investigators[1-5]. Kinetic studies, however, have scarcely
been carried out because the reaction is too fast to be measured
by ordinary methods. Only a kinetic study on the proton
adsorption-desorption at the TiO$_2$-water interface has been
reported using the pressure-jump technique by the present
authors[6].

W. J. Gettins and E. Wyn-Jones (Eds.). Techniques and Applications of Fast Reactions in Solution. 587–596.
Copyright © 1979 by D. Reidel Publishing Company.

The purpose of the present study is to clarify the dynamics of adsorption-desorption in γ-Al_2O_3-Pb^{2+} system, where Pb^{2+} ion adsorbs strongly on the γ-Al_2O_3 surface.

2. THEORY

Let us consider the following scheme for the adsorption-desorption reaction of Pb^{2+} ion on γ-Al_2O_3 surface, where step 1 is an adsorption-desorption of Pb^{2+} on the hydrous oxide surface group $\equiv Al$-OH, and step 2 the deprotonation-protonation induced by Pb^{2+} adsorbed:

$$\equiv Al\text{-}OH \underset{k_{-1}}{\overset{k_1}{\rightleftharpoons}} \equiv Al\text{-}O\overset{\cdots Pb^{2+}}{\underset{H}{\diagdown}} \underset{k_{-2}}{\overset{k_2}{\rightleftharpoons}} \equiv Al\text{-}OPb^+ \tag{1}$$

$$Pb^{2+} \qquad\qquad\qquad\qquad\qquad H^+$$

$$\text{step 1} \qquad\qquad\qquad \text{step 2}$$
$$\tau_1 \qquad\qquad\qquad\qquad \tau_2$$

where $\equiv Al\text{-}O\overset{\cdots Pb^{2+}}{\underset{H}{\diagdown}}$ is the encounter complex formed by adsorption of Pb^{2+}, $\equiv Al\text{-}OPb^+$ the surface complex, τ_1 and τ_2 are the relaxation times of the steps 1 and 2, respectively, and k_i are the rate constants. At the constant ionic strength the following rate equations are given for these steps:

$$-\frac{d[Pb^{2+}]}{dt} = k_1 C_p (\Gamma_t^\infty - \Gamma_t)[Pb^{2+}] - k_{-1} C_p \Gamma_1 \tag{2-a}$$

$$-\frac{d[H^+]}{dt} = k_{-2} C_p \Gamma_2 [H^+] - k_2 C_p \Gamma_1 \tag{2-b}$$

with

$$\Gamma_t = \Gamma_1 + \Gamma_2 = \Gamma_t^\infty \frac{K_1 (1 + \frac{K_2}{[\overline{H^+}]})[\overline{Pb^{2+}}]}{1 + K_1 (1 + \frac{K_2}{[\overline{H^+}]})[\overline{Pb^{2+}}]} \tag{3}$$

where [ ] denotes the bulk concentration, C_p the concentration of particle, K the equilibrium constant, t the time, Γ, Γ^∞, and Γ_t are the equilibrium, saturated, and total amounts of adsorbed Pb^{2+}, respectively, and the subscripts 1 and 2 refer the states $\equiv Al\text{-}O\overset{\cdots Pb^{2+}}{\underset{H}{\diagdown}}$ and $\equiv Al\text{-}OPb^+$, respectively. Thus, the reciprocal relaxation times can be expressed as

$$\tau_{1,2}^{-1} = \frac{1}{2}\{(a_{11}+a_{22})\pm\sqrt{(a_{11}+a_{22})^2 - 4(a_{11}a_{22}-a_{12}a_{21})}\} \quad (4)$$

with

$$a_{11} = k_1\{C_p(\Gamma_t^\infty-\overline{\Gamma}_t)+[\overline{Pb^{2+}}]\} + k_{-1}$$

$$a_{12} = k_{-1}$$

$$a_{21} = k_2$$

$$a_{22} = k_{-2}(C_p\overline{\Gamma}_2+[\overline{H^+}]) + k_2$$

Equation 4 is simplified under special conditions

$$\tau_1^{-1} \gg \tau_2^{-1} \quad (a_{11} \gg a_{22})$$

$$\tau_1^{-1} = a_{11} = k_1\{C_p(\Gamma_t^\infty-\overline{\Gamma}_t)+[\overline{Pb^{2+}}]\} + k_{-1} \quad (5\text{-a})$$

$$\tau_2^{-1} = a_{22} - \frac{a_{12}a_{21}}{a_{11}} = k_{-2}(C_p\overline{\Gamma}_2+[\overline{H^+}]) + k_2$$

$$- \frac{k_{-1}k_2}{k_1\{C_p(\Gamma_t^\infty-\overline{\Gamma}_t)+[\overline{Pb^{2+}}]\} + k_{-1}} \quad (5\text{-b})$$

The apparent and true overall equilibrium constants $K_{o,ap}$ and K_o, respectively, are related by the following equation:

$$K_{o,ap} \equiv \frac{\overline{\Gamma}_t[\overline{H^+}]}{(\Gamma_t^\infty-\overline{\Gamma}_t)[\overline{Pb^{2+}}]} = K_o + K_1[\overline{H^+}] \quad (6)$$

with

$$K_o = K_1K_2$$

3. EXPERIMENTAL SECTION

Apparatus.   (a)Pressure-jump apparatus   The Pressure-jump apparatus used is a modification of that described by Knoche and Wiese[7]. A pressure jump of 100 atm was realized within 80 μsec. An advantage of this apparatus is that chemical relaxation induced by pressure jump can be observed simultaneously by both changes in electric conductivity and optical absorbance.
(b)Electric field pulse apparatus   The electric field pulse apparatus used has already been reported[8]. The electric field

intensity in the cell was 16 kV cm^{-1}. The duration of the high
voltage pulse applied was 25 μsec, where the rise and decay
times of the pulse were much faster than 0.1 μsec.

Materials.   A commercial product "Aluminium Oxides C" supplied
by Japan Aerosil Co. was produced by the hydrolysis of AlCl$_3$ in
a flame process. The X-ray diffraction pattern confirmed that
the sample was predominantly γ-Al$_2$O$_3$. The zero point of charge
was pH 9.0. The specific surface area determined by the BET
adsorption measurements was 100 ± 15 m^2 g^{-1}. The particle size
was smaller than 1 μm and apparently uniform.
    Prior to use, the sample was electrodialyzed until the
electric conductivity became equal to that of the distilled
water. The sample dried at 100°C was redispersed by ultra-
sonication, and was aged for a few days. The particle in the
suspension did not sediment as observed during the period of one
hour.

Analytical method.   In order to measure colorimetrically the
bulk concentration of lead ion in the aqueous suspension of γ-
Al$_2$O$_3$ containing lead nitrate, the interfering particles were
filtered with Millipore filter (Millipore Filter Corp. GSWP
04700: with pore size of 0.22 μm). The colorimetric determination of lead ion was performed at λ = 510 nm by the dithizone
extraction method.

4. RESULTS AND DISCUSSION

    The kinetic measurements were carried out in the aqueous
suspension of γ-Al$_2$O$_3$ containing Pb(NO$_3$)$_2$ at 20°C. A typical
relaxation curve obtained by means of the pressure-jump technique
with conductivity detection is shown in Figure 1(a) in which the
electric conductivity increases with the pressure. As seen from
Figure 1(a), at the beginning of the relaxation a very fast
change in conductivity was also found. To resolve this, measurements were performed by the electric field pulse technique.
Here, a single relaxation was found as shown in Figure 2. Additional measurements were carried out by both the pressure-jump
and electric pulse methods in the γ-Al$_2$O$_3$ suspension, the
aqueous solution of Pb(NO$_3$)$_2$, the supernatant solution of the
γ-Al$_2$O$_3$-Pb(NO$_3$)$_2$, and the γ-Al$_2$O$_3$-NaNO$_3$ suspension, but no
relaxation effect was observed. Since the γ-Al$_2$O$_3$-Pb(NO$_3$)$_2$
suspension was slightly acidic (pH 5.4), the experiment in the
γ-Al$_2$O$_3$-HNO$_3$ suspension was also performed in order to clarify
the contribution of proton to the relaxation. But, no relaxation was observed. On the other hand, similar relaxation to that
observed in the γ-Al$_2$O$_3$-Pb(NO$_3$)$_2$ suspension could also be seen
in the γ-Al$_2$O$_3$-PbCl$_2$ system. These facts indicate that two kinds
of relaxations occur in the γ-Al$_2$O$_3$ suspension containing Pb^{2+}.

The electric conductivity changed with increasing pressure in the γ-Al$_2$O$_3$-Pb(NO$_3$)$_2$ suspension. The plausible reactions, therefore, may be an aggregation-dispersion of the γ-Al$_2$O$_3$ particles caused by Pb^{2+} and an adsorption-desorption of Pb^{2+} on the surface of the γ-Al$_2$O$_3$ particles. If the former reaction occurs, a change in scattered light intensity should be expected in the spectrophotometric sample cell[6]. As seen from Figure 1(b), however, the oscilloscope trace with the turbidity detection did not reveal any relaxation on the same time scale as that with the electric conductivity detection. This result, consequently, suggests that the relaxation observed is associated with the adsorption-desorption of Pb^{2+}.

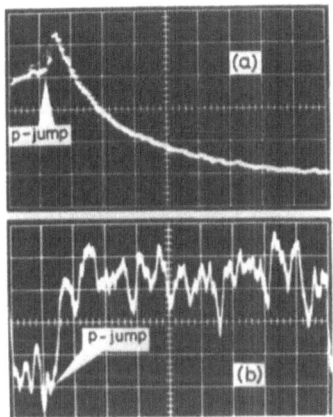

Figure 1. Typical relaxation curves in aqueous γ-Al$_2$O$_3$-Pb(NO$_3$)$_2$ suspension observed by the pressure-jump method with (a) electric conductivity and (b) turbidity detection. Concentration of Al$_2$O$_3$: $C_p = 15$ g dm^{-3} at 20°C. Sweep: 2 msec/div, wave length in (b): 525 nm.

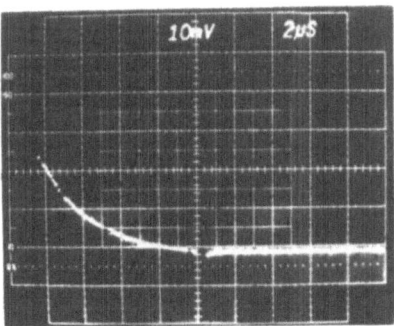

Figure 2. Typical relaxation curve in aqueous γ-Al$_2$O$_3$-Pb(NO$_3$)$_2$ suspension observed by the electric field pulse method. $C_p = 25$ g dm^{-3} at 20°C. Sweep: 2 µsec/div.

The Pb^{2+} concentration dependence of the reciprocal slow and
fast relaxation times, τ_s^{-1} and τ_f^{-1}, respectively, were measured
at various ionic strengths and at various pH, and the results are
shown in Figures 3 and 4. As seen from the figures, the value of
τ_s^{-1} is dependent on the concentration of Pb^{2+} and pH, but is
independent of the ionic strength. On the other hand, the value
of τ_f^{-1} decreases with increasing the concentration of Pb^{2+}.

The adsorption isotherms of Pb^{2+} at various pH were obtained
in order to decide the total concentration of adsorption site and
the equilibrium constant K associated with the adsorption-
desorption reaction. At each pH, the adsorption isotherms obtained
were of the Langmuir's ones. Consequently, the values of Γ^∞ and
K were obtained from the intercepts and the slopes of the Langmuir
plots.

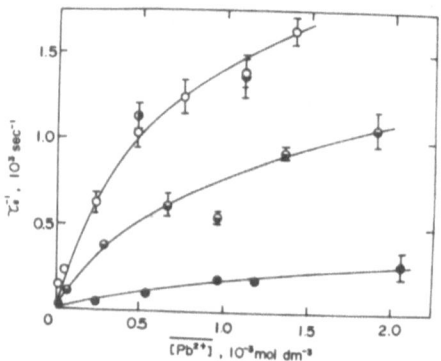

Figure 3. The Pb^{2+} concentration dependence of τ_s^{-1} in aqueous
γ-Al_2O_3-$Pb(NO_3)_2$ suspension of $C_p = 25$ g dm^{-3} at $20^\circ C$. (o)pH 5.4,
(o)pH 5.4 : ionic strength of $I = 6 \times 10^{-3}$, (o)pH 5.0, (•)pH 4.5.

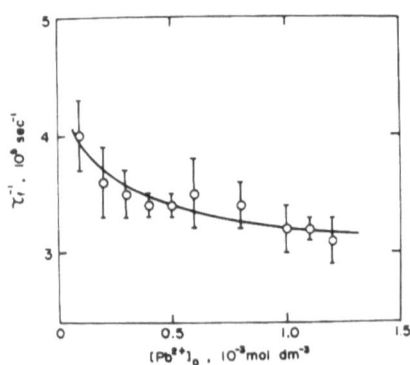

Figure 4. The Pb^{2+} concentration dependence of τ_f^{-1} in aqueous
γ-Al_2O_3-$Pb(NO_3)_2$ suspension of $C_p = 25$ g dm^{-3} at $20^\circ C$ and pH 5.4.

Slow process.   In order to interpret the concentration
dependences of τ_s^{-1} shown in Figure 3, various plausible mechanisms for the adsorption-desorption reaction of Pb^{2+} on $\gamma-Al_2O_3$
surface were examined.

(I) The Langmuir-type adsorption-desorption reaction.

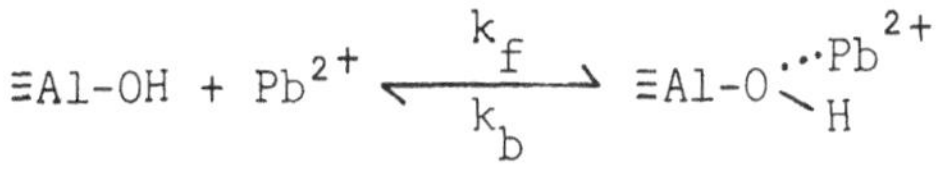

$$\equiv Al-OH + Pb^{2+} \underset{k_b}{\overset{k_f}{\rightleftharpoons}} \equiv Al-O\underset{H}{\overset{\cdots Pb^{2+}}{\diagup}}$$

(II) The adsorption-desorption reaction involving the ion-exchange
with two protons.

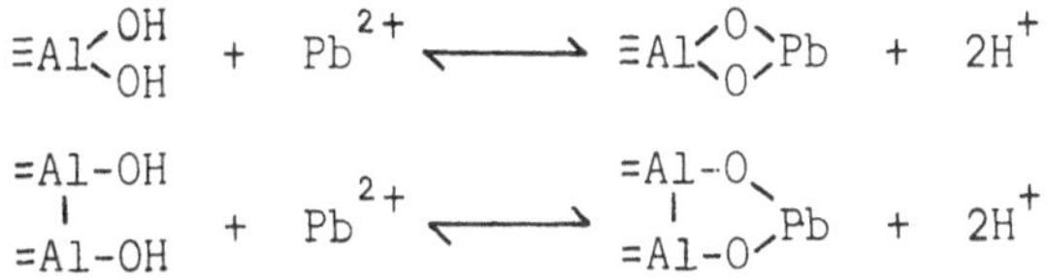

$$\equiv Al\underset{OH}{\overset{OH}{\diagdown}} + Pb^{2+} \rightleftharpoons \equiv Al\underset{O}{\overset{O}{\diagdown}}Pb + 2H^+$$

$$\begin{matrix} =Al-OH \\ | \\ =Al-OH \end{matrix} + Pb^{2+} \rightleftharpoons \begin{matrix} =Al-O \\ | \quad\diagdown \\ =Al-O\diagup \end{matrix}Pb + 2H^+$$

(III) The adsorption-desorption reaction involving the ion-exchange
with proton.

$$\equiv Al-OH + Pb^{2+} \underset{k_b}{\overset{k_f}{\rightleftharpoons}} \equiv Al-OPb^+ + H^+$$

(IV) The adsorption-desorption reaction in the presence of proton
catalyst.

$$\equiv Al-OH + Pb^{2+} + H^+ \rightleftharpoons \equiv Al-OPb^+ + 2H^+$$

We found that all of them could be ruled out.

(V) The two-step adsorption-desorption reactions involving the
ion-exchange with proton.   The existence of two relaxations
in Figures 1(a) and 2 suggests that the adsorption-desorption
reaction of Pb^+ consists of two processes where one of them is
very fast and the other one is slow. As a possibility, mechanism
III can be devided into two elementary steps as follows:

$$(V-a) \quad \equiv Al-OH \underset{H^+}{\overset{\diagdown}{\rightleftharpoons}} \equiv Al-O^- \underset{Pb^{2+}}{\overset{\diagup}{\rightleftharpoons}} \equiv Al-OPb^+$$

step 1   step 2

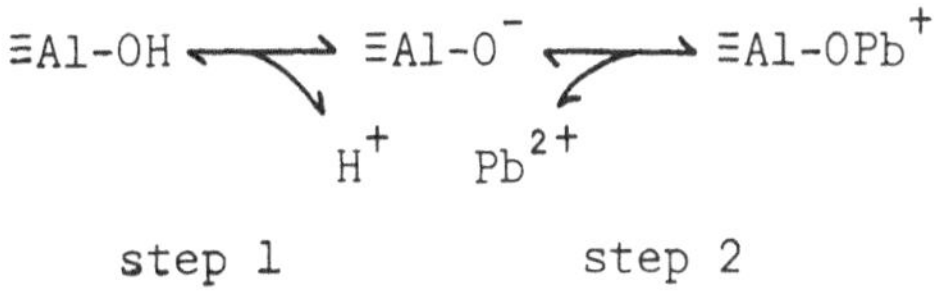

(V-b)   Another possible two-step mechanism is given by
         scheme 1.

In the case where step 1 is very fast compared to step 2 in V-a,
or in the converse case, the value of τ_s^{-1} calculated from the
rate equations for these two steps is not proportional to the
amount of adsorbed Pb^{2+}. This prediction contradicts the
experimental fact that the value of τ_s^{-1} measured is proportional
to the amount of adsorbed Pb^{2+}.

Next, the experimental results of τ_s^{-1} are analized using the theoretical equation 5-b for mechanism V-b. In the case of $\tau_1^{-1} \ll \tau_2^{-1}$ the value of τ_s^{-1} should not be proportional to the amount of absorbed Pb^{2+}, but it should be according to eq 5-b. Comparing these predictions with the experimental facts, eq 5-b remains as a possible description of the slow relaxation. Under the present experimental conditions, eq 5-b can be approximated as

$$\tau_s^{-1} \simeq k_{-2} \frac{C_p \overline{\Gamma}_t}{1 + K_2^{-1}[\overline{H^+}]} \tag{7}$$

Here, the value of K_2 can be obtained from both the slope and intercept of the straight line of Figure 5 where the dependence of $K_{o,ap}$ on $[H^+]$ is given by eq 6. Γ_t^∞ in eq 6 can be obtained from eq 3. The experimental plots of τ_s^{-1} vs. $C_p \overline{\Gamma}_t (1 + K_2^{-1}[\overline{H^+}])^{-1}$ are shown in Figure 6. As seen from this figure, the data points fall on a straight line through an origin. This fact verifies that the slow relaxation can be ascribed to step 2 in mechanism V-b, given by scheme 1. The value of k_{-2} was obtained from the slope of the straight line in Figure 6, and the value of k_2 was calculated using the values of K_2 and k_{-2}. The results are listed in Table I.

Fast process.   The Pb^{2+} concentration dependence of τ_f^{-1} shown in Figure 4 was analized by eq 5-a. The experimental plot of τ_f^{-1} vs. the expression between braces in eq 5-a are shown in Figure 7. As seen from this figure, the data points are on a straight line. From the slope and intercept of this line, the values of k_1 and k_{-1} were obtained, which are also listed in Table I. The value of K_1 calculated from k_1 and k_{-1} is $(1.4 \pm 1.0) \times 10^3$ mol^{-1} dm^3, and is in agreement with the K_1 value obtained statically using eq 6 as seen in Table I. These facts leads to the conclusion that the fast relaxation can be attributed to the step 1 in the mechanism V-b, given by scheme 1.

Similar kinetic investigations are in progress for aqueous suspensions of metal oxides containing various positive and negative ions, and the results will be reported in the near furture.

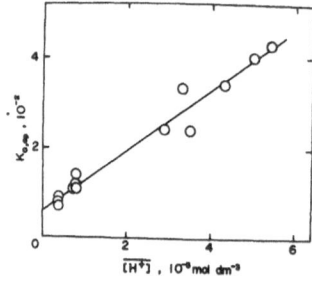

Figure 5.   Plot of $K_{o,ap}$ vs. $[H^+]$.

TABLE I   Kinetic Parameters of Adsorption-desorption of Pb^{2+} on γ-Al_2O_3 Particles at 20°C

Step 1		
k_1 10^8 mol^{-1} dm^3 s^{-1}	k_{-1} 10^4 s^{-1}	K_1 10^2 mol^{-1} dm^3
1.4±0.4	1.0±0.4	14±10
		6.7±0.3[a]
Step 2		
k_2 10 s^{-1}	k_{-2} 10^6 mol^{-1} dm^3 s^{-1}	K_2 10^{-6} mol dm^{-3}
1.3±0.6	1.5±0.1	9±4

[a] This value was obtained from the slope of the straight line in Figure 5.

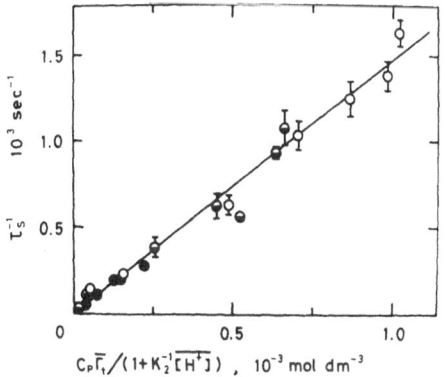

Figure 6.   Plot of τ_s^{-1} vs. $C_p\overline{\Gamma}_t(1+K_2^{-1}[\overline{H}^+])^{-1}$ in eq 7 .
(o)pH 5.4, (◓)pH 5.0, (●)pH 4.5.

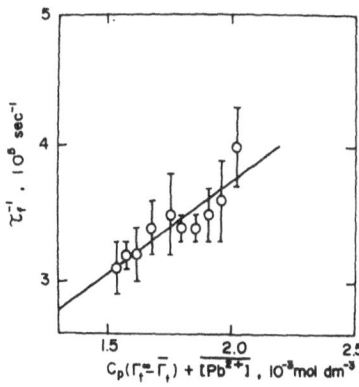

Figure 7.   Plot of τ_f^{-1} vs. the expression between braces in eq 5-a.

REFERENCES AND NOTES

1. H. Hohl and W. Stumm, J. Colloid Interface Sci., 55, 281
    (1976).
2. P. W. Schindler and H. Gamsjäger, Kolloid-Z. Z. Polym., 250,
    759 (1972).
3. P. W. Schindler, B. Fürst, R. Dick, and P. U. Wolf,
    J. Colloid Interface Sci., 55, 469 (1976).
4. J. B. Perri, J. Phys. Chem., 69, 211, 220 (1965).
5. C. P. Huang and W. Stumm, J. Colloid Interface Sci., 43,
    409 (1973).
6. M. Ashida, M. Sasaki, H. Kan, T. Yasunaga, K. Hachiya, and
    T. Inoue, J. Colloid Interface Sci., (in press).
7. W. Knoche and G. Wiese, Chem. Instru., 5, 91 (1973-74).
8. Y. Tsuji, T. Yasunaga, T. Sano, and H. Ushio, J. Am. Chem.
    Soc., 98, 813 (1976).

LIST OF PARTICIPANTS

A. Those who attended the full NATO Advanced Study
Institute at the University College of Wales.

D. Attwood, Department of Pharmacy, The University of
    Manchester, Manchester M13 9PL, U.K.
J. Aubard, Laboratoire de Chimie Organique Physique,
    Université Paris VII, 1 rue Guy de la Brosse,
    75005 Paris, France.
D. Bauernschmitt, Institut für Physicalische Chemie,
    Universität Bayreuth, Am Birkengut,
    Postfach 3008, 8580 Bayreuth, W. Germany.
O. Bensaude, Laboratoire de Chimie Organique Physique,
    Université Paris VII, 1 Rue Guy de la Brosse,
    75005 Paris, France.
F.J. Bermejo-Barrera, Institute for Chemical Physics,
    "Rocesolano", Madrid, Spaim.
C.F. Bernasconi, Department of Chemistry, Division of
    Natural Sciences, Thimann Laboratories,
    University of California, Santa Cruz,
    California 95064, U.S.A.
A. Beyer, Institut für theoretische Chimie und
    Strahlenchimie, Universität Wien,
    Wahringerstrasse 38, 1090 Wien, Austria.
D. Bloor, Department of Chemistry and Applied Chemistry,
    University of Salford, Salford M5 4WT, U.K.
M. Braithwaite, Corporate Laboratory, I.C.I. Ltd.,
    P.O. Box 11, The Heath, Runcorn, Cheshire
    WA7 4QE, U.K.
R. Brouillard, Laboratoire de Chimie Organique Physique,
    Université Paris VII, 1 rue Guy de la Brosse,
    75005 Paris, France.
E.F. Caldin, University Chemical Laboratory, The
    University of Kent, Canterbury, Kent CT2 7NH, U.K.
F. Ceuterick, Laboratorium voor Chemische en
    Bioloogische Dynamica, Katholiek Universiteit
    te Leuven, Celestijnenlaan 200D, B-3030
    Heverlee, Belgium.
S.P. Dagnall, University Chemical Laboratory, University
    of Kent, Canterbury, Kent CT2 7NH, U.K.

J.R. Darwent, The Dyson Perrins Laboratory, The
        University of Oxford, South Parks Road,
        Oxford OX1 3QY, U.K.
A. Dawson (Miss), Department of Chemistry and Applied
        Chemistry,University of Salford, Salford M5 4WT
        U.K.
K. Dela, The Royal Danish School of Pharmacy, Chemical
        Laboratory AD, Universitetsparken 2,
        DK 2100, Copenhagen ø, Denmark.
D. Devia, Max-Planck-Institut für Biophysikalische
        Chemie, Postfach 968, D-3400 Gottingen, B.R.D.
G. Dewey, Department of Chemistry, University of
        Rochester, Rochester, New York 14627, U.S.A.
E.D. Dio, Istituto de Chimica-Fisica, Università
        di Palermo, 26 via Archirafi, 90123 Palermo,
        Italy.
H.B. Dunford, Department of Chemistry, University of
        Albertam Edmonton, Alberta, Canada T6G 2G2.
A.E. Durno, Corporate Laboratory, I.C.I. Ltd., The
        Heath, Runcorn, Cheshire WA7 4QE, U.K.
J. Everaert, Laboratorium voor Chemische en Bioloogische
        Dynamica, Katholiek Universiteit te Leuven,
        Celestijnenlaan 200D, B-3030 Heverlee, Belgium.
E.M. Eyring, Department of Chemistry, The University
        of Utah, Salt Lake City, Utah 84112, U.S.A.
M.M. Farrow, Staff Chemist, I.B.M. Corporation,
        P.O. Box 1900, Boulder, Colorado 80302, U.S.A.
J.B. Field, University Chemistry Laboratory, The
        University of Kent, Canterbury, Kent CT2 7NH, U.K.
P.D.I. Fletcher, University Chemical Laboratory,
        University of Kent, Canterbury, Kent CT2 7NH, U.K.
W. Frisch, Fritz-Haber-Institut der Max-Planck-
        Gesellschaft, Faradayweg 4-6, 1 Berlin 33
        Dahlem, W. Germany.
S. Gabrielson, Chemistry Department, University College
        P.O. Box 78, Cathays Park, Cardiff CF1 1XL, U.K.
J. Garcia-Rosas, Chemistry Department, Stirling University
        Stirling FK9 4LA, U.K.
W.J. Gettins, Department of Chemistry and Applied
        Chemistry, University of Salford, Salford
        M5 4WT, U.K.
B. Gibson, Department of Chemistry, University of
        Durham, Durham DH1 3LE, U.K.
J. Gormally, Department of Chemistry and Applied
        Chemistry, University of Salford,
        Salford M5 4WT, U.K.
C.M. Gould, Department of Chemistry and Applied Chemistry
        University of Salford, Salford M5 4WT, U.K.
J. Grandjean, Institut de Chimie Organique (B6),
        Université de Liège, Sart-Tilman par 4000,
        Liège I, Belgium.

R.C. Greenwood, University Chemical Laboratory,
University of Kent, Canterbury, Kent CT2 7NH, U.K.

L. Grozinski, 1639 Walters Street, Campbell,
California 95008, U.S.A.

P. Hemmes, Olson Laboratories, Department of Chemistry,
Newark College of Arts and Sciences,
Newark, New Jersey 07102, U.S.A.

B. Henry, Laboratoire de Chimie Physique Organique,
Université de Nancy I, C.O. 140,
54037 Nancy Cedex, France.

M.S. Henty, Research and Development Department, Mond
Division, I.C.I. Ltd., P.O. Box 8, The Heath,
Runcorn, Cheshire WA7 4QD, U.K.

J.F. Hinton, Department of Chemistry, University of
Arkansas, Fayetteville, Arkansas 72701, U.S.A.

J. Holzwarth, Fritz-Haber-Institut der Max-Planck-
Gesellschaft, 1 Berlin 33, Faradayweg 4-6,
W. Germany.

O.W. Howarth, Department of Molecular Sciences,
University of Warwick, Coventry CV4 7AL, U.K.

M. Jensen, Chemistry Laboratory AD, The Royal Danish
School of Pharmacy, Universitetsparken 2,
DK 2100, Copenhagen Ø, Denmark.

P.L. Jobling, Department of Chemistry and Applied
Chemistry, University of Salford,
Salford M5 4WT, U.K.

G.H. Jones, Office 039, Corporate Laboratory,
I.C.I. Ltd., P.O. Box 11, The Heath,
Runcorn, Cheshire, U.K.

Y. Konstantatos, Greek Atomic Energy Commission,
Nuclear Research Centre, Democritos,
Athens, Greece.

W. Knoche, Max-Planck-Institut für Biophysikalische
Chemie, Postfach 968, D-3400 Gottingen, B.R.D.

D. Labuda, Department of Biochemistry, Adam Mickiewicz
University at Poznan, 61-701 Poznan, Fredry 10,
Poland.

H.H. Limbach, Institut für Physikalische Chemie der
Universität Freiburg, Albertstrasse 21,
D-7800 Freiburg I. BR., B.R.D.

D.M. Lynch, Chemistry Department, University College,
P.O. Box 78, Cathys Park, Cardiff CF1 1XL, U.K.

M. De Maeyer, Laboratorium voor Chemische en Bioloogische
Dynamica, Katholiek Universiteit te Leuven,
Celestijnenlaan 200D, B-3030 Heverlee, Belgium.

E. Morris, Unilever Research, Colworth House,
Sharnbrook, Bedford MK44 1LQ, U.K.

R. Natarajan, Department of Chemistry and Applied
Chemistry, University of Salford,
Salford M5 4WT, U.K.

I.T. Norton, Department of Chemistry, University of
        York, Heslington, York YO1 5DD, U.K.
O. Ortona, Istitute Chemicao, Università di Napoli,
        via Mezzocannone 4, 80134 Napoli, Italy.
R.A. Pethrick, Thomas Graham Building, Department of
        Pure and Applied Chemistry, University of
        Strathclyde, 295 Cathedral Street,
        Glasgow G1 1XL, U.K.
M.C. Pereira, Department of Chemistry and Applied
        Chemistry, University of Salford,
        Salford M5 4WT, U.K.
B.F. Peterman, Department of Chemistry, University of
        Ottowa, 365 Nicholas Street, Ottawa, Ontario,
        Canada K1N 9B4.
S. Petrucci, Department of Chemistry, Plolytechnic
        Institute of New York, 333 Jay Street,
        Brooklyn, New York 11201, U.S.A.
G. Platz, Institute für Physikalische Chemie,
        Universität Bayreuth, Am Birkengut, Postfach
        3008, 8580 Bayreuth, B.R.D.
R. Ramaswamy, Department of Chemistry, Princeton
        University, Princeton, New Jersey 08540, U.S.A.
J.E. Rassing, Chemical Laboratory AD, The Royal Danish
        School of Pharmacy, Universitetsparken 2,
        DK 2100, Copenhagen Ø, Denmark.
A. Riches (Miss), University Chemical Laboratory,
        The University of Kent, Canterbury,
        Kent CT2 7NH, U.K.
J. Ridge, Department of Biochemistry, Stanford University
        School of Medicine, Stanford, California 94305,
        U.S.A.
L. Rodehuser, Max-Plank-Institute für Biophysikalische
        Chemie, Postfach 968, D-3400 Gottingen, B.R.D.
B.H. Robinson, University Chemical Laboratory, University
        of Kent, Canterbury, Kent CT2 7NH, U.K.
O. Rogne, Division for Toxicology, Norweigian Defence
        Research Establishment, P.O. Box 25,
        N-2007 Kjeller, Norway.
C. Sbriziolo, Istituto de Chimia-Fisica, Università di
        Palermo, 26 via Archirafi, 90123 Palermo, Italy.
Z.A. Schelly, Department of Chemistry, The University
        of Texas at Arlington, Arlington, Texas 76109,
        U.S.A.
G. Serratrice, Laboratoire de Chimie Physique Organique,
        Université de Nancy I, C.O. 140, 54037
        Nancy Cedex, France.
S.B. Sidais, Chemistry Department, University College,
        University of Wales, Cathays Park, P.O. Box 78,
        Cardiff CF 1 1XL, U.K.

H. Sindermann, Medizinische Hochschule Hannover,
        Org. Nr. 4350, Postfach 61 01 80,
        D-3000 Hannover-61, W. Germany.
F. Strohbusch, Institut für Physikalische Chemie der
        Universität Freiburg, Albertstrasse 21,
        78 Freiburg, W. Germany.
Y.P. Tan, Max-Planck-Institut für Biophysikalische
        Chemie, Abt. Molekular Biologie, D-3400
        Gottingen, W. Germany.
G.J.T. Tiddy, Basic Research Laboratories, Unilever Ltd.,
        Port Sunlight, The Wirral, Merseyside L62 4XN,
        U.K.
J.A. Tolley (Miss), Department of Pharmacy, University
        of Manchester, Manchester M13 9PL, U.K.
F. Turan (Mrs) Chemistry Department, Eczacilik
        Fakultesi, Hacettepe Universitesi, Kimya
        Bilim Dali, Ankara, Turkey.
M.V. Twigg, Agricultural Division, I.C.I. Ltd.,
        P.O. Box 6, Billingham, TessideTS23 1LD, U.K.
V. Vitagliano, Istituto Chimico, Univiversità di Napoli,
        via Mezzocannone 4, 80134 Napoli, Italy.
I. Wagner, Max-Planck-Institut für Physikalische Chemie,
        Postfach 968, D-3400 Gottingen, W; Germany.
K. Wallenhauer (Ms), Institut für Physikalische
        Chemie, Universität Bayreuth, Am Birkengut,
        Postfach 3008, 8580 Bayreuth, W. Germany.
M.F. Walsh, Department of Chemistry and Applied
        Chemistry, University of Salford,
        Salford M5 4WT, U.K.
B. Wojak, Lehrstuhl für Physikalische Chemie,
        Universität Bayreuth, 8580 Bayreuth, Postfach
        3008, W. Germany.
E. Wyn-Jones, Department of Chemistry and Applied
        Chemistry, University of Salford,
        Salford M5 4WT, U.K.
T. Yasunaga, Faculty of Science, Hiroshima University,
        Hiroshima, Japan.
R. Zana, Centre de Recherches sur les Macromolecules,
        C.N.R.S., 6 rue Boussingault, 67083-Strasbourg
        Cedex, France.
D.A. Sweigart, Department of Chemistry, Swarthmore
        College, Swarthmore, PA 19801, U.S.A.

B. Those who attended the two-day meeting on "The
        Current State of the Art", Fast Reactions in
        Solution Group Meeting, 14-15 September, 1978.

E.A.C. Aniansson, Department of Physical Chemistry,
        University of Gottingen, LTH Fack. 5-402 20
        Goteborg 5, Sweden.

P. Bayley, National Institute for Medical Research,
          The Ridgeway, Mill Hill London NW7 1AA, U.K.
R. Bird, Department of Chemistry and Metallurgy,
          The Royal Military College of Science,
          Shrivenham, Swindon, Wilts., U.K.
R. Bisby, Department of Chemistry and Applied Chemistry,
          University of Salford, Salford M5 4WT, U.K.
M. Booy, University Chemical Laboratory, The University
          of Kent, Canterbury, Kent CT2 7NH, U.K.
L.J. Brubacher, Department of Chemistry, University of
          Waterloo, Waterloo, Ontario, Canada N2L 3G1.
G. Brunton, Shell Research Ltd., Thornton Research
          Centre, P.O. Box 1, Chester CH1 3SH, U.K.
R. Cerf, Laboratoire d'Acoustique Moleculaire,
          Université Louis Pasteur, 4 rue Blaise Pascal,
          67070 Strasbourg, France.
P.W. Couldrey, I.C.I. Corporate Laboratory, P.O. Box 11,
          The Heath, Runcorn, Cheshire, U.K.
B.G. Cox, Biokemisk Institut, Odense Universitet,
          Campusvej 55, DK-5230 Odense M, Denmark.
Kathleen Damps, The Royal Institution of Great Britain,
          31, Albemarle Street, London W1X 4BS, U.K.
R. Day, University Chemical Laboratory, The University
          of Kent, Canterbury, Kent CT2 7NH, U.K.
H.E. Edwards, North East Wales Institute, Kelsterton
          College, Connah's Quay, Clwyd, U.K.
H. Erne, Dia-Log G.m.b.H. & Co. K.G., Harfstrasse 34,
          6000 Dusseldorf 13, W. Germany.
A.G. Evans, Chemistry Department, University College,
          University of Wales, Cathays Park, P.O. Box 78,
          Cardiff CF1 1XL, U.K.
R. George, Durrum Instruments Corp., 77 Tudor Drive,
          Yateley, Camberley, Surrey GU17 70B, U.K.
D.M. Goodall, Department of Chemistry, University of
          York, Heslington, York YO1 5DD, U.K.
B. Gruenewald, Biozentrum, Klingelbergstrasse 70,
          CH-4056 Basel, Switzerland.
L. Hellemans, Lab. Chem. and Biol. Dynamica,
          Katholiek Universiteit te Leuven, Celestijnenlaan
          200D, B-3030, Heverlee, Belgium.
K. Heremans, Lab. Chem. en Biol. Dynamica, Katholiek
          Universiteit te Leuven, Celestijnenlaan 200D,
          B-3030 Heverlee, Belgium.
U. Hilverkus, Universität Bielefeld, 4800 Bielefeld, B.R.D.
T.C. Jenkins, The Institute for Cancer Research, Physics
          Department, Clifton Avenue, Sutton, Surrey
          SM2 5PX, U.K.
J.G. Jones, New University of Ulster, Coleraine, N. Ire.
P. Jones, University of Newcastle upon Tyne, Radiation
          and Biophysical Chemistry Laboratory, School

of Chemistry, The University, Newcastle upon
Tyne, NE1 7RU, U.K.

I.D. Johnson, Department of Chemistry and Applied
Chemistry, The University of Salford,
Salford M5 4WT, U.K.

A. Jost, Fakultät für Chemie, Universität Bielefeld,
Postfach 8640, 48 Bielefeld 1, W. Germany.

R.W. King, National Institute for Medical Research,
The Ridgeway, Mill Hill, London NW7 1AA, U.K.

U.K. Klaning, Aarhus University, Department of
Chemistry, 140 Langelandsgade, DK-8000
Aarhus C, Denmark.

K.T. Leffek, Dalhousie University, Chemistry Dept.,
Halifax, Nova Scotia, Canada.

J.S. Littler, University of Bristol, School of
Chemistry, Cantocks Close, Bristol BS8 1TS, U.K.

W.J. Louw, Institut für Physikalische Chemie, Frankfurt
am Main, Robert-Mayerstrasse 11, 6000
Frankfurt/Main, W. Germany.

L.I.A. Martins, Department of Chemistry and Molecular
Sciences, University of Warwick, Coventry
CV4 7AL, Warwickshire, U.K.

J.J. McGarvey, Department of Chemistry, David Keir
Building, Queen's University of Belfast,
Stranmillis Road, Belfast BS9 5AG, U.K.

G. Meier, Universität Bielefeld, 4800 Bielefeld 1,
Universitätsstrasse, W. Germany.

R. Morris, Durrum Instrument Corp. 1228 Titan Way,
Sunnyvale, CA94088, U.S.A.

P. O'Neill, Institute of Cancer Research, Physics
Department, Clifton Avenue, Sutton, Surrey
SM2 5PX, U.K.

J.E. Packer, Biochemistry Dept. Brunel University,
Kingston Lane, Uxbridge, Middlesex UB8 3PH, U.K.

B. Parsons, Kelsterton College, Connah's Quay,
Clwyd CH5 4BR, U.K.

A. Persoons, Lab. Chem. en Biol. Dynamica, Katholiek
Universiteit te Leuven, Celestijnenlaan 200 D,
B-3030 Heverlee, Belgium.

C.R. Rabl, Max-Planck-Institut für Biophysikalische
Chemie, Karl-Friedrich-Bonhoeffer Institut,
D-3400 Gottingen-Nikolausberg, Postfach 968,
W. Germany.

N.H. Rees, Chemistry Dept., University College,
University of Wales, Cathays Park,
Cardiff CF1 1XL, U.K.

P.E. Sorensen, Chemistry Dept. A, The Technical University of Denmark, DK-2800 Lyngby, Denmark.

P. Suppan, Nortech Labs. Ltd., Brunel Rd., Churchfields
Estate, Salisbury, Wiltshire, U.K.

K. Takagi, Institute of Industrial Science, University
        of Tokyo, 22-1 Roppongi 7 Chome, Minato-Ku,
        Tokyo, Japan.
P.J. Thomas, I.C.I. Ltd., Mond Division, Research and
        Development Dept., P.O. Box 8, The Heath,
        Runcorn, Cheshire WA7 4QD, U.K.
R.N.F. Thorneley, ARC Unit of Nitrogen Fixation,
        University of Sussex, Brighton, Sussex
        BM 9QJ, U.K.
C.J. Tredwell, TheRoyal Institution of Great Britain,
        21 Albemarle St. London W1X 4BS, U.K.
D.H. Turner, Department of Chemistry of Chemistry,
        University of Rochester, Rochester, N.Y.
        14647, U.S.A.
P. Wardman, Cancer Research Campaign, Gray Laboratory,
        Mount Verneon Hospital, Northwood, Middx.
        HA6 2RN, U.K.
P. Warwick, University Chemical Laboratory, The
        University of Kent, Canterbury, Kent CT2 7NH,
H. Winkler, Max-Planck-Institut für Physicalische
        Chemie, Karl-Friedrich-Bonhoeffer-Institut,
        D-3400 Gottingen-Nikolausberg, Am Fasberg,
        Postfach 968, W. Germany.
M. Wyn-Jones, Department of Chemistry and Applied
        Chemistry, The University of Salford,
        Salford M5 4WT, U.K.